SOILS IN ARCHAEOLOGICAL RESEARCH

SOILS IN ARCHAEOLOGICAL RESEARCH

Vance T. Holliday

OXFORD
UNIVERSITY PRESS
2004

OXFORD
UNIVERSITY PRESS

Oxford New York
Auckland Bangkok Buenos Aires Cape Town Chennai
Dar es Salaam Delhi Hong Kong Istanbul Karachi Kolkata
Kuala Lumpur Madrid Melbourne Mexico City Mumbai Nairobi
São Paulo Shanghai Taipei Tokyo Toronto

Published by Oxford University Press, Inc.
198 Madison Avenue, New York, New York 10016

www.oup.com

Library of Congress Cataloging-in-Publication Data
Holliday, Vance T.
Soils in archaeological research / by Vance T. Holliday.
p. cm.
Includes bibliographical references and index.
ISBN 0-19-514965-3
1. Soil science in archaeology. I. Title.
CC79.S6 H65 2004
930.1′028—dc21 2003005784

9 8 7 6 5 4 3 2 1
Printed in the United States of America
on acid-free paper

To my pedologic mentors: B. L. Allen and Peter W. Birkeland

(Left) B. L. Allen in the field on the High Plains, 2002. (Right) Pete Birkeland at the Geological Society of America Penrose Conference on Paleosols, Oregon, 1987.

Preface

This book is a discussion of the study of soils as a component of earth science applications in archaeology, a subdiscipline otherwise known as geoarchaeology. The volume focuses on how the study of soils can be integrated with other aspects of archaeological and geoscientific research to answer questions regarding the past. To a significant degree, the book approaches soils as a function of and as clues to the factors of soil formation; that is, the external or environmental factors of climate, organisms, relief, parent material, and time (making up the well-known CLORPT formula of Jenny, 1941; discussed in chapter 3) that drive the processes of soil formation. Reconstructing the factors is important in reconstructing the human past. The book outlines the many potential and realized applications of soil science, especially pedology and soil geomorphology, in archaeology. This approach contrasts with earlier systematic, single-author volumes on the topic (Cornwall, 1958; Limbrey, 1975). The older works tend to emphasize human impacts on soils, particularly from an agricultural perspective, which is not surprising given their focus on northwest Europe. Moreover, soil geomorphology was essentially unrecognized when Cornwall's classic study was published and was just beginning to come into its own as a subdiscipline when Limbrey's book appeared.

The volume is designed for use by students and professionals with backgrounds in both archaeology and earth science, particularly pedology, geomorphology, and Quaternary stratigraphy. The target audience is the archaeologists and geoarchaeologists who want to know how soils can be used to aid in answering archaeological questions. In addition, I hope this book will help pedologists and soil geomorphologists understand more about investigating the human past.

A few basic concepts and principles in pedology are presented as necessary. More attention is devoted to theoretical, conceptual, and especially practical issues in soil geomorphology because few students or professionals in archaeology and in the geosciences have access to training in soil geomorphology and because a variety of issues in soil geomorphology are of direct relevance to geoarchaeology. However, this book is not an introductory text to pedology or soil geomorphology. Some of the world's leading investigators in these disciplines have already prepared good introductions, including Buol et al. (1997) and Fanning and Fanning (1989; for U.S. views of pedology); Birkeland (1999) and Daniels and Hammer (1992; for North American approaches to soil geomorphology); Fitzpatrick (1971), Duchaufour (1982), Gerrard (2000), and Van Breemen and Buurman (2002; for British/European perspectives on pedology); and Gerrard (1992; for a British/European view of soil geomorphology). These summaries, and for that matter this volume, are no substitute for formal instruction and practical field experience, however. Pedology, soil geomorphology, and geoarchaeology are all "hands-on" field disciplines.

Field experience and instruction applies to geoscientists interested in archaeology as well as to archaeologists who want to use soils in their research, a point raised in one of the earliest papers on soils in archaeology (Cornwall, 1960, p. 266). Such training is an essential key to mutual understanding. Lack of communication or, more typically and specifically, the inability to communicate between archaeologists and geoscientists (or any other scientists outside of mainstream archaeology), despite the best of intentions, is a frequent source of frustration and tension on interdisciplinary archaeological projects. A personal experience illustrates the point. I was briefly involved in an archaeological survey that included a well-respected soil scientist who had just retired from the Soil Conservation Service (now the National Resource Conservation Service). The archaeologist in charge was quite excited at the prospect of having this veteran pedologist on the team, though was vague when I asked what results were expected of the pedologist. The pedologist was, in private conversations with me, equally bewildered in regard to his duties and the larger archaeological efforts, but decided he would just do what he knew best. The end result was a frustrated archaeologist with an excellent soil map of the project area, but a map containing little information of archaeological or geomorphological significance. I hope this volume serves to facilitate communication between archaeologists and soil scientists or other geoscientists and will help investigators minimize or avoid similarly frustrating situations.

Geoarchaeologists must understand the questions asked in archaeology and must also understand that, unfortunately, geoscience training is not a common component of most archaeology degree programs. Archaeologists, in turn, must understand that the utility of soils in archaeology goes beyond knowing how to describe or classify them and goes beyond knowing some laboratory techniques. I have worked with archaeologists—good ones—who could identify an A or Bt horizon in the field and who could tell me that their site area was mapped as a Haplustalf, but who were otherwise clueless as to the stratigraphic, chronologic, or geomorphic implications of these soil characteristics. Field description and classification are simply means to an end.

In an attempt to resolve some of these problems, I have written a book that pulls together my own ideas and those of many others regarding the role that soil science and particularly pedology can play in archaeological research. This approach is based on my own training and experience as well as that of colleagues in soil geomorphology, geoarchaeology, and archaeology. Some of the examples are not related to archaeological research because so little of this type of soils work has been done in archaeological contexts, but these examples illustrate the principles and the potentials for archaeology.

The first three chapters of the volume present introductory discussions of soils in geoarchaeology and basic concepts (chapter 1), basic terminology and methods of studying soils (chapter 2), and theoretical or conceptual aspects of soil genesis, including further discussion of the CLORPT approach to soil geomorphology (chapter 3). The next three chapters deal with two fundamental applications of soils in geoarchaeological research: soil surveys (chapter 4) and soil stratigraphy (chapters 5 and 6). In a sense, soil survey involves the landscape or relief factor and soil stratigraphy the parent material factor, though both components of soil investigation involve aspects of the other factors. Chapters 7 through 9 are more explicitly organized around the soil-forming factors: the concept of time in pedogenesis and soils as age indicators (chapter 7); soils as indicators of past climate and vegetation (chapter 8); and soils as related to and indicators of relief and landscape evolution (chapter 9). The final two chapters discuss soils in the context of investigations that have been more commonly an explicit component of archaeological research: site-formation processes (chapter 10) and land use and human impacts on the landscape (chapter 11). Three appendixes are also provided: 1, on variations to the standard U.S. Department of Agriculture soil-horizon nomenclature useful in soil geomorphic and geoarchaeological research; 2, on comparisons of some common laboratory methods for analysis of soils in archaeological contexts; and 3 (with coauthors Julie Stein and Bill Gartner), on comparisons of some common laboratory methods for analysis of soils in archaeological contexts.

This book is written from a geoscience perspective. Conventions regarding age estimates and chronostratigraphy, therefore, follow geologic standards. Ages of less than 100,000 yr are expressed in "yr B.P." as are uncalibrated radiocarbon ages unless otherwise noted. Ages of 100,000 yr or older are expressed as "ka" (thousands of years) or "Ma" (millions of years). The age of the Plio-Pleistocene boundary is placed at 1.8 Ma (Harland et al., 1990; Pasini and Colalongo, 1997) and the age of the Pleistocene-Holocene boundary is 10,000 yr B.P. (after Hageman, 1972). The early-middle Pleistocene boundary (equivalent to the early-middle Quaternary boundary) is placed at the Brunhes-Matuyama polarity reversal, 788 ka (after Harland et al., 1990, p. 68, sec. 3.21.2). The middle-late Pleistocene boundary (equivalent to the middle-late Quaternary boundary) is placed at the beginning of marine oxygen isotope stage 5e (after Harland et al., 1990, pp. 68–69, sec. 3.21.2), which represents the beginning of the last interglacial period before the Holocene, dated to ca. 125 ka (following Winograd et al., 1997).

This book began to take shape when I was a Visiting Professor at the Alaska Quaternary Center at the University of Alaska–Fairbanks (spring 1994). Jim

Dixon and Mary Edwards helped significantly in arranging my stay in Fairbanks. The next phase of writing began during a sabbatical leave granted by the College of Letters and Sciences of the University of Wisconsin–Madison (fall 2000).

I appreciate the help of many individuals who supplied line drawings and photographs that appear in this book and who allowed the photographs to be reproduced: Art Bettis, John and Bryony Coles, Jonathan Damp, Rick Davis, Ed Hajic, John Jacob, Jim Knox, Rolfe Mandel, Charlie Schweger, Marc Stevenson, Mike Wiant, and Don Wyckoff.

The line drawings and most of the photographs were prepared with support from the Cartography Laboratory of the Department of Geography at the University of Wisconsin–Madison. My gratitude to Onno Brouwer, director of the Cartography Laboratory, for his generous support. This chore was patiently and expertly carried out by Rich Worthington and Erik Rundell. Laura Pitt (University of Wisconsin) prepared many of the tables. Dirk Harris (University of Arizona) helped prepare some of the photo scans. Additional support for preparation of the artwork was provided by the Office of the Provost of the University of Arizona.

This book has its roots in my initial experience with and thoughts about soils in archaeological contexts in the 1970s and in a few subsequent attempts to organize my thinking on the subject (Holliday, 1989a, 1990a). Many people, some who became good friends and close colleagues, have directly or indirectly influenced my experiences and ideas regarding soils in archaeology, and I take great pleasure in acknowledging them here. My initial exposure to soils came when I started working on the Lubbock Lake Project (run under the auspices of the Museum of Texas Tech University) as a research assistant (1974–1978) in the Museum Science graduate program. As I became familiar with the remarkable soils record at Lubbock Lake and took my first soils courses, my budding interests were encouraged by Chuck Johnson and especially by Eileen Johnson, who were codirecting the project. However, the person key to pushing me in the direction I took was B. L. Allen, Professor (now Emeritus) of Soil Science at Texas Tech: one of this country's great pedologists, an outstanding teacher and mentor and one of Texas's fine, decent gentlemen. I took all of my basic soils training from B. L., but more than that, he shared an interest in archaeology and in the record of the past that soils contain. We began work together on the soils at Lubbock Lake, and he enthusiastically encouraged me to pursue these investigations for a Ph.D. As a result, I entered the graduate program in Geological Sciences at the University of Colorado (1978) to work on a doctoral dissertation under Peter W. Birkeland (now Professor Emeritus).

My four and half years at the University of Colorado were one of the great experiences in my professional career. The faculty and students in the department, and Pete Birkeland in particular, instilled and inspired my approach to soil geomorphology, Quaternary geology, and the academic life. Pete is an amazing individual, both as a scientist and a friend, with his laid-back style, deep concern for students and teaching, and substantial research productivity. Studying with him is one of my proudest accomplishments.

After graduate school I spent two years at Texas A&M University in the departments of Geography and Anthropology. There I had the opportunity to get to know two other great pedologists: Larry Wilding and Tom Hallmark. Discussions with both of these men, and some enjoyable fieldwork with Tom, provided valuable insights into soil-forming processes and how they might be important in archaeological research.

Most of my postgraduate career until 2002 was in the Department of Geography at the University of Wisconsin–Madison. My approach to soil stratigraphy, soil geomorphology, and soil investigations in archaeological research gelled during my 16 years at the UW. I benefited greatly from many discussions with my colleagues there: Jim Knox, Tom Vale, Karl Zimmerer, and the late Francis Hole (all in Geography), and Kevin McSweeney and Jim Bockheim (both in Soil Science). The real learning came in teaching classes and seminars and working in the field with graduate students. Those particularly interested in soils and geoarchaeology and who expanded my pedoarchaeological horizons include John Anderton, Mike Daniels, Bill Gartner, Peter Jacobs, Jim Jordan, Samantha Kaplan, David Leigh, Joe Mason, James Mayer, Jemuel Ripley, Garry Running, Ty Sabin, and Catherine Yansa (in Geography); Danny Douglas, Jeff Monroe, and Jesse Rawling (in Geology); Steve Cassells, Pat Lubinski, Bill Middleton, Megan Partlow, Jeff Shockler, and Tina Thurston (in Anthropology); and David Brown (in Soil Science).

Over the years I've met many other colleagues who share my interests in using soils to unravel the human past. We've talked and corresponded, coauthored papers, coedited books, worked in the field, and in some fortunate situations become friends. All have influenced my thinking about this topic, and with great pleasure I acknowledge and thank them: Art Bettis, Andrei Dodonov, Bill Farrand, Paul Goldberg, Ed Hajic, Rich Macphail, Les McFadden, Rolfe Mandel, Dave Meltzer, Dan Muhs, Lee Nordt, Julie Stein, and Dan Yaalon. A number of colleagues very kindly and very helpfully reviewed chapters: Art Bettis (chapters 1 through 6), Paul Goldberg (chapters 1 and 11), Jeff Homburg (chapter 11), Rich Macphail (chapter 11), Rolfe Mandel (chapters 1, 2, 5, and 6), Lee Nordt (chapters 1, 7, 8, 9, and 10), Mike Schiffer (chapter 10), and Bill Woods (chapters 1 and 10). James Mayer helped with statistical analyses of the radiocarbon ages (chapter 7). Thanks also to Julie Stein and Bill Gartner for collaborating on appendix 3. Additional information, commentaries, or data were provided by Jesse Ballenger, Pete Birkeland, Glen Doran, Charles Frederick, Bill Johnson, Don Johnson, Rob Kemp, Mike Kolb, Mary Kraus, Johan Linderholm, Randy Schaetzl, Russell Stafford, Julie Stein, Gregory Vogel, and Don Wyckoff. Jim Burton, Phil Helmke, Tina Thurston, and Bill Woods also helped me out on the ticklish topic of soil phosphorus.

Finally, my deep gratitude to two lovely ladies—my wife Diane and my daughter Cora—for their patience during this long writing process.

Contents

SOILS IN ARCHAEOLOGICAL RESEARCH

1

Introduction

Soils are a potential source of much information in archaeological studies on site- and feature-specific scales as well as on a regional scale. Soils are a part of the stage on which humans have evolved. As an integral component of most natural landscapes, soils also are an integral component of cultural landscapes. "Soils are active components of functioning ecosystems that reflect the spatial variability of ecological processes and at the same time have varying degrees of suitability for different kinds of human behavior" (Warren, 1982b, p. 47). Beyond physically supporting humans and their endeavors, however, soils are indicators of the nature and history of the physical and human landscape; they record the impact of human activity, they are a source of food and fuel, and they reflect the environment and record the passage of time. Soils also affect the nature of the cultural record left to archaeologists. They are a reservoir for artifacts and other traces of human activity, encasing archaeological materials and archaeological sites. Soil-forming processes also are an important component of site formation processes. Pedogenesis influences which artifacts, features, and environmental indicators (floral, faunal, and geological) are destroyed, which are preserved, and the degree of preservation.

Those involved in field archaeology (as archaeologists, geoscientists, or bioscientists) routinely deal with soils—probably more so than most soil scientists or geologists (Birkeland, 1994, p. 143). However, what the soils or a soil scientist can tell archaeologists about the site and about the archaeological record is not always clear. In part, the integration of soil science in archaeology has been hampered by ambiguities in use of the term "soil" and confusion over what a soil is or is not. The bigger issue is that pedological research, particularly in the United

States, has not traditionally been a component of geoarchaeology (the application of the earth science in archaeology) until recent years, in comparison with applications of other aspects of geoscience such as stratigraphy, sedimentology, or geomorphology. This situation evolved in large part because the academic study of soils typically is located in the agricultural sciences rather than the earth sciences. Students of archaeology and the geosciences, therefore, often have no access to courses in soil science because agriculture programs are considerably less common than schools of arts and sciences. As Tamplin (1969, p. 153) noted, most archaeologists are well trained in the principles of stratigraphy and the "Law of Superposition" long before they learn about soils and soil formation. Further compounding the problem is the focus of most soil science training and research, which is on mapping, contemporary land use, soil quality, and plant productivity and not on reconstructing the past (Tandarich and Sprecher, 1994; Bronger and Catt, 1998a; McFadden and McDonald, 1998; Holliday et al., 2002). Soil scientists are often unfamiliar with questions of concern to archaeologists, geologists, and geographers—questions of stratigraphy, landscape evolution, and paleoenvironments. In addition, U.S. pedologists seldom gain experience in dealing with extensively altered soils such as middens and plaggens because they are rare or of limited extent in North America and are therefore of limited interest in terms of mapping and land use.

Soil Science, Soils, and Soil Horizons

Before continuing into the substance of this chapter, some fundamental disciplinary and conceptual issues must be reviewed. This book is an application of subfields of soil science in archaeology and geoarchaeology. Soil science is the study of soils as a natural resource on the Earth's surface and includes the study of soil formation, classification and mapping, soil chemistry, soil physics, soil biology, and soil fertility (Soil Science Society of America, 1987, p. 24). The principal subfields of soil science that are the focus of this book are pedology and soil geomorphology, both of which overlap with the disciplines of geology and physical geography. Pedology is the study of soils as three-dimensional bodies intimately related to the landscape, focusing on their morphology, genesis, and classification. Soil geomorphology is the study of relationships between soils and landscapes (e.g., Ruhe, 1956, 1965; Daniels and Hammer, 1992; Gerrard, 1992; Birkeland, 1999). In its broadest sense, soil geomorphology is the investigation of soils as they were influenced by climate, flora, fauna, topography, and geologic substrate operating over time (e.g., Birkeland, 1999).

What Is a Soil?

The word "soil" is used by different individuals in different ways. To the farmer, the agricultural scientist, and some soil scientists, it is simply the medium for plant growth. To the engineer, some geologists, and probably many archaeologists, it is unconsolidated sediment including loose or weathered rock or regolith. To the pedologist and soil geomorphologist, however, soil has a very specific definition that is not always properly understood or appreciated. Using this definition, a soil is a natural three-dimensional entity that is a type of weathering phenomena occurring at the immediate surface of the earth in sediment and rock, acting as a medium for plant growth, and the result of the interaction of the climate, flora, fauna, and landscape position, all acting on sediment or rock through time (modified from Soil Science Society of America, 1987). The medium for soil development (i.e., the rock or sediment in which the soil forms) is referred to as "parent material."

Key concepts in the pedologic and soil geomorphic view of soils are that, first, soils form in or represent an alteration by physical, chemical, and biomechanical weathering of sediments and rocks over time (i.e., soils are a type of surface weathering phenomena); second, pedogenesis includes interaction with flora and fauna and accumulation of organic matter; third, there is some movement or redistribution (typically downward, but also upward) of clastic, biochemical, and ionic soil constituents (e.g., clay, organic carbon, iron, aluminum, and manganese compounds, and calcium carbonate in ionic solution); fourth, soils are an intimate component of the landscape, form on relatively stable land surfaces, and are approximately parallel to the land surface; fifth, soils are dynamic and are components of the ecosystem representing the interface of the atmosphere, the biosphere, and the geosphere; and sixth, soils are extremely complex systems.

Soils are laterally extensive across the landscape. They form across various landforms and in a variety of parent materials and vary in a predictable manner because of changes in erosion, deposition, drainage, vegetation, fauna, and age of the landscape. Soils also vary as the microclimate and macroclimate varies. This predictable variability is referred to as the "constancy of relationships" (Brewer, 1972, p. 333) and is unique to soils among geomorphic phenomena. This characteristic of soils in buried contexts allows them to be traced in three dimensions over varying paleotopography. Individual layers of sediment, in contrast, will be confined to particular depositional environments and will thin to nothing away from that environment (Mandel and Bettis, 2001b, p. 180).

Soil Horizons

"Soil horizons" are zones within the soil (i.e., subdivisions of the soil) that parallel the land surface and have distinctive physical, chemical, and biological properties (table 1.1; fig. 1.1). Soil horizons result from mineral alteration, biogenic activity, additions of organic matter, leaching of soluble materials, and translocation of fine particles, humus, and chemical compounds (table 3.1; fig. 3.1). Together, a set of genetically related horizons produce a "soil profile." A soil profile is the vertical arrangement of soil horizons, typically seen in a two-dimensional exposure down to and including the parent material (fig. 1.1), similar to a standard archaeological profile—which may exhibit a soil profile. Soil profiles vary because of the complex interaction of climate, the biota living on and in the soil, the nature of the soil parent material, the landscape position, and the age and evolution of the landscape (i.e., the soil-forming factors, discussed in chapter 3). The "solum" is the upper and most weathered part of the soil profile, the A, E, and B horizons. A "sequum" is an eluvial horizon (e.g., E) and an

Table 1.1. General definitions of soil horizons used in the United States

Soil Master Horizons

O horizon or layer: Horizons or layers dominated by organic material. Some are saturated with water for long periods or were once saturated but are now artificially drained; others were never saturated. Some O horizons consist of undecomposed or partially decomposed litter, such as leaves, needles, twigs, moss, and lichens, that were deposited on the surface; they may be on top of either mineral or organic soils. Other O layers are organic materials that were deposited under saturated conditions and have decomposed to varying stages.

A horizon: Mineral horizon that formed at the surface or below an O horizon and that exhibits 1) obliteration of all or much of the original rock structure and 2) an accumulation of humified organic matter intimately mixed with the mineral fraction.

E horizon: Mineral horizon in which the main characteristic is loss of silicate clay, iron, aluminum, or some combination of these, leaving a concentration of sand and silt particles. This horizon exhibits obliteration of all or much of the original rock structure. An E horizon is usually lighter in color than an overlying A horizon and an underlying B horizon. In some soils the color is that of the sand and silt particles, but in many soils coatings of iron oxides or other compounds mask the color of the primary particles.

B horizon: Horizon that forms below an A, E, or O horizon and is dominated by obliteration of all or much of the original rock structure and shows one or more of the following: 1) illuvial concentration of silicate clay, iron, aluminum, humus, carbonates, gypsum, or silica, alone or in combination; 2) evidence of removal of carbonates; 3) coatings of sesquioxides that make the horizon conspicuously lower in value, higher in chroma, or redder in hue than overlying and underlying horizons without apparent illuviation of iron; 4) alteration that forms silicate clay or liberates oxides or both and that forms granular, blocky, or prismatic structure; or 5) brittleness.

C horizon or layer: Horizon or layer, excluding hard bedrock, that is little affected by pedogenic processes and lack properties of O, A, E, or B horizons. The material of C layers may be either like or unlike that from which the solum formed. The C horizon may have been modified even if there is no evidence of pedogenesis. Included as C layers are sediment, saprolite, unconsolidated bedrock, and other geologic materials that commonly are uncemented.

R layers: Hard (minimally weathered) bedrock.

Horizons dominated by properties of one master horizon but having subordinate properties of another: Two capital letter symbols are used: AB, EB, BE, or BC. The master horizon symbol given first designates horizon whose properties dominate the transitional horizon (e.g., an AB horizon has characteristics of both an overlying A horizon and an underlying B horizon, but it is more like the A than like the B).

Horizons in which distinct parts have recognizable properties of the two kinds of master horizons indicated by the capital letters: The two capital letter are separated by a virgule(/): E/B, B/E, or B/C. Most of the individual parts of one of the components are surrounded by the other.

Subhorizons or Subordinate Horizons of Master Horizons

a *Highly Decomposed Organic Material:* Used with "O" to indicate the most highly decomposed of the organic materials. The rubbed fiber content is less than about 17 percent of the volume.

b *Buried Soil or Horizon:* Used in mineral soils to indicate identifiable buried horizons with major genetic features that were formed before burial. Genetic horizons may or may not have formed in the overlying material, which may be either like or unlike the assumed parent material of the buried soil.

c *Concretions or Nodules:* Indicate a significant accumulation of cemented concretions or nodules. The cementing agent is not specified except it cannot be silica. This symbol is not used if concretions or nodules are dolomite or calcite or more soluble salts, but it is used if

Table 1.1. (*cont.*)

	the nodules or concretions are enriched in minerals that contain iron, aluminum, manganese, or titanium.
d	*Physical Root Restriction:* Indicates root-restricting layers in naturally occurring or manmade unconsolidated sediments or materials such as dense basal till, plow pans, or other mechanically compacted zones.
e	*Organic Material of Intermediate Decomposition:* Used with "O" to indicate organic materials of intermediate decomposition. Rubbed fiber content is 17 to 40 percent of the volume.
f	*Frozen Soil:* Indicates that the horizon or layer contains permanent ice. Symbol is not used for seasonally frozen layers or for "dry permafrost" (material that is colder than 0°C but does not contain ice).
g	*Strong Gleying:* Indicates either that iron has been reduced and removed during soil formation or that saturation with stagnant water has preserved a reduced state. Most of the affected layers have chroma of 2 or less and many have redox concentrations. The low chroma can be the color of reduced iron or the color of uncoated sand and silt particles from which iron has been removed. Symbol "g" is not used for soil materials of low chroma, such as some shales or E horizons, unless they have a history of wetness. If "g" is used with "B," pedogenic change in addition to gleying is implied. If no other pedogenic change in addition to gleying has taken place, the horizon is designated Cg.
h	*Illuvial Organic Matter:* Used with "B" to indicate the accumulation of illuvial, amorphous, dispersible organic matter-sesquioxide complexes. The sesquioxide component coats sand and silt particles. In some horizons, coatings have coalesced, filled pores, and cemented the horizon. The symbol "h" is also used in combination with "s" as "Bhs" if the amount of sesquioxide component is significant but the value and chroma of the horizon are 3 or less. This horizon is not to be confused with the "Ah" used to designate human impacts (appendix 1).
i	*Slightly Decomposed Organic Material:* Used with "O" to indicate the least decomposed of the organic materials. Rubbed fiber content is more than about 40 percent of the volume.
k	*Carbonates:* Accumulation of calcium carbonate.
m	*Cementation or Induration:* Continuous or nearly continuous cementation. The symbol is used only for horizons that are more than 90 percent cemented, although they may be fractured. The layer is physically root restrictive. If the horizon is cemented by carbonates, "km" is used; by silica, "qm"; by iron, "sm"; by gypsum, "ym"; by both lime and silica, "kqm"; by salts more soluble than gypsum, "zm."
n	*Sodium:* Accumulation of exchangeable sodium.
o	*Residual Sesquioxides:* Residual accumulation of sesquioxides.
p	*Plowing or Other Disturbance:* Disturbance of the surface layer by mechanical means, pasturing, or similar uses. A disturbed organic horizon is designated Op. A disturbed mineral horizon is designated Ap even though it was clearly once an E, B, or C horizon.
q	*Silica:* Accumulation of secondary silica.
r	*Weathered or Soft Bedrock:* Used with "C" to indicate root restrictive layers of soft bedrock or saprolite, such as weathered igneous rock; partly consolidated soft sandstone; siltstone; and shale. Excavation difficulty is low or moderate.
s	*Illuvial Accumulation of Sesquioxides and Organic Matter:* Used with "B" to indicate the accumulation of illuvial, amorphous, dispersible organic matter-sesquioxide complexes if both the organic matter and sesquioxide components are significant and the value and chroma of the horizon is more than 3. The symbol is also used in combination with "h" ("Bhs") if both the organic matter and sesquioxide components are significant and the value and chroma are 3 or less.
ss	*Slickensides:* Presence of slickensides, which result directly from the swelling of clay minerals and shear failure, commonly at angles of 20 to 60 degrees above horizontal.
t	*Silicate Clay:* Accumulation of silicate clay translocated within the horizon or moved into the horizon by illuviation, or both. At least some part should show evidence of clay accumulation in the form of coatings on surfaces of peds or in pores, or as lamellae ("clay bands"), or bridges between mineral grains.

Table 1.1. (*cont.*)

v	*Plinthite:* Presence of iron-rich, humus-poor, reddish material that is firm or very firm when moist and that hardens irreversibly when exposed to the atmosphere and to repeated wetting and drying. This horizon is not to be confused with the "Av" used to designate a vesicular horizon in arid environments (appendix 1).
w	*Development of Color or Structure:* Used with "B" to indicate the development of color or structure or both, with little or no apparent illuvial accumulation of material (see appendix 1 for additional usages).
x	*Fragipan:* Genetically developed layers that have a combination of firmness, brittleness, very coarse prisms with few to many bleached vertical faces, and commonly higher bulk density than adjacent layers.
y	*Gypsum:* Accumulation of gypsum.
z	*Salts More Soluble than Gypsum:* Accumulation of salts more soluble than gypsum.
Combinations of Symbols:	A B horizon that is gleyed or that has accumulations of carbonates, sodium, silica, gypsum, salts more soluble than gypsum, or residual accumulation or sesquioxides carries the appropriate symbol—g, k, n, q, y, z, or o. If illuvial clay is also present, "t" precedes the other symbol: Btg.

Modified from Soil Survey Division Staff (1993, pp. 118–126). These symbols are used for describing soils in the field. For more complete definitions see Buol et al. (1997), Birkeland (1999), Schoeneberger et al. (1998), or Soil Survey Division Staff (1993). Some alternative horizon designations, including those developed outside of the United States, are presented in appendix 1.

underlying B horizon. Two sequums in a vertical sequence are a "bisequum" (common in some podzolizing environments; discussed below).

Soil horizons are the most obvious features of soils in the field because of their unique physical, biological, and chemical characteristics such as structure and color (fig. 1.1). Moreover, the development of soil horizons is a characteristic of soils that is unique among geomorphic processes and features. The ability to recognize soil horizons is a key first step in developing the ability to recognize soils. The visual distinctness of soil horizons and soils is one of the principal reasons they have long been used as stratigraphic markers. Careful scrutiny and description of soil profiles and horizons (chapter 2) are critical elements of pedology and require considerable training and practice.

Soil horizon nomenclature includes a few master or major horizons (the well-known A-B-C sequence), a considerable number of subhorizon symbols that act as modifiers of the master horizons, and additional descriptive terminology (table 1.1; appendix 1). The soil horizon nomenclature commonly used in the United States was developed largely by the U.S. Department of Agriculture (USDA) to meet the requirements of a standardized, nationwide soil survey. This system is fully explained by the Soil Survey Division Staff (1993; available at http://soils.usda.gov) and Schoeneberger et al. (1998). Excellent summaries are provided by Buol et al. (1997), Birkeland et al. (1991), and Birkeland (1999). Vogel (2002) and Reed et al. (2000) have prepared very handy booklets on soil description for archaeologists. The Soil Science Society of America also has a very useful glossary of soil science terms (http://www.soils.org/sssagloss/). Canadian terminology is presented by Soil Classification Working Group (1998), and the Australian nomenclature is described by McDonald et al. (1998). For Europe,

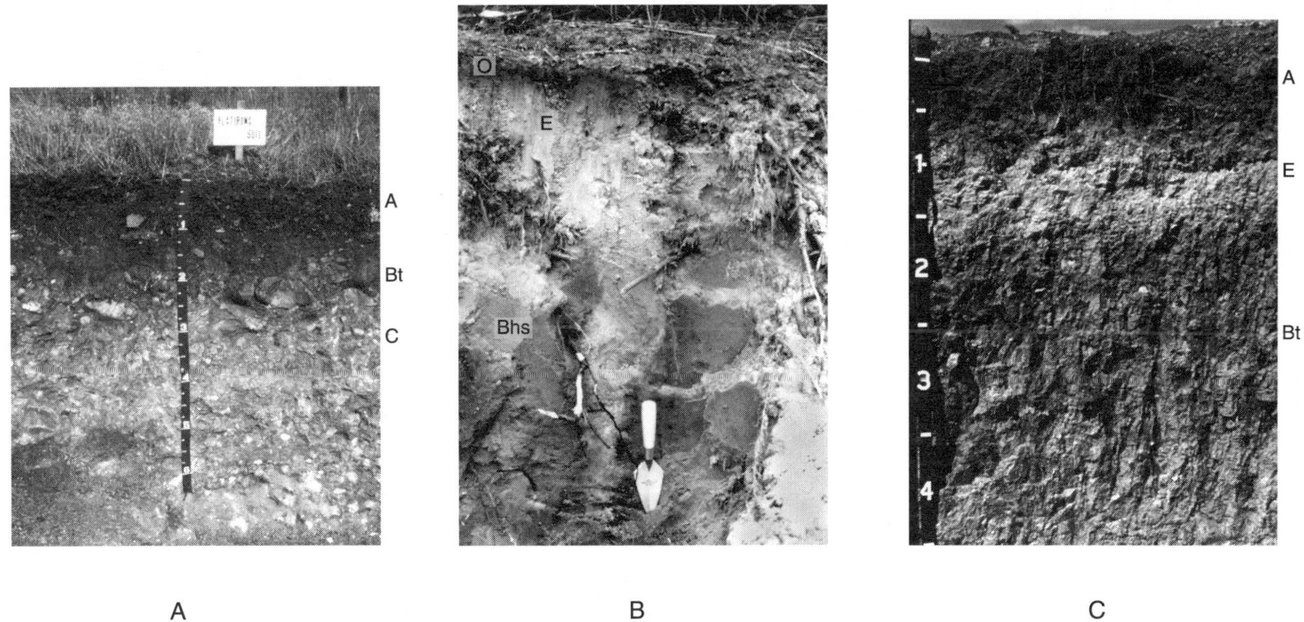

Figure 1.1 Examples of various soil types and profile morphologies from North America. (A) Paleustoll (Flatirons series) formed in alluvium on an early Pleistocene pediment in the Colorado Piedmont, just east of the Rocky Mountain front. The soil has a thick, dark, surface horizon high in organic matter (a mollic epipedon), classifying it as a Mollisol. The current climate is semiarid, with a spring-summer rainy season (ustic moisture regime). The soil has a very well expressed (deep reddish-brown, thick, clay-rich) argillic (Bt) horizon (with intensely weathered gravel), placing it in the "pale" Great Group. The scale is in feet. (B) Spodosol from the Upper Peninsula of Michigan, illustrating development of the E and Bhs horizons in sandy, glacial outwash. (C) Alfisol (Hapludalf) from southern Michigan illustrating development of A-E-Bt horizonation typical of postglacial soils in the area developed on loess and till (slide 1–6 from the Marbut Memorial Slide set, Soil Science Society of America; reproduced with permission of the Soil Science Society of America). Scales are in feet.

Hodgson (1997) presents the standards used in Great Britain (see also Catt, 1990), and Duchaufour (1998, pp. 146–147, 148) and van Breemen and Buurman (2002, pp. 141, 365–366) summarize the Food and Agriculture Organization of the United Nations (FAO-UNESCO) system (to be updated and superceded by the FAO World Reference Base for Soil Resources [FAO-WRB]; see Duchaufour, 1998, pp. 151–152) employed throughout continental Europe (see Driessen and Dudal, 1991). All of these sources also provide additional specifics on the terminology and data necessary to describe soils. Some additional non-standard (i.e., non-USDA-approved) horizon nomenclature, developed by soil geomorphologists and Quaternary geologists, or by pedologists in other countries, is also provided in table 1.1 and discussed in appendix 1. The USDA horizon nomenclature, with modifications described in appendix 1, is used throughout this volume. Older or foreign nomenclatures used in sources for figures and tables were converted, unless noted.

To fully understand late-20th-century and contemporary USDA-based pedology, the concept of the "pedon" must be noted. The pedon is the smallest body of one kind of soil large enough to represent the nature and arrangement of horizons (Soil Survey Division Staff, 1993, p. 18). Essentially, the pedon is the soil profile in three dimensions; a conceptual unit of soil defined for sampling purposes (see Schelling, 1970, p. 170; Buol et al., 1997, pp. 36, 43–44; Soil Survey Staff, 1999, pp. 10–14). Whereas the pedon is conceptual, the "soil individual" (or "polypedon") is a real body of soil on the landscape (essentially, more than one pedon; Schelling, 1970, pp. 170–171; Buol et al., 1997, pp. 36–37). These terms and definitions are obscure and somewhat unfathomable, and the concepts have little relevance to geomorphology or geoarchaeology; they are mentioned because they are key concepts in USDA soil mapping (chapter 4).

Soil Horizons versus Geologic Layers

Soil horizons are not the same as geologic layers. Soil horizons form in geologic layers. Learning how to distinguish between soil horizons and unaltered sediments is another important step in learning how to recognize soils (Stein, 1985, p. 6; Mandel and Bettis, 2001b, p. 175). The use of the term "layer" interchangeably with "horizon" in some literature (including soil science publications) is particularly unfortunate and further confuses the issue of differentiating soils from sediments (Wilson, 1990, pp. 61–62, 71). Geologic layers follow the Law of Superposition: they are deposited one atop the other, with the bottom layer being the oldest and the top layer the youngest. Soil horizons are superimposed over, and thus postdate, the geologic materials in which they form (their parent material), and in general, horizons develop from the top of their parent materials downward (see chapter 5 and Cremeens and Hart, 1995). The boundaries between soil horizons, therefore, typically have no relationship to geological layering (discussed further in chapter 5). An individual soil horizon can form through several depositional layers, and conversely, several horizons can form within a single deposit.

Distinguishing between horizons and geologic layers sometimes is difficult (discussed further in chapters 5 and 10), particularly given the heavy emphasis on stratification and superposition in the training of archaeologists as well as geologists (e.g., Tamplin, 1969, pp. 153–154; Wilson, 1990, pp. 61–62, 71). For example, a layer of organic-rich sediment subjected to bioturbation and burial may be confused with a buried A horizon (fig. 5.7). A zone of pedogenically translocated humus (Bh horizon), common in podzolizing environments, likewise may be misidentified as a buried A horizon. Conversely, Rutter (1978) recalled an incident in which an archaeologist described an A-Bw-Bk profile, subsequently identified by a pedologist as layers of peat, loess, and volcanic ash, respectively. Confusion of horizons for layers, and vice versa, can have profound consequences in interpreting landscape evolution, site formation, or cultural stratigraphy, as discussed in chapters 9, 10, and 11.

Soils in Geoarchaeological Research

Soil science has its roots in both the geological and agricultural sciences (Tandarich, 1998a,b). This book is written largely from the geosciences perspective but also includes agricultural aspects that bear on the interpretation of the human past. Much of the volume deals with pedology. Though traditionally taught in agricultural schools, pedology is an earth science by virtue of its focus on soil in the context of landscape, surficial processes, and surficial deposits. The subdiscipline of pedology and geoscience that is most directly related to archaeology is soil geomorphology. In particular, much soil geomorphic research involves the study of soils in an attempt to reconstruct paleoenvironments and paleolandscapes or for dating (e.g., Ruhe, 1965; Boardman, 1985b; Jungerius, 1985; Richards et al., 1985; Knuepfer and McFadden, 1990; Gerrard, 1992; Birkeland, 1974, 1984, 1999) and thus has obvious archaeological implications. Geoarchaeologists K. Butzer (1982) and T. Van Andel (1994), a geographer and a geologist, respectively, suggest that one of the most significant aspects of geoarchaeological research is the analysis of landscapes, especially in terms of the changing options they present to their human occupants (Van Andel, 1994, p. 32; see also Fedele, 1976, and Gladfelter, 1977). Few aspects of the environment are as intimately linked to the landscape as soils, and thus, the appreciation and study of soils should be an integral component of geoarchaeology.

Increased interest in soils and soil science applications in archaeology has followed the growth of geoarchaeology, which began more or less when the term "geoarchaeology" was coined (Butzer, 1973; Rapp et al., 1974). For example, in 1977 the Geological Society of America (GSA) established an Archaeological Geology Division; in 1986 the journal *Geoarchaeology* was inaugurated; in 1990 the GSA published a volume on the subject (Lasca and Donahue, 1990) as part of its Centennial series; and in 1992 M. R. Waters published the first single-author volume devoted exclusively to geoarchaeology (Waters, 1992). The study of soils as a component of geoarchaeology similarly evolved, though lagging somewhat behind the broader geoscientific aspects of archaeology. The late 1980s saw publication both of an edited volume devoted to anthropogenic soils (Groenman-van Waateringe and Robinson, 1988) and of the first volume on soil micromorphology in archaeology (Courty et al., 1989), followed by an issue of

World Archaeology (1990, v. 22, n. 1) on "soils and early agriculture" and the first edited volume on soils in archaeology (Holliday, 1992b). This trend continued through the 1990s and into the early 2000s. Almost half of the 35 papers in Lasca and Donahue (1990) deal directly or indirectly with soils. They are also a prominent component of subsequent edited volumes on geoarchaeology (Barham and Macphail, 1995; Goldberg et al., 2001; Stein and Farrand, 2001), including historical treatments (Mandel, 2000).

In the meantime, archaeology emerged in mainstream soil science. Two international conferences on "pedoarchaeology" were held in 1992 (Orlando, Fla.; Foss et al., 1993b) and 1994 (Columbia, S.C.; Goodyear et al., 1997). Archaeology was featured prominently in two symposia at the annual meeting of the Soil Science Society of America in 1993, resulting in publication of an edited volume on pedology in archaeology (Collins et al., 1995). Furthermore, the role of soils in archaeological research (Holliday, 1994) was highlighted in a symposium honoring the 50th anniversary of Jenny's (1941) "Factors of Soil Formation" (Amundson, 1994). Scudder et al. (1996) also produced a lengthy review paper on soil science and archaeology in an agronomy monograph series, which introduced the term "archaeopedology" (p. 6; along with Reitz et al., 1996, p. 5) without defining it.

A telling example of the wide interest in the subject of soils in archaeology is in the disciplinary backgrounds of the investigators dealing with soils in Lasca and Donahue (1990), Holliday (1992b), Collins et al. (1995), and Goldberg et al. (2001): The authors include archaeologists, pedologists, geologists, and geographers. Such cross-disciplinary interests and interdisciplinary approaches significantly advance the field.

Soil science, particularly pedology, and archaeology are closely allied in their temporal and spatial scales, and among the earth sciences, pedology is most similar to archaeology in scales of operation and process (Holliday et al., 1993). These similarities in scale are apparent in both regional and site-specific studies. At large (regional) scales, soil stratigraphy has long been used in archaeology for correlating sites and for dating (e.g., Leighton, 1937; Albritton and Bryan, 1939; Bryan, 1941a; Movius, 1944). Soil geomorphic investigations also are compatible in scale to regional archaeological investigations, focusing on dating, environmental reconstruction, and late-Quaternary landscape evolution (e.g., Dan et al., 1968; Dan and Yaalon, 1971; Gile et al., 1981; Pope and Van Andel, 1984; Grolier, 1988; Overstreet and Grolier, 1988, 1996; Blair et al., 1990; Fedele, 1990; Wells et al., 1990; Mandel, 1994; Brinkmann, 1996; Wilkinson, 1997; Belcher and Belcher, 2000). Soil micromorphology (soil petrography; see chapter 2) is also useful for regional geomorphic and archaeologic studies, including investigations of sediment provenance, landscape evolution, environmental reconstructions, and agricultural development (e.g., Courty et al., 1989; Courty, 1992, 2001).

At small (site-specific) scales, the focus of pedology—the soil profile—is similar in scale to many archaeological sites (tens of centimeters to a few meters thick), and the scale of many pedological features is similar to that of archaeological features (a few millimeters to tens of centimeters thick; Holliday et al., 1993). Soil variability at small scales as a function of slope, drainage, or lithologic change is a common theme in pedology and is also of archaeological significance for stratigraphic correlation and interpretation of site formation processes. Temporal scales of formation of individual pedogenic versus anthropic features are disparate (centuries to millennia versus days to decades, respectively), but overall, processes of site formation and cultural evolution operate at temporal scales similar to those of soil formation (decades to millennia). The scalar compatibility of archaeology and pedology strongly argues for pedologists and pedologic perspectives to be involved in all phases of archaeological research (Holliday et al., 1993).

Beyond the issue of scale, pedologists and archaeologists also share another important perspective: understanding the soil as a resource, now and in the past (Jacob, 1995b, p. 54). The agriculture tradition in pedology provides a unique perspective for archaeologists who are trying to understand the origins, evolution, and characteristics of agriculture. Pedologists can also provide important insights into understanding human effects, such as soil erosion, on soils.

Soil stratigraphy and soil chemistry, rather than pedology and soil geomorphology, are perhaps the best-known and oldest applications of soil science in archaeology and also will be discussed in this volume. These two applications have very different disciplinary traditions, however. Soil stratigraphy has its origins in Quaternary geology, in which soils have long been recognized as stratigraphically and paleoenvironmentally significant (Leverett, 1898; Leighton, 1937, 1958; Bryan, 1941a, 1948; Bryan and Albritton, 1943; Ruhe, 1965; Haynes, 1968; Valentine and Dalrymple, 1976). Quaternary geologists and geomorphologists working with archaeologists were quick to recognize soils in stratified archaeological contexts and to use the soils as clues to past environments (e.g., Leighton, 1936; Bryan, 1941a; Haynes, 1968; Antevs, 1941). Paleontology and paleobotany (especially palynology) were important components of such research, but the physical and chemical characteristics of the soils themselves seldom were discussed or dealt with. Soil chemistry has long been studied both by soil scientists and archaeologists for clues to human impact on the landscape, especially for reconstructing agricultural activity and for detecting human occupation (e.g., Arrhenius, 1931, 1963; Solecki, 1951; Cornwall, 1958; Berlin et al., 1977; Eidt, 1977, 1984, 1985). Soils as pedologic entities, however, often are unrecognized or not dealt with in these studies. R. C. Eidt (1984, 1985), a geographer, is one of the few scientists to combine traditional pedologic approaches with soil chemistry to investigate anthropogenically modified soils ("anthrosols"; further discussed in chapter 11).

The historic dichotomy in the use of soils for stratigraphic or paleoenvironmental purposes versus their use as keys to anthropogenic impacts on the landscape also has a rough geographic separation. Most research on soils in archaeological contexts has taken place in North America and Europe. Many of the North American studies focused on soils as stratigraphic markers and as age and paleoenvironmental indicators, as is the case in the loess-rich areas of Europe. Outside these areas in Europe, however, and most notably in Great Britain, the research emphasis tended to be on the use of soils as indicators of human impact or human paleoenvironments (though rapid growth in these areas of interest began in the 1970s in North, Central, and South America, e.g., Eidt and Woods, 1974; Woods, 1975, 1977, 1984; Eidt, 1984, 1985; Sandor et al., 1986;

Sandor, 1987; Dunning, 1994). This regional variation probably is a result of differences in the nature of the archaeological record in the two regions and in the history of geoarchaeological research in each. In North America, archaeological sites with long records of human occupation in thick, well-stratified deposits with intercalated buried soils are relatively common, especially in the central and western regions. The impact of prehistoric peoples on soils and on the landscape was minimal, however. In Europe, in contrast, humans exerted a profound effect on the landscape for thousands of years. These effects have long attracted the attention of archaeologists and geoscientists, whose geographically distinct approaches to soils applications in archaeology can be seen in some of the earliest systematic treatments of soils as clues to the past (compare the North American perspectives of Bryan [1948] and Bryan and Albritton [1943] with the British approach of Cornwall [1958, 1960]) and are readily apparent by comparing the papers on North American archaeological sites in Lasca and Donahue (1990), Holliday (1992b), and Collins et al. (1995) with the studies from northern Europe and Great Britain in the papers assembled by Groenman-Van Waateringe and Robinson (1988) and Barham and Macphail (1995) and in the comprehensive works of Cornwall (1958) and Limbrey (1975).

Soil geomorphology, with the landscape at its core, provides an integrative link between soil stratigraphy, pedology, and soil chemistry and their applications in archaeological and geoarchaeological research. In this volume, a soil geomorphic approach is used to assess soil surveys (chapter 4) and to link studies of soils as stratigraphic markers (chapters 5, 6), as age and paleoenvironmental indicators (chapters 7 and 8), as clues to landscape evolution and site formation processes (chapters 9 and 10), and for the study of anthropogenic soils and human effects on the environment (chapter 11). Each chapter includes a discussion of basic principles and their archaeological implications and a presentation of case histories.

2

Terminology and Methodology

The long history of soil science (e.g., well over 100 yr in North America) and its bureaucratic institutionalization as a component of agricultural research in many countries resulted in the evolution of a substantial vocabulary and methodology for the discipline. A wide array of methods for the field and laboratory investigation of soils also is available to geoarchaeologists. The first part of this chapter is a discussion of some basic terms and definitions used in pedology and soil geomorphology. Some specific terms (e.g., soil stratigraphic nomenclature) are discussed as necessary elsewhere in the text and in appendix 1. There is a sizable body of nomenclature in pedology and soil geomorphology for describing and classifying soils. Indeed, there is a tendency in soils research toward an overabundance of nomenclature and jargon (e.g., Fastovsky, 1991). All scientific fields necessarily have a specialized nomenclature, however. Researchers in any field, and especially interdisciplinarians such as archaeologists working with soils and soil scientists working with archaeology, should be aware of the nomenclature, jargon, and lingua franca of the new fields they enter. A pedologist who becomes involved with North American archaeology would have to become familiar with terms and concepts such as "Paleoindian" or "Archaic" or "site." Likewise, archaeologists and geoscientists interested in understanding soils for geoarchaeological purposes must learn some basic soil science terminology and the principles behind issues of proper use (or misuse) of some terms. This fosters communication and problem solving and avoids ambiguities.

The rest of this chapter is a discussion of some of the more widely used approaches in the field and in the laboratory, especially in archaeological contexts. Key points to be made are that, first, investigators select the methods that best suit the field situation and the research questions being posed; second, if comparisons are made to other research, the comparable methods should be used; and third, all field and laboratory methods should be referenced in publications and deviations from standard practices or procedures should be described.

Nomenclature and Definitions

Some terms introduced below and elsewhere are well defined and generally agreed on, whereas others are variously or vaguely defined. There are also variations in terms and nomenclature from country to country because much of the jargon was devised by or developed under the direction of the agricultural agencies of national governments for the mapping, classification, and management of farm land or other aspects of land use. The history of some of the terms (and the ensuing confusion over meanings) is discussed by Johnson and Hole (1994), and varying applications (and definitions) of the terms are well illustrated among the papers collected and edited by Follmer et al. (1998). Some of the more useful international terminology is discussed in the text (especially chapter 5) and in appendix 1.

Soil Classification

An important component of pedology is soil classification, which is the categorization of soils into groups at varying levels of generalization according to their morphological and chemical properties and sometimes their assumed genesis (Buol et al., 1997, p. 5). The purpose of classification is systematizing knowledge about soils and determining the processes that control similarity within a group and dissimilarities among groups (Birkeland, 1999, p. 29). The classification system used in the United States is the U.S. Comprehensive Soil Classification System, or "soil taxonomy," published as *Soil Taxonomy* (Soil Survey Staff, 1975, 1999; www.soils.usda.gov). This system often is incorrectly referred to as the "7th Approximation" (the title of an earlier version of the classification system [Soil Survey Staff, 1960; see Soil Survey Staff, 1975, preface]). The U.S. system was developed in the 1950s and 1960s and was a revolutionary concept in soil classification. Most earlier schemes were based in large part on the presumed genetic history of the soils (e.g., red desert soil, brown forest soil)—which is almost never immediately apparent—and on nonsoil characteristics (e.g., local vegetation or groundwater level) rather than on properties of the soils themselves. In addition, many of the terms used in the earlier systems derived from various foreign languages, folk terms, and coined names; were generally poorly defined; and were not always mutually exclusive (Butler, 1980, pp. 72–74; Buol et al., 1997, pp. 222–223). The newer U.S. system is an approach to soil classification based entirely on the soils as they are, relying on properties observable in the field or measurable in the laboratory (Soil Survey Staff, 1975, 1999; Bartelli, 1984).

To appreciate the applications and limitations of soil taxonomy in archaeological and geoscientific research, the purpose of the system must be fully understood. An explanation of soil taxonomy is facilitated by contrasting it with what it is not. Soil taxonomy was designed to facilitate classification for soil survey and land-use purposes (Soil Survey Staff, 1975, pp. 7, 8; Bartelli, 1984) and is geographically biased toward the agriculturally productive soils of the midlatitudes. It was not designed to be a tool in soil geomorphic or other geoscientific research. The Soil Survey Staff (1975, p. 7; 1999, p. 15) describes the primary objective of soil taxonomy as having "hierarchies of classes that permit us to understand, as fully as existing knowledge permits, the relationships between soils and also between soils and the factors responsible for their character." Bartelli (1984, p. 9), in discussing the development of soil taxonomy, further notes that observable or measurable soil properties were selected to group soils of similar genesis.

Soil taxonomy is arguably at odds with these objectives, however. The system does not provide a means of understanding relationships between soils beyond their spatial relationships on soil maps, and it is also largely divorced from the factors of soil formation (viewed as both an advantage and disadvantage of the system; see Morrison, 1978; Birkeland, 1999; Holliday et al., 2002). The development of soil taxonomy involved essentially no research into the genetic relationships among soils, and very little soil survey research in the United States has focused on the genesis of soils or soil mapping units (Holliday et al., 2002). Examples of these aspects of soil taxonomy as manifested in soil surveys are explored in chapter 4. Furthermore, and of particular significance to geoarchaeology, soil taxonomy is not well suited for application to buried soils (discussed below and in chapter 5) and inadequately deals with soils heavily altered by human activity (so-called "anthrosols," discussed below and in chapter 11). Finally, soil taxonomy is not an exhaustive inventory of all known soils or pedogenic relationships. This is an important consideration when using soils, either at the surface or buried, for reconstructing the past. Some assume or imply that soil taxonomy represents the universe of soils and that interpretation is simply a matter of picking out the correct soil or soils from those listed (e.g., Smith and McFaul, 1997, p. 130; Retallack, 2001), but most of the soils listed in taxonomy at the suborder level are soils that have been investigated, to some degree, largely in the United States. Undoubtedly, many more variants (both in the United States and, particularly, around the world) await description and classification.

The U.S. soil classification system is based on a variety of differentiating characteristics including diagnostic horizons (table 2.1) and related properties, such as soil moisture and soil temperature. The terms for the different characteristics were derived from Greek and Latin roots. The diagnostic horizons and other characteristics are strictly defined and based on measurable soil properties and, therefore, convey a wealth of data. The diagnostic horizons are not the same as the more generally defined A-B-C horizon symbols used in field descriptions, although there is often a general correlation (table 2.1). The mollic epipedon, for example, is a surface horizon identified on the basis of thickness, Munsell color, organic carbon content, and citrate-extractable phosphorus content, among other characteristics. It may or may not be the equivalent to the A horizon (e.g., it may include the A and upper B horizon). In contrast, the A horizon is a field designation for a horizon found at the surface or below an O horizon and characterized by humified organic matter mixed with mineral material.

Table 2.1. General concepts for selected diagnostic horizons in soil taxonomy

Epipedons (Diagnostic Surface Horizons)	
Anthropic	Mollic epipedon high in phosphorous content
Histic	Surface horizon very high in organic matter (O)
Mollic	Deep, dark, humus-rich surface horizon with abundant cations (A, A&B)
Ochric	Surface horizon that does not meet the qualifications of any other epipedon (A)
Plaggen	An artificially made surface layer produced by long-term manuring
Diagnostic Subsurface Horizons	
Albic	Light-colored horizon with significant loss of clay and free iron oxides (E)
Argillic	Horizon of significant clay accumulation (Bt)
Calcic	Horizon of significant accumulation of calcium carbonate (Bk)
Cambic	Some reddening or structural development; reorganization of carbonates if originally present (Bw)
Kandic	Heavily weathered, clay-rich horizon low in bases (Bt)
Natric	Argillic horizon high in sodium (Btn)
Oxic	Intensely weathered horizon virtually depleted of all primary minerals and very low in bases
Petrocalcic	Calcic horizon strongly cemented by calcium carbonate (km or K)
Spodic	Horizon of significant accumulation of aluminum and organic matter with or without iron (Bh, Bs, Bhs)

These are very general definitions of terms used in the soil classification system of the U.S. Department of Agriculture (based in part on Wilding et al., 1983a). Considerable field and laboratory data are necessary to determine diagnostic horizons. For a complete list and criteria see *Soil Taxonomy* (Soil Survey Staff, 1999). Diagnostic horizons are not exact equivalents of field designations (e.g., not all Bt horizons are argillic horizons), although there is a general relationship. Some probable field equivalents are given in parentheses.

The classification system is hierarchical with six categories (from general to specific): order (table 2.2), suborder, great group, subgroup, family, and series. A formative element of the term used at each higher category is carried down through successive lower categories to the great group level (fig. 1.1A). For example, the Flatirons series is a Mollisol in an ustic moisture regime with a well-expressed argillic horizon, classifying it in the Paleustoll great group (fig. 1.1A). The subgroup is written as two words. In the case of the Flatirons soil, it is in a dry setting and therefore identified as an Aridic Paleustoll. Families differentiate the subgroups on the basis of physical and chemical characteristics such as texture, mineralogy, and temperature. The Flatirons soils are clayey-skeletal, smectitic, mesic Aridic Paleustolls. The soil series is the basic unit of soil mapping (further discussed in chapter 4). The series represents the grouping of soils with similar profile characteristics within a given region. Soil series are typically named after places, and usually a town. For soil geomorphological and geoarchaeological research, an understanding of the classification to the great group or possibly subgroup level probably is sufficient and, in any case, more meaningful than the family and series nomenclature (further discussed in chapter 4).

The terms used in U.S. soil taxonomy are many and are strictly defined (tables 2.1, 2.2). Many of these terms and their definitions appear unusual and confusing at first, but with experience, researchers should find them quite usable and

Table 2.2. General concepts of the soil orders in soil taxonomy

Term	Definition
Alfisols	Soils with argillic horizon, but no mollic (A-Bt), that are lower in bases than Mollisols; typically found in humid, temperate regions
Andisols	Soils formed in volcanic ash and related volcanic parent materials (A-C, A-Bw)
Aridisols	Soils formed in desert conditions (Entisols can also be found in deserts) or under other conditions restricting moisture availability to plants (high salt content; soils on slopes); with or without argillic horizon, but commonly with calcic, gypsic or salic horizons (A-Bw-Bk; A-Bt-K; A-By)
Entisols	Soils with little evidence of pedogenesis (A-C, A-R); very few diagnostic horizons
Gelisols	Permafrost soils; very common in high latitudes
Histosols	Organic soils, such as peats
Inceptisols	Soils exhibiting more pedogenic development than Entisols, with appearance of diagnostic surface and subsurface horizons that are not as well developed as in most other orders (A-Bw)
Mollisols	Soils with a mollic epipedon and high in bases throughout; typical of continental grasslands
Oxisols	Soils with an oxic horizon; found in tropical regions and include many soils formerly termed Laterites and Latosols
Spodosols	Soils with spodic horizons (O-A-E-Bh/Bs/Bhs); typical in cool, humid climates under coniferous forests
Ultisols	Highly weathered soils that have argillic horizons and that are very low in bases (A-Bt); typically found on older landscapes in warm, humid climates
Vertisols	Soils high in clay content in climates with distinct wet and dry seasons and that shrink and swell markedly

For a complete list and criteria see *Soil Taxonomy* (Soil Survey Staff, 1999). To properly classify a soil one must follow the guidelines and criteria for diagnostic horizons and classification in soil taxonomy (Soil Survey Staff, 1999). This table presents only the principal characteristics of the soil orders.

useful. Buol et al. (1997) provide a good introduction to soil classification, but formal instruction is especially recommended.

The U.S. soil classification system has been compared to hierarchical biological classification (e.g., Retallack, 1990, p. 91). As pointed out by Fitzpatrick (1971), however, most soil classification systems are hierarchical, yet there is no reason to assume that soils are genetically related in the manner that they are grouped into hierarchies. Soil classifications are developed for a variety of purposes and are imposed on a natural system. Most soil classification schemes, such as soil taxonomy, are designed for agricultural and other types of land use, not for interpreting landscape evolution or human history. This particular aspect of soil taxonomy seems to be poorly understood by many geoscientists.

Soil taxonomy has been criticized by some—geologists and soil geomorphologists in particular (e.g., Hunt, 1972; Morrison, 1978; Holliday et al., 2002), but also pedologists (e.g., Fitzpatrick, 1971, 1979). Hallberg (1984) provides an excellent overview and critique of soil taxonomy from the perspective of a geologist. Most problems in the classification system stem from the need for arbitrary rules or decisions inherent in any attempt to categorize and classify parts of a continuum. As Hallberg (1984, p. 53) notes, "classification involves . . . the Tyranny of the Pigeonhole." He further points out that, "the institutional, or bureaucratic implementation of the U.S. system of soil taxonomy . . . has often had the effect of making [it] inflexible; its implementation often rigid and legalistic" (Hallberg, 1984, p. 57). In other geosciences, in contrast, there are a number of "scientific codes or guidelines put forth by professional societies, which are freely debated in the scientific literature [and at] professional meetings," (Hallberg, 1984, p. 57) such as the *Code of Stratigraphic Nomenclature* (e.g., North American Commission on Stratigraphic Nomenclature [NACOSN], 1983), in geology (further discussed in chapter 5). Because many soil geomorphologists are trained in geology or physical geography they often alter terms from the soil taxonomy to suit their needs. For example, they provide adjectives such as "weak argillic horizon" (Btj or juvenile Bt in field nomenclature) for horizons that barely meet argillic criteria (Birkeland, 1999). This can be a very useful approach in soil geomorphic and geoarchaeological research, and a number of nonstandard terms are used in this volume (see further discussion and appendix 1).

More specifically, much of the criticism of soil taxonomy is aimed at the terminology introduced, the absence of genetic information in the system, and the difficulties in applying the system to buried soils. The last point is of concern in soil geomorphic studies, but much of the basic terminology of soil taxonomy remains useful in such circumstances even if full classification is not. Some of the classificatory terms do appear odd at first, but they are simply not that difficult to learn. Once the basic diagnostic terms are understood, a vast number of classificatory words can be put together, and a single word will then carry a large amount of qualitative and quantitative information (fig. 1.1A). Moreover, the system is used by most individuals doing the basic soils research (in both academic and governmental contexts) in the United States, and therefore any investigator interested in soils in the United States must become familiar with the system to understand the literature.

Beyond the pros and cons of the basic concepts behind soil taxonomy and its terminology, the system is difficult to apply to buried soils (Mack et al., 1993; Nettleton et al., 1998; Holliday et al., 2002; see also the discussion of buried soils in chapter 5). Applying both the diagnostic horizon nomenclature and taxonomic classification to buried soils is problematic because of either the erosion of near-surface horizons or the postburial alteration of the soils (chapters 5, 10), both of which are greater the longer the soil or sediment has been buried, and because soil taxonomy is explicitly designed for surface soils. Components of buried soils can be described in terms of diagnostic horizons, but the characteristics of the horizon may be different from its preburial state. Erosion or compaction changes horizon thickness, for example, which is a significant component of the requirements for a mollic epipedon and a calcic horizon (table 2.1). The color of a mollic epipedon, also a classificatory requirement, usually changes after burial because of oxidation of organic matter. Furthermore, pedogenesis in the deposits that bury a soil may modify the buried soil in a process known as "soil welding" (Ruhe and Olson, 1980; also see chapter 5): a calcic or argillic horizon can be superimposed over a buried mollic epipedon or argillic horizon, for example.

Taxonomic classification of a buried soil often is possible if a complete buried profile is preserved and the burial was recent, but the classification will differ from the preburial classification over the long term. Burial changes characteristics of the diagnostic horizons and almost always changes soil moisture and soil temperature—environmental characteristics necessary for much classification. Significant changes in soil and water chemistry can also accompany or follow burial and can significantly affect classificatory characteristics such as base saturation. Further problems can be encountered when classifying buried soils in terms of horizons and taxonomy for paleoenvironmental interpretations because few types of horizons or taxonomic categories are associated with unique environmental conditions (e.g., Dahms and Holliday, 1998), an issue further explored in chapter 8.

Because of the difficulty of applying soil taxonomy to the classification of buried soils, especially pre-Quaternary soils, alternative classifications have been proposed (Mack et al., 1993; Nettleton et al., 1998, 2000). Interestingly, they use terms, concepts, and a structure from or similar to that of soil taxonomy. The determination of which system, if any, becomes the lingua franca in studies of buried soils awaits extensive field testing.

Knowing how to describe or classify a soil is only the first step in using soils in archaeology or soil geomorphology. Such knowledge is only a tool for communication and interpretation. The number of archaeological site reports with descriptions and classifications of soils, but no further discussion of them, indicates that this aspect of recognizing and classifying soils is poorly understood by many archaeologists (and some collaborating soil scientists and geologists). Recognizing soil horizons or classifying a soil is a very basic first step in the geoarchaeological interpretation of a site, akin to learning pottery types or recognizing flaking patterns on stone tools as a step toward reconstructing human behavior.

Working outside of the United States, researchers may want to become familiar with other classification systems. Some are similar to soil taxonomy, but others are not. Lof (1987) provides a useful correlation and comparison, with color photos, of the FAO classification scheme with those of the United State, Canada, England and Wales, France, Germany, and Australia. Comparisons of the structure, philosophy, advantages, and disadvantages of a variety of soil classification systems, including soil taxonomy, are provided by the Soil Survey Staff (1975, pp. 437–455), Butler (1980, pp. 72–122), and Buol et al. (1997, pp. 195–233). The International Institute for Geo-Information Science and Earth Observation (ITC), in The Netherlands, prepared a very useful "Compendium of On-Line Soil Survey Information" including information on and comparisons of the major national soil classification systems (www.itc.nl/~rossiter/research/rsrch_ss_class.html).

Other Terms

Beyond the "official" governmental soils nomenclature, a variety of informal terms is used in the more geologically focused studies of soils such as soil stratigraphy and soil geomorphology, and in the more archaeologically focused work in geoarchaeology. The following section is a discussion of the more widely used terminology from these fields. Some other, specific soil geomorphic terms are introduced in chapter 3. The specifics of soil stratigraphic terminology are reviewed in chapter 5.

The term "paleosol" is widely used in archaeology and other Quaternary studies and is variously defined. These definitions include soils of obvious antiquity (Morrison, 1967, p. 10), ancient soils (Butzer, 1971, p. 170), soils formed on a landscape of the past (Ruhe, 1965, p. 755; Yaalon, 1971c, p. 29; Gerrard, 1992, p. 202; Catt, 1998) or under an environment of the past (Yaalon, 1983), a soil with distinct evidence that the direction of soil development was different from that of the present (Catt, 1998), a soil formed during an earlier period of pedogenesis (Allaby and Allaby, 1991), or soils formed under conditions generally different from those of today (Plaisance and Cailleux, 1981, p. 702).

Specific types of paleosols include "buried soils," which are soils covered by sediment (figs. 2.1–2.3, 6.4, 6.14, 6.18, 6.20, 6.21, 7.6, 7.11, and 8.4–8.6); "relict soils," which are soils formed on past landscapes or under past environments and never buried; and "exhumed soils," which are soils that were buried and subsequently re-exposed (Ruhe, 1965; Valentine and Dalrymple, 1976; see Johnson and Hole [1994] and chapter 5 for further discussion of these terms). Among these terms, "buried soil" is probably the least ambiguous, although a distinction between a buried soil and a buried paleosol is proposed (Catt, 1998, p. 84). The former term is used for "soils buried by deposits too thin to seal them from present pedogenesis and not showing evidence of development in a direction different from the present," and the latter term is proposed for soils that are buried and "isolated from present pedogenesis" or soils that are buried and exhibit "distinct evidence that the direction of soil development was different from that of the present."

Otherwise, among the definitions and types of paleosols, exactly what constitutes a past landscape or past environment or how old the soil has to be was never defined. Because landscapes always are being subjected to some modification and the environment is never static, and because all soils take some time to form, arguments have been made that all soils are paleosols and all unburied soils are relict soils, making such terms redundant (see also Bos and Sevink, 1975; Fenwick, 1985; Johnson et al., 1990; Bronger and Catt, 1998a,b; Follmer, 1998; D. L. Johnson, 1998; Johnson and Hole, 1994). Moreover, these definitions require that the history of the soil, the landscape, or both be known before the term can be applied. In any case, there seems to be no reason to differentiate soils based on their relevance to the past or present. In geology, a gravel layer is a gravel layer, whether it was deposited in this century or in the Permian. If the gravel is lithified then it is a conglomerate, but otherwise no distinction is necessary. The term "paleosol" seems to have some utility, however, judging from its widespread use (e.g., Follmer et al., 1998), especially in dealing with soils in the rock record—so-called pre-Quaternary soils (e.g., Retallack, 1990).

The following criteria and definitions, based largely on the author's experience and views, may clarify the differentiation of soils, buried soils, and paleosols. A soil, as defined previously, can be a recently formed unburied soil or deeply buried soil from the Triassic. Age, genesis, and stratigraphic position are irrelevant. A "ground soil" or "surface soil" refers specifically to unburied soils. A "buried soil" is any soil in which sediments cover a clearly recognizable horizon sequence, regardless of the thickness of the cover sediments (see also the definition of an "isolated paleosol" by Schaetzl and Sorenson [1987]). A paleosol is a

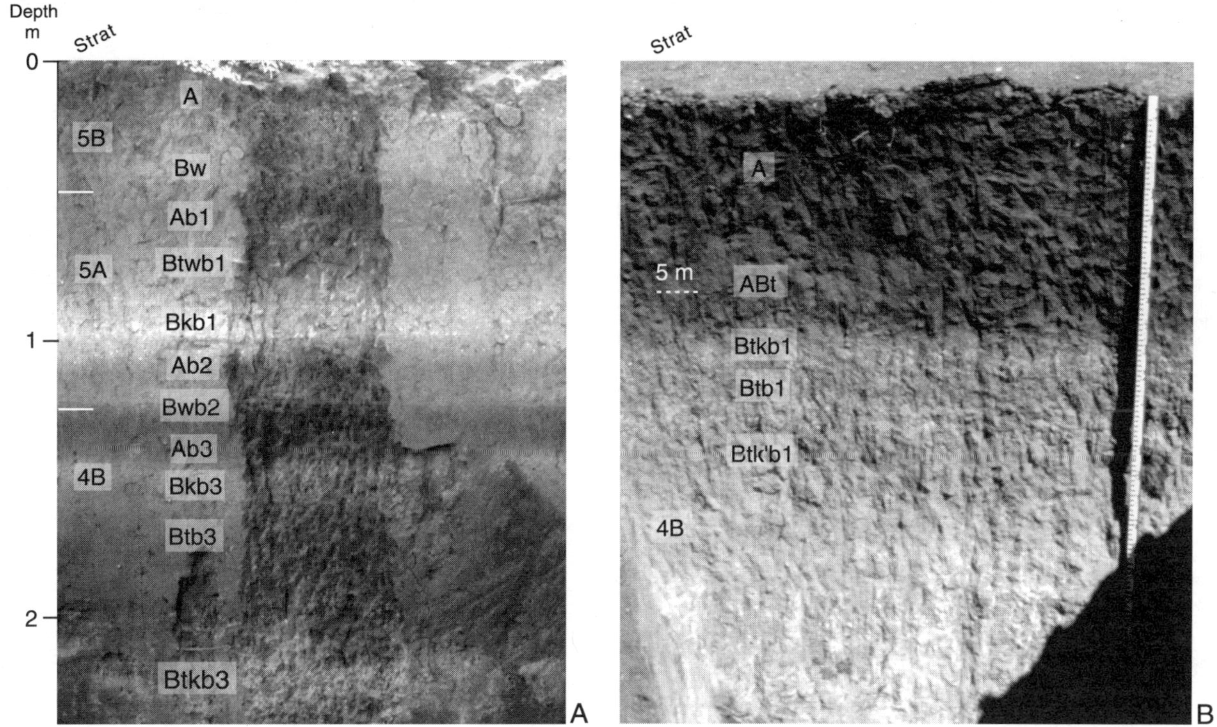

Figure 2.1 Buried soils at the Lubbock Lake Site, Texas. (A) Good example of the late Holocene soil sequence (Trench 95, with a profile picked out by a knife). The b3 is the Lubbock Lake soil, formed in stratum 4B ~4500–1000 yr B.P., then buried by stratum 5A. The carbonate in the Bkb3 horizon probably is from the b2 soil (lower 5A), providing a nice illustration of soil welding. The Apache soil (b1) formed in upper 5A. The weakly expressed surface soil is the Singer Soil, formed in stratum 5B. (B) Cumulic facies of the Lubbock Lake soil in a lowland valley-axis position (Trench 141). The Btkb1, Btb1, and Btk'b1 horizons are soil stratigraphic equivalents of the Bkb3-Btb3-Btkb3 in (A). The lower ABt probably was the original A horizon. It was cumulized by slowly aggrading mud of stratum 5 m (a lowland, muddy, valley axis facies of 5A) and eventually evolved into a Bt horizon. Thus, the cumulic A-ABt horizons in the lowland setting are a facies of the b2, Apache, and Singer soils more common in valley margin positions (fig. 2.1A).

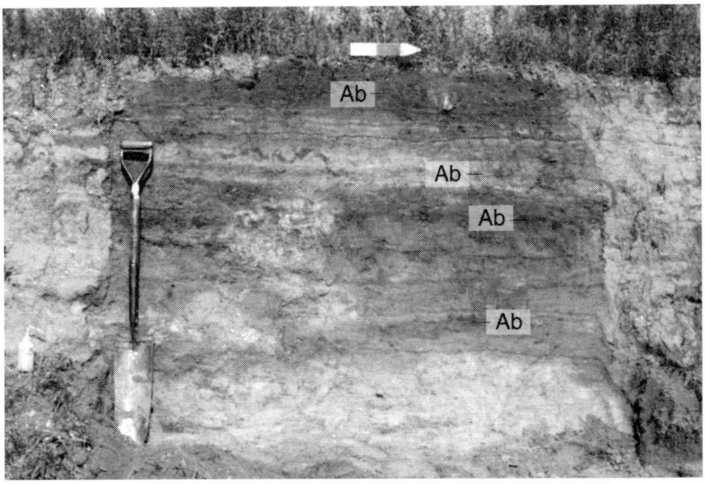

A

B

Figure 2.2 Buried soils in alluvium. (A) Fluvent in the floodplain of the Brazos River, north-central Texas. Multiple thin, dark, buried A horizons (Ab) are apparent, in contrast to the lighter color of the unaltered alluvium. (B) Buried soils developed in alluvium at the Alum Creek site, central Kansas (Mandel and Bettis, 2001b, fig. 7.3; from *Earth Sciences and Archaeology*, by P. Goldberg, V. T. Holliday, and C. R. Ferring, © 2001, Kluwer Academic/Plenum Publishers. Reproduced with permission of Kluwer Academic/Plenum Publishers and R. D. Mandel). A relatively thin and weakly expressed buried A horizon is apparent above the shovel handle. The shovel is resting against a dark, cumulic A horizon. Note the sharp upper boundaries of the two buried A horizons in contrast to their more gradual lower boundaries (and compare with the abrupt lower boundary of the organic-rich deposit illustrated in fig. 5.7). Photo provided by and reproduced with permission of R. D. Mandel.

Figure 2.3 A very strongly expressed Bt horizon (with Bk horizon immediately below, and a less well expressed Bt below that) buried in fine-grained Pleistocene alluvium in southeastern Arizona. The A horizon is missing, probably because of oxidation and perhaps erosion. Note the very strong development of prismatic structure and the sharp upper boundary, but more gradual lower boundary.

soil that is both buried and lithified. This definition is relatively unambiguous and generally easy to apply in the field on visual examination. A related term that has some utility but is now out of favor is "fossil soil." The term was essentially synonymous with buried soil until the mid-20th century (Johnson and Hole, 1994). Subsequent usage of the term was somewhat muddled, however, and it has largely disappeared from the literature. Fossil soil is a useful concept, though, as it refers to something that was once living and is now dead and buried or, using a broader definition of fossil (as an adjective), as something that existed in the geologic past and of which there is still evidence (Bates and Jackson, 1980, p. 243).

Morrison (1978, p. 100) seems to equate buried soils with bisequums. A bisequum, however, refers to a pair of genetically related sets of illuvial B horizons with overlying eluvial horizons—typically an E-Bs sequum overlying an E′-Bt sequum (Schaetzl, 1996, p. 23). By definition, a bisequum is a single soil.

The terms "monogenetic" and "polygenetic" soils were coined by Bryan and Albritton (1943) based on geoarchaeological research in west Texas. A monogenetic soil was initially defined as "one developed in one climatic regime," and a polygenetic soil was defined as having "developed in more than one climatic regime" (Bryan and Albritton, 1943, table on p. 477). Catt (1998, p. 84) provides more up-to-date definitions. A monogenetic soil "formed in a period when the variation in environmental factors was too small to produce detectably different assemblages of soil features . . . (i.e., the direction of soil development was constant)." A polygenetic soil "formed in two or more periods when the environmental factors were sufficiently different to produce detectably different assemblages of soil features . . . (i.e., the directions of soil development were different in the periods involved)." Similar to the criticism of the term and concept of paleosol (vs. soil), some argue that there is no such thing as a monogenetic soil and that all soils are polygenetic because environmental factors are always changing to some degree (e.g., Yaalon, 1971a, p. 155; Johnson et al., 1990; Johnson and Hole, 1994; D. L. Johnson, 1998). With the emphasis now placed on the direction of pedogenesis and the detectability of evidence for these changes, the terms may have merit and utility. This is clearly indicated by the concept of the "vetusol," described next.

Vetusol is a more clearly defined term that provides an important addition to concepts of buried soils and what they may tell us about environmental change. It refers to soils that evolved under the same or similar processes from their inception to the present time (Cremaschi, 1987). The term can be applied to a surface soil or a buried soil or both. For example, Busacca and Cremaschi (1998) describe a surface soil and as many as six buried soils that comprise a single vetusol. The point is that vetusols developed "under the influence of a single set of major pedogenic processes throughout their history and that they reacted only slightly to Pleistocene environmental changes" (Busacca and Cremaschi, 1998, p. 96). A vetusol, then, could be considered a kind of monogenetic soil that also represents long-term pedogenesis.

The study of paleosols/buried soils often is referred to as "paleopedology" (Yaalon, 1971b; Retallack, 1990; Follmer, 1998). As with "paleosol," this term seems to be very popular and is widely used, but its real utility is questionable.

In a broad, holistic sense, the study of buried soils, including paleosols, is a component of pedology, soil-geomorphology, and soil stratigraphy. As noted, however, pedology as traditionally taught in the United States does not deal with buried soils, and thus, some workers have found it necessary to use the term paleopedology, essentially to differentiate a geoscientifically based study of soils from an agricultural one.

"Catena" is another important term in soil geomorphology because it describes a fundamental and significant aspect of soils (noted in chapter 5 and more fully discussed in chapter 9). As originally proposed (Milne, 1935a,b), a catena is a mapping unit defining a group or pattern of soils along a slope formed as a result of differential drainage on the slope, solute leaching down-slope, and erosion and deposition along the slope. In the United States, however, the term came to be applied to a sequence of soils developed in uniform parent material on a hill slope that vary according to drainage (fig. 2.4; Soil Survey Staff, 1951, p. 160). The term is often used interchangeably with "toposequence" (a group of soils that vary only in their topographic position; they otherwise formed in the same parent material, under the same climate and vegetation, and for the same amount of time; discussed in chapter 3) (Jenny, 1980, p. 280; Hall, 1983, p. 124; Buol et al., 1997, pp. 154–155; Birkeland, 1999, p. 235). The processes of erosion and deposition inherent in the formation of catenas mean that the age of the landscapes will vary, however (figs. 2.4 and 9.1). So even following the Soil Survey Staff approach, only some catenas are toposequences, and only some toposequences are catenas. In this volume the term catena is used more in the sense of Milne's original concept.

Related to catena is the concept of the "soilscape." A soilscape is the pedologic portion of the landscape (Buol et al., 1997, p. 383) or the pattern and dis-

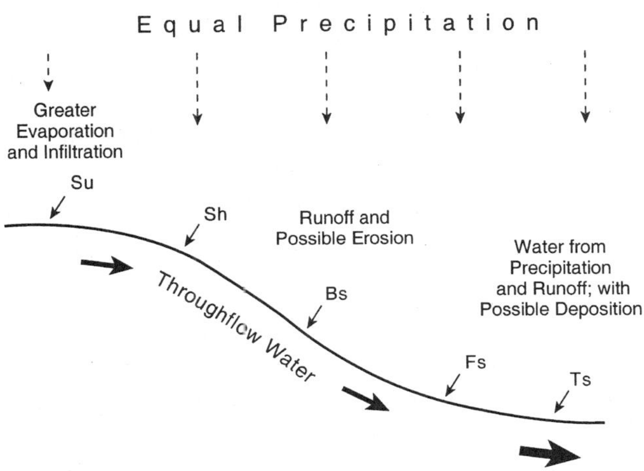

Figure 2.4 Schematic diagram illustrating geomorphic and pedogenic processes along a hypothetical catena (based on Birkeland et al., 1991, fig. 5.1, and Birkeland, 1999, fig. 9.2). Slope positions are: Su = summit, Sh = shoulder, Bs = backslope, Fs = footslope, and Ts = toeslope.

tribution of soils across the landscape. In a sense, soilscapes represent the expansion or distribution of catenas across the landscape. Catenary relationships among soils on a slope and the distribution of catenas across soilscapes are unique characteristics of soils as they form on and across landscapes and serve to distinguish soils from other geologic phenomena.

The terms "anthrosol," "anthropic soil," or "anthropomorphic soil" have been around since at the least the mid-20th century (e.g., Soil Survey Staff, 1951, p. 133; Bunting, 1965, p. 131), but became popularized by Eidt (1977, 1984, p. 23). He initially defined anthrosols as "settlement-affected soils" (Eidt, 1977, p. 1327), and later as soils "whose native traits have been significantly altered by human activities" (Eidt, 1984, p. 23). Simple physical changes of a "native soil" resulting from human activity do not seem to be enough to qualify the soil as an anthrosol. In the examples discussed below and in most of the other literature on this topic, soils identified as anthrosols typically had significant chemical inputs as well as obvious physical changes resulting from human activity. In any case, the term "anthrosol" is now part of the general parlance and is used more broadly than Eidt probably intended. Some refer to purposely altered soils as anthrosols and use the term interchangeably with anthropogenic (e.g., Smith, 1980). In contrast, in the FAO-UNESCO (1994, p. 61) soil classification system, anthrosol is one of the major soil groupings for a broad array of soils "in which human activities have resulted in profound modification or burial of the original soil horizons." Specific types of anthrosols in the FAO-UNESCO scheme include those subjected to deep cultivation, to long-term manuring, and to burial from long-term irrigation and from urban spoil. The FAO definition does not seem entirely satisfactory, however, because it could also include soils buried under mounds or tells or soils buried because of human-induced soil erosion. Few geoarchaeologists would consider those soils as anthrosols. Yet another use of the term is for soils that formed in anthropogenic sediments, but that do not necessarily display any indication of human effects on pedogenesis (e.g., Overstreet et al., 1988; Brinkmann, 1996; Overstreet and Grolier, 1996). This is not a widely used definition and is perhaps overly broad. The heart of the other definitions is that the soil itself is altered by human activity.

Eidt (1984, p. 23) further subdivides anthrosols into "anthropogenic" soils, which were intentionally altered, and "anthropic" soils, which were unintentionally altered. Establishing the intent of ancient occupants of a region is difficult to do, but there are several kinds of soils that almost certainly were purposefully modified (e.g., the "plaggen" and "terra preta" discussed in chapter 11), as well as soils on many different landscapes that were artificially altered, but not with the intent of making specific changes. "Anthropogenic" literally refers to having an origin stemming from human factors. An anthrosol, therefore, is the result of anthropogenic processes. Distinguishing an "anthropogenic soil" as a type of anthrosol seems to serve no useful purpose. The term "anthropic" also has meanings that vary from Eidt's definition. "Anthropic" is most widely used as a component of the U.S. soil taxonomy as a particular type of surface horizon. The "anthropic epipedon" is similar to the mollic in requirements for color, thickness, and organic carbon content, but it has a higher content of P_2O_5 because of "long-continued use of the soil by humans, either as a place of residence or as a site for growing irrigated crops" (Soil Survey Staff, 1999, p. 22), although little chemical or archaeological data are available for these horizons (e.g., Pettry and Bense, 1989). In this volume the term "anthrosol" is used to refer to soils that exhibit anthropogenic physical and chemical alterations. Soils subjected solely to chemical or physical alterations generally are not considered anthrosols, though some exceptions could be made.

Describing Soil Development

In discussing soils, particularly in terms of their age or for making stratigraphic comparisons or correlations, describing the relative degree of soil development or pedogenic expression often provides a useful shorthand. Formal classification is often cumbersome and overly detailed for general comparisons, and many classification schemes are not designed to distinguish between degrees of soil development. For example, in soil taxonomy, horizons that qualify as argillic or calcic can vary significantly in their degree of development and morphological expression. In view of this situation a variety of informal terms are used to distinguish between soils based on overall degree of development. The following is one of the more widely used qualitative schemes for describing the degree of soil development (from Retallack, 1990, table 13.1; Birkeland, 1999, pp. 19–20), incorporating the well-known system for describing stages of carbonate accumulation (after Gile et al., 1966, and Machette, 1985; see fig. 3.3):

> A *very weakly developed soil* is one with an A horizon and perhaps a Cox horizon. If pedogenic carbonate is present, carbonate morphology is stage I.
>
> A *weakly developed soil* is one with an A/Bw/Cox or Bk horizon sequence. Accumulation of pedogenic constituents is not sufficient to qualify the horizons as argillic, spodic, or calcic. If carbonate horizons are present, morphology is probably stage I.
>
> A *moderately developed soil* is one with an A/Bt/Cox, A/E/Bs/Cox, or A/Bt/Bk/Cox horizon sequence. The Bt, E, Bs, and Bk horizons meet the criteria for argillic, albic, spodic, and calcic horizons. Carbonate accumulations would display stage II morphology.
>
> A *strongly developed soil* is similar to a moderately developed one except that the B horizon in the strongly developed profile is generally thicker and redder, contains more clay, and has a more strongly developed structure (possibly a kandic horizon); if carbonates are present, they would form a K horizon (stage III and higher morphology).

The distinctions between a moderately and a strongly developed profile are qualitative, but they could be quantified on color and texture. In comparing texture, however, the parent materials should be similar. For example, comparing soil development on a loess with 20% primary clay with that on a gravelly out-wash with 5% primary clay is difficult. As regards color, there is no fast rule. Each development rank might be accompanied by a different hue. Thus, if a moderately developed soil has a 10YR hue, a strongly developed soil would have a redder color, perhaps a 5YR hue.

Methodology

Soil science in general and pedology in particular are unique among the geosciences in having sets of standardized and detailed methods for field investigations and laboratory analyses. This stems from the necessity to standardize methods used for soil mapping and soil characterization in the USDA. Hundreds if not thousands of pedologists work for the USDA, and standardization of methods is needed to produce soil maps, soil surveys, and soil data that are comparable in execution. Field methods are prescribed by the Soil Survey Division Staff (1993; replacing Soil Survey Staff, 1951), and the laboratory methods used for sample analyses are detailed by the Soil Survey Laboratory Staff (1996). The lab methods are updated infrequently to reflect new advances in soil analysis. The 1996 monograph supercedes several earlier versions (e.g., Soil Survey Staff, 1972, 1984; Soil Survey Laboratory Staff, 1992). The American Society of Agronomy also publishes a set of volumes on *Methods of Soil Analysis* (Page et al., 1982; Klute, 1986; Sparks et al., 1996; Dane and Topp, 2002). Soil geomorphology, however, melds the standard USDA and soil science approaches with methods developed in the geological sciences. The result is a broad array of field and lab methods for geomorphic and Quaternary stratigraphic research (e.g., Singer and Janitzky, 1986; Birkeland et al., 1991; Birkeland, 1999). Geoarchaeology has its own large and growing battery of methods (e.g., Shackley, 1975; Goldberg et al., 2001), including a large literature on the laboratory study of soils in archaeological contexts (e.g., Courty et al., 1989).

The availability of the many different methods in the study of soils provides geoarchaeologists with a formidable array of techniques. Because this volume is based largely on a soil geomorphic approach to the study of soils, much of this chapter emphasizes soil geomorphic methods. Other approaches are presented, however, particularly laboratory methods developed in geoarchaeology. Specific applications of individual techniques are provided throughout the rest of this volume, particularly in the discussions on identifying buried soils (chapter 5) and on human effects on soils (chapter 11) and in appendixes 2 and 3. Step-by-step details of individual methods are not presented here, however. For that the reader is directed to the references.

Field Methods

As in most field sciences, geoarchaeological interpretations of a site are rarely better than the field data on which they are based. "Post excavation analyses cannot overcome deficiencies in . . . observations, recording, or sampling" (Dincauze, 2000, p. 293). Careful observation and recording are essential for understanding and interpreting soils as well as other geoarchaeological phenomena. A number of authors describe basic geoarchaeological methods (e.g., Limbrey, 1975, pp. 254–280; Shackley, 1975, pp. 10–32; Gladfelter, 1977; Butzer, 1982, pp. 35–97; papers in Goldberg et al., 2001, and Stein and Farrand, 2001) and pedologic and soil geomorphic techniques (Limbrey, 1975, pp. 254–280; Catt, 1990, pp. 4–17; Retallack, 1997a, pp. 113–115; Birkeland, 1999, pp. 347–448) beyond the description of soil profiles, discussed above. Most of the nuts and bolts will not be repeated, but a few comments are offered based on personal experiences.

First and foremost, the geoarchaeologist should be involved in all phases of any archaeological project from the beginning with the initial planning and proposal stage (Rapp, 1975; Butzer, 1982, pp. 35–97; Goldberg, 1988; Luff and Rowley-Conwy, 1994; Quine, 1995; Dincauze, 2000, pp. 502–503, 510–511; papers in Goldberg et al., 2001, and Stein and Farrand, 2001). Many of us have had the experience of being brought into a project as afterthoughts to look at exposures (i.e., after much of the archaeology was removed) to answer questions such as "Is that a soil?" (or, in more hushed, almost reverential tones, "Is that a paleosol?"). And if it is, "what can the soil tell us about the archaeology?" Such issues are better dealt with if they are raised and addressed at the outset of any project.

The plea for involving geoarchaeologists at all phases of an archaeological project has been around for several decades (e.g., Rapp, 1975; Gladfelter, 1977; Goldberg, 1988) and now may not be as urgent as it once was. At this time, geoarchaeologists are much more commonly and intimately involved in field archaeology. In part this is probably because there are more university faculty who teach and conduct geoarchaeological research. There are also interdisciplinary professional organizations devoted to the topic (the Archaeological Geology Division of the Geological Society of America and the Geoarchaeology Interest Group of the Society for American Archaeology). However, the advent of Cultural Resources Management (CRM) projects in archaeology and the requirements for interdisciplinary evaluation and study of threatened archaeological sites and resources, and demands for predicting site locations, may be the biggest factors in establishing geoarchaeological studies as a routine component of many archaeological investigations (e.g., Bettis, 1992; Ferring, 1992; Stafford, 1995; Green and Doershuk, 1998).

Even though geoarchaeologists are now commonly involved in archaeological research, the role of soil studies in geoarchaeological research still seems unclear to many, including geoarchaeologists with little or no training in pedology (a common problem for geoarchaeologists trained in geology programs). Too many of us have had to deal with "cigar box geoarchaeology" (Thorson and Holliday, 1990), a term the author coined to refer to situations in which an archaeologist shows up on the doorstep of a geoscientist with a cigar box full of dirt clods and asks for help in interpreting their site (e.g., Farrand, 2001, p. 37). This is an almost meaningless exercise, in which geoarchaeological considerations clearly are an afterthought. Most archaeologists have some basic familiarity with stratigraphic principals and basic concepts of sedimentology and geochronology. Many are less familiar with pedology. In particular, identifying and interpreting soils on an archaeological site sometimes seems a bit mysterious. An important role for the geoarchaeologist is to familiarize the archaeologist with soils and their role in interpreting the site and to make plans for field sampling (e.g., Goldberg, 1988; Courty, 2001; Macphail and Cruise, 2001).

Once the fieldwork is underway, geoscientists often are amazed at the length, depth, and general availability of exposures in archaeological excavations. Most pedologists, stratigraphers, sedimentologists, and geomorphologists are accustomed to working with small exposures in road cuts, quarries, trenches, or cores.

Indeed, the standard soil profile in pedology and soil geomorphology is an individual pit or trench. Archaeological excavations often provide an unusual degree of exposure, allowing examination of (usually unavailable) microstratigraphic or facies details (e.g., fig. 7.10). The level of detail can be sometimes overwhelming and even obscure broader and more pertinent issues of human activity, site formation processes, or landscape evolution. This situation can usually be resolved by lengthy or repeated visits to a site. Familiarity with the site situation, the stratigraphy and geomorphology, and the archaeological and geoarchaeological questions being posed often help in resolving the more important aspects of the record available in exposure.

A crucial aspect of any geoarchaeological fieldwork is off-site investigation (Rapp, 1975; Gladfelter, 1977; Butzer, 1982; Quine, 1995). The soils, stratigraphy, sedimentology, and geomorphology of any site must be placed in a regional landscape context. There are several reasons for this. Archaeological sites were created in a regional context, and this context must be understood before the site can be fully understood. At a general level this means understanding the regional environment and its evolution, which also aids in reconstructing site formation processes. More specifically, understanding the "natural" setting (i.e., that undisturbed or minimally disturbed by human activity) around an archaeological site provides a control sample (Stein, 1985, 2001a; Quine, 1995; Courty, 2001, p. 211) for assessing human effect on the soils, sediments, and landscapes in the site. Another important though often overlooked aspect of comparing both landscape and site-specific aspects of geoarchaeology is to understand the unique character, if any, of an archaeological site. A variety of factors may contribute to decisions by ancient people on where habitations should be located. Some of these factors may be unique on the landscape (e.g., a spring or availability of other resources), and the resulting stratigraphic record in the site may not reflect the regional record—and vice versa.

There are many different settings for the examination and recording of soils and stratigraphy, both on site and off site. As noted, archaeological excavation profiles often provide outstanding exposures. Recording the profiles is usually a component of the archaeological research, and every effort should be made to have the geoarchaeologist involved in this important data collection, including preparation of field drawings, field descriptions, and photography of the sections. Beyond the excavation units, however, opportunities for examining the stratigraphy may be limited, and artificial exposures must be created. Excavation of backhoe trenches is a common means of stratigraphic prospecting in geoarchaeology (e.g., Stafford, 1981; Mandel, 1992; Stafford, 1995, pp. 85–86; Garrison, 2003, pp. 104–105). They can provide a good look at the stratigraphy, can be excavated and covered up relatively quickly, and backhoes are widely available on a rental basis. Effectively, however, they provide exposures of only a few meters depth. These trenches also can be extremely dangerous, and when working in or around them all safety precautions should be taken (e.g., shoring or stepping of walls, wearing hard hats, and working from ladders) (Bergman and Doershuk, 1995; Green and Doershuk, 1998, p. 131).

Another means of stratigraphic prospecting is by examination of cores extracted from the ground using any one of several powered, truck- or trailer-mounted soil-coring machines (fig. 2.5A; e.g., Giddings, Geoprobe, PowerProbe; McManamon, 1984; Stein, 1986; Stafford, 1995, pp. 83, 86–87; Mandel and Bettis, 2001b, pp. 183–184; Garrison, 2003, pp. 104–117). These probe machines yield continuous cores in the range of 5–9 cm in diameter and typically 120 cm in length, providing the sample is easily recovered (fig. 2.5B). Core extraction sometimes is a problem, however. Further, cores do not provide as good a view of stratigraphic continuity or lateral variability of stratigraphic units as a trench, but they are much less destructive. They can also penetrate significant thickness of sediment; as deep as 10–15 m, depending on the nature of the deposits. Soils (surface and buried) and sedimentary structures are easily recognized in cores. Cores can also be recovered relatively quickly and can be easily archived (e.g., using plastic sleeves). Samples for radiocarbon dating can also be taken from cores if sufficient organic matter is present, which is determined by the organic matter content and the thickness of the sample itself, the diameter of the core, and the dating method (e.g., radiocarbon by conventional vs. AMS techniques). Many geoarchaeologists also have easier access to coring devices (through university departments, government agencies, or consulting firms) than they do to backhoes. Coring machines also provide a means of rapid stratigraphic investigation of relatively large areas.

Hand-operated augers and probes have been used for many years in both archaeological and stratigraphic studies (e.g., Gifford, 1969; McManamon, 1984; Stein, 1986, 1991; Lippi, 1988; Schuldenrein, 1991; Erikson, 1995; Stafford, 1995; Weston, 1996; Cannon, 2000). The most common device is the bucket-auger, 8–10 cm in diameter, than can recover samples at approximately 10-cm intervals. With extension rods they can penetrate deposits to depths of 3 m or more, again depending on the nature of the sediments and soils. The augers can also be used sideways for sampling from sections. These bucket augers are cheap, lightweight, and easy to use. They can also be taken into areas not accessible to backhoes and power-corers (e.g., Lippi, 1988). Their portability is also a drawback, however: They are labor-intensive devices. The principal drawback to bucket augers, however, is that they mix the sample. The augering process obscures sedimentary and soil structures, depositional contacts, and soil boundaries. Bucket augers provide generalized stratigraphic information but are not useful for microstratigraphy or detailed stratigraphic mapping. Hand probes of the "Oakfield" variety will yield undisturbed cores. They are simply pushed down into the ground using the weight of the operator. They are very useful in coring shallow (~≤1 m) sites in fine-grained sediment. Thicker sediments can be difficult to force, though, unless the sediment is particularly soft, such as loess. Hand-operated probes with a larger and heavier sampling barrel, driven with a sliding drop hammer, can go deeper through relatively coarse material and yield intact cores (Cannon, 2000).

Intermediate between the powered, mounted coring machines and hand augers are portable power drills capable of taking cores. Fuel must be carried, but initial cost is significantly less than for truck- or trailer-mounted rigs. Gordon (1978) illustrated the utility of portable core drills for coring frozen sites. He was able to map soil stratigraphy as well as identify occupation zones and site boundaries by examining frozen cores. In Britain, Canti and Meddens (1998) and Bates

A

B

Figure 2.5 Examples of soil coring. (A) A trailer-mounted Giddings soil-coring rig in use at the Miami (Clovis) site in the Texas Panhandle (Holliday et al., 1994). It was towed into position to extract a series of cores from playa sediments that contained mammoth bone and Paleoindian artifacts. (B) An example of a 3-inch diameter core extracted with a Giddings rig. Buried soils and sediments show well in these cores (arrow at 30 cm indicates top of Btb horizon).

et al. (2000) describe several relatively portable and inexpensive commercially available devices capable of extracting cores. Bates et al. (2000) also discuss the use of geotechnical borehole data in geoarchaeological contexts.

Coring and augering devices both have been employed to prospect for deeply buried archaeological sites and to recover microdebitage (McManamon, 1984; Stein, 1986; Michlovic et al., 1988; Stafford, 1995). These approaches raise a variety of questions and issues of sampling and representativeness (reviewed and assessed by Stein, 1986, and Stafford, 1995), but with proper caution and an awareness of the potential problems, they can be useful for directly linking buried occupation zones to the stratigraphy. In the study reported by Michlovic et al. (1988), samples of buried A horizons found in cores were specifically targeted for screening and examination under a microscope to detect microartifacts as indicators of buried sites.

A combination of trenching, coring, and augering techniques probably is the most valuable and informative approach for tracing stratigraphy over large areas. Which method or methods to emphasize, however, depends on local field conditions, thickness of the deposits, size of the area to be investigated, and current land use (Stafford, 1995, p. 88), as well as the time and funds available. Zangger (1993, pp. 10–15), for example, discusses the integration of a variety of field strategies, including coring, for a regional geoarchaeological study on the Argive Plain of Greece.

Archaeologists and geoscientists (especially those with training in pedology) often have very different approaches to preparing a profile for study and recording. Archaeologists invariably prepare smooth, straight vertical walls using sharp trowels as part of a basic unit-by-unit gridded excavation strategy. In addition, the clean wall often displays color contrasts and bedding well. Geoscientists, especially pedologists, on the other hand, often poke at walls and sections with a field knife or small pick to allow the soil structure to be expressed (this "plucking" procedure is well described by Vogel, 2002, pp. 3–4; the technique sometimes allows sedimentary structures to be expressed as well; e.g., Turner et al., 1982; see fig. 2.1A). Archaeologists and geoscientists tend to be quite surprised when first encountering the technique of the other field (and archaeologists often are aghast at having their straight walls poked), but both approaches are very helpful in deciphering the subtleties of microstratigraphy and soil horizonation. Ideally, exposures should be troweled alongside a section that is worked out for structure. Both should be included in photographs. Cores will produce similar kinds of information. If not smeared, the smooth exterior of a core, perhaps after light troweling, should be measured and described. Then the core should be opened. If it is a moist core high in fines (silt and clay), it will probably have to be cut open (a trowel works well for this), which inhibits the expression of structure. Otherwise, a core can be opened to expose the structure by taking a field knife, holding it above and parallel to the core, and carefully striking the core with the sharp edge down.

The description of soil horizons and profiles (i.e., the description of what can be called soil macromorphology, to differentiate between descriptions in the field and descriptions of thin sections in the lab for micromorphology) is based almost entirely on qualitative features of the soil observed in the field. The standard

Table 2.3. The interpretive value of soil morphology at archaeological sites

Soil property	Potential interpretative characteristics
Texture	Lithologic and pedologic discontinuities (erosion or hiatus in deposition); sediment type and sediment origin; porosity; relative energy of deposition for alluvial sediment; presence of Bt or argillic horizons
Structure	Relative abundance of macropores and potential artifact movement; degree of development (weak to strong) as indicator of soil age; degree of development of clay or organic coatings on ped faces in Bt or argillic horizons as indicator of soil age
Color	Indicator of organic matter and free iron content; sediment type; delineation of horizons; drainage and water table fluctuations (mottles or gley colors)
Boundary	Abrupt boundary as indicator of plowing (Ap horizon) or erosional disconformity; more diffuse boundary as indicator of horizonation (vs. depositional contact)
Consistence	Indicator of structural development, cementation, or consolidation, e.g. recent alluvium very friable or loose and noncompacted; indicator of porosity; indicator of primary (e.g., lacustrine) carbonate vs. pedogenic carbonate
Clay coatings	On peds or pores an indicator of degree of Bt or argillic development
Carbonate	Secondary $CaCO_3$ coatings, bodies, pore filling, and cementation as indicator of degree of development; primary $CaCO_3$ as indicator of lacustrine carbonate; removal as indicator of degree or depth of leaching
Horizon identification	Implications of major soil-forming processes, e.g., A = organic matter accumulation—stable surface, E = leached zone, Bt = clay illuviation; distinguish natural vs. artificial horizons; horizon thickness as indicator of degree of development

Modified from Foss et al. (1993b, table 1).

properties noted in U.S. Soil Survey work include color (using Munsell nomenclature) and presence or absence of mottles, texture (relative content of sand, silt, and clay), consistence (degree of cohesiveness of soil peds), structure (or lack thereof—the characteristic shape and arrangement of the peds), cutans (or coatings) on grains or ped faces, presence or absence and types of nodules or concretions (e.g., carbonates), voids (number, size, and shape), reaction to dilute hydrochloric acid (to indicate presence or absence of carbonates), and boundary characteristics of each horizon (Soil Survey Division Staff, 1993; Schoeneberger et al., 1998). The same basic characteristics have proven useful in soil geomorphic and geoarchaeologically-focused soils research (table 2.3; e.g., Limbrey, 1975, pp. 254–270; Foss et al., 1993a; Quine, 1995; Birkeland, 1999; Reed et al., 2000; Vogel, 2002; Garrison, 2003, pp. 95–103). Additional properties that may be locally significant include the nature of organic remains, the nature and degree of biologic activity (e.g., worm casts or krotovinas), the degree of stoniness, and the size, shape, and orientation of clasts. Most of these properties can also be used to identify and describe unlithified soil parent material; that is, unweathered geologic deposits.

An important point made by Vogel (2002, p. 2) regarding soil description is the difference between observation and interpretation. Determining characteristics such as soil color and soil texture are observations. Identifying soil horizons (or for that matter identifying the origins of soil parent material), however, are interpretations. Descriptions of field characteristics should be clearly distinguished from interpretations of horizons or soil genesis.

Good photography of soil profiles and stratigraphic exposures is a fundamental aspect of data recovery but requires some experience and patience. In this writer's experience (working mainly in the semiarid and arid midlatitudes) there are no hard-and-fast rules regarding photography of sections in direct light versus shade because so many variables are involved: the direction the exposure faces, time of day, time of year, degree and type of shading (cloud cover? vegetation? buildings?), and soil moisture. Some of these variables can be controlled (e.g., if soil pits are being excavated they can be oriented to take advantage of lighting conditions), but most cannot. Some field scientists argue that sections should never be photographed in direct light (e.g., Shackley, 1975, p. 19; Garrison, 2003, p. 105), but this writer has had excellent results under such conditions. Ideally, sections should be studied and photographed under a variety of lighting conditions. Wetting of exposures with a fine spray of water can help bring out color contrasts, but the utility of this method will depend on the texture of the sediment (clayey deposits will hold more moisture and stay darker longer than sandier layers).

Photographs should be in both color and black and white (BW). Color should be either color slides or digital images. Color prints should be avoided because they typically use secondary colors, whereas slides use primary colors. Slides, therefore, tend to produce "truer" colors, although color can vary among types of slide film. Film such as Ektachrome tends to emphasize blues and greens, whereas Kodachrome emphasizes reds and browns. Kodachrome, therefore, produces excellent stratigraphic photographs if soils and sediments tend to display brownish or reddish hues. Cameras with self-developing film can be used for quick reference or for storing with field notes, though the colors often are not reliable as "true." BW photographs can be made from color slides or from digital images, but the best approach is to have a second camera carrying BW film. The advent of digital cameras provides new possibilities for storing, recovering, and publishing photos, though many publishers prefer to work with glossy BW prints from negatives.

Sampling for subsequent laboratory analyses usually is the final step in fieldwork. A sampling strategy should be developed during the course of the field investigation. In a well known and widely used manual for archaeological field techniques, archaeologists were urged to collect "soil samples . . . from every site excavated" (Heizer, 1966, p. 78; Heizer and Graham, 1967, p. 106), but with no discussion of what should be sampled or why the samples should be collected (and no discussion of what "soil" is), an important omission criticized by both pedologists (Tamplin, 1969, p. 155) and archaeologists (Binford, 1968). Subsequent editions mention the collection of soil samples to detect evidence for human activity (Hester et al., 1997, p. 136), but otherwise provide no substantive discussion of soils or soil sampling (Hester et al., 1997, pp. 136–137, 246, 250–251).

There are many reasons for sampling and subsequent laboratory analyses. If the degree of pedogenic development is of interest, then all soil horizons should be sampled along with unweathered sediment from below the soil (the parent material) and, in the case of buried soils, from above. Samples can be collected by horizon or in specified intervals (e.g., every 5 or 10 cm) within the horizon. If the research questions focus on site formation processes such as sedimentation rate or depositional environment, then samples should be collected at close intervals from throughout each stratigraphic unit. "Close" is a relative term depending on the thickness of the deposits, but sampling a few to every ~10 cm in a vertical column would be appropriate. Sampling at 1 to 2 cm intervals might be more appropriate if the research focuses on variation in occupation intensity through time via phosphorus analysis at a stratified site. Sampling laterally across a stratigraphic unit is appropriate if sedimentological or pedological facies variation is an issue or to aid in stratigraphic correlation.

Sample size will depend on many variables. In sampling from cores, the size of the core will pose obvious limitations on sample size. The logistics of getting samples from the field to the laboratory, such as a long hike with limited carrying capacity, may also be a consideration. More generally, the samples should be representative of the soils and sediments under scrutiny and be appropriate to the research question being asked (Garrison, 2003, pp. 118–119). Ideally samples should be at least 50–100 g. This is usually enough to run most standard analyses and still provide spare sample in case reruns are necessary. Larger samples are useful for both backup and archival purposes. A somewhat unusual kind of sampling is recovery of soil "monoliths" and "peels" (Dumond, 1963; Garrison, 2003, pp. 106–107). Monoliths are solid, intact columns of sediment and soil representing preserved sections of a profile. They are on the order of 10 to 20 cm wide, and one to two m high or higher. They are made by carving the section from a stratigraphic exposure. The carved sections are removed and preserved in one of two ways: They can be cut to fit long containers of wood or metal open on one side and constructed for the purpose of housing the monolith (Garrison, 2003, pp. 106–107), or the carved section can be mounted on a plank before detaching from the section, impregnated with a resin, and then removed (Dumond, 1963). Peels are preserved sections of a profile that contrast with monoliths by being thin (<1 cm thick) but tens of centimeters (or more) wide and high. Monoliths and peels preserve bedding, soil color, structure, and boundaries, and stratigraphic contacts, and they can be further sampled for laboratory analyses if they are not impregnated. Monoliths and peels are particularly useful as archives of site stratigraphy if the site is in danger of complete destruction.

Ultimately the options for sampling strategies are almost limitless and require careful consideration by the geoarchaeologist and archaeologist (e.g., Goldberg and Macphail, 2003; French, 2003, pp. 45–46). Courty (2001, p. 212) provides a particularly honest and well-taken observation about sampling: "The novelty of archaeological sites and soil/depositional-related archaeological problems makes irrelevant a standardized sampling procedure for routine soil investigations. To the contrary, sampling requires flexibility, intuition, and the ability to accept the fact that errors will be later revealed . . ." Her remark pertains to micromorphology, but can be applied to any sort of geoarchaeological sampling.

Development Indices

A number of investigators have attempted to quantify field characteristics as a means of assessing the relative degree of pedogenesis for stratigraphic correlation or relative dating. The more successful of these "semiquantitative" indices (in which a number is assigned to a qualitative feature or characteristic) are summarized by Birkeland (1999, pp. 19–23; see also table 7.4). Color indices convert the Munsell notation (for a horizon or profile) to a single number as a means of comparing the degree of reddening. Profile development indices assign numbers or a "rating" to physical properties such as grade and type of structure and boundary distinctness, incorporates horizon thickness, and then compares the properties to the parent material or C horizon as an indicator of degree of soil development (or relative age). An attraction of these various indices is that they require no special equipment and cost nothing to perform beyond an investment of time.

Laboratory Methods

The study of soils in pedology, soil geomorphology, and geoarchaeology often includes analysis of physical and chemical characteristics in the laboratory. These characteristics have proven useful in identifying and interpreting natural and human-induced pedogenic processes and in assessing and comparing the degree of soil development—characteristics of archaeological significance discussed in subsequent chapters. Soil chemistry, especially phosphate analysis, has been used for decades to detect evidence of human occupation and to decipher prehistoric land use and modification of the soil (e.g., Arrhenius, 1931, 1963; Solecki, 1951; Woods, 1977; Berlin et al., 1977; Eidt, 1977, 1984; Gurney, 1985; Lippi, 1988; Schuldenrein, 1995; see chapter 11). In pedology and soil geomorphology some of the more commonly measured and useful soil attributes include particle-size distribution (relative percentages of sand, silt, and clay; also known as grain-size distribution or granulometry in geoarchaeology), bulk density (weight per unit volume of soil), clay mineralogy, and the content of organic matter (or organic carbon), calcium carbonate, iron, aluminum, and phosphorus (Singer and Janitzky, 1986; Catt, 1990; Birkeland, 1999). Aspects of laboratory analysis of soils are further discussed in appendixes 2 and 3.

Another powerful analytical tool in soil analysis is micromorphology, the study of soils and unconsolidated sediments in thin section using a petrographic microscope (Brewer, 1976; Wilding and Flach, 1985; Bullock et al., 1985; Fitzpatrick, 1993). The technique was developed largely in Europe early in the 20th century and essentially is soil petrography. There were a few early attempts at applying micromorphology to soils in archaeological sites (e.g., Dalrymple, 1958), but initially most early applications were in the petrographic analysis of pottery (Goldberg, 1983). However, the use of soil micromorphology in archaeological contexts matured in the last two decades of the 20th century (e.g, Goldberg, 1983, 1992; Fisher and Macphail, 1985; Courty et al., 1989; Macphail et al., 1990a; Courty, 1992; Davidson et al., 1992; Macphail and Goldberg, 1995; Macphail and Cruise, 2001; French, 2003). The method in archaeological contexts is still most commonly employed in Europe, but it is gaining wider application (e.g., Goldberg, 1979, 1992, 1998, 2000; Courty and Fedoroff, 1985; Goldberg and Sherwood, 1994; Goldberg and Whitbread, 1993; Schiegl et al., 1996; Courty and Weiss, 1997; Weiner et al., 1998; Bookidis et al., 1999; Goldberg and Arpin, 2000; Courty, 2001; French, 2003).

Though often viewed as an ancillary tool in soil science and archaeology, micromorphology, as the term implies, provides data that complements those gathered from the more traditional macromorphological field studies and laboratory analyses. The method provides a microscopic look at pedologic features such as cutans (coatings), voids, grain shape, and weathering (figs. 2.6, 8.3, 8.8, 11.1, 11.2, 11.8). Examining thin sections also aids in differentiating soils from unweathered sediments and natural from anthropogenic inputs in soils (figs. 11.1 and 11.2; see chapter 11) not otherwise discernable from analyses of bulk samples. The primary drawback of micromorphology is that individual thin sections represent a very small and essentially two-dimensional sample of a soil. Thin-section analysis is tedious work, and therefore, an adequate thin section sample of a soil is rarely attempted. Another potential problem that can be encountered in the microscopic study of soils is equifinality. Not all microscopic characteristics of soils have unique origins, which can result in more than one interpretation for an individual micromorphological feature (Carter and Davidson, 1998; Kemp, 1998). For best results, micromorphology must be used in tandem with traditional field-based macromorphology (Wilding and Flach, 1985) and bulk analyses.

The basic preparation and analysis of soil thin sections is similar to that employed in traditional rock petrography (specific step-by-step discussions of thin section preparation are provided by Brewer, 1976, pp. 447–449, and Murphy, 1986). One significant exception is that soft sediments and soils must first be hardened—that is, turned to rock—before a thin section is made. A soil thin section is generally prepared from a block (~10 × 5 × 7 cm) brought in from the field and oven-dried for several days at ~60°C. The block is then impregnated with either polyester resin or epoxy under vacuum and sliced with a rock saw, using kerosene. The block is finally mounted on a glass slide and polished to the standard thickness of 30 μm. The section can then be viewed on a petrographic microscope under plain or polarized light (fig. 2.6) and described accordingly.

A wide array of laboratory methods is available for characterizing and quantifying various physical and chemical parameters of soils. Compendia of many of the more widely used and useful analytical techniques in pedology and geoarchaeology are provided by Jackson (1958, 1969), Shackley (1975), McKeague (1978), Page et al. (1982), Klute (1986), Singer and Janitzky (1986), Catt (1990), Gale and Hoare (1991), the Soil Survey Laboratory Staff (1996), Sparks et al. (1996), Dane and Topp (2002), and Garrison (2003). The preparation, description, and interpretation of soil thin sections is discussed by Brewer (1976), Bullock et al. (1985), Kemp (1985c), Murphy (1986), and Courty et al. (1989).

Choosing a particular technique depends on a variety of factors, including what questions are being asked about the soils, the degree of precision required, whether one wants to compare results with those of another worker, and of course, the cost in time and money. Beyond finances, two basic considerations should be kept in mind when the question of analysis subsequent to fieldwork is raised: the necessity of the analysis and the type of analysis required. In the author's experience laboratory work is carried out on soils from many archaeological sites with no problem orientation; it is done simply because someone felt

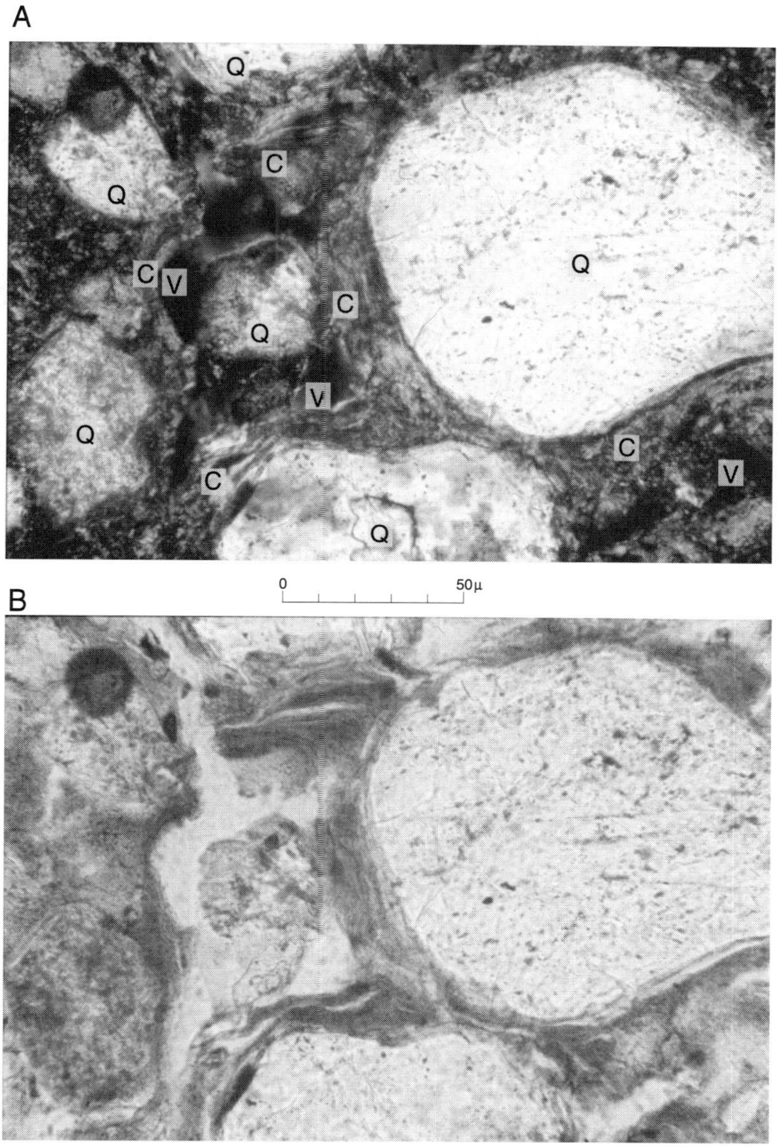

Figure 2.6 Thin section photomicrographs of the Bt horizon from the Rocky Flats soil (eastern Colorado), which has been forming since the early Pleistocene. (A) Taken under crossed polarized light (or "crossed nicols"), quartz grains (Q), clay coatings (C), and voids (V) show well. The quartz grains are sand-size and are part of the parent material. The clay coatings are illuvial and can be seen to consist of very fine layers. The quartz and clay transmit colored light that flickers and darkens as the stage on the microscope is rotated. The voids are the intergranular spaces and are black. Color characteristics, among other properties, can be used to differentiate, for example, clay from quartz, or host of other minerals or other geogenic, pedogenic, and anthropogenic features in the soil. (B) This is the same view as (A), but taken under plane light. The clay laminations show well.

it should be done (see also Limbrey, 1975, pp. 277–280). Analytical data can be of the utmost importance in interpreting soils and extending those interpretations to the archaeological problems being addressed, but there are occasions when the fieldwork alone could solve the problems or when laboratory work is simply of no use or will not answer the question or questions being asked.

Once the decision is made to proceed with laboratory work, selection of particular techniques is critical. For most routine analyses there are a variety of terms and techniques, and the results of each analytical procedure are not always comparable. For example, in determining particle-size distribution there are different definitions of what constitutes the size ranges of each category (e.g., Shackley, 1975; Briggs, 1977; Blatt et al., 1980); there are also various pretreatment procedures (such as removal of organic material or carbonates), and whether or not they are used can profoundly affect the resulting particle-size data. Different approaches to particle-size analysis as well as organic carbon and calcium carbonate determination in archaeological contexts and the varying results are discussed in appendix 3. Phosphorus determination, commonly employed in archaeological research, also provides an example. Phosphorus occurs in a variety of forms in the soil (chapter 11 and appendix 2). Different laboratory procedures extract different forms, and the results are not always comparable (chapter 11 and appendixes 2 and 3). In summary, the types of analyses and techniques employed to analyze soils should be chosen on the basis of the research objectives and on the analyses and techniques others have used to reach the same objectives. Also, when comparing analytical results, all terms and techniques must be clearly defined and referenced.

3

Conceptual Approaches to Pedogenesis

To fully appreciate and apply pedologic principals in archaeology, some of the theoretical underpinnings of pedology and especially soil geomorphology must be outlined. Pedologists and soil geomorphologists have attempted to describe, if not model, the processes of soil formation, the factors that drive the processes, and the evolution of soils as landscapes evolve (summarized by Smeck et al., 1983; Johnson and Watson-Stegner, 1987; and Gerrard, 1992, pp. 1–50, 217–220). The task is a difficult one, however, because of the complex and variable sets of processes responsible for soil development. Several of the resulting approaches have proven useful for conceptualizing pedogenesis and, more important, for interpreting soils. In addition to understanding soil-forming processes for interpreting soil profiles, understanding soil formation is important for understanding site formation. The conceptual approaches particularly useful in soil geomorphic and geoarchaeological research are summarized below. Soil-forming processes as components of site formation are discussed more fully in chapter 10.

The following discussions of conceptual approaches to pedogenesis are roughly arranged in order of increasing complexity. The "multiple-process model" is essentially a categorization of soil-forming processes. It does not explain pedogenesis but is a useful way to sort and group the many soil-forming processes. The "state factor" approach and the "K-cycle" concept do not deal directly with soil formation, but instead focus on important external factors and processes that drive or affect pedogenesis such as climate and geomorphic evolution. The "soil evolution" model and the "new global view of soils" attempt to integrate pedogenic process with landscape evolution, climate, and other factors. This section closes with discussion of two important aspects of pedogenesis and pedogenic pathways that offer caveats in the use of soils for reconstructing the past.

Multiple Process Model

Soils are the result of biogeochemical processes determined and driven by the ecosystem (following Buol et al., 1997). This relationship is more simply described as "internal soil-forming processes" driven by "external soil-forming factors" (fig. 3.1; after Buol et al., 1984). A useful approach to categorizing the many and varied internal soil-forming processes responsible for pedogenesis is the multiple-process model of Simonson (1959, 1978). Simonson groups the internal pedogenic processes into four categories: additions (e.g., water and organic matter), losses (e.g., leaching of carbonate), transfers (e.g., clay and iron illuviation), and transformations (e.g., organic residues decomposing and forming humus) (fig. 3.1). The many soil-forming processes and their respective quadripartite grouping are presented in table 3.1. The external soil-forming factors that drive the internal processes are the subject of the following section and, indeed, form the framework for much of this volume.

State Factors

The presence or absence or degree of activity of the internal processes is a function of the complex interaction of external environmental factors. Consideration of these factors produced the "state factor" approach to soil genesis (Jenny, 1941, 1980), which is probably the most widely known theoretical framework in pedology. This approach is successfully applied in soil geomorphology, has many applications to archaeology, and is used throughout much of this book. Jenny (1941, 1980), following the work of many others (see Johnson and Hole, 1994; and Tandarich and Sprecher, 1994), defined the factors of soil formation as climate, organisms (flora and fauna), relief (or landscape setting), parent material, and time, often written as the equation

$$S = f(cl, o, r, p, t \ldots),$$

where the uppercase S is the whole soil. This equation defines the state of the soil as a function of the five factors (the state factors, or CLORPT factors) and other, unspecified factors of local or minor importance ("..."; e.g., dust fall, groundwater fluctuations). The equation as a whole has never been solved, but Jenny (1941, 1980) proposed solving the equation by studying the variation in a soil as a function of one factor, keeping the others constant or accounted for. For example, variation in soils resulting from differences in climate could be studied by keeping all factors except climate constant. Variations in any soil property or properties can then be attributed to variations in climate. This is written as

$$s = f(\underline{cl}, o, r, p, t),$$

where the lowercase s denotes a soil property or properties. Qualitative statements about soils forming as a function of any one factor are called sequences

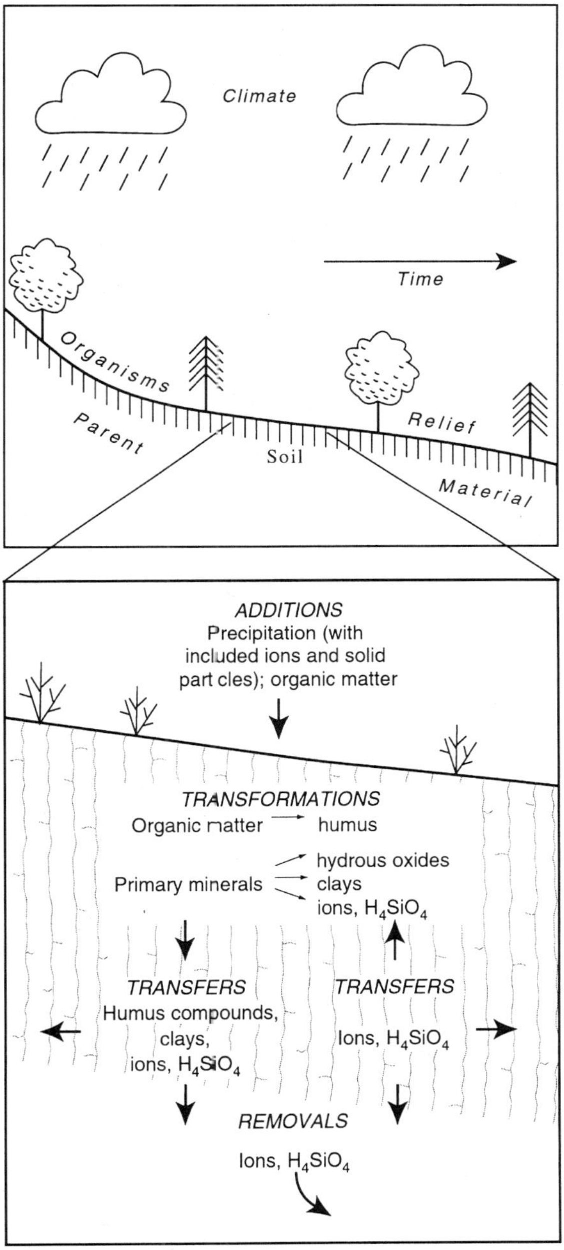

Figure 3.1 Schematic illustration showing the relationship of the five external (or environmental) factors of soil formation (top; from Jenny, 1941) to the internal processes of soil genesis (bottom; based on Birkeland, 1999, fig. 5.1; from Simonson, 1959, 1978).

Table 3.1. Soil-forming processes, grouped according to Simonson's fourfold classification

Process	Fourfold categorization[1]	Brief definition
Eluviation	3 hz	Movement of material out of a portion of a soil profile as in an albic horizon
Illuviation	3 hz	Movement of material into a portion of soil profile as in an argillic or spodic horizon
Leaching (depletion)	2	General term for washing out or eluviating soluble materials from the solum
Enrichment	1	General term for addition of material to a soil body
Erosion	2 hp	Removal of material from the surface of a soil
Cumulization	1 hp	Eolian, hydrologic, and human-made additions of mineral particles to the surface of a soil solum
Decalcification	3 hz	Reactions that remove calcium carbonate from one or more soil horizons
Calcification	3 hz	Processes including accumulation of calcium carbonate in Bk and possibly other horizons of a soil
Salinization	3 hz	The accumulation of soluble salts such as chlorides, sulfates, and bicarbonates of sodium, calcium, magnesium, and potassium in salic horizons
Desalinization	3	The removal of soluble salts from salic soil horizons
Alkalization (solonization)	3 hz	The accumulation of sodium ions on the exchanges sites in a soil
Dealkalization (solodization)	3	The leaching of sodium from natric horizons
Lessivage	3 hz	The mechanical migration of small mineral particles from the A to the B horizon of a soil, producing a B horizon relatively enriched in clay as in an argillic horizon
Pedoturbation	3 hp	Biologic, physical (freeze–thaw and wet–dry cycles) churning and cycling of soil materials, thereby homogenizing the solum in varying degrees
Podzolization	3, 4 hz	The chemical migration of aluminum and iron or organic matter, resulting in the concentrations of silica (i.e., silication) in the layer eluviated
Desilication (ferrallitization, ferritization, allitization)	3, 4 hz	The chemical migration of silica out of the soil solum and thus the concentration of sesquioxides in the solum (e.g. goethite, gibbsite), with or without formation of ironstone (laterite; hardened plinthite) and concretions
Resilication	4	Formation of kaolinite from gibbsite in presence of excess $Si(OH)_4$ in solution or formation of smectite from kaolinite in presence of large amounts of $Si(OH)_4$ at higher pH values
Decomposition	4	The breakdown of mineral and organic materials
Synthesis	4	The formation of new particles of mineral and organic species
Melanization	1, 3 hz	The darkening of light-colored, unconsolidated, initial materials by admixture of organic matter (as in a dark colored A horizon)
Leucinization	3 hz	The paling of soil horizons by disappearance of dark organic materials either through transformation to light-colored ones or through removal from the horizons
Littering	1	The accumulation on the mineral soil surface of organic litter and associated humus to a depth of less than 30 cm
Humification	4	The transformation of raw organic material into humus

Table 3.1. (*cont.*)

Process	Fourfold categorization[1]	Brief definition
Paludization	4 hp	Processes regarded by some workers as geogenic rather than pedogenic, including the accumulation of deep (>30 cm) deposits of organic matter as in mucks and peats (Histosols)
Ripening	4 hz	Chemical, biological, and physical changes in organic soil after air penetrates previously waterlogged material
Mineralization	4	The release of oxide solids through decomposition of organic matter
Braunification; Rubification; Ferrugination	3, 4 hz	Release of iron from primary minerals and the dispersion of particles of iron oxides or oxyhydroxides in increasing amounts. Progressive oxidation or hydration of the iron colors the soil mass brownish, reddish brown, and red, respectively
Gleization	3, 4 hp	The reduction of iron under anaerobic soil conditions, with the production of bluish to greenish-gray matrix colors, with or without yellowish brown, brown, and black mottles and ferric and manganiferous concretions
Loosening	4	Increase in volume of voids by activity of plants, animals, and humans and by freeze–thaw or other physical processes and by removal of material by leaching
Hardening	4	Decrease in volume of voids by collapse and compaction and by filling of some voids with fine earth, carbonates, silica, and other materials

From Buol et al. (1997, table 3.1).

[1] Soil-forming processes: 1 = addition; 2 = removal; 3 = translocation; 4 = transformation; "hz" indicates processes that promote horizonation; "hp" indicates processes that promote haploidization (see chapter 10).

(climosequence, biosequence, toposequence, lithosequence, chronosequence), and quantitative statements, in which functions have been solved for any one factor, are called functions (climofunction, biofunction, topofunction, lithofunction, chronofunction).

Birkeland (1974, 1984, 1999) took Jenny's CLORPT approach for describing and quantifying the five factors and recast it in order to reconstruct the factors as they acted in the past or over time. For example, by looking at a group of soils that vary in age but otherwise formed in the same parent material in the same landscape position under the same climate and biota (i.e., by examining a chronosequence), rates of soil development can be established.

The state factor approach to the study of soil genesis is not without criticism, which is summarized by Birkeland (1999, pp. 144–145), Johnson and Watson-Stegner (1987), and Gerrard (1992, pp. 3–7). In particular, Paton et al. (1995) attempt to overturn what they see as the "clorpt paradigm." Their "new global view" is presented and discussed below. Several other criticisms should be noted, however. At the most general level, the state factors do not deal with soils and pedogenic process, but with external factors that affect the soil. In archaeology and other areas of Quaternary research, though, the external factors often are the object of concern, and soils can be a means of reconstructing them; the soils are environmental proxies (e.g., Birkeland, 1999). The state factor approach cannot describe landscapes, however, and soils are selected on a landscape for their ability to solve the equation. The state factor approach also tends to treat the factors individually as independent variables, although they often act together, such as climate and biota. The time factor is the only truly independent variable, but the passage of time in and of itself does not form a soil; it simply allows the other factors to operate. For the most part, the general validity of the state factor approach has been upheld, especially in soil geomorphic research (e.g., Yaalon, 1975; Beecham, 1980; Birkeland, 1999), and applied in related fields (e.g., Major, 1951). Other theoretical approaches deal directly with soils and better describe pedogenic processes (as discussed elsewhere in this section and by Smeck et al., 1983, and Johnson and Watson-Stegner, 1987), but no other conceptual approach to soils has the proven utility in direct pedologic research, especially in soil geomorphology, as does Jenny's.

The Jenny–Birkeland approach to understanding pedology is particularly useful "from the point of view of a field-oriented geologist-pedologist, working with a wide variety of soils at the earth's surface" (Birkeland, 1984, p. 166). Because this is the same point of view taken by many geoarchaeologists, the state factor approach is useful in archaeological pedology and is one approach used in subsequent chapters. The factors of soil formation generally of most concern in archaeology are environment (i.e., climate and vegetation) and time. State factor analysis examines soils to determine the effects of climate and organisms and to determine how soils vary as a function of time. The resulting data then can be applied to soils in similar contexts as a means of environmental reconstruction or for estimating the age of deposits. An understanding of soil variability as a function of topographic position or parent material variation, which also can be investigated via state factor analysis, also may be important in reconstructing local environments and in soil stratigraphy. Chapters 6, 7, 8, and 9 include discussion of the archaeological applications of the state factor approach to soil stratigraphy, dating, and environmental and landscape reconstructions.

The state factor approach in a geoarchaeological context is furthered by consideration of human activity as another factor of soil formation. Bidwell and Hole (1965) discuss the role of human activity in modifying the traditional five factors (table 11.3). For example, people have altered the biotic factor by removing forests and replacing them with farm crops, and they have modified the topographic factors by digging canals, constructing terraced fields, or forming tells. In contrast, Amundson and Jenny (1991) place humans and human activity as separate factors. They symbolize independent anthropogenic state factors by their inheritable attributes (genotypes), a subset of the organism factor (o_h), and by their culture (c; products of human work or thought). For the purposes of reconstructing human behaviors using soils, they may have complicated the equation more than necessary (though geoarchaeological applications of soils were not their purpose). The symbol c could stand for the various kinds of human behavior that change the physical, chemical, and biological characteristics of the soil (e.g., clearing vegetation, dumping refuse, building fires, growing crops), and theoretically, the state factor equation could be solved for c. These issues of human effects on the state factors and on soil-forming processes are explored in chapter 11.

K-Cycles

Beyond the processes and factors that create soil horizons and make soil profiles, a key concept in soil geomorphology (and soil stratigraphy in particular) is the relationship between soils and landscape stability. Most simply, soils form on stable landscapes; that is, landscapes with no or little erosion or aggradation. There are exceptions to this concept, such as slowly aggrading floodplains or eolian landscapes (see chapter 5) or regions of continuous, intense weathering (e.g., tropical settings), but the basic idea has proven useful in the interpretation of buried soils (and is further discussed in chapters 5 and 6). This view of soils and landscapes as "periodic phenomena" was formalized by Butler (1959, 1982) in his K-cycle concept. Each cycle includes an unstable phase of erosion and deposition (Ku), followed by a stable phase (Ks) and formation of a "groundsurface" with concomitant pedogenesis. A sequence of buried soils therefore represents both, first, a sequence of groundsurfaces and, second, recurrent cycles of landscape stability and instability. Butler (1959, 1982) also emphasizes the lateral variability of soils on both modern and buried soilscapes and thus, without saying so, describes catenas and "paleocatenas." In many ways the K-cycle concept is an oversimplified approach for interpreting soil geomorphic and soil stratigraphic relationships, but it is a useful conceptual framework for understanding the relationship between pedogenesis and other earth surface processes (Brewer, 1972; Catt, 1986, p. 166; Gerrard, 1992, pp. 216–220). In geoarchaeology the concept of soils as periodic phenomena is very useful for interpreting and understanding the evolution of archaeological sites and landscapes (chapter 9). The K-cycle concept is also important because of its focus on landscapes, groundsurfaces, and stratigraphy and because it emphasizes soils, buried and unburied, as three- and four-dimensional entities.

Soil Evolution Model

An approach to modeling soil formation that incorporates aspects of several other models (including Simonson's and Jenny's approaches as well as the concept of intrinsic thresholds, discussed later) and also the concept of landscape evolution (though not expressly the idea of K-cycles) is the "soil evolution" model of Johnson and Watson-Stegner (1987; see also Johnson et al., 1987, 1990). The authors recognize that soil formation is not a linear, unidirectional process. Soils do not simply exhibit more horizons and get thicker over time, for example. They are affected by processes (external and internal) that promote or inhibit horizonation (or both) or promote or inhibit profile thickening (or both; e.g., fig. 3.2). Johnson and Watson-Stegner recognize two basic components to soil formation: "progressive pedogenesis" and "regressive pedogenesis," expressed as $S = f(P, R)$. Progressive pedogenesis includes processes and factors that promote

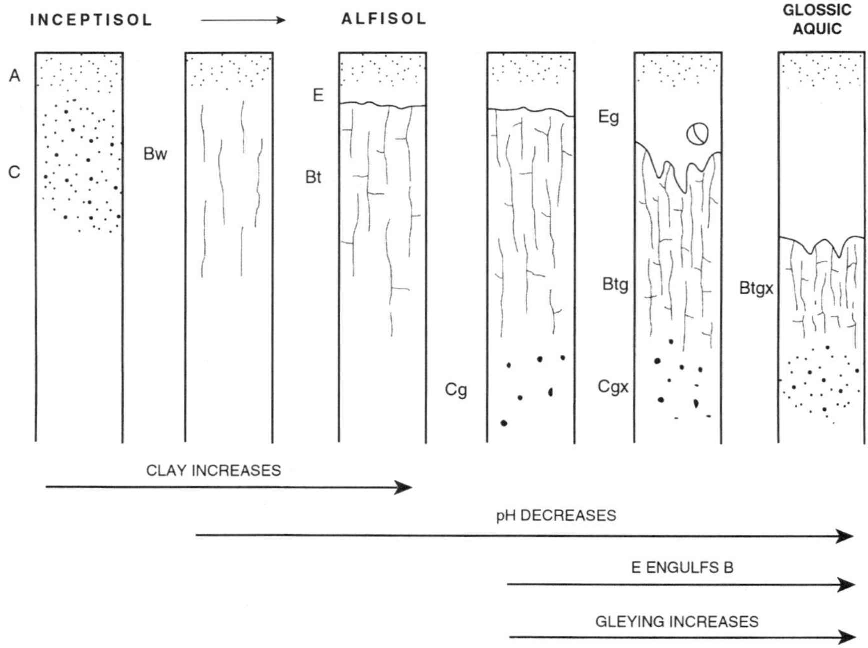

Figure 3.2 Soil profile changes through time in a calcareous loamy parent material, illustrating progressive pedogenesis (processes favoring horizonation such as eluviation, clay translocation, and leaching), followed by regressive pedogenesis (processes favoring homogenization or destructing of horizons such as deep eluviation or gleying; modified from Birkeland, 1999, fig. 8.42). The gleying is the result of continued clay accumulation, which eventually impedes drainage. This change in drainage also represents an intrinsic pedogenic threshold.

horizonation, the deepening of the soil, and the incorporation of material added to the surface (table 3.1). Regressive pedogenesis includes processes and factors that promote haploidization and simplified, more homogeneous profiles with time (table 3.1).

Progressive and regressive pedogenic processes can and usually do operate together. At any given site or in any given region, a subset of these processes will operate, and the resulting soil usually reflects the dominance of one group of processes over the other. The dominance of one group could shift to the other group over time as either internal or external processes or factors change. Such changes could include shifts in climate, vegetation, animal communities, human activity, or geologic events and processes. The soil evolution model is the most sophisticated of the several models that have been proposed, in its holistic view of soil formation. It also provides considerable explanatory power in discussing soils once they have been studied. The model does not provide the *a priori* means of investigating soil development that the Jenny–Birkeland factorial approach holds, but it is a useful conceptual approach that links the factors to the soil itself.

The soil evolution model and its components have some explicitly geoarchaeological applications, particularly in understanding site-formation processes (e.g., Johnson, 1989, 1993; Johnson and Watson-Stegner, 1990). Of particular significance are the concept of the biomantle and the role of biomechanical processes (Johnson, 1990, 2002). This idea focuses on the role of fauna in creating the upper portion of soil profiles—typically the A horizon. Johnson (1993, 1994a, 2002) argues that many models of soil formation, in particular the state factor approach of Jenny (1941), tend to focus only on the role of plants and, mostly, on their biochemical roles in soil genesis. These models did not, for a variety of reasons, fully take into account the dynamic nature of soil fauna in producing soil horizons. The impact of animals in processes of soil bioturbation and creating of stone-lines and artifact-lines is of particular significance in the evolution of archaeological sites and is more fully discussed in chapter 10.

The "New Global View" of Soils

The soil evolution model of Johnson and his colleagues emphasizes geomorphic processes more heavily than most of the other well-known models of pedogenesis. Much of the research that went in to developing the soil evolution model was in the tropics and led Johnson (1993, 1994a,b) to question the applicability in the tropics of the traditional vertical-process-oriented factorial approach to interpreting and describing soil-forming processes. This issue was more forcefully and sweepingly presented by Paton et al. (1995) in their "new global view" of soils, in which they essentially reject the concept of soil "zonalism"—the idea that the distribution of soil zones is determined by the five factors, particularly climate and vegetation—and the corollary concept that the factors drive soil-forming processes vertically down through the soil, resulting in A-B-C profiles. Schaetzl (2000, p. 772) succinctly summarizes their critique and their own concept of global pedogenesis: "Paton and his coauthors take soil genesis out of its traditional five-factor paradigm, with its accompanying A-B-C horizon model, to a more geomorphic- or earth science-based approach. The authors argue that 'zonal soils' or soil zonalization is flawed, and propose to replace this concept with one involving much more geomorphology: surficial and biological processes acting upon a weathering and inherently more mobile mantle of sediment. As such, their new model is much better at explaining the *spatial patterns of the soil mantle*, especially in tropical landscapes, than are some of the existing models" (described earlier; emphasis in original citation). Schaetzl (2000, pp. 772–773) goes on to summarize: "Bioturbation is stressed as a soil geomorphic vector, always churning the upper horizons and rendering them downwardly mobile (down*slope*, that is!). Eolian and sheetwash processes contribute in this arena as well. Downward-moving water, long stressed in the northern hemisphere as the overriding vector in soil-genesis processes, is strongly deemphasized for soils on older landscapes. . . . This model, developed in the dry and subhumid tropics, operates well there, and indeed may operate anywhere, given enough time" (emphasis in original citation).

The "new global view" has come under heavy criticism (e.g., Catt, 1996). In particular, Beatty (2000), Courchesne (2000), Johnson (2000), and Schaetzl (2000) forcefully argue that the traditional "CLORPT, A-B-C" model does successfully account for the genesis and distribution of soils on younger landscapes in the middle latitudes. Johnson (2000) also notes that the "new global view" is not really "new" (see references cited by Johnson, 2000, pp. 778, 780), though Paton et al. (1995) present the argument more fully and in more detail. Courchesne (2000, p. 784) suggests that Paton et al. (1995) purposely overstate their case to make a point, which Paton and one of his coauthors acknowledge (Paton and Humphreys, 2000, p. 787). The authors of the book are in danger of "throwing the baby out with the bath water," however (Beatty, 2000, p. 787). Johnson (2000, p. 780) is probably correct in his view that the old CLORPT, A-B-C model "formulated for soils in the plainlands of European Russia and North America, conceptually promotes an atypical (but not aberrant) style of pedogenesis relative to most of the nonglaciated and loess-free rest of the world, an atypical style that has been applied, unfortunately, as the world standard." Most soil geomorphologic research and, more to the point of this volume, most geoarchaeological research, however, has been carried out in the Quaternary glacial, alluvial, eolian, and desert regions of the middle latitudes, where the traditional model works and, generally, works well. The traditional model, therefore, is the point of view of this volume.

Caveats: Intrinsic Thresholds and Equifinality

There are many potential pitfalls in using soils to reconstruct the archaeological and geologic past, as discussed throughout this volume. Two general issues, related directly to pedogenic processes and their results, are intrinsic thresholds and equifinality. Because they can be significant in the interpretation of soils in most any context, these issues are summarized here.

In thinking about soil development, an important consideration is that the genetic pathways of soil development can shift as either factors or processes or both change. A traditional view of pedogenesis is that changes in internal processes reflect shifts in external factors. For example, evidence that moisture conditions in a soil changed from well drained to poorly drained is often taken to indicate a rising water table caused by climate changes. However, changes in soil morphologic stability can be caused by internal changes in soil morphology, chemistry, or mineralogy alone. These "intrinsic thresholds" are discussed by Muhs (1984). In the example of the change in drainage conditions, this could be brought about simply through steady clay illuviation in the B horizon of a well-drained soil (i.e., formation of an argillic horizon) until pore spaces are clogged and downward movement of water is impeded, creating poor drainage conditions. This condition, in turn, can inhibit subsoil leaching and result in a subsoil buildup of salts and a rise of pH. Similarly, zones of carbonate accumulation (calcic horizons) can eventually become plugged with calcium carbonate so that downward movement of carbonate-bearing water is not possible. The water then moves laterally over the top of the plugged calcic horizon, significantly changing its mor-

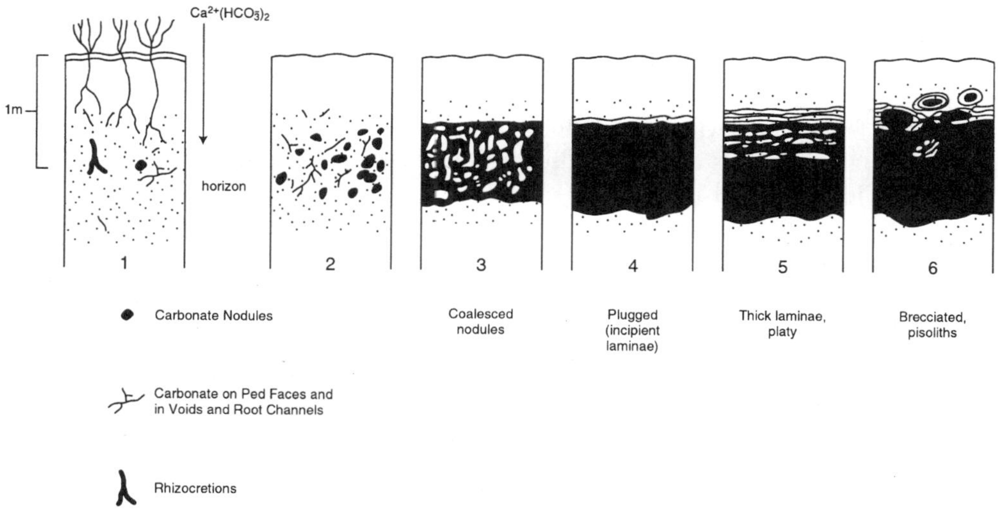

Figure 3.3 Illustration of an intrinsic pedogenic threshold in the evolution of calcic horizons (modified from Allen and Wright, 1989, fig. 1.4; based on Gile et al., 1966, fig. 5; Machette, 1985; published with permission of V. P. Wright and J. R. L. Allen). Initial stages of carbonate accumulation are characterized by development of nodules and coatings on gravel (1–2). With time the carbonate becomes more pervasive (3–4). Stage 4 is characterized by development of a laminar zone, which represents the crossing of an intrinsic threshold as the calcic horizon becomes plugged, impeding downward movement of carbonate-laden water. The laminar zone becomes thicker (5) and eventually brecciated (6).

phology (fig. 3.3). In both examples (and others provided by Muhs, 1984) no external environmental change is needed.

Understanding the effect of intrinsic thresholds is important in the geoarchaeological study of soils. Changes in soil morphology, chemistry, or mineralogy are not necessarily the result of changes in geoarchaeologically significant external factors of soil formation such as climate or vegetation. Failure to recognize or consider intrinsic thresholds as a component of soil evolution could mislead or misdirect interpretations of landscape age, environmental change, or landscape evolution (chapters 7, 8, and 10).

A final point concerning soil genesis is the issue of "equifinality," which is the "convergence to similar forms despite variations in processes and controls" (Phillips, 1997, p. 1). Specifically in the case of soils, equifinality may be thought of as a variety of pedogenetic pathways producing soils with similar morphology. Ruhe (1965, p. 762), Pawluk (1978, pp. 62–64), Phillips (1997), and Leigh (2001) are some of the few workers who discuss aspects of equifinality and soils, but Ruhe's and Pawluk's discussions are brief (and they doesn't use the term "equifinality"), Phillips's argument is theoretical and has just a few examples, and Leigh's is very specific. Otherwise there is little literature on the topic, and there are relatively few cases in which equifinality has clearly been shown. Intuitively, however, given the almost infinite combinations of external environmental conditions but the relatively limited number of weathering processes, there likely are soils with varying genetic environmental histories that are similar at least in general appearance. So in the middle latitudes, for example, most soils will redden through the hues 10YR-7.5YR-5YR and probably will illuviate clay. Thus, reddish-brown soils with argillic horizons can be found under a variety of environmental conditions and on landscapes that vary significantly in age (e.g., the soils in figs. 1.1A and 1.1C).

The point here is to simply offer a cautionary note in the use of soils to reconstruct the past. They can provide valuable clues, particularly when combined with other data, but a variety of factors must be considered in any interpretation.

4

Soil Surveys and Archaeology

Soil survey and mapping is one of the most fundamental and best-known applications of pedology. The preparation of soil maps began in the 19th century (Yaalon, 1997), but systematic county-based soil surveys began in the 20th century in the United States (Simonson, 1987, p. 3). The production of soil maps based on systematic soil surveys has been one of the primary driving forces in pedologic research in both academic and governmental settings in the United States and worldwide through much of the 20th century (Simonson, 1987, 1997; Yaalon and Berkowicz, 1997). For example, soil survey and mapping has been a primary function of the USDA since 1899 (Simonson, 1987, p. 3; Soil Survey Division Staff, 1993, p. 11). Soil maps have been prepared for a variety of uses at scales ranging from a few hectares to those of continental and global magnitude. Published soil surveys contain a wealth of data on landscapes as well as soils, but are generally an underused (and likely misunderstood) resource in geoarchaeology, probably because of their agricultural and land-use orientation. This chapter presents a discussion of what soil surveys are (and are not) and potential as well as realized applications in archaeology. Much of the discussion focuses on the county soil surveys published by the USDA because they are so widely available, although applications of other kinds and scales of soil maps that have been applied in archaeology or that have archaeological applications also are discussed.

The Soil Survey

Many countries in the world have national soil surveys whose primary mission is the mapping and inventorying of the nation's soil resource. In the United States, soil survey is a cooperative venture of federal agencies, state agencies (including the Agricultural Experiment Stations), and local agencies, coordinated by the National Cooperative Soil Survey (Soil Survey Division Staff, 1993, p. 11). The principal federal agency involved in soil survey is the National Resource Conservation Service (NRCS; formerly the Soil Conservation Service, SCS) of the USDA. The mapping of soils by the NRCS/USDA is probably the agency's best-known activity. Its many published county soil surveys are its most widely known and widely used product. The mapping of the United States is not complete, but the surveys are available for many counties; they contain a wealth of information regarding soils and land-use capability, and many include descriptions of soil properties of interest to engineers. In some cases, they also contain data of relevance to soil geomorphic studies. For example, soil surveys in the midwestern United States often characterize loess deposits, and these data have been used in soil geomorphologic studies of loess thickness and origin, soil erosion, and textural controls on soil geography (e.g., Fehrenbacher et al., 1986). Maps of bedrock geology and landslides in the Valley and Ridge country of Virginia were compiled in part from soil surveys (Lindholm, 1993, 1994a,b). The surveys also are a remarkable source of maps and air photographs. Most archaeologists are aware of this resource, but its usefulness sometimes seems to be overestimated or misunderstood in some cases and underused in others (Voight and O'Brien, 1981), as illustrated below. The county soils surveys have potential for archaeological applications, but their purpose and limitations must first be recognized.

The USDA county soil surveys and those of most nations are prepared for land-use planning, especially for soil-conservation programs, planning agriculture programs, financial credit, zoning, construction and engineering purposes, and land evaluation (Butler, 1980, pp. 1–7; Broderson, 1994; Buol et al., 1997, pp. 412–413; King, 1983, pp. 102–103). They are produced at a variety of scales and at various levels of detail. Most USDA county soil surveys are in the range of 1 :12,000 to 1:31,680 (Soil Survey Division Staff, 1993, pp. 47–46). More generalized soil maps of individual states in the United States typically are in the range of 1:250,000 to 1:2,000,000. In contrast, King (1983) describes soil maps for a part of Cyprus produced at two scales: 1:25,000 and 1:200,000. Globally, FAO-UNESCO (1974) produced a 10-volume set of soil maps of the world at 1: 500,000.

Limitations of Soil Surveys

Soil surveys have been extolled for their use in archaeological research (Saucier, 1966; Limbrey, 1975, pp. 246–250; Olson, 1981, pp. ix, 1–6; Scudder et al., 1996), but their purpose and preparation impose a variety of limits on their utility. The following subsections discuss the principal problems in applying soils surveys, particularly the USDA county surveys, in archaeological research.

Scale and Map Generalization

A key feature of soil surveys that imposes limits on their use for site-specific interpretations are the generalizations inherent in defining mapping units. Indeed, soil variability and how to deal with it in mapping is a significant issue in the pedologic literature (e.g., Wilding and Drees, 1983; Mausbach and Wilding, 1991). The distribution of soils across a landscape may be too complex to map accurately either because of complexities in their evolution (i.e., variability in the factors or processes of soil formation) or because of variability in their taxonomic classification. The reasons for and problems of map generalizations are succinctly stated by Hole and Campbell (1985, p. 113):

> Because limitations imposed by scale and legibility do not permit exact representation of all detail observed (or inferred) in the field, *mapping units cannot correspond exactly to the taxonomic or genetic units that they represent*. . . . Mapping units therefore typically include areas of soils other than those specified in the legend. . . . Mapping units named for one soil taxon may encompass more than one taxon; the inclusions of unnamed soils may be too small to map, or occur in a pattern too complex to specify. An important practical problem in most national soil surveys is that *the map user has no way to learn of the amount, character, and pattern of these variations, because he or she must reply completely upon the map, its legend, and the accompanying report, which usually presents little detail on such matters.* The typical mapping unit includes an unknown amount of variation, usually said to be insignificant in respect to management of the soil." (emphasis added)

For example, soil taxonomy (Soil Survey Staff, 1975, pp. 408–409) permitted a mapping unit based on a soil series to include a single contrasting soil series as an inclusion if it does not exceed 10% of the area of the mapping unit. If the inclusions are similar to the named series, they are permitted up to 50% of the area of the mapping unit. The minimum size area permitted for delineation on most county soil surveys is 1 ha. Depending on the scale of the map, the minimum size permitted may be as large as 4 ha (Soil Survey Division Staff, 1993, pp. 52–55, table 2–1). Therefore, the scale of many archaeological sites is smaller than most soil mapping units and the variability inherent in those mapping units.

The problem of scale in dealing with soil mapping units and archaeological sites has been confronted in archaeological surveys around the world. Jones (1990, p. 271), working on the North Island of New Zealand, presents a good summary of the situation: "The main difficulty that arises is that of converting pedological classifications and mapping done on a small scale [i.e., covering a large area], for gross land valuation and modern horticultural and cropping purposes, to the fine detail required for site-specific considerations. The small areas [of many archaeological sites] . . . require much more detailed and expensive large-scale soil mapping than is typically carried out in general soil surveys." King (1983, pp. 103–104) described a very similar situation in trying to apply published soil maps for Cyprus in an archaeological survey. The smallest area that could be delimited on the 1:25,000-scale soil maps is 3 ha, but the archaeological survey parties plotted their finds on 1:5,000-scale maps and located sites as small as 0.06 ha. As a result, there were considerable problems in abstracting soil data from the available soil surveys maps that had any archaeological significance (King, 1983, p. 103).

More generally stated, a fundamental problem of soil survey is that the soils in a survey area are grouped into a limited number of soil individuals on the basis of selected soil properties (King, 1983, p. 102). Selection of the properties is determined by the objectives of the soil survey. The choice of soil properties is therefore a highly selective procedure.

The boundaries of a particular mapping unit do not always represent actual soil boundaries because soils may grade from one type into another rather than having sharp contacts like many geologic units. Soil boundaries are rarely as distinct as the lines used to symbolize them (Hole and Campbell, 1985, p. 101). "The conventional soil map implicitly employs a model in which soil bodies form discrete, internally uniform, units, with abrupt discontinuities at their edges [i.e., sharp boundaries indicated by a line]. . . . [T]his model of soil variation must be regarded as a useful expedient, but one that generally provides only a rough approximation of the actual landscape" (Hole and Campbell, 1985, p. 102). The fundamental lesson here is that the soils that can be seen at an archaeological site may not necessarily be those indicated on the published soil survey of the area.

The Soil Series

The fundamental mapping unit in county soil surveys is the "soil series." Soil series are defined on the basis of one or more of the following characteristics: kind, thickness, and arrangement of soil horizons, along with the color, texture, structure, consistence, pH, content of carbonates and other salts, content of coarse fragments, and mineralogy. A "soil association" is a more generalized mapping unit (e.g., used on county-scale or statewide soil maps) typically based on two or more similar series plus inclusions (see chapter 5 on soil stratigraphy for discussion of a more generic kind of soil association).

The focus on the soil series further inhibits the soil geomorphic and geoarchaeologic applications of county soil surveys. The soil series, though the lowest level of soil classification, typically is the highest level of grouping "soil individuals" (polypedons; chapter 1; Swanson, 1990a); that is, it is the most homogeneous grouping of soils made in soil taxonomy (Buol et al., 1997, p. 381). Soil series are defined strictly as subdivisions within the classificatory system (see chapter 1) and are intended to "record pragmatic distinctions, i.e., to be keyed to soil usefulness" (Simonson, 1997, p. 80). The emphasis on specific physical or chemical characteristics to define soil series results in mapping units that are rarely related to one another except taxonomically. To illustrate, "the soil classification relates an Aquept in a certain map unit to other Inceptisols throughout the world. But it does not address the relationship between the map unit containing the Aquept and other map units that occur adjacent to it in the survey area" (Swanson, 1990a, p. 52).

The heavy emphasis on the soil series, combined with the lack of information regarding genetic relationships among series (discussed later), results in a "strong component of geographic isolation involved in series definition" (Hallberg, 1984,

p. 51). This approach, combined with the use of place names for series (rather than connecting them to the higher levels of classification via the hierarchical root-word system that revolutionized soil classification) and the view of soils as "isolated points" or "soil individuals," tends to promote a view of the soil landscape as if the soil series are independent segments, like tiles in a tile floor. For these reasons as well as others discussed below, the soil series should not be emphasized in regional geoarchaeological or soil geomorphic investigations. The great group or perhaps subgroup level of soil taxonomy probably is more meaningful and, in any case, more informative, though investigators will need to be familiar with series names to read soil maps.

Other Problems

Further limiting the applicability of soil surveys in geomorphic or stratigraphic studies (geoarchaeological or otherwise) is the different spatial perspective of geoscientists versus pedologists (Holliday et al., 2002). Geomorphologists and stratigraphers, on the one hand, tend to view the world horizontally, emphasizing features such as geomorphic surfaces and land forms, as well as three-dimensionally, studying stratigraphic units. Many pedologists, on the other hand, traditionally tend to view soils as "independent entities occurring at specific points" (Daniels and Nelson, 1987, p. 289) and focus on the vertical dimension of soils. Pedology field training and field experience often deals with soil pits and soil profiles rather than soil landscapes (Daniels and Hammer, 1992, p. xv–xvi), probably because, first, so much of training and research in pedology involves digging soil pits or taking soil cores for mapping and studying profiles for taxonomic classification (Swanson, 1990b; Paton et al., 1995, p. 1); second, the soil taxonomy defines soils as single points (Daniels and Hammer, 1992, p. 77); and third, many soil-forming processes promote downward movement (i.e., emphasize the vertical dimension). Soil survey and soil classification historically emphasized the soil profile and the "soil individuals."

Typically, soil surveys are not designed for use as guides to local geology or geomorphology (though there are exceptions, discussed later in this chapter) or to aid in interpreting or reconstructing the past, and few contain substantive data on local soil-forming processes and factors. In particular, this is the case for the USDA county surveys (Holliday et al., 2002). As soil surveys developed through the 20th century, genetic and factorial relationships among series as well as geologic and other "physiographic" aspects were expressly deemphasized, despite work that showed a good relationship (Simonson, 1997; Holliday et al., 2002). The USDA surveys reflect local bedrock and land forms to some degree. Indeed, a key tenet in the philosophy of soil mapping is that soil–landscape relationships should be predictable given enough field data (Wilding and Drees, 1983; Hall and Olson, 1991; Hartung et al., 1991). However, the degree to which the USDA county surveys accurately depict soil geomorphic relations will vary tremendously from survey to survey, depending on the size of the features of interest, the area, and the training and experience of the mappers. During the course of the survey, soil scientists may learn a lot about the "nonsoil" characteristics of their mapping units (Swanson, 1990b), but in general, little emphasis is placed on the origin or historical development of the soils. As a result, data on regional geomorphology and geology in the surveys often are secondary and skimpy at best, and in most cases are very minor components of the survey. Discussion of all five factors of soil formation typically occupies no more than two to three pages out of perhaps 50–100 pages of text in a soil survey, with only a few paragraphs spent on the most general characteristics of the geology and geomorphology.

Though landscape position is an important component of field investigations of soils, soil surveys in recent decades have deemphasized viewing or investigating soilscapes or soils as components of landscapes; that is, dealing with soils as three-dimensional, contiguous bodies (Daniels and Nelson, 1987; Daniels and Hammer, 1992, p. 77; Swanson, 1993; Hall and Olson, 1991; Paton et al., 1995, pp. 5–8). "Much effort has been expended on taxonomic classification of soils . . . but the importance of proper representation of landscape relations within and between soil mapping units has been virtually ignored" (Hall and Olson, 1991, p. 21). Many soil mapping units have been observed to occur in a variety of landscape positions (Hall and Olson, 1991, p. 21), which violates the basic soil-mapping principal of understanding soil–landscape relationships and increases the variability within a given map unit. Guy Smith, the "father" of soil taxonomy, correctly noted that, "The identification of a single series in two or three different landscape positions suggests that neither the genetic nor use relationships of the soil have been sufficiently studied" (Forbes, 1986, p. 49). Put another way, the heavy emphasis placed on developing soil taxonomy over the past few decades, in both governmental and academic settings, deemphasized research by pedologists into understanding soil genesis or soil geomorphic relationships (Holliday et al., 2002).

Soil surveys place even less emphasis on the soil parent material or on soil evolution through time (i.e., as four-dimensional bodies formed in sediment or rock; Daniels and Hammer, 1992, pp. 10, 77; Simonson, 1997) than they do on geomorphology. The underemphasis on the landscape and soil parent materials probably is because taxonomic classification emphasizes soils as units unto themselves (Daniels and Hammer, 1992, p. 77) or "soil individuals," a view that developed along with the arbitrary subdivision of soils into pedons and the resulting separation of soils and pedons from natural landscapes (Knox, 1965; Swanson, 1990a,b; Daniels and Hammer, 1992, p. 77). There are exceptions to this view of soils, however, particularly in areas of substantial soil geomorphic research or where soil surveys were conducted by geologically trained investigators (e.g., Ruhe et al., 1967; Balster and Parsons, 1968; Parsons et al., 1970; Gile et al., 1981; summarized in Holliday et al., 2002).

The general lack of emphasis on the genesis or soil geomorphic relationships of soil series, combined with the necessity to generalize mapping units, results in problems of particular geoarchaeological significance. Some series include a variety of soils that are genetically and geomorphically distinct. For example, soil series and standard soil survey maps are of limited utility in valley landscapes because of the inherent variability, over short distances, of valley fills and geomorphic surfaces (Daniels and Hammer, 1992, pp. 57–58), compounded by the lack of adequate landscape models in use by soil mappers. Indeed, soils in alluvial settings have received relatively little attention by soil geomorphologists

Table 4.1. Soils of the Lubbock Lake area, Texas

Setting and code[1]	Soil series	Classification
Upland		
1, 3	Acuff	Paleustoll
5	Amarillo	Paleustalf
18	Estacado	Paleustoll
30	Olton	Paleustoll
38	Potter	Ustollic Paleorthid
26	Midessa	Ustochrept
Playa/Lunette		
8	Arch	Calciorthid
16, 17	Drake	Ustorthent
24, 25	Mansker	Paleustoll
38	Potter	Ustollic Paleorthid
26	Midessa	Ustochrept
Draw		
24, 25	Mansker	Paleustoll
38	Potter	Ustollic Paleorthid
14, 15	Bippus	Cumulic Haplustoll
10, 11	Berda	Ustochrept
19	Estacado	Paleustoll
42	Randall	Pellustert
9	Arents and Pits	—

From Blackstock (1979).

[1] Coded to numbers on figure 4.1.

(Ferring, 1992; Foss et al., 1995), although, of course, these settings have a high likelihood of containing archaeological features and sites. Stratigraphic and geomorphic distinctions in these settings may occur at spatial scales much smaller than the mapping scale of the soil survey (discussed above; Brown, 1997; Ferring, 1992, 2001). Sorting out these distinctions also may be beyond the scope of the mappers' training or duties. For example, the floors of dry valleys or "draws" on the High Plains of Texas are mapped as one unit, the Berda Loam (classified as an Aridic Ustochrept; table 4.1, fig. 4.1). Geoarchaeological field studies show, however, that this mapping unit can include a group of geomorphically and stratigraphically distinct soils and sediments (fig. 4.2; Holliday, 1985d, 1995), locally containing a high potential for late Prehistoric, proto-Historic, and Historic archaeological sites (Johnson and Holliday, 1995). The scale of the geomorphic and stratigraphic variability is significantly smaller than the scale of the mapping unit but is still very important geoarchaeologically.

In contrast, other series may relate to the same stratigraphic or geomorphic unit but are taxonomically differentiated because of the hierarchical and often arbitrary distinctions of soil taxonomy. A specific example of this problem can be found among surface soils across the High Plains of Texas and New Mexico,

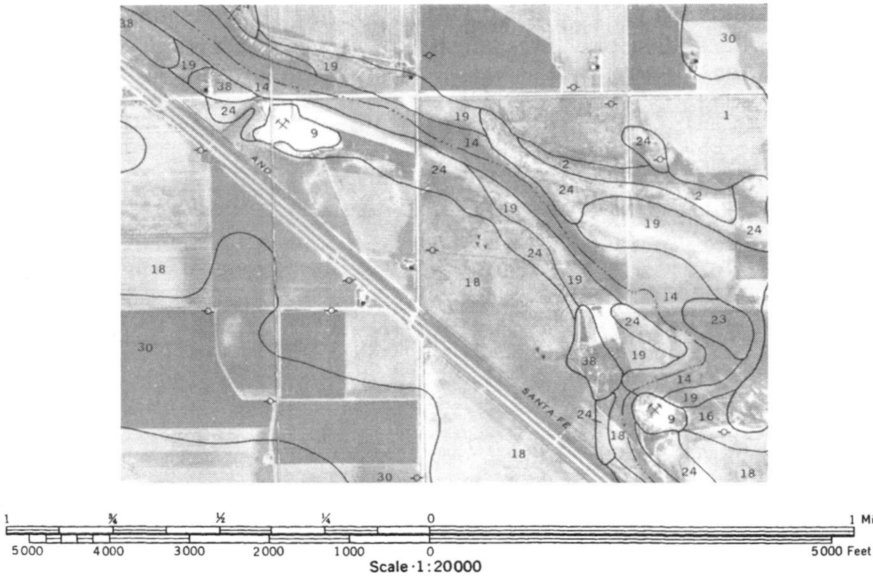

Figure 4.1 Portion of a soil survey map from Lubbock County, Texas, in the area of the Lubbock Lake site (numbers are coded to soil series, table 4.1; Blackstock, 1979, sheet 29). The strongly developed Ustalfs and Ustolls of the uplands (map units 1, 3, 5, 18, and 30) contrast with the more weakly expressed Haplustolls and Ustochrepts of the draw (map units 14, 15, 10, and 11) and the lunette (map units 16 and 17).

where a group of well-developed soils formed in Pleistocene eolian sediments (Holliday, 1989b, 1990b). Some of the more common upland surface soils are the Acuff, Amarillo, Arvanna, Brownfield, Mansker, Olton, Patricia, and Pullman series, all of which have thick, well-expressed argillic and calcic horizons (table 4.1). One of the most significant differences between these soils is the thickness of the A horizon: the Amarillo, Arvanna, Brownfield, and Patricia soils have a thinner A; the Acuff, Mansker, Olton, and Pullman have a thicker A (Holliday, 1990b). Otherwise, they are the same soil geomorphically and stratigraphically. These differences in A-horizon thickness probably are the result of wind erosion (Holliday, 1990b), but this single difference between otherwise identical soils results in their classification in two orders: Paleustolls (Acuff, Mansker, Olton, Pullman) and Paleustalfs (Amarillo, Arvanna, Brownfield, Patricia; table 4.1).

Archaeological Applications of Soil Surveys

The above-described characteristics of soil surveys have resulted in the minimal usage of the surveys in geoarchaeologic and soil geomorphic research. Nevertheless, the surveys have some utility in these fields in addition to providing maps and aerial photos. Several writers have noted that some surveys record site locations (Saucier, 1966; Almy, 1978). For example, the soil survey for Manatee

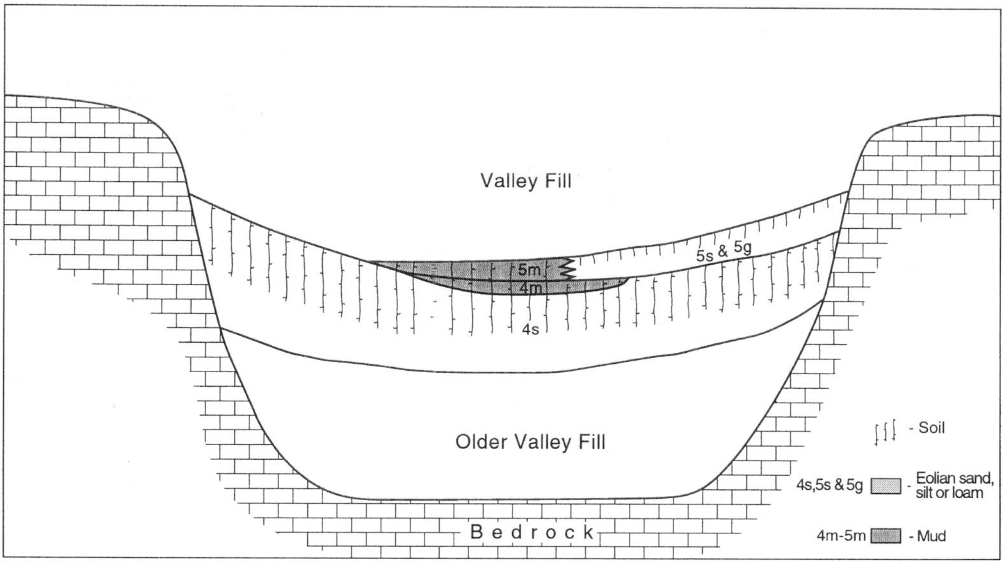

Figure 4.2 Stratigraphy of middle and late Holocene fill in draws of the Southern High Plains showing variation in surface soils as a function of age and parent materials (modified from Holliday, 1995, fig. 6; reproduced with permission of the Geological Society of America). The soil formed in stratum 4s is ~4500 years old with A-Bt-Bk morphology where unburied, but ~3500 years old, with similar morphology, where buried (fig. 2.1A). The soil in 5m is a cumulic ABg formed in dark gray mud for the last few hundred years (fig. 2.1B), but the soils in 5s and 5g exhibit both A-C and A-Bw profiles (fig. 2.1A) and are also a few hundred years old. The designations 4s, 5s, and 5g refer to lithostratigraphic units. Stratum 4s is equivalent to 4B at Lubbock Lake. Strata 5A and 5B at Lubbock Lake (fig. 2.1A) formed in both 5s and 5g.

County, Florida, describes shell middens as a land type: "This miscellaneous land type consists of large heaps of oyster, clam, and other shells . . . most areas are small though some cover as much as 10 acres and are 2 to 20 feet thick" (Caldwell et al., 1958, p. 20). Symbols on soil maps indicating features such as chert fragments, rock outcrops, sandy spots, or blowouts may also be indicative of archaeological sites (Saucier, 1966; Almy, 1978). These examples are rare, however, and county soil surveys should not be assumed to have any data relevant to specific archaeological site locations.

The advent of cultural resources management (CRM) work in archaeology since the 1970s produced a demand for readily available information on regional geomorphology, which the county soil surveys can supply to some extent. Specific soil series sometimes are associated with specific land forms and, thus, can provide clues to archaeologically significant settings. Almy (1978) illustrates how much more data on specific land form types and contemporary environmental settings can be gathered from a county soil survey than from a standard U.S. Geological Survey topographic map, although data on vegetation and soil characteristics for specific areas were taken from the soil survey rather more literally than may seem wise. For example, 49 soil series are mapped in the study area, which is taken to mean that 49 soil "types" are present, that is, reflecting some sort of natural reality (Almy, 1978, p. 78), without recognizing the purpose of soil series or of soil survey. Moreover, an assumption was made that all soil "types" (except modern soils resulting from historic activity) were present in the prehistoric past (Almy, 1978, p. 78); that is, there seemed to be no understanding that the soils evolve through time. Along the lower Mississippi River, archaeological sites are associated with natural levees of the current and abandoned courses of the Mississippi River. A few specific soil series are associated with the levees and therefore can be used as guides to likely site locations (Saucier, 1966). On the High Plains of the central United States, small, circular basins with seasonal lakes or "playas" and adjacent dunes or "lunettes" are associated with a few specific soil series and show up well on county soil surveys (table 4.1, fig. 4.1). These settings clearly were attractive to humans in the region throughout the late Quaternary and are useful guides to archaeological survey and site prediction (Litwinionek et al., 2003).

More broadly, drainage characteristics of soils, emphasized on soil surveys, can be useful for archaeological site prediction, with better-drained soils more likely to contain sites (i.e., prehistoric occupants probably preferred better-drained settings for most of their activities; Mandel and Bettis, 2001b, p. 182). This soil–site relationship has been applied or at least considered in developing strategies for archaeological site survey and site prediction (Plog, 1971; Lovis, 1976; O'Brien et al., 1982). This was one aspect of delineating natural levees (well-drained soils) from backswamp areas (poorly drained soils) along the lower Mississippi (Saucier, 1966). In southwestern Florida, drainage characteristics among soil series seemed to be one of several important factors in determining the likelihood of finding archaeological sites, with slightly better-drained soils having more sites (Almy, 1978). Poorly drained floodplains and other marshy settings, however, can be important settings for some archaeological sites (e.g., Warren, 1982a).

There are several examples of using the distribution of specific soil series to predict the presence or absence of sites by simply looking at the correlation between soil series and site frequency (Almy, 1978; Voight and O'Brien, 1981; Warren et al., 1981). "Distributions of cultural locations can be compared readily with distributions of soil properties to detect patterns of association or to test implications of locational hypotheses" (Warren et al., 1981, p. 36). The broad theoretical approach here is that "[r]egional distributions of soil characteristics can be useful for interpreting distributions of human settlement because (1) the process of soil formation is influenced by environmental factors that often leave distinguishable traces in soils, and (2) human settlement patterns often are influenced by these same environmental factors" (Warren et al., 1981, p. 47). This approach can be useful if the basic environmental conditions that influenced settlement have not changed since the period or periods of human occupation of interest. In marshy southwestern Florida, Almy (1978) reports a high statistical correlation between high site frequency and well-drained soils. The drainage conditions have not changed significantly except that late Holocene sea-level rise increased the distribution of poorly drained soils.

There are a number of reasons, however, why the distribution of modern soil series may be poor indicators of site distributions (following Warren et al., 1981). First, the factors and processes of soil formation can and do change over time, and the more time that passes following occupation, the more likely that the environment changed and the more likely that the change was significant in terms of either settlement or soil formation or both. Second, human perceptions of soil quality or environmental quality change through time; that is, successive groups of occupants may have different criteria for what constitutes "good" or attractive soils or environments even though there were no substantive changes in them. Finally, soils are not necessarily the best indicator of past environments, which is discussed more fully in chapter 8. Other kinds of data (e.g., pollen) may provide a better picture of the environmental conditions that influenced settlement.

In the Cannon Reservoir area of Missouri, soils data from county surveys were effectively integrated with other environmental information for a variety of purposes. Soil associations reflected both modern and early-19th-century drainage characteristics and were useful for characterizing and classifying prehistoric and historic site settings and for designing sampling strategies (Warren and O'Brien, 1981; Warren, 1982b). Soil maps in comparison to descriptions of early-19th-century vegetation also revealed some "discontinuities" (e.g., the soil types did not match vegetation types; soils typical of grasslands had tree cover). This was attributed to late Holocene climate change that shifted vegetation communities (Warren and O'Brien, 1982).

The soils themselves can have some interpretive value for archaeological research. Assessing relative landscape age based on soil classification is perhaps the most useful yet most underused aspect of soil surveys in geoarchaeology and geomorphology. Soil taxonomy was not designed to be a means of estimating soil age or to make other sorts of genetic interpretations, but with some experience and caution, age assessments can be offered based on horizon characteristics, particularly the B horizon, and based on classification. As discussed more fully in chapter 7 (and well reviewed by Birkeland, 1999), through time, translocation and transformation processes cause B horizons to become thicker and redder, and they can accumulate soil materials such as clay, carbonate, and oxidized iron. These soil characteristics can be determined from soil survey data. For example, the presence or absence and the thickness and color of the argillic horizon are well-known pedogenic characteristics that can be indicative of relative soil age (chapter 7; Birkeland, 1999). In addition, the presence or absence of an argillic horizon is a key component of soil taxonomy at a high level of classification, distinguishing Alfisols and Ultisols from the other orders and differentiating a variety of Mollisols and Aridisols at the suborder and great group levels. Thus, the classification of soil series can provide clues to relative ages, especially if other factors of formation for the series under study were roughly similar. This approach is perhaps best applied to river terraces. For example, terraces with soils classified in the "Pale" great group (high content of illuvial clay and reddish brown or redder hues, or a petrocalcic horizon; e.g., Paleustalf, Paleudoll), likely took tens of thousands of years to form. In the Americas, therefore, they probably have no subsurface archaeological sites, but on the surface they could contain sites spanning all ages of occupation since the late Pleistocene. Lower terraces of the same river with more typical (i.e., less developed) argillic horizons in the soils (e.g., Haplustalfs) may well contain buried archaeological sites depending on the local rates of clay translocation. Lower terraces with no argillic soils are likely to be Holocene in age, if not late Holocene, and thus have a high potential for occupation debris.

This approach was taken by Artz (1985) in a search for late Archaic sites (~5000–2000 yr B.P.) in a small stream system in northeast Oklahoma. He determined that sediments or soils old enough to be associated with late Archaic occupations should have soils characterized by moderately developed Bt horizons, based on data from other parts of the central United States. On county soil surveys, three soil series were mapped on the alluvial valley fill: a Haplaquoll (Osaga series), a poorly drained floodplain soil; a Hapludoll (Verdigris series), a moderately drained floodplain soil; and an Argiudoll (Mason series), a well-drained soil away from the floodplain. The Mason series, therefore, was the only soil mapped in the drainage that contained an argillic horizon and therefore was used as a key to finding late Archaic sites. The other soils were more weakly expressed and therefore were considered too young to contain Archaic occupations. The subsequent archaeological survey produced artifacts and radiocarbon ages that verified the premise that only soils with Bt horizons would be old enough to be associated with late Archaic sites.

In Iowa, soils surveys along the Des Moines and Racoon rivers (e.g., Andrews and Dideriksen, 1981) have proven useful in delineating alluvial terraces (Bettis, 1992; E. A. Bettis, personal communication, 2001). Late Wisconsin surfaces have Mollisols with Bt horizons and Alfisols. Early-middle Holocene surfaces are dominantly Mollisols with Bt horizons or Inceptisols. Late Holocene surfaces have Mollisols without Bt horizons and Entisols. These relationships have proven useful in mapping stream valleys in the region for purposes of archaeological site prediction (Bettis, 1992; discussed further in chapter 7).

Research on soils, geomorphology, and geoarchaeology along the Kansas River (Johnson and Logan, 1990; Johnson and Martin, 1987; Mandel, 1995; Sorenson et al., 1987) provide a good illustration of the relationship between traditional soil survey mapping, soil series, and soil geomorphic relationships along terraces (table 4.2). The land forms most likely to contain *in situ* archaeological sites are the Newman and Holliday terraces (Johnson and Logan, 1990, pp. 275–277). The Newman terrace aggraded during the early to middle Holocene and thus is likely to contain Paleoindian and early-middle Archaic sites. The principal soils on the Newman surface are moderately expressed Mollisols of the Muir-Reading-Wabash association (Sorenson et al., 1987, p. 96). Most common are well-drained Mollisols (Haplustolls) with a cumulic A horizon (caused by occasional flooding). Some of the Mollisols exhibit argillic horizons (the Reading Argiudolls), which formed in parent material higher in clay. Soils formed in very clayey parent materials typical of floodplain backswamp areas are poorly drained Mollisols with some characteristics of Vertisols (Haplaquolls of the Wabash series).

The Holliday terrace is younger, inset against the Newman and composed of sediment that accumulated in the late Holocene. The alluvium of the Holliday terrace is thus likely to contain late Archaic and late Prehistoric sites. The youth-

Table 4.2. Soils on Holocene terraces of the Kansas River

Terrace	Age of terrace and alluvium	Principal soil series	Classification
Holliday[1]	Late Holocene	Eudora	Fluventic Hapludoll
		Kimo	Aquic Hapludoll
		Sarpy	Typic Udipsamment
Newman[2]	Early-Middle Holocene	Muir	Cumulic Haplustoll
		Reading	Typic Argiudoll
		Wabash	Vertic Haplaquoll

From Sorenson et al. (1987).

[1] Other soil series on the Holliday Terrace include Carr, Haynie (Typic Udifluvents); Ivan (Cumulic Hapludoll); Humbarger, Muir (Cumulic Haplustoll); and Solomon (Vertic Haplaquoll).

[2] Other soil series on the Newman Terrace include Carr, Haynie (Typic Udifluvents); Eudora (Fluventic Hapludoll); Judson, Kennebec (Cumulic Hapludolls), Kimo (Aquic Hapludoll), Chase (Aquic Argiudoll); Zook (Cumulic Haplaquoll); Tully (Pachic Argiustoll); Geary, Hastings (Udic Argiustoll); and Sutphen (Udertic Haplustoll).

fulness of the terrace is indicated by the soil series. The principal soils are poorly expressed Mollisols of the Eudora-Kimo-Sarpy association (Sorenson et al., 1987, p. 96). Most common are Mollisols with some characteristics of Fluvents formed in overbank deposits (Hapludolls of the Eudora series). Other Mollisols are poorly drained (Hapludolls of the Kimo series). Sandy Entisols are mapped on historic flood deposits (Udipsamments of the Sarpy series). During archaeological surveys the soils on the terrace surface provide a good first approximation of the age of the land form.

The soils of the Kansas River terraces appear to exhibit considerable variability, especially when considering the minor soil associations (table 4.2). This led Sorenson et al. (1987, p. 100) to caution that "a simple linkage of key soils and major alluvial terraces is inadequate as an aid in understanding landscape evolution in the Kansas River valley." This is an important caveat given the variability in parent material characteristics that are produced by a large meandering stream. Their work, however, was based on an examination of published soil surveys. Another cautionary note in the form of a question must be raised: How well do the soil surveys reflect pedologic and geomorphic reality? The answer must rely on field checking.

In a few cases, archaeologists attempted to use the distribution of specific soil types or pedologic characteristics, based on data from county soil surveys, as indicators of past environments. This can be a misleading exercise, however, involving simplistic notions of soil–environment relationships. The relationship between plant types or local environments and soil morphology is complex because of environmental changes through time, because of the complex interaction of local environmental factors and pedogenic processes, and because few pedogenetic characteristics are related to specific types of plants or environments (Pawluk, 1978; Birkeland, 1984, pp. 268–271, 304–317; 1999, pp. 293–306; more fully discussed and illustrated in chapter 8).

In an archaeological site-catchment analysis in Iowa, the accuracy of the soil maps, and statements about soil–vegetation relationships and soil evolution in the county soil surveys, were taken at face value: "In the distribution of soil types, one has a reasonably accurate record of the distribution of the various types of vegetation. One can similarly use variation in aspect, drainage, and other land form features identified on soil maps to construct micro-vegetative variation within plant communities using detailed vegetative ordinations. For example, one can distinguish wet prairie from mesic prairie. The systematic changes in soil properties across a toposequence, therefore, serve as proxy data for detailed vegetative reconstruction. The process of mapping and identification is simplified by the fact that descriptions of soil types state the past vegetation responsible for the formulation of that particular soil type" (Tiffany and Abbott, 1982, pp. 315–316). The investigators are not considering map accuracy relative to map scale, the problems of map generalization, the lack of attention on the part of soil surveyors to the reconstruction of factors driving local soil genesis, and that soils are historical bodies affected by environmental changes. The time depth of the soil–environment relationship is unknown. In particular, reconstructing past vegetation is not a part of the county soil survey program, and in any case, such data are not available for most counties in the United States. Therefore, statements in county soil surveys concerning environmental changes must be regarded with skepticism unless they are well documented.

Voight and O'Brien (1981) attempted to use county soil surveys to aid in environmental reconstructions of the Mississippi River floodplain in southeastern Missouri, and they present the following cautionary tale:

> Properties of soils developed in a dynamic flood plain environment provide clues as to the degree and kind of interactions that have occurred among environmental variables. While soil distributions reflect both past and current processes involved in flood plain formation, the mapped soils are superficial sediments [sic] that have formed on land forms of variable age. Thus, vegetation maps generated through the use of SCS soil survey maps cannot be used as analogs of past spatial configurations of plant communities. As an example, soil survey maps of soil series and phases were used in conjunction with data from soil series interpretation sheets to construct a vegetation map of the study area.... From previous discussion it is evident that while soils properties are in part a function of the kind of vegetation under which a soil developed, important changes in community composition and distribution caused by habitat disruption can occur and have little effect on development of soil properties. Furthermore, the present configuration of soil in relation to land forms in the Ste. Genevieve flood plain is not analogous to prehistoric or historical period configurations. Thus, a vegetation map based on modern soils distributions is simply that—a map of modern vegetation distribution. (p. 31)

The kinds of geologic and geomorphic data that can be gleaned from county soil surveys often are limited, as discussed above, but there are some examples of soil surveys prepared with the type of background information that permits their use as guides to archaeologically significant soil geomorphic relationship. For example, in the 1960s the USDA sponsored an intensive investigation of soil geomorphic relationships on late Pleistocene and Holocene landscapes in the Willamette River valley of northwestern Oregon (Balster and Parsons, 1968; Parsons et al., 1970). The model of fluvial landscape evolution that resulted from this work has been used for mapping soils along rivers in the region (e.g., Gerig,

1985) and on the Pacific coast (e.g., Shipman, 1997). These county soil surveys should be useful in predicting the likely presence or absence of or age of archaeological sites, either buried or at the surface.

Likewise, the USDA soil surveys have been used for some geomorphic mapping, which has geoarchaeological potential. Areas with strongly contrasting parent materials and landforms are a key component in linking soil series to specific rock types, deposits, and geomorphic features (e.g., Lindholm, 1993, 1994a,b). Brevik and Fenton (1999) used soil surveys to map late Pleistocene strandlines of Glacial Lake Agassiz in eastern North Dakota. The strandlines are commonly preserved in the area as beach ridges composed of sand and gravel, in contrast to the silt and clay of the Lake Agassiz lake beds. Further, the strandlines are low but distinct ridges on the otherwise flat lake plain. The soils formed on the ridges, therefore, are better drained because of their coarser parent material and topographic setting, in contrast to soils on the surrounding, poorly drained lake plain.

Beyond the county soil surveys produced by the USDA, a wide variety of soil maps exist, produced for an array of purposes by a wide variety of agencies in many different scales. Many of these maps probably have archaeological applications, especially the larger-scale ones (i.e., maps covering small areas). In any case, the purpose of and intended audience for the maps and the manner in which they were prepared must be determined before archaeological applications can be made. In some countries, geomorphologists and Quaternary geologists work closely with pedologists to produce large-scale soil maps (i.e., maps of relatively small areas). The result is maps with soil types related to specific lithologies and depositional environments and also linked to landform types and landscape age. In The Netherlands, a soil survey produced at 1:10,000 proved useful in predicting likely site locations (beach ridges and natural levees) and site age because it was prepared by linking soil types to specific late Quaternary deposits and land forms (Dekker and De Weerd, 1973).

The Soil Survey for Scotland (1981) was put to good use in a geoarchaeological context as part of a study of ancient land management in Orkney (Simpson, 1997). The work focused on the Bilbster soil series, described as a "freely or imperfectly drained podzol developed on drift" (Davidson and Simpson, 1984, p. 75). This series includes a "deep top phase" (Davidson and Simpson, 1984, p. 75) or facies characterized by a surface horizon >75 cm thick. Davidson and Simpson (1984) showed that this thicker phase of the Bilbster is a plaggen, which is an anthropogenic soil resulting from repeated manuring of soils and is common in northwest Europe (see chapter 11). The distribution of the overthickened Bilbster series could be delineated on the soil maps of Orkney (Soil Survey for Scotland, 1981), allowing Simpson (1997) to conduct an in-depth study of the origins of plaggen soils and their implications for land use in the region in the 13th–19th centuries.

Soil surveys also have been carried out for expressly archaeological purposes with considerable success. In the midwestern United States, soil surveys along alluviated valleys have been important components of some archaeological site prediction studies (e.g., Bettis, 1992; Mandel, 1992, 1994; more fully discussed in chapters 7 and 9). The work entails looking at the distribution of soil types (taxonomic great groups and soil series) and their parent materials (i.e., the sediments comprising the alluvial fill); dating the sediments, soils, and associated land forms (typically stream terraces); modeling the environmental evolution of the sediments and soils; and using the resulting data to predict the location and age of archaeological sites, buried and at the surface, and to interpret the distribution of sites once surveys are completed. Not only are these surveys important for locating sites but they can also indicate likely areas in which sites will be absent, either because the sediments and landscapes likely to be associated with archaeological sites are missing or because the extant deposits and land forms are too young.

A good example of this approach is provided by Stafford and Creasman (2002), working in the lower Ohio River valley. Their work attempts to explain the "hidden record" of the Woodland and later prehistoric occupations on the floodplain and offers a model for future site prospecting. Data on soil geomorphic relations in the valley were determined using county soil surveys as well as other kinds of maps and extensive trenching by the geoarchaeologists. Landforms with Alfisols are underlain by Pleistocene fill and likely contain very few archaeological sites. Most of the archaeology represents Woodland and later occupations, which are buried in extensive deposits of late Holocene alluvium. Prehistoric sites typically are in deposits associated with Inceptisols (A-Bw profiles) and Mollisols (cumulic A-Bw profiles). These deposits and some late Prehistoric sites are buried beneath Historic alluvium characterized by Entisols (A-C profiles) and Mollisols (cumulic A-C profiles). The distribution of Entisols, Inceptisols, and Mollisols, therefore, can be used to predict the distribution of late Holocene alluvium, including Historic deposits, as well as potential areas with Woodland and other late Prehistoric sites. A very similar pattern of late Holocene soil geoarchaeological relationships is reported in the Upper Midwest and used as the basis for a similar model of site prediction (Bettis, 1992; further discussed in chapter 7).

Soil geomorphic mapping was carried out to predict site locations and to interpret archaeological finds in the desert terrain of Ft. Bliss in south-central New Mexico and far western Texas (Monger, 1995). The work built on the considerable soil geomorphic research of the USDA "Desert Project" in the Rio Grande Valley just to the west (e.g., Gile et al., 1965, 1966, 1981). Maps were prepared of soil geomorphic surfaces, which span the Pleistocene and Holocene (fig. 4.3). The degree of soil development on the surfaces was used to identify and locate late Quaternary soils and associated sediments, which are those that might contain archaeological sites; that is, the maps provided a method for subdividing land forms according to age (Monger, 1995, p. 38). The maps also aided in locating old surfaces with no late Quaternary deposits where occupation debris from a variety of periods is mixed together. Finally, the maps were also used to identify areas of wind erosion, which increases artifact visibility but destroys stratigraphic relationships.

In southwestern Greece, soil mapping was a component of long-term investigations in and near Nichoria (McDonald and Rapp, 1972; Rapp and Aschenbrenner, 1978). The mapping was a traditional soil survey and provided relatively little data of archaeological significance, but the surveys did show significant differences in soil morphology (Alfisols and Inceptisols) in and near the field area

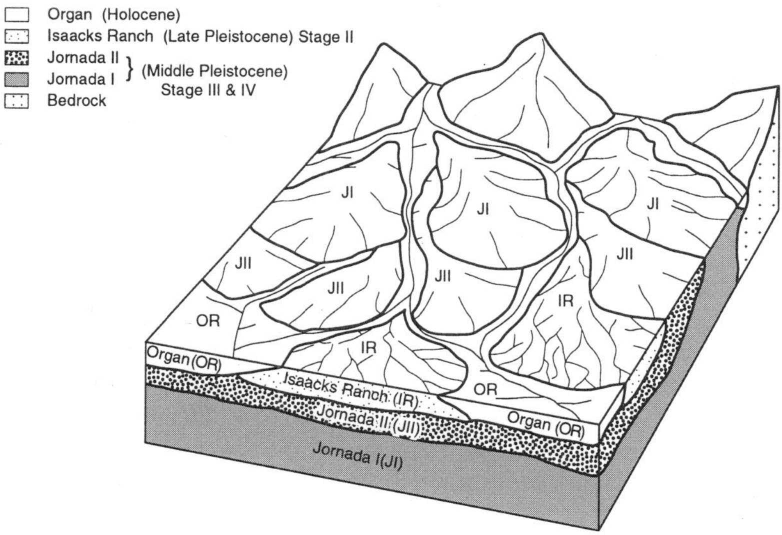

Figure 4.3 Block diagram based on mapping showing the geomorphic and stratigraphic relationships of soils, sediments, and landforms in south-central New Mexico (modified from H. C. Monger, 1995, "Pedology in arid lands archaeological research: An example from southern New Mexico–western Texas," fig. 3.3. *In Pedological Perspectives in Archaeological Research*, M. E. Collins, B. J. Carter, B. G. Gladfelter, and R. J. Southard, Eds., pp. 35–50. Soil Science Society of America, Special Publication no. 44; reproduced with permission of the Soil Science Society of America). The "Stages" refer to stages of carbonate development defined by Gile et al. (1966; see fig. 3.3). The surface soils are useful guides to locating and interpreting the archaeology. The Jornada I and II sediments predate human presence in the region, and archaeological remains would therefore be found as a palimpsest on these surfaces. The Isaacks Ranch (15,000–8000 yr B.P.) and Organ (<8000 yr B.P.) alluvium, however, could contain buried sites.

(Yassoglou and Nobeli, 1972; Yassoglou and Haidouti, 1978). This difference in soil types was attributed to differences in landscape stability and soil age (Alfisols indicate an older, more stable landscape, Inceptisols a younger, disturbed landscape) caused by human activity.

A survey designed expressly to identify relict arable soils was carried out on the Lofoten Islands of Norway (Simpson et al., 1998). Traditional field survey was combined with analysis of pollen cores to detect the onset of agriculture and the examination of thin sections to complement field evidence for agriculture or other human impact (see also chapter 11). The regional uncultivated soils are dominantly sandy soils and peats resting on sand. Two distinctly different and more localized soils were also identified: loamy sands with scattered bone, charcoal, and pottery shards, and sandy peats with some bone and charcoal. These two soils are considered to be ancient cultivated fields, with mixed textures caused by plowing, and the bone, charcoal, and pottery represent amendments derived from midden debris used as fertilizer.

Site-specific soil surveys have proven helpful in solving particular archaeological research questions. At the Altofts site, West Yorkshire, in the United Kingdom, an area identified as a henge (a circular or oval area enclosed by a bank and an internal ditch) was investigated to determine whether in fact it was a henge (Weston, 1996). The studies included a soil survey built around a transect using a bucket auger. The augering revealed that the heaps of sediment thought to be a henge were composed of coal slag, charcoal, and burned sediment and rock. This was a clear indication that the heaps were relicts of the Industrial Revolution, probably from a coal mine. The degree of leaching of the mining debris, the formation of a Bw horizon in the debris, and the translocation of coal dust into the underlying pre-Industrial soil suggested that the mounds were formed in the late 19th century.

In western Belize, soils provided insights into the location of Maya cacao orchards (Muhs et al., 1985). Cacao was one of the most important crops of the lowland Maya, but evidence of where the cacao was grown is rarely evident. In addition, cacao macrofossils are rarely preserved and cacao sheds little pollen. Muhs et al. (1985) addressed the problem by mapping and analyzing soils at an ethnohistorically documented cacao-growing center. A 1:10,000 map of soils and geomorphic surfaces in the study area was prepared. The conclusions of Muhs and colleagues regarding the best soils for cacao growth were based on the distribution of soils with suitable physical and chemical characteristics (floodplain Argiudolls and Hapludolls). Supporting evidence included buried Maya-period stone walls that may have marked field boundaries.

Several site-focused soil surveys have helped resolve issues of site location and age (also discussed in chapter 7). A soil survey around the ancient Bronze Age city of Harappa in Pakistan (Pendall and Amundson, 1990; Amundson and Pendall, 1991) used a soil geomorphic approach to reconstruct the Holocene evolution of the landscape in the site area and to understand the relationship of the site to the adjacent Ravi River. Unraveling the fluvial history of the site area was a key to resolving longstanding debates over the effect of floods and climate change on Harappa. The survey showed that the site was built on the oldest and highest terrace in the area but was also adjacent to a river channel active at the time the site was occupied (fig. 4.4). A dry channel of the Ravi River lies just north of Harappa, but a series of parallel, semiconcentric bands formed by the soil mapping units to the east, south, and west of the site (fig. 4.4) shows that an older meander once flowed around the south side of the ancient city. These data were interpreted to indicate that the site location was carefully chosen to avoid most floods but also was located to take advantage of the river for transportation and floodplain cultivation. Subsequent soil and topographic survey work (Belcher and Belcher, 2000) substantiated these conclusions. The numerical age of the abandoned channel was not determined, but the topographic position of this alluvial landscape and degree of soil development in the channel fill relative to the alluvial surface beneath the site showed that it existed as an active channel before or during occupation of the site, but not after. Belcher and Belcher (2000) also suggest that site selection had a significant influence on the development of the Indus Valley Tradition. In a somewhat similar approach, Jones (1990) conducted soil surveys at a series of sites on the North Island of New Zealand. Using

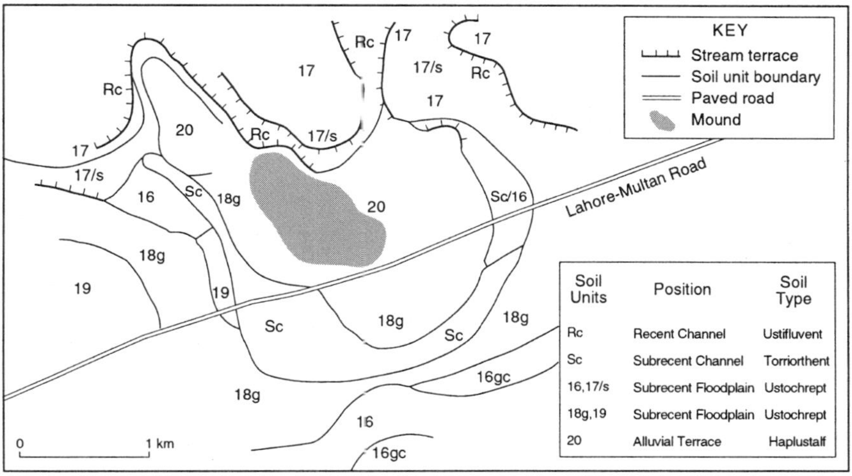

Figure 4.4 Soil map of the Harappa area, Pakistan (based on Pendall and Amundson, 1990, fig. 2; Belcher and Belcher, 2000, figs. 7 and 8). The pattern of stream meanders is clearly apparent around the site (soil units Rc, Sc, 16, 17, 18, and 19).

published, regional soil surveys combined with data from studies of local fluvial geomorphology, he was able to apply local soils data to estimate the maximum ages for construction of fortified settlements.

The demands of CRM archaeology in the United States resulted in several innovative approaches to acquiring and applying soils data. In a study within the Rum River basin, near Mille Lacs Lake in Minnesota, Kolb (2000) incorporated a 1927 soil survey into a geoarchaeological investigation. The older survey emphasized the surface texture of the soils. The textural data were used to estimate the thickness and distribution of a sheet of eolian silt and sand that blanketed the area. The textural data were sufficiently detailed to permit identification of predominate sediment-moving wind direction and the location of source areas for the eolian sediment. In a study at Wright-Patterson Air Force Base, Ohio, archaeological data were used to construct a soil map for geoarchaeological purposes (Kolb, 1996). Much of the archaeological fieldwork consisted of "shovel probing" at 10-m intervals. Following an initial study by the geoarchaeologists, the archaeological crews were trained to recognize and record a few significant soil variables during the shovel work. The result is a detailed soil map (Kolb, 1996, fig. 35) that can be used to identify disturbed areas that require no further study or areas of recent fill in which archaeological sites might be buried. The soil map also indicated drainage conditions, which were used to assess the potential for archaeological sites: Better-drained settings were considered more likely to have sites than poorly drained settings.

5

Soil Stratigraphy

Soils have been employed in archaeological stratigraphy since at least the 1930s, including topical discussions of the significance of soils in stratified deposits (e.g., Leighton, 1936, 1937; Bryan and Albritton, 1943). This apparently was for several reasons. The unique physical and chemical properties that distinguish soils from sediments make soils quite useful for stratigraphic subdivision and correlation. In particular, pedologic features, most notably soil horizons, are often the most visually prominent features in stratified deposits. Furthermore, much of the early archaeological pedology was done by individuals trained in Quaternary geology (e.g., Leighton, 1937; Bryan, 1941a; Bryan and Albritton, 1943; Movius, 1944, pp. 49–62), in which soils have been recognized as stratigraphically important since the late 19th century (Bowen, 1978, pp. 10–56; Finkl, 1980; Tandarich, 1998a).

The recognition of soils and the differentiation of soils from sediments in archaeological contexts is one of the most fundamentally significant aspects of geoarchaeological stratigraphy. This initial step in stratigraphic interpretation is crucial to most of the applications of pedology and soil geomorphology discussed in subsequent chapters. Because soils indicate periods of stability or hiatuses in deposition, the identification of soils or the lack thereof in a stratigraphic sequence provides information on the number of depositional episodes and intervals of stability. The identification of specific soil horizons also provides clues to the degree and duration of soil development, the nature of the soil-forming environment, and the kinds of soil-forming processes that may affect the archaeological record. Further, tracing of soils from exposure to exposure is a key aspect of correlating strata and interpreting the evolution of archaeological landscapes.

This chapter presents a discussion of some principals of soil stratigraphy, and the following chapter focuses on the archaeological significance of soils as stratigraphic units. This chapter begins with a discussion of basic stratigraphy, which is one of the fundamental components of field-based geoscience. That section is followed by a closer look at soil stratigraphy, including a summary of both formal and informal soil stratigraphic nomenclature as well as a discussion of the unique characteristics of soils when used as stratigraphic markers and their archaeological implications. The recognition of buried soils and the processes of soil burial are then discussed. A number of terms are introduced in this chapter, including discussion of their appropriateness, use, or misuse. These issues may seem picayune, but understanding them is vital to intra- and interdisciplinary communication and understanding. "Stratigraphic nomenclature, like systematic taxonomy, is a tedious means to interesting ends" (Hopkins, 1975, p. 10).

Stratigraphy

Stratigraphy is the study of the sequence, age, and correlation of sediments and rocks and their interpretation regarding mode of origin and geologic history (Bates and Jackson, 1980; p. 615; Salvador, 1994, p. 13). Stratigraphic classification "systematically arranges and partitions bodies of rock or unconsolidated materials of the Earth's crust into units based on their inherent properties or attributes" (North American Commission on Stratigraphic Nomenclature [NACOSN], 1983, p. 847). The principles of stratigraphic classification are spelled out by Salvador (1994), and excellent discussions of stratigraphy from the perspective of Quaternary studies are provided by Bowen (1978, pp. 84–104) and Lowe and Walker (1997, pp. 298–323). In geoarchaeological contexts, general stratigraphic principles are reviewed by Stein (1987, pp. 341–346; 2000), Waters (1992, pp. 60–88), and Thorson (1990, pp. 400–406), and some basic concepts of soil stratigraphy are illustrated by Brewer (1972) and Cremeens and Hart (1995).

The grouping of sediments or soils into stratigraphic units can be based on a variety of criteria. "Lithostratigraphy" is the grouping of sediments or rocks on the basis of lithological characteristics such as sediment or rock type (e.g., limestone or alluvium). Lithostratigraphy is perhaps the most common type of stratigraphy and is what most people probably think of when the term "stratigraphy" is used. There are other types of stratigraphic organization and correlation, however. "Chronostratigraphy" is the grouping of sediment or rocks based on their age (e.g., Cretaceous or Pleistocene or some specific numerical interval of time). "Biostratigraphy" is the grouping of rocks or sediments based on their content of plant or animal remains—usually fossilized remains in most of geology, but not necessarily in Quaternary stratigraphy (e.g., vertebrate faunas or pollen assemblages). Lithostratigraphic units and biostratigraphic units can have diachronous or time-transgressive boundaries, but chronostratigraphic units always (by definition) have synchronous boundaries (fig. 5.1). Moreover, in a sequence of rocks or sediment, the boundaries of different stratigraphic units may not coincide; that is, the rocks or sediment may be grouped in several different ways depending on the stratigraphic approach taken (fig. 5.2).

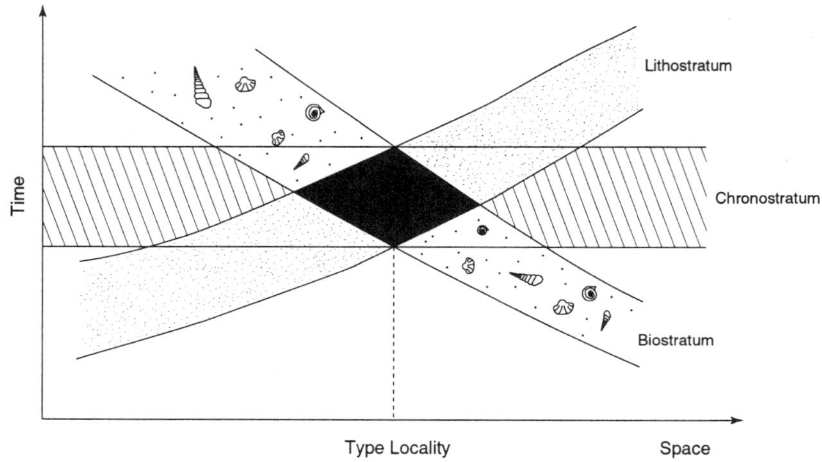

Figure 5.1 The three principal kinds of stratigraphic units, illustrating the time-transgressive nature of boundaries of lithostratigraphic and biostratigraphic units in comparison to the synchronous boundaries of chronostratigraphic units (Holliday, 2001a, fig. 1.1a; from *Earth Sciences and Archaeology*, P. Goldberg, V. T. Holliday, and C. R. Ferring, Eds., © 2001, Kluwer Academic/Plenum Publishers. Reproduced with permission of Kluwer Academic/Plenum Publishers).

"Soil stratigraphy" or "pedostratigraphy" is another way of grouping and correlating sediments and rock (fig. 5.2). It has been defined in several ways. Finkl (1980, p. 171), focusing on the temporal aspects of stratified sequences of soils, defines soil stratigraphy as the "chronological ordering of pedological episodes" and the aim of soil stratigraphers to "recognize significant soil-stratigraphic units and to rank them chronologically." A much broader view is offered by Catt (1990, p. 169), who defines soil stratigraphy as "the study of different soil associations formed in an area during past periods of varied soil-forming conditions." (Here the term "association" refers to a group of related soils that vary laterally because of variation in soil-forming factors [discussed further in the text]. This is a more generic use of the term compared to the soil-mapping parlance of the USDA, discussed in chapter 2.) This approach is attractive from a geoarchaeological perspective because it emphasizes the landscape and not just time, providing a conceptual framework that includes landscape reconstruction as well as correlation and dating.

Soil Stratigraphic Units

More germane to this discussion are definitions of formal stratigraphic units, which are classificatory terms based on any of the various physical, chemical, biological, or chronological attributes that rocks or sediments may possess. Geoarchaeologists should be aware of some of the basic concepts and controversies behind this terminology to better understand the literature and to make their

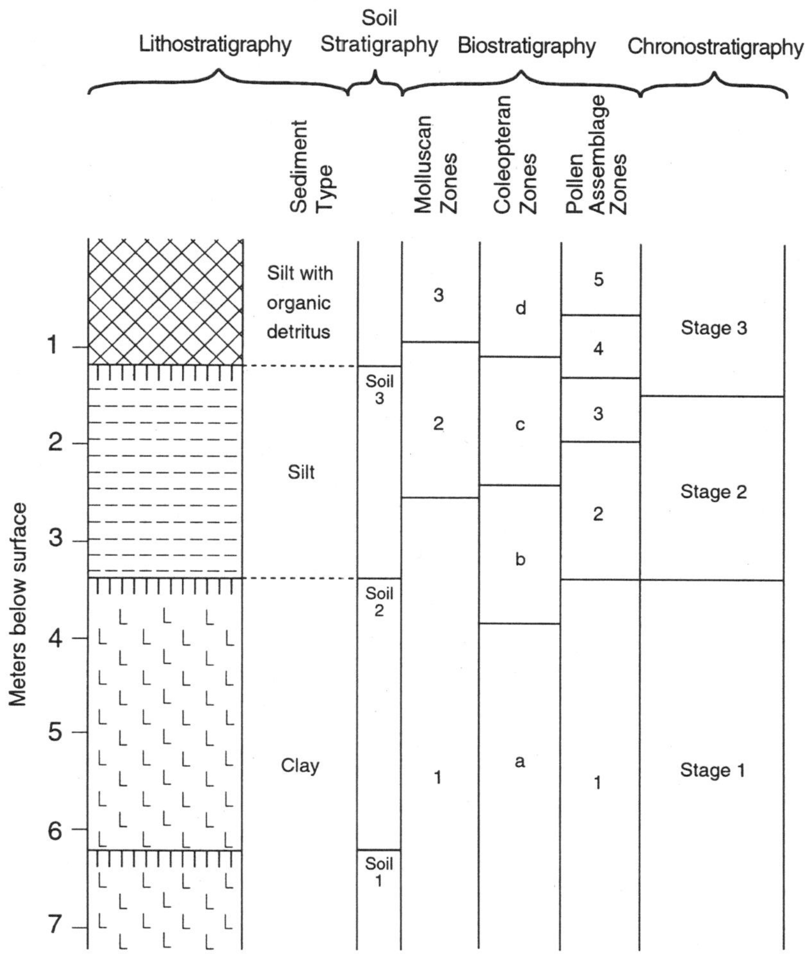

Figure 5.2 Subdivisions of a sedimentary sequence showing different positions of stratigraphic boundaries depending on different stratigraphic criteria (Holliday, 2001, fig. 1.1b; from *Earth Sciences and Archaeology*, P. Goldberg, V. T. Holliday, and C. R. Ferring, Eds., © 2001, Kluwer Academic/Plenum Publishers. Reproduced with permission of Kluwer Academic/Plenum Publishers).

own decisions on which terms to use and how to use them. Soil stratigraphic units have been defined in several ways. They are formally referred to as "pedostratigraphic units" in the current North American Stratigraphic Code (NACOSN, 1983, pp. 850, 864). A pedostratigraphic unit is a "buried, traceable, three-dimensional body of rock [or sediment] that consists of one or more differentiated pedologic horizons" developed in and overlain by "one or more formally defined lithostratigraphic or allostratigraphic units" (NACOSN, 1983, p. 864). The definition of "pedologic horizons" in the North American Code excludes the O horizon and need not include the C horizon. The base of most soil stratigraphic units, following these criteria, is therefore likely to be the base of the B horizon.

An earlier version of the code defined a soil stratigraphic unit as, "a soil with physical features and stratigraphic relations that permit its consistent recognition and mapping as a stratigraphic unit" (American Commission on Stratigraphic Nomenclature [ACSN], 1961, p. 654).

The 1983 approach to pedostratigraphic units is more specific than the 1961 approach, which some view as an advantage over the earlier definition (Cremeens and Hart, 1995, p. 18), but there are several disadvantages to the 1983 definition, particularly in geoarchaeological contexts. The requirement that the soils be buried is illogical and, in any case, impractical in many field situations. The processes that govern whether or not a soil is buried can be variable and highly localized, particularly in Holocene deposits such as dunes. The 1983 definition can result in a situation in which a clearly traceable soil that is locally buried or locally exposed at the surface is a pedostratigraphic unit in some places but not others. Likewise, the requirement that pedostratigraphic units be developed in and overlain by formally defined lithostratigraphic and allostratigraphic units (discussed later) is impractical and unrealistic because formally defined Holocene strata are rare and formally defined Quaternary units are only locally common. Pedostratigraphic units, which formed as a result of pedogenesis, cannot include O horizons according to NACOSN (1983) because these organic-rich zones are not considered products of pedogenesis. This approach depends on one's definition of what constitutes pedogenic processes and creates problems in differentiating buried Oa horizons (zones of high organic-matter content with little recognizable plant remains) from the rest of the soil profile (Cremeens and Hart, 1995, p. 18). The requirement for burial and the restriction of the O horizon in the definition of the pedostratigraphic unit represent seemingly arbitrary rules and hinder the field applications of the concept of the pedostratigraphic unit. Our professional vocabulary—unlike the NACOSN definition of pedostratigraphic units—should be based on practical utility.

An important new component of the 1983 code, with implications in soil geomorphology and soil stratigraphy, is the definition of the "allostratigraphic unit." It is a "mappable stratiform body of sedimentary rock [or sediment] that is defined and identified on the basis of its bounding disconformities" (NACOSN, 1983, p. 865). The definition of these units was a means of mapping and formally defining layers of rock or sediment that are lithologically similar (i.e., that could not otherwise be differentiated as lithostratigraphic units) but are clearly separable on the basis of disconformities. The provision for the allostratigraphic unit has been especially useful in Quaternary mapping and stratigraphic research (Autin, 1992; Blum and Valastro, 1994; Birkeland, 1999, pp. 308–309). For example, allostratigraphic units could be used to differentiate otherwise lithologically similar stream terraces, alluvial fans, or glacial tills. Soils, surface or buried, represent a kind of disconformity and therefore can be used to identify the upper boundary of an allostratigraphic unit (NACOSN, 1983, p. 866). The formal definition of a sequence of lithologically similar sediments with intercalated buried soils can, therefore, be done in terms of allostratigraphy as well as soil stratigraphy.

Several terms have been proposed as the "fundamental unit" of pedostratigraphic classification (i.e., a term equivalent to "formation" as the fundamental unit of lithostratigraphic classification). In the North American Stratigraphic

Code, the "Geosol" was adopted as the fundamental unit, following Morrison (1967, 1978). For example, in the midwestern United States, the Sangamon Geosol is a formal pedostratigraphic unit (e.g., Johnson et al., 1997; Jacobs, 1998). The above-cited definition of a pedostratigraphic unit followed Morrison's concept of "Geosol." "Geosol" was chosen over "soil" apparently because of the many different definitions and connotations of "soil." "Geosol" has not been shown to have any more utility than "soil," however. "Geosol" is clearly defined in the code, so "soil" likewise could be defined. There is rarely any confusion over the meaning of "soil" among geoarchaeologists, geomorphologists, or Quaternary stratigraphers. Moreover, from the time "Geosol" was first proposed (Morrison, 1967) until its adoption in the Code (NACOSN, 1983), it was seldom applied, indicating that it lacked clear utility and was not superior to "soil." Other terms with a connotation more or less similar to geosol include "groundsurface" (Butler, 1959; see chapter 1), "pedolith" (Crook and Coventry, 1967), "pedomorpholith" (Van Dijk et al., 1968), and "pedoderm" (Brewer et al., 1970). Using "soil" to refer to pedostratigraphic units still seems the most sensible and straightforward approach, however. Ultimately, field investigators are free to choose the terminology that best suits them and their field situation.

Soil Stratigraphic Terminology

A wide variety of sometimes confusing terminology has been proposed and used to describe the many possible stratigraphic relationships between and within buried soils and soil stratigraphic units. Most of the terms are seldom used, largely because they lack utility in illuminating depositional or landscape history. A review of these terms, however, is a good way to systematically examine and explain possible stratigraphic relationships among buried soils.

Partial or localized burial of a soil produces complexities in soil stratigraphic relationships. If a soil is buried in some places but not in others, and the buried soil is taken out of the soil-forming environment (if it is buried deep enough), pedogenesis will then begin in the new sediment, whereas soil development will continue uninterrupted in the unburied soil (fig. 5.3; e.g., fig. 4.2). The unburied soil will be exposed to more variation in the pedogenic factors than either the buried soil or the soil formed in the new sediment. For example, the unburied soil will undergo a longer period of pedogenesis than either of the other two soils (fig. 5.3). Moreover, the unburied soil may also be exposed to more variation in the environment. In this situation, the unburied soil is most commonly referred to as a "polygenetic soil" (fig. 5.3; Bryan and Albritton, 1943; Catt, 1998, p. 84; see also chapter 1). Equivalent to polygenetic soil are the terms "polycyclic soil" (Duchaufour, 1982, p. 144) and "polyphased soil" (Courty et al., 1989, pp. 13, 15). Other generally equivalent terms include "superimposed soil" (Hunt, 1972, p. 199) and "polymorphic soil" (Simonson, 1978, p. 20), but these terms do not necessarily refer to soils subjected to climate changes, whereas polygenetic and polycyclic do. The term "relict soil" has long been used synonymously with the term "polygenetic soil" (e.g., Thorp et al., 1951; Catt, 1998; discussed by Johnson and Hole, 1994, pp. 117–118), but in a well-known definition, Ruhe (1965, p. 755) defined relict soils as "soils that formed on pre-existing landscapes but were not

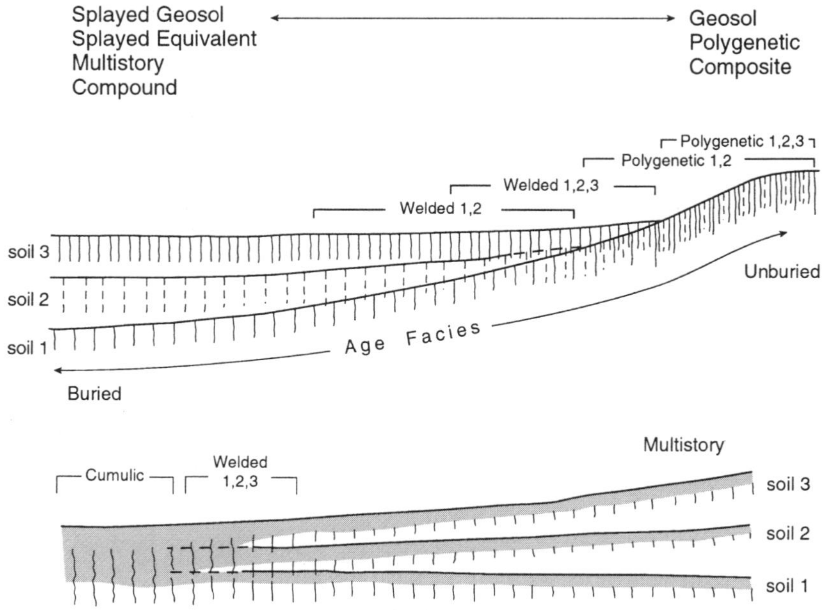

Figure 5.3 Schematic diagrams illustrating the soil stratigraphic and facies relationships of buried, unburied, and cumulic soils and associated terminology (see also fig. 6.1). In the upper diagram, duration of pedogenesis in the unburied, polygenetic soil is equal to combined duration of pedogenesis of soils 1, 2, and 3, plus time of sedimentation.

buried by younger sediments." This definition does not deal with the issue of both past and current pedogenesis (see Catt, 1986, pp. 169, 171; Johnson and Hole, 1994, pp. 117–118), and the term is considered obsolete by the Working Group on Definitions used in Paleopedology (Catt, 1998).

Closely spaced multiple buried soils (the "multistory soils" of Morrison, 1967, 1978, pp. 83–84) are described by a number of stratigraphic terms (fig. 5.3). Buried soils separated by unweathered or minimally weathered sediment are described as "compounded soils" by Morrison (1967, p. 25; 1978, pp. 83–84). If pedogenic processes from an upper soil extend down into or "overlap" with a lower soil, the process is referred to as soil "welding" and produces a set of "welded soils" (figs. 2.1A, 5.3, 5.10, 6.17, 7.7, 8.6, 8.8, 10.12, and 10.13) (Ruhe and Olson, 1980). Soil welding is an important process of postburial alteration of soils (see chapter 10). Welded soils are also referred to as "polypedomorphic soils" (Bos and Sevink, 1975; Duchaufour, 1982, p. 144) or "composite soils" (Morrison, 1967, p. 25; 1978, pp. 83–84). Bos and Sevink (1975, pp. 224–225) refer to a set of welded soils as a "complex profile." In Europe the term "pedocomplex" often is used to describe multistory soils, but there is confusion over the definition of the term. Catt (1990, p. 18; 1991, p. 174) emphatically defines a pedocomplex as a set of welded soils, whereas Finkl (1980, p. 181) and Catt (1998; representing the Working Group on Definitions used in Paleopedology) specifically define the term to refer to closely spaced buried soils with no "overlap." On the basis of the usage of the term "pedocomplex" in the literature by European workers (e.g., Kukla, 1970, 1975, 1977, 1987; Lazarenko, 1984; Dodonov, 1991; Bronger et al., 1995; Kemp, 2001), it seems to be applied to closely spaced sets of either welded or unwelded soils and also to overthickened or cumulic soils (see following).

Multiple buried soils not welded to one another may be accorded individual soil stratigraphic status if they are stratigraphically significant. If a set of buried soils can be traced to a single soil (fig. 5.3), then (1) the buried soils are described as "subdivided soils" by Morrison (1967, p. 26; 1978, p. 84), whether or not they are welded; (2) the buried soils are termed "compound soils" by Duchaufour (1982, p. 144), "splayed soils" by Catt (1990, p. 18), and a "composite profile" by Bos and Sevink (1975, pp. 224–225) if they are not welded; (3) the buried soils are described as a "pedocomplex equivalent" to the single soil by Catt (1990, p. 18) if the buried soils are welded; and (4) the single soil that is equivalent to the multiple buried soils may be polygenetic or composite. The subdivided equivalents of a single soil stratigraphic unit are identified as a "subdivided geosol" by Morrison (1967, p. 26; 1978, p. 84) and a "splayed equivalent" by Catt (1990, p. 18; figs. 5.3 and 6.1). The subdivided or multistory equivalents of a single soil could also be considered facies of one another (soil facies are discussed further in the text).

The various pedostratigraphic terms and concepts discussed above indicate that buried soils come in neat "multistoried" sets of either welded or compound soils (e.g., Bos and Sevink, 1975, fig. 2; Morrison, 1978, figs. 1–3). In the field, however (and illustrated further in this chapter), workers will find any combination of welded soils, unwelded soils, pedocomplexes of both, subdivided polygenetic soils, deeply buried soils, and shallowly buried soils. Knowing the terms for the stratigraphic relationships is not the point. What is important is understanding how various soil stratigraphic relationships evolve, what they mean in terms of landscape evolution, and what they mean geoarchaeologically.

Much of the discussion of buried soils in soil stratigraphic literature, including the above, emphasizes the soils and not the burial processes. Terms such as "subdivided" and "splayed" suggest that some process split the soils apart, when in fact the main process in operation was episodic and often local sedimentation. This is a good place to reemphasize the notion that soil formation is a process on relatively stable landscapes that probably represents the long-term norm, whereas sedimentation (i.e., soil burial processes) often represents relatively brief processes and, in many situations, a brief departure from the norm (further discussed in chapter 7). Indeed, a focus on soil stratigraphic units tends to place an undue emphasis on the soil-forming conditions and "soil-forming intervals" (discussed in chapter 8; McFadden and McDonald, 1998) rather than on the processes of erosion and sedimentation that produce individual buried soils and that may, in fact, be more environmentally significant because they represent a shift away from the norm, or prevailing conditions.

Soil Facies

"Soil facies" are lateral variations in soils caused by variations in one or more of the soil-forming factors (Morrison, 1967, pp. 13, 15; 1978, pp. 86–87). Soil facies are also known as "soil variants" (Leamy et al., 1973) or "soil associations" (Catt, 1986, p. 168). The term "soil association" follows the connotation of formally defined "associations" used in soil mapping (chapter 2), and "pedofacies" are defined as corresponding approximately with soil series (Catt, 1998, p. 84). Some groups of soil series can be interpreted as representing variations in a soil stratigraphic unit across a landscape, but as discussed in chapter 4, many variables are involved in defining and differentiating soil series and soil associations and they may or may not have anything to do with the factors influencing the soil. The Working Group on Definitions used in Paleopedology (Catt, 1998) defines a pedofacies as essentially the same thing as a soil facies. However, as originally introduced (Bown and Kraus, 1987; Kraus and Bown, 1988), a pedofacies specifically referred to variability in the "maturity" or degree of development of buried soils, typically on floodplains, as a function of their distance from areas of relatively high sediment accumulation (fig. 6.1; discussed further in chapter 6). Areas of slower sediment accumulation will display more strongly expressed soils. A pedofacies, therefore, according to the original definition, is a very specific kind of soil facies.

Regardless of the terminology chosen, a significant aspect of soil stratigraphic units is that they vary as the factors that influenced their development varied. This is a very useful concept in soil stratigraphy because buried soils, similar to surface soils, can vary laterally. For example, a buried soil that formed in several different parent materials will exhibit different lithological facies (fig. 6.1). A soil stratigraphic unit that is buried in some places but not in others will exhibit facies variations caused by differences in the age of the soil (fig. 5.3)—that is, in the amount of time the soil had to form, and possibly by environmental variations.

Of particular significance when using soils as age indicators is that buried soils, similar to lithostratigraphic units and contemporary surface soils, are diachronic or time transgressive. The age of the upper boundary of the pedostratigraphic unit can vary as the age of the paleo-landsurface associated with the soil varied. Rapid burial over a large area can produce a buried soil with an isochronous upper boundary (at least within the limits of most dating methods). More typically, however, minor or slow sedimentation on one part of the landscape and erosion on another part produces a soil surface of variable age. Put another way, "Soil formation is discontinuous when . . . erosion and deposition alternate with periods of greater surface stability. Some parts of a land surface may persist through the climatic and other environmental changes that cause erosion and deposition elsewhere; pedogenesis on them is continuous, though the changes usually result in partial or complete replacement of some soil-forming processes by others" (Catt, 1986, p. 169).

Several examples illustrate the time-transgressive nature of some soil strata. At the Lubbock Lake site in Texas, a prominent soil stratigraphic marker is a moderately expressed soil (the Lubbock Lake soil) formed on middle Holocene eolian valley fill (Holliday, 1985c, 1988). Sedimentation slowed and pedogenesis began ~4500 yr B.P. The soil underwent local burial beginning ~1000 yr B.P. Thus the age of the surface varies from one buried 1000 yr ago to one still exposed today (see also chapters 6, 7, 9). At a much broader temporal and geographical

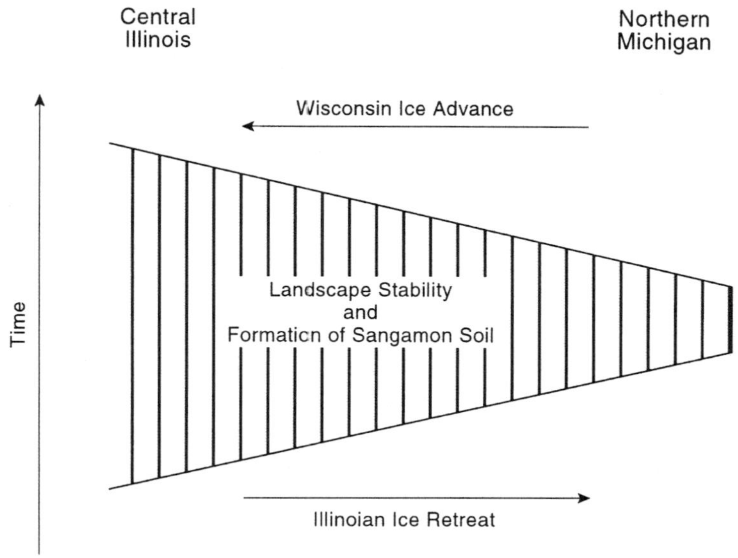

Figure 5.4 The time-transgressive nature of soil development illustrated by the Sangamon Soil of the midwestern United States (based on W. H. Johnson et al., 1997, fig. 1).

scale, in the midwestern United States, one of the best known soil stratigraphic units in the world, the Sangamon Soil (or Sangamon Geosol) formed during the last interglacial period (the Sangamon Interglacial) on the landscape created as the penultimate ice sheet (of the Illinoisan Episode in midwestern stratigraphic parlance) retreated out of the Great Lakes region into Canada (Follmer, 1978, 1982, 1983; Johnson et al., 1997; Hall, 1999). The landscape was subsequently buried as ice advanced during the Wisconsinan Episode. The Sangamon Soil, therefore, is much older in terms of the time it had to form to the south (in the lower Midwest) than it is to the north (in the upper Midwest and into Canada) because the landscape was formed first in the south and buried last in the south (fig. 5.4).

Soil facies are also produced as the parent material, topography, vegetation, and biota vary. The above-mentioned Lubbock Lake soil, in addition to local burial, formed in parent material that includes both sandy valley-margin facies and more loamy valley-axis facies and in topographic positions such as well-drained steeper slopes near the valley margin and on more poorly drained flat surfaces along the valley axis (fig. 9.3; Holliday, 1985a,d, 1988). At a much larger spatial scale, soils on the Sangamon landscape exhibit considerable facies variation (Hall, 1999). The soil formed in till, glacial outwash, loess, alluvium, slopewash, and freshly exposed bedrock. The soil also developed in a variety of topographic settings, from well-drained uplands to waterlogged lowlands and closed basins. Sangamon pedogenesis also was variously affected by a wide array of environments as plant communities and climate varied locally and in a south-to-north gradient.

Variation in a buried soil or soil stratigraphic unit resulting from variation in the topography and drainage conditions during soil genesis produces a "paleocatena." Most typically, a paleocatena is simply a buried catena. The concept of a buried catena or paleocatena is another important one, however, because this is a unique aspect of soils relative to other kinds of stratigraphic units and is a key to the recognition of buried soils (discussed in chapter 9). Put another way, soil stratigraphic units should and do vary as soils at the surface vary today. Components of the buried soil stratigraphic units at Lubbock Lake and the buried Sangamon Soil comprise paleocatenas.

Soil Stratigraphy versus Soil Horizonation

The many variables in local and regional soil-forming factors mean that the number and thickness of soil horizons in a soil stratigraphic unit and the position of the lower boundary of the soil can vary significantly. The individual soil horizons and the lower boundary of the soil, therefore, have no stratigraphic significance whatsoever other than to define the soil stratigraphic unit. The lower boundary, similar to the upper one, is time transgressive or diachronic. It represents the lowermost significant alteration of the parent material by pedogenic processes (fig. 5.5).

Soil horizons are not geologic layers; horizons represent an alteration of the layers. Cremeens and Hart (1995, p. 26) further note that the Law of Superposition does not apply to the vertical relationships of individual soil horizons such as the typical A-B-C horizon sequence. This is an important point. The concept of soil horizonation and the specific spatial succession of soil horizons is a significant pedologic paradigm with important stratigraphic ramifications. Soil

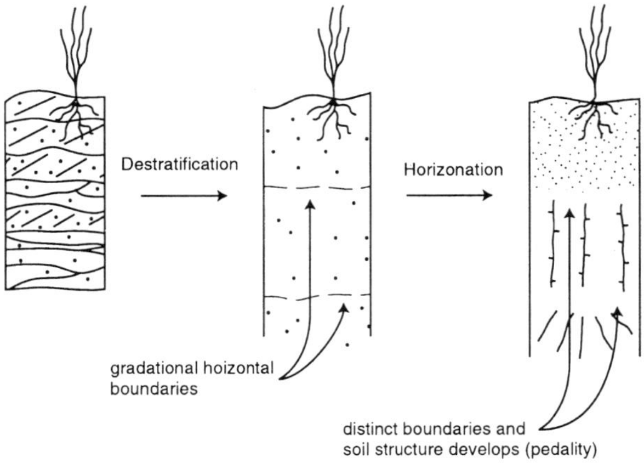

Figure 5.5 Pedogenic modification of alluvium (modified from Allen and Wright, 1989, fig. 1.3; published with permission of V. P. Wright and J. R. L. Allen). Primary stratification is gradually destroyed, along with or followed by development of soil horizons.

horizonation extends from the surface downward, as a function of pedologic processes (fig. 5.5). Lithostratigraphic units in which soils form do obey the Law of Superposition, as do pedostratigraphic units; that is, in a sequence of multiple buried soils, the uppermost soil is the youngest and the deepest one is the oldest (fig. 5.3). Individual soil horizons should not be used as individual stratigraphic entities, however. Soil horizons are separated in space but are not necessarily separated in time, though some horizons may evolve sequentially after others form (e.g., an A-C profile evolving into an A-Bw-C soil and then into an A-Bt-C sequence; see chapter 7). Moreover, soil horizons, particularly the A horizon, can be thoroughly churned because of the effects of floral and faunal mixing (discussed in chapter 10).

There are several specific exceptions in which the Law of Superposition does apply to soil horizonation and in which bedding and lithostratigraphic contacts are preserved. Aggrading floodplains typically have multiple, weakly expressed buried soils composed of an A and C horizon (the classic Fluvents of soil taxonomy; chapter 7). They are essentially stacks of alluvial sediment with some organic matter accumulation at contacts and with bedding preserved (Ferring, 1992, 2001; Cremeens and Hart, 1995). Archaeological materials in these Fluventic deposits should likewise be stacked up oldest to youngest (further discussed in chapter 7; Cremeens and Hart, 1995). Soil O horizons, which are essentially accumulations of organic matter (e.g., in a peat bog; further discussed in chapter 10), also build vertically and can be stratified. Associated archaeology likely would be in correct stratigraphic order, though some bioturbation is possible (chapter 10). Slowly aggrading "cumulic" or "pachic" A horizons (see below) also may have stratified artifactual debris, but again, bioturbation could be a significant problem.

The formation of soil horizons tends to complicate lithostratigraphic interpretations because soil horizons can obscure lithostratigraphic contacts. Flegenheimer and Zárate (1993) note this as a distinct geoarchaeological problem in loess on the Argentine Pampas. Individual layers of loess typically are less than 2 m thick (Imbellone and Terugi, 1993) and can be fully altered by pedogenic processes. Though not mentioned by Flegenheimer and Zárate (1993), soil welding probably complicates the stratigraphic interpretations. The archaeological pitfalls of confusing soil horizons with stratigraphic units are well illustrated at Lubbock Lake. Work at the site in 1939 and 1941 exposed the Holocene fill from valley axis to valley margin (Wheat, 1974). The stratigraphic sequence that was recognized included buried soils, which was unusual for the time and reflects the archaeologists' unusual degree of appreciation for the geologic history of the site. In any case, the stratigraphic subdivisions included several soil horizons, and, therefore, archaeological materials that spanned a significant amount of time were lumped together (Johnson and Holliday, 1986). For example, the calcic horizon of the Lubbock Lake soil (mentioned earlier) was a stratigraphic marker. In most areas it formed near the base of the middle Holocene eolian deposits (stratum 4), but in the area of the early excavations the calcic horizon formed in early Holocene eolin sediments (stratum 3; fig. 5.6; Johnson and Holliday, 1986). Artifacts and features from that horizon therefore come from two time periods.

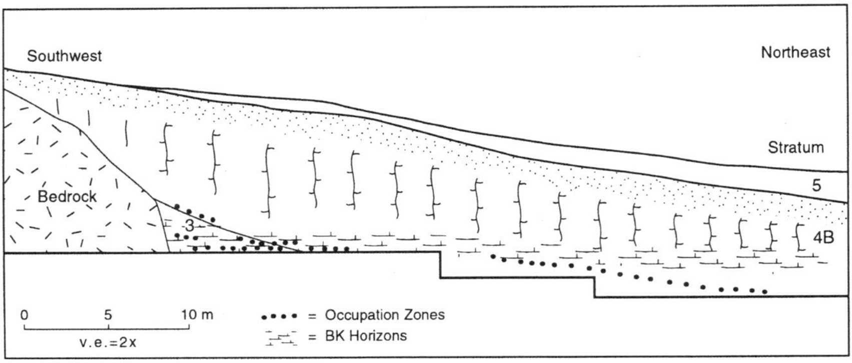

Figure 5.6 Stratigraphy of strata 3 and 4 at the Lubbock Lake site, Texas, showing the position of the calcic horizon of the Lubbock Lake soil (largely in stratum 4B) superimposed over archaeological features in both strata (based on Johnson and Holliday, 1986, fig. 5).

Buried Soils

Soil stratigraphy deals with buried soils, which is to say it deals with buried landscapes. Because we usually see strata in two dimensions, there is a tendency to think of them as two-dimensional entities. Strata are, of course, three-dimensional bodies, and because soils require some time to form, they can be considered four-dimensional entities. Waters (1992, pp. 82–83) provides a good description of the evolution of strata, one particularly appropriate to the visualization of buried soils (see also Kraus, 1999; Kraus and Bown, 1986; Kraus and Aslan, 1999):

> The nature and structure of a stratigraphic sequence in any environment is determined by the number, timing, areal extent, magnitude, and duration of individual periods of deposition, degradation, and stability. . . . Over a given interval of time, individual episodes of deposition, stability, and erosion may be of long duration, with changes occurring infrequently, or they may be of short duration and alternate frequently. Furthermore, these episodes may affect a large region or only a small area. Consequently, different . . . sequences of lithostratigraphic units, pedostratigraphic units, and erosional contacts are created.
>
> The total duration of deposition, degradation, and stability determines how much of the time continuum of a stratigraphic sequence is recorded by . . . depositional units (lithostratigraphic units) versus the combined time represented by . . . surfaces of erosion (erosional unconformities) and surfaces of stability (pedostratigraphic units). Episodes of landscape degradation [or] stability create gaps in any depositional sequence and thus leave an incomplete geologic record of lithostratigraphic units. The ratio of the total amount of time represented by nondepositional and erosional contacts between lithostratigraphic units produces a measure of the completeness of the stratigraphic sequence. . . . Generally this ratio is low because the contacts between and within lithostratigraphic units represent the passage of more time than the physical sediments themselves. (Waters, 1992, pp. 82–83)

An important point raised in this quotation is that soils represent gaps in the sedimentological record; that is, periods of generally stable landscapes because there is little or no deposition or erosion. More broadly, Waters's succinct summary supports the conclusion that recognizing buried soils and establishing soil stratigraphic relationships is important for a broad array of geoarchaeological questions and interpretations, which are dealt with in this and following chapters.

Recognition

Recognizing buried soils seems to be something of an art. Attitudes on the recognizability of buried soils range from high confidence (e.g., Ruhe, 1965; Retallack, 1990, pp. 20–54) to considerable skepticism (e.g., Ruellan, 1971, p. 9; Valentine and Dalrymple, 1976, pp. 209–210; Dormaar, 1987; Catt, 1990, p. 5). The ability to recognize buried soils depends on the degree of preservation of the soil (which in turn depends on the nature of the burial processes and postdepositional alterations) and on the experience of the investigator, including experience working with surface soils and with stratigraphic studies in the given field area.

Fundamentally, buried soils can be recognized using the same characteristics used to recognize surface soils (Yaalon, 1971a, pp. 154, 157; Valentine and Dalrymple, 1976, pp. 209–213; Fenwick, 1985, pp. 5–11; Jenkins, 1985; Catt, 1986, pp. 173–174; 1990, pp. 6–7; Birkeland, 1999, pp. 24–28; French, 2003, pp. 41–43). In the field these characteristics include absence of geologic bedding, recognizable soil horizons, horizons in a typical vertical sequence, typical soil horizon boundaries (sharper upper boundary, more gradual lower boundary), and the formation of stone lines in some environments. Bettis (1992, p. 129) provides a useful summary of criteria for differentiating soil A horizons from organic-rich alluvium, criteria also applicable to distinguishing other kinds of soil horizons: a gradual or clear lower boundary for A horizons as opposed to an abrupt lower boundary in geologic deposits, absence of bedding in A horizons and its presence in geologic deposits, and the presence of granular, or crumb, soil structure in A horizons and its absence or weaker development in geologic deposits.

One significant problem in identification of buried soils is that a number of sedimentary, hydrogeologic, and pedogenic processes can produce zones that looks like buried soils (Pyddoke, 1961, pp. 37–39; Tamplin, 1969, pp. 153–154; Brewer, 1972, p. 332; Rutter, 1978; Catt, 1986, p. 173; Thorson, 1990, pp. 402–403, 405–406; Ito et al., 1991; Tandon and Kumar, 1999, pp. 131–135; Mandel and Bettis, 2001b, pp. 175–180). These processes and deposits include deposition of organic-rich sediment in lacustrine, palustrine, or alluvial settings (confused with A horizons); deposition of red or reddish-brown or brown sediment (confused with Bw or Bt horizons); deposition of reddish or reddish-brown clay-rich sediment (confused with Bt horizons); deposition of gray sediment (confused with a redoximorphic horizon); deposition of volcanic ash (confused with Bk or in particular E horizons); illuviation of humus in a Bh or Bhs horizon below a leached, sandy E horizon (confused with an A horizon buried below unweathered sand); and deposition of calcium carbonate, soluble salts, or iron oxides in lacustrine, palustrine, alluvial, spring, or groundwater settings (confused with Bk, K, By, Bz, or Bs horizons). Postburial processes (discussed in chapter 10) can also produce zones mistaken for soil horizons.

Figure 5.7 Stratified alluvium in north-central Iowa showing a dark, organic-rich flood drape just above the north arrow (Mandel and Bettis, 2001b, fig. 7.2; from *Earth Sciences and Archaeology*, P. Goldberg, V. T. Holliday, and C. R. Ferring, Eds., © 2001, Kluwer Academic/Plenum Publishers. Reproduced with permission of Kluwer Academic/Plenum Publishers and R. D. Mandel). This zone could be confused with a buried A horizon. Note the abrupt lower boundary of the drape (in contrast to the gradual lower boundaries of the A horizons illustrated in figs. 1.1, 2.1, and 2.2). Photo provided by and reproduced with permission of R. D. Mandel.

The difficulty and importance of distinguishing sedimentary and pedogenic features is illustrated in several geoarchaeologically focused examples. In establishing a Holocene soil/weathering chronosequence for use by archaeologists, Bettis (1992, p. 129) notes that an essential component of its application is distinguishing between dark-colored, organic-rich A horizons of soils and dark-colored, organic-rich late Holocene alluvial deposits (fig. 5.7). Gladfelter (1992) discusses an approach to differentiating paludal marls from pedogenic calcic horizons in Upper Paleolithic sites in the southern Sinai. Zones of carbonate accumulation are intercalated with thick alluvial sequences, and distinguishing the origins of these carbonates was critical to understanding Paleolithic subsistence. A floodplain with paludal or lacustrine marls would be indicative of a very different setting from a floodplain in which calcic soils formed. Following the properties listed by Fenwick (1985) for the identification of buried soils, Gladfelter determined that the carbonate is marl. Key indicators that the carbonate accumulation is not pedogenic include little enrichment by organic matter, no increased redness through individual profiles, no changes in clay content indicative of illuviation, no indication of mineral weathering, no disruption of primary bedding, and no pedogenic structure recognized (Gladfelter, 1992, p. 184).

At the Wasden site on the Snake River Plain of Idaho, multiple occupation zones associated with buried soils are reported (Moody and Dort, 1990). A sequence of buried A-Bw, A-E-Bw, and A-E-Bt soils spanning the Holocene were described and analyzed. The exact nature of the purported buried soils is not clear, however. Some of the E and Bt horizons are described as microstratified, although true E and Bt horizons should exhibit little if any stratification. Some E horizons also have higher contents of organic matter and clay than horizons immediately above and below them, which is the reverse of that expected for true E horizons. Clearly there is a problem in reconciling the soil horizon nomenclature with the field descriptions and laboratory data.

Some sediments may even derive from soil horizons, such as high clay, reddish-brown deposits eroded from a well-developed Bt horizon. In the Thessaly region of Greece, for example, differentiating between *in situ* red Mediterranean soils and redeposited red clay (itself a weathering product) has profound implications for correlating and interpreting Paleolithic sites (Van Andel, 1998). The red clay or "redbeds" represent active erosion and sedimentation, whereas buried, well-developed, red Bt horizons represent prolonged landscape stability. Moreover, the identification of a buried soil within the redbeds aided in dating a hand axe to the Lower Paleolithic, a very significant find. Failure to recognize buried soils within the redbeds was a significant problem in attempts to understand the regional stratigraphy and its archaeological significance (Runnels and Van Andel, 1993a, p. 194).

Recognition of soil horizons can still result in stratigraphic and environmental misinterpretation if the genetic relationships of the horizons are not recognized. Buried soils are common in layers of tephra (Dalrymple, 1967; Kato and Matsui, 1979; Campbell, 1986; Shoji et al., 1993; Baümler and Zech, 2000), as are archaeological sites (Sheets and Grayson, 1979; Olson, 1983; Sheets, 1983; Davis, 1986; Sheets and McKee, 1994; further discussed in chapter 6). Soil horizons in tephra pose some problems of recognition, however. Shoji et al. (1993, p. 9) discuss the potential difficulties in differentiating spodic horizons (and Bh horizons) from buried A horizons. Laboratory analyses may be required.

Buried soils in loess on the coastal plain of Israel pose problems of recognition and interpretation that are crucial for correlating and dating Paleolithic archaeology (e.g., Ronen, 1975; Tchernov et al., 1994; Tsatskin et al., 1995; Gvirtzman et al., 1999; Porat et al., 1999; Tsatskin and Ronen, 1999; the geoarchaeology of some of these Israeli coastal plain soils is discussed further in chapters 6, 8, and 9). The loess includes multiple buried soils with two dominant horizon types: clay-rich Bt horizons alternating with somewhat coarser Bk horizons. These two horizon types had been accorded independent stratigraphic status and separate paleoenvironmental interpretations; that is, they were considered two different types of soils developed in two different parent materials (Bruins and Yaalon, 1979). The Bt horizons were believed to have formed in finer-grained parent material, and the Bk horizons were interpreted as forming in coarser parent material. Further, the two horizon types were believed to have different paleoenvironmental implications: the Bt horizons formed under a wetter semiarid climate, and the Bk horizons formed under a drier semiarid climate.

Wieder and Gvirtzman (1999) reexamined the soils using both macromorphology and micromorphology. Their work showed that both kinds of horizons formed in the same parent material, a sandy loess. The "clayey" nature of the Bt horizon is the result of illuviation of clay, probably deposited on the soil surface as dust. The "sandy" texture of the calcic horizon is the result of pervasive reprecipitation of secondary calcium carbonate. The investigators concluded that the parent materials are the same in all soils, deposition was a gradual process and the stratigraphy does not represent distinct cycles of coarser- and finer-textured sediment, and sets of Bt-Bk horizons are genetically related (i.e., they formed together). Therefore, four sets of Bt horizons and four sets of Bk horizons do not represent eight layers and eight buried soils, as proposed by Bruins and Yaalon (1979), but rather four cycles of deposition and soil formation (Wieder and Gvirtzman, 1999).

A particularly important characteristic that can be key to identifying a buried soil is predictable, lateral variability in morphology (the "constancy of relationships" of Brewer, 1972, p. 333) caused by topographic variability (i.e., it is part of a "paleocatena"; see chapter 9) and by parent material variability, discussed earlier. The topographic and catenary variability of soils is a unique characteristic that is particularly important in the identification of buried soils—one not likely mimicked by any other geological phenomena (Brewer, 1972; Valentine and Dalrymple, 1975, 1976; Finkl, 1980; Catt, 1986, pp. 168–169). This characteristic of soils allows them to be traced in three dimensions over varying paleotopography. Individual layers of sediment, in contrast, will be confined to particular depositional environments and will thin to nothing away from that environment (Mandel and Bettis, 2001b, p. 180).

A variety of laboratory techniques can be used to help identify buried soils. Laboratory data indicative of buried soils include depth functions in the content of organic carbon or organic matter, clay, and calcium carbonate. Fecal pellets in buried A horizons may be apparent in thin sections. Remnants of a truncated A horizon may be indicated by higher organic carbon or organic matter at the top of a suspected buried soil, with a rapid decline down the profile. In an archaeological context, Foss et al. (1995, p. 7, table 1–6) note that some trace elements can be used to identify buried A horizons if there are textural contrasts between the buried zone and overlying deposits. Clay content will be highest in the upper part of a Bt horizon and decrease with depth, in contrast to a zone with primary clay, which should be distributed more evenly through the zone or perhaps be layered. Secondary carbonate in a Bk or calcic horizon may, but not necessarily, be higher in the upper part of the zone, especially in well-developed calcic horizons. In thin section, illuvial clay will be well oriented (microlaminated), have high birefringence, exhibit optical continuity with straight extinction and strong anisotropy, and coat sand grains, pores, and ped faces (Bullock et al., 1985, p. 114; Courty et al., 1989, pp. 156–159). Pedogenic carbonate in thin section will appear as fine calcite concentrations of micrite and microsparite in voids (Courty et al., 1989, pp. 175–176).

Kemp (1985b) presents a good example of using laboratory data, primarily micromorphology, to determine the presence or absence of a buried soil at the famous Paleolithic site of Swanscombe on the Lower Thames. The site is in a terrace of the Thames and produced a variety of finds including the skull of "Swanscombe Man" and Clactonian and Acheulian flint artifacts (Wymer, 1985, table 14; Gibbard, 1994, pp. 167–168, table 6). The stratigraphy is divided into three gravel units: lower, middle, and upper. The Paleolithic artifacts are in the lower (Clactonian artifacts) and middle (Acheulian artifacts with hominid remains) gravels. At issue is the "Lower Loam" at the top of the lower gravels, deposited during the Hoxnian interglacial stage (either oxygen isotope stage 7 or 11, the dating is uncertain; see Gibbard, 1994, pp. 182–183). The upper 50–60 cm of the Lower Loam is a noncalcareous reddish clay, interpreted as either an interglacial soil or the result of postburial diagenesis (see Kemp, 1985b, for a review of the issue). Thin-section study revealed that decalcification of the upper part of the altered zone was associated with formation of secondary carbonate lower in the profile; that is, there was evidence of downward translocation of calcium carbonate. Furthermore, the decalcification in the Lower Loam had no relationship to the deposits that buried the zone in question. If decalcification was the result of water moving down through both the overlying sediments and the Lower Loam, then pH in the upper unit should be as low or lower than that in the Lower Loam, but it is not. Therefore, the decalcification process occurred before burial. Root channels in both the zones of decalcification and carbonate accumulation show that the translocation occurred while plants grew at the top of the zone. The zone was also subjected to an increasingly higher water table before burial, as indicated by iron and manganese coatings over the calcite precipitate and the decalcified root channels.

Laboratory data can also be used to identify geologic layers confused with soil horizons. For example, at the site of Berekhat Ram in the Golan Heights, Acheulian artifacts occur in a "red clay" between two layers of basalt (Goldberg, 1987; Courty et al., 1989, pp. 228–234). The red clay was variously interpreted as the result of colluviation and pedogenesis, but its origin, of significance to the interpretation of a rare Acheulian assemblage, was not clearly documented. Micromorphology showed the base of the layer to be colluvial with some pedogenic modification, but most of the zone is colluvial, modified by groundwater weathering. That is, the artifacts were in colluvium that was weathered long after the occupation. In a very different setting, excavations at Edinburgh Castle, Scotland, revealed an organic-rich zone separating Iron Age from early Medieval deposits (Davidson et al., 1992). The zone was identified in the field as a buried Ah horizon and was believed to represent a hiatus in deposition. Thin-section analysis revealed, however, that the zone was composed almost entirely of plant stems, fragments of wood charcoal, and coprolites, probably representing a layer of plant refuse or perhaps straw for animal bedding. There is no buried soil in the section, therefore, and the entire sequence probably represents gradual accumulation of refuse.

An important method for recognizing and especially for correlating buried soils is magnetic susceptibility (e.g., Maher and Thompson, 1991; Singer et al., 1992; Antoine et al., 1999), which is proportional to the amount of strongly magnetic minerals in soils and parent materials. The reasons for the high magnetic susceptibility of soils are many, but ultimately the reason seems to be related to the pedogenic formation of ultrafine ferrimagnetic grains (magnetite and maghemite), mineral grain size, and concentration. The strength of magnetic susceptibility is related to soil formation and therefore can be an indicator of buried soils. Comparisons must be made to the parent materials in which the soil formed, however. The most successful results for using magnetic susceptibility to identify and correlate buried soils are where the parent materials have been uniform through time, such as the thick, stratified sequences of soils buried within loess (e.g., Maher, 1986; Kukla, 1987; Maher and Thompson, 1991; Derbyshire et al., 1995a,b).

Catt (1990, p. 4) strongly recommends a combination of careful field study and laboratory investigation in the identification of buried soils (see also Courty, 2001). Ideally, this approach entails a field assessment, followed by laboratory analyses and then a return to the field with the laboratory data. Realistically, however, constraints on time or funding or both may prohibit this approach in many geoarchaeological contexts. If questions arise in the field regarding the status of a stratigraphic unit as a buried soil, however, laboratory investigation may be the only way to solve the issue, whether or not a return to the section is possible.

Burial Processes

The recognition and interpretation of buried soils can be complicated by a number of processes that take place before, during, or after burial. As a landscape is buried, some or all of the following processes can operate, depending on landscape position and the nature of the burial process or processes:

1. Rapid burial that leaves a complete soil profile preserved under younger sediment
2. Erosion before burial, resulting in a truncated soil profile preserved under younger sediments
3. Slow burial that allows pedogenesis to keep pace with sedimentation

Once a soil is buried, it can still be subjected to a variety of weathering processes that bring about further alterations. Soil parent materials can also be subjected to alterations that mimic pedogenesis. Awareness and recognition of these various processes of burial and postburial alteration is essential in the interpretation of buried soils.

The literature on buried soils commonly emphasizes the incompleteness of many if not most of these soils (Ruellan, 1971, pp. 10–11; Finkl, 1980, p. 186; Catt, 1990, pp. 4, 5–6; Gerrard, 1992, p. 207), in particular, the absence of an A and E horizon (fig. 2.3). Certainly some components of a number of buried soils were removed by erosion before burial. The A and the E horizons generally are friable and noncohesive and, therefore, are easily removed by wind or water. In contrast, B horizons often are more consolidated, particularly those characterized by illuviation of clay, carbonate, or sesquioxides, rendering them more difficult to erode, especially in contrast to the A and E horizons. The apparent absence of an A horizon in a buried soil, however, may be the result of postburial processes that can make A horizons difficult to recognize (Johnson, 1977; McDonald and Busacca, 1992); a problem further discussed in chapter 10. Erosion of the A horizon before burial or postdepositional modifications can increase the difficulty of identifying a buried soil because of the loss of the distinctive darker colors of the A. Moreover, even in the absence of training in pedology, most archaeologists are aware of the relationship between organic matter accumulation and buried surfaces and, thus, the presence of an A horizon is more likely to be recognized than a truncated B horizon. Removal of the A horizon also has important archaeological implications because of the higher probability that artifacts and occupations zones will be concentrated there (see chapter 7).

Buried soils with complete or nearly complete pedons are relatively common in late Quaternary stratigraphic and geoarchaeological records (Hoyer, 1980; Holliday, 1985b,d, 1995; Hajic, 1990; Mandel, 1992, 1994, 1995; Ferring, 1990, 1992, 1995b; Zangger, 1993). They tend to be most common in aggrading low-energy depositional settings such as floodplains, some eolian environments, and some coastal plains, as described in chapter 6. The best clue to the nature of the contact between the soil and the sediments that bury it is the relative degree of abruptness. A surface that was eroded prior to burial will be knife-edge sharp and sometimes irregular. A surface buried rapidly but without erosion will produce a relatively abrupt boundary, but it will be gradational at a scale of some millimeters. Plant macrofossils may be preserved in both the Ab horizon and protrude up into the overlying sediment. Another clue to a partially eroded Ab is that it may seem thinner than usual relative to the thickness of the underlying horizon, but this sort of assessment requires experience with the local surface and buried soils.

Soil burial can also be a slow process in which pedogenesis keeps place with sedimentation (figs. 5.8 and 5.9). The result is "upbuilding" (Johnson and Watson-Stegner, 1987) or "overthickening" or "cumulization" of a soil (Buol et al., 1997, pp. 111, 113; figs. 2.1B, 2.2B, 5.3, 5.8–5.10, 6.1, 6.4, 6.6, 6.10, 9.6, and 11.7). Overthickening means that a particular horizon literally becomes thicker than average

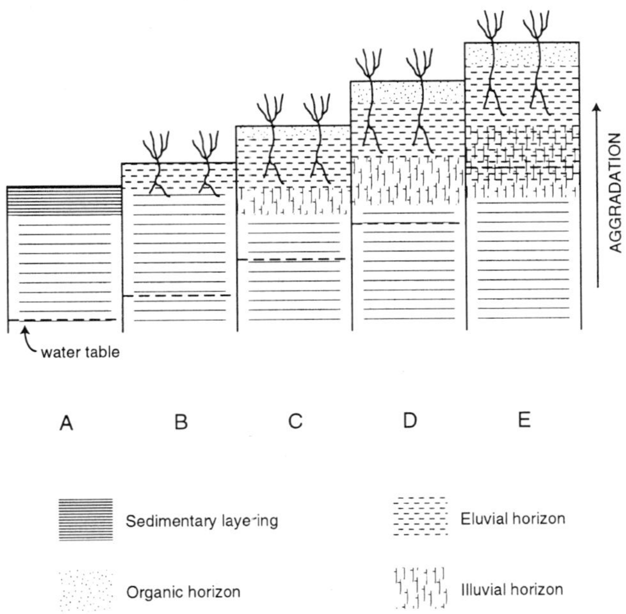

Figure 5.8 Upbuilding or cumulization of a floodplain soil (modified from Allen and Wright, 1989, fig. 2.4; published with permission of V. P. Wright and J. R. L. Allen). As sediment is added, the initial surface of the soil eventually becomes part of the eluvial zone, then the illuvial zone, and finally is submerged if the water table rises.

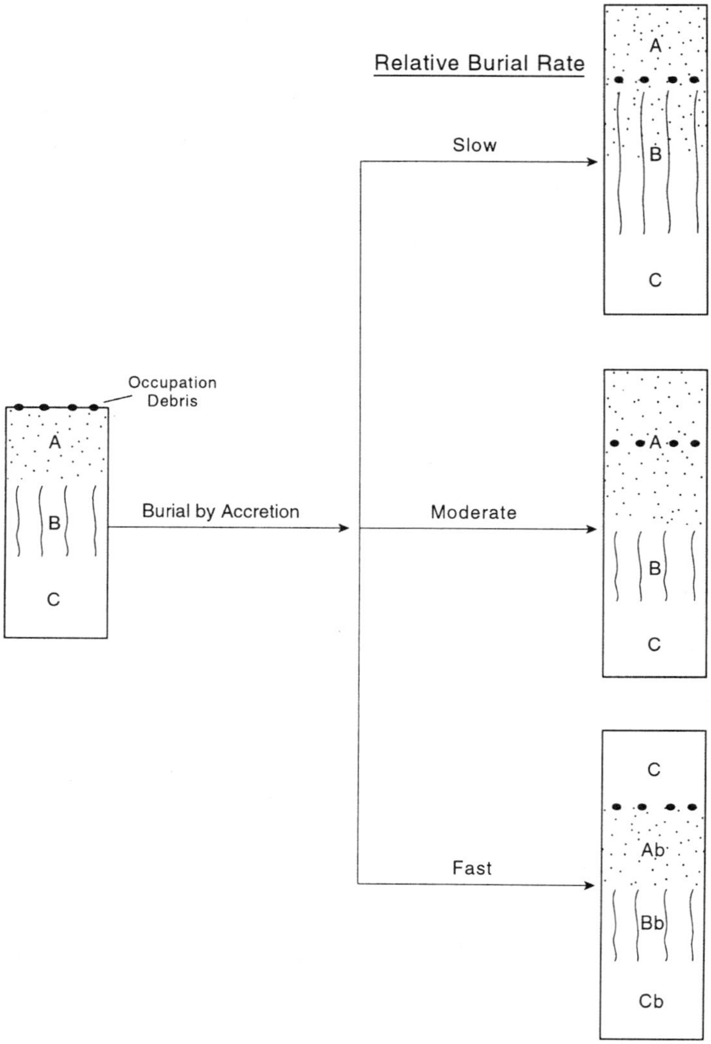

Figure 5.9 The potential effect of varying rates of sedimentation on buried soils and the resulting soil horizonation and geoarchaeology (modified from D. L. Cremeens and J. P. Hart, 1995, "On chronostratigraphy, pedostratigraphy, and archaeological context," fig. 2–4. *In Pedological Perspectives in Archaeological Research*, M. E. Collins, B. J. Carter, B. G. Gladfelter, and R. J. Southard, Eds., pp. 15–33. Soil Science Society of America, Special Publication no. 44; reproduced with permission of the Soil Science Society of America). Slow and moderate rates of burial result in cumulization of the soil, whereas the rapid burial produces a typical buried soil. The effects of varying rates of burial on the same soil are illustrated by fig. 2.1B (slow burial) and fig. 2.1A (rapid burial) See also fig. 7.2.

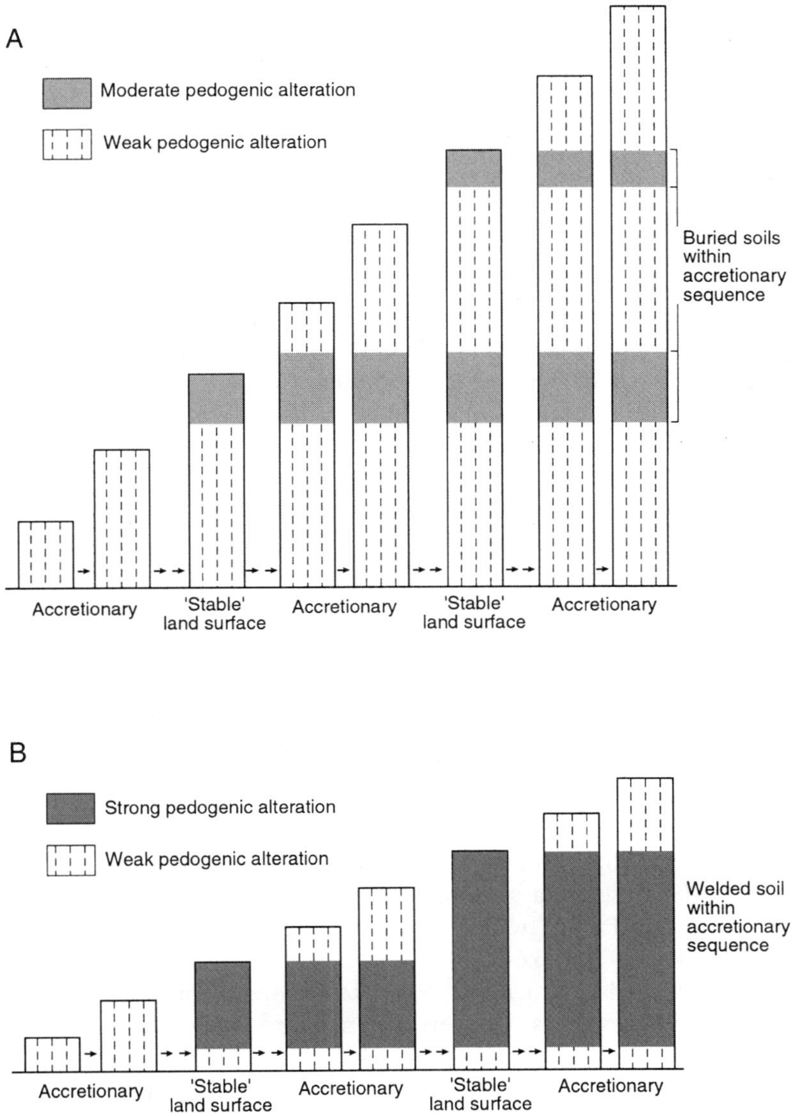

Figure 5.10 Two pathways for soil stratigraphic development on the Loess Plateau of China (modified from Kemp and Derbyshire, 1998, fig. 8). (A) Cumulization of soils results from phases of rapid deposition of loess and weak soil development, alternating with periods of low sedimentation and more intense weathering. (B) Welding of soils results from slower deposition or greater depth of pedogenic alteration or both. As rates of sedimentation vary across a region, these soil sequences may represent facies of one another (see also fig. 6.17).

(relative to more typical horizons associated with similar soils in similar settings in the region). Overthickened soils are referred to as "accretionary" (Catt, 1990, p. 6) or "cumulative" (Nikiforoff, 1949, pp. 227–228; Birkeland, 1999, pp. 165–167), and overthickened horizons are called "cumulic" in soil taxonomy (Soil Survey Staff, 1999, p. 155). Soil taxonomy further distinguishes between "cumulic" and "pachic" horizons: Cumulic horizons can exhibit evidence for stratification, whereas a pachic horizon is overthickened with no evidence of stratification. In general parlance, however, the term "cumulic" is usually applied to both situations.

In describing the processes of overthickening, some investigators distinguish between the effects of slow rates of burial and moderate rates of burial (fig. 5.9; Follmer, 1982, p. 120; Cremeens and Hart, 1995, pp. 23–24). The difference between an overthickened soil and several welded soils, therefore, is related to local rates of sedimentation (figs. 5.9 and 5.10). A set of welded soils and a single overthickened soil could be facies of one another (figs. 2.1, 5.3, and 6.1; e.g., Kemp et al., 1997). This facies relationship could be at the scale of a floodplain or valley (figs. 6.1 and 6.6) or at a regional scale (figs. 5.10 and 6.17). Geoarchaeologically, variation in rates of burial of soils and occupation debris will significantly affect where in a soil an archaeological horizon will occur (figs. 5.9, 6.6, and 7.2).

Most typically, overthickening affects the A horizon and results from moderate rates of sediment accumulation and burial, where organic matter additions and bioturbation more or less keep pace with sedimentation (figs. 2.2B and 5.9). Overthickened A horizons are archaeologically significant because they can preserve stratigraphic relationships among artifacts, features, and living surfaces that are otherwise obscured or destroyed in the more typical A horizon. Some accretionary A horizons have weakly expressed stratification, which further aids in archaeological correlation and dating. According to Follmer (1982, p. 120) slow rates of burial allow both the A and B horizons to grow upward, resulting in overthickening of the B horizon (figs. 2.1B, 5.9, and 5.10). Stevenson (1969, p. 474), Johnson (1977, p. 196), Holliday (1988, pp. 601–602), and Mandel and Bettis (2001a, pp. 18–19) describe aggrading landscapes in which upward development of a B horizon overlapped with the original portion of an overthickened A horizon (figs. 2.1B, 5.8, and 6.6).

Overthickened soils, particularly B horizons, are common in loess. Kemp (2001, pp. 152–154) describes the overthickening processes as representing shifts in the balance between sediment accumulation and pedogenesis (fig. 5.10A). During accretionary phases, sediment accumulates relatively rapidly and undergoes weak pedogenic alteration. When dust input slows, the loess undergoes more substantial pedogenic alteration. Over time, these processes result in a pedocomplex consisting of an overthickened soil with some zones exhibiting stronger pedogenic expression than others (e.g., Kemp et al., 1996).

Zangger (1993, p. 50) describes a distinctive buried A horizon making up the "Pleistocene surface" and marking the Pleistocene-Holocene boundary along the coast of the Argolid in southern Greece. This horizon also contains archaeological material, the earliest in the region, making it an important geoarchaeological stratigraphic marker. The buried soil is described as "ca. 1 m thick, homogeneous, dark brown [7.5YR 3/4, 10YR 3/2], clay rich" (Zangger, 1993,

p. 50) with common roots and root casts. The soil apparently formed along the low-lying coastal plain and represents the remains of a "once well developed A horizon formed under thick vegetation cover" (Zangger, 1993, p. 50). The soil was modified by the addition of archaeological charcoal and phosphates and also by slow marine sedimentation, resulting in overthickening. The soil was finally buried by the early Holocene marine transgression.

In cold, aggrading environments the effects of cumulization may be much less obvious (Thorson, 1990, p. 403). Loess accumulating in colder higher latitudes or altitudes or in proximity to glaciers probably fell on a slowly aggrading, weakly expressed ochric A horizon. Little organic matter accumulated, and that which did decomposed, resulting in a thick loess with no obvious evidence of soil horizonation. Thorson and Hamilton (1977) used this model of loess deposition and soil formation to explain the vertical distribution of artifacts at the Dry Creek site, Alaska.

Pedogenic features are clearly the primary clues to the presence of buried soils, but the morphologic and chemical characteristics typical of soils can undergo moderate to profound changes following burial. Yaalon (1971c) grouped soil horizons and features according to their relative persistence after burial (table 5.1). Generally speaking, the likelihood of preservation or persistence is related to the rate of development of the horizon. Soil features that form rapidly, such as the A horizon, tend to be the least persistent after burial, whereas soil horizons that form more slowly, such as the Bt horizon, tend to be more persistent in buried soils. Specific aspects of postburial alterations are discussed in chapter 9.

Table 5.1. Soil horizons and features grouped according to their relative persistence

Altered easily, generally $<10^3$ years to reach steady state (Properties acquired by reversible, largely self-regulating processes)	Relatively persistent, slowly adjusting, generally $>10^3$ years to reach steady state (Mostly steady-state near-equilibrium features or metastable state)	Persistent features (Features produced by essentially irreversible, self-terminating processes)
Mollic horizon (f)[1]	Cambic horizon (f)	Oxic horizon (f)
Slickensides (c)	Umbric horizon (f)	Placic horizon (f)
Salic horizon (c)	Spodic horizon (f)	Plinthite (f)
Gypsic horizon (o)	Fragipan (f)	Durinodes (f)
Mottles (c)	Mottles (o)	Petrocalcic horizon (f)
Gilgai features (c)	Argillic horizon (c)	Gypsic crust (f)
Cambic horizon (o)	Natric horizon (o)	Argillic horizon (c)
Spodic horizon (o)	Calcic horizon (f)	Natric horizon (c)
	Gypsic horizon (c)	Albic horizon (o)
	Histic horizon (c)	Fragipan (o)
		Histic horizon (o)

From Yaalon (1971c, table 1).

[1] Letters in parenthesis are semiquantitative estimates of the frequency or persistence of the feature within the group: f = frequently; c = commonly; o = occasionally; r = rarely.

The process of soil burial and then postburial alterations can effectively obscure a buried soil. To summarize, the following questions can be posed as a means of identifying the partial or modified remains of a buried soil (modified from Catt, 1990, pp. 6–7):

1. Do the layers concerned contain fossil remains of plants (e.g., roots, charcoal), indicating the former presence of a land surface?
2. Do the layers display a vertical sequence of horizons that partly or wholly resembles that typical of known types of present surface soils?
3. If they are traced laterally, do the layers transgress bedding planes or other rock structures? If so, they are likely to be soil horizons, though in some soils, horizons develop parallel to bedding planes.
4. If they are traced laterally, do the transgressive layers change in character, not only in relation to changes in the underlying sediments or rock type but also in response to changes in the slope of the associated buried land surface? That is, do they form a paleocatena?

6

Soil Stratigraphy in Geoarchaeological Contexts

The various kinds and states of soils, surface and buried, discussed in the previous chapter can be found in an almost infinite variety of combinations, and most can also be found in archaeological contexts. Furthermore, most soil stratigraphic relationships and conditions of soil burial can form a continuum through time or space or both, depending on local and regional variations in rates and depth of burial (i.e., rates and thickness of sedimentation). The most common and most extensive depositional environments with buried soils that illustrate these relationships are alluvial and eolian. These are the settings for much research on buried soils and soil stratigraphy. Alluvial settings likewise have been the loci of considerable archaeological and geoarchaeological research. Tephra—airfall deposits from volcanic eruptions—also commonly contain buried soils because of the episodic nature of eruptions. Though not as extensive as alluvial or other kinds of eolian deposits, tephra stratigraphy is locally important. Archaeological sites are also common in tephra layers from a variety of settings and regions. This chapter illustrates geoarchaeologically significant soil stratigraphic relationships in a variety of alluvial and eolian settings and at various spatial and temporal scales.

Alluvial Soil Stratigraphy

Alluvial systems probably have been the site of more geoarchaeological research than any other type of depositional environment because they have always attracted occupants who left archaeological sites. A significant amount of archaeological research has also focused on riverine settings owing to "rescue" or "salvage" archaeology. In the United States, for example, this work included the federally funded River Basin Surveys of the 1940s, 1950s, and 1960s, followed by CRM studies beginning in the 1970s and continuing into the 21st century. The importance of alluvial stratigraphy in interpreting the archaeological record of alluvial settings has been recognized throughout most of this work (e.g., Mandel, 2000). Furthermore, the significance of soils in alluvial stratigraphic records has long been recognized; for example, soils were an important component of Haynes's (1968) classic geoarchaeological model of an "alluvial chronology" for the central and western United States. Alluvial soil stratigraphy per se is more poorly known, however, being underrepresented in the traditional pedology or even traditional soil stratigraphic literature. Most substantive research on alluvial soil stratigraphy comes from the sedimentological literature (e.g., Kraus and Bown, 1986; Kraus, 1999; Kraus and Aslan, 1999) or from geoarchaeological research (e.g., Gladfelter, 1985, 2001; Hassan, 1985; Ferring, 1992, 2001; Brown, 1997; Foss et al., 1995; Waters, 2000; Huckleberry, 2001).

Floodplains, Draws, Arroyos, and Alluvial Valleys

Floodplains and other alluvial settings provide all manner of possible soil stratigraphic situations because of their dynamic and variable nature. For the same reasons, alluvial settings also exhibit considerable soil stratigraphic variability over short scales in both time and space. Soil stratigraphy is a key component of alluvial geoarchaeology, therefore, and takes on considerable significance in view of the importance of alluvial settings in human history. Soils are especially important components of the stratigraphy in aggrading floodplains because of the episodic deposition over large areas. These processes result in multiple and laterally extensive buried soils. The following summary of floodplain soil stratigraphy follows Kraus and Aslan (1999, pp. 306–308). Soil variability at the scale of the floodplain is closely related to alluvial processes such as lateral channel migration, crevassing, and overbank flooding; topography and catenary relationships (described in more detail in chapter 9); and water-table fluctuations. Floodplain sedimentation usually is sporadic, with relatively long periods of inactivity between episodes of deposition or erosion. Depending on short-term sedimentation and erosion, a variety of floodplain soils can form. Aggradation is rapid in proximity to the channel and decreases toward flood basins, resulting in formation of a natural levee next to the channel. If erosion is insignificant and sedimentation is rapid but unsteady, multistory soils form in the levee (fig. 6.1). Typically they will be weakly developed and separated by minimally weathered alluvium. These soils are the classic Fluvents in soil taxonomy (fig. 2.2A). The parent materials are usually sandy. Archaeological sites will be well stratified and typically contain well-preserved and discrete occupation zones (fig. 7.2; e.g., Stafford et al., 1992).

Moving away from the channels, sediments become finer and thinner. Individual layers of sediment and discrete buried soils will be difficult to discern. Depending on the rates of deposition and rates of pedogenesis, some buried soils may be more strongly expressed than those in the levee and may be welded, and

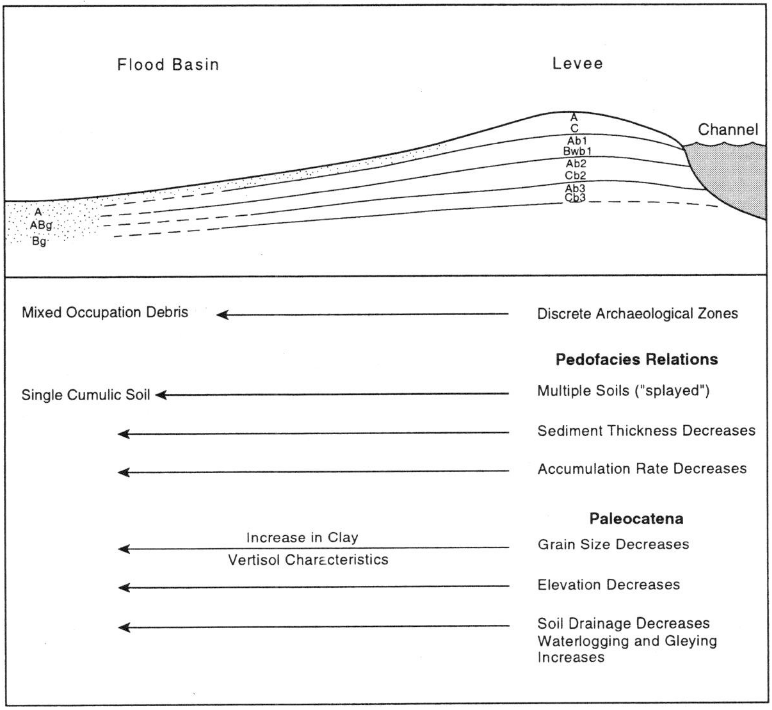

Figure 6.1 Schematic illustration of floodplain soil stratigraphy and pedogenic processes in a floodplain catena (see also fig. 5.3). Modified from Kraus (1999, fig. 9) and Kraus and Aslan (1999, fig. 2).

others may exhibit cumulic profiles (figs. 5.3 and 6.1). Occupation zones may be stratified and relatively discrete, though mixing via bioturbation will obscure some sites or features. Where sediments pinch out and soils merge, occupation zones will likewise merge. In some settings, some or all of the occupation zones may merge and become mixed.

The morphology of a soil (surface or buried) will also vary because of topographic position and drainage; that is, the soils will form a catena (figs. 2.4 and 6.1; see also chapter 9). Soils formed in higher topographic positions such as a levee may have thinner A horizons and perhaps oxidized Bw horizons. At lower landscape positions, the soils show increased gleying, thicker and more organic-rich A horizons, and locally even O horizons.

In addition to thinning of individual layers, soil strata may be difficult to trace across a floodplain because of the effects of sediment fining and poor drainage (see chapters 9 and 10). With distance from the channel, sediments tend to be finer, with a high clay content (fig. 6.1). In regions with a distinct dry season, Vertisols (shrinking–swelling soils) will form, which can obscure stratigraphy and mix occupations (discussed further in chapter 10). The soils in the levees will be better drained and have more distinctive horizonation and strata because of the coarser texture of the sediment as well as the topographic position. Flood basin soils will

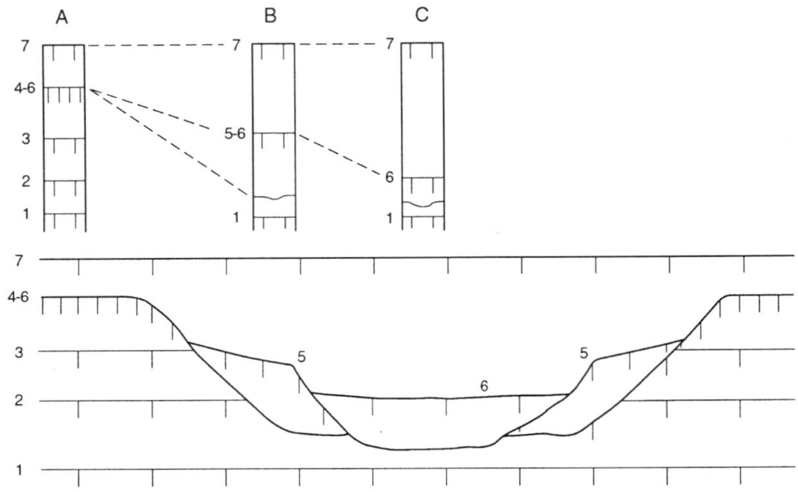

Figure 6.2 Schematic diagram illustrating complex soil stratigraphy resulting from a simple cut-and-fill sequence. Numbers represent geomorphic surfaces; A–C are sections. Spacing of vertical lines is indicative of relative duration of pedogenesis (from Allen and Wright, 1989, fig. 2.3; published with permission of V. P. Wright and J. R. L. Allen).

be more poorly drained because of the finer texture and lower topographic setting, producing gleying and mottling, which obscure stratification and horizonation.

Rivers and streams characterized by entrenchment, cut-and-fill, and terrace formation have soil stratigraphic records even more complex than those found on aggrading floodplains. In such settings the usefulness of buried soils for stratigraphic correlation diminishes, although they remain important for identifying buried landscapes that may contain occupation zones (fig. 6.2).

There are a number of examples of multiple archaeological occupations zones sandwiched among multistory floodplain soils (Stevenson, 1985; Chatters and Hoover, 1986; Bobrowsky et al., 1990; Stewart et al., 1991; Pincha and Gregg, 1993; Jing et al., 1995; Cremeens et al., 1998; Mandel and Bettis, 2001a,b; May, 2002). These stratigraphic sequences typically contain a series of A-C soil profiles; the C horizons are overbank sediments, and the A horizons denote accumulation of organic matter on temporarily stable floodplain surfaces. Not surprisingly, therefore, the archaeological material is typically associated with the A horizons rather than the C horizons. These A horizons can yield discrete occupation zones, but they may be difficult to employ for even local stratigraphic correlation because they are not particularly distinctive and because localized sedimentation in different areas through time results in buried soils whose A horizons merge and bifurcate, hampering intersite comparisons.

In river valleys of the Southern Plains of western Oklahoma and north-central Texas, episodic alluviation produced a series of geoarchaeologically significant multistory buried soils (fig. 6.3; Ferring, 1990, 1992, 1995b). These include weakly expressed soils with thin A-C profiles in middle and late Holocene alluvium. Most

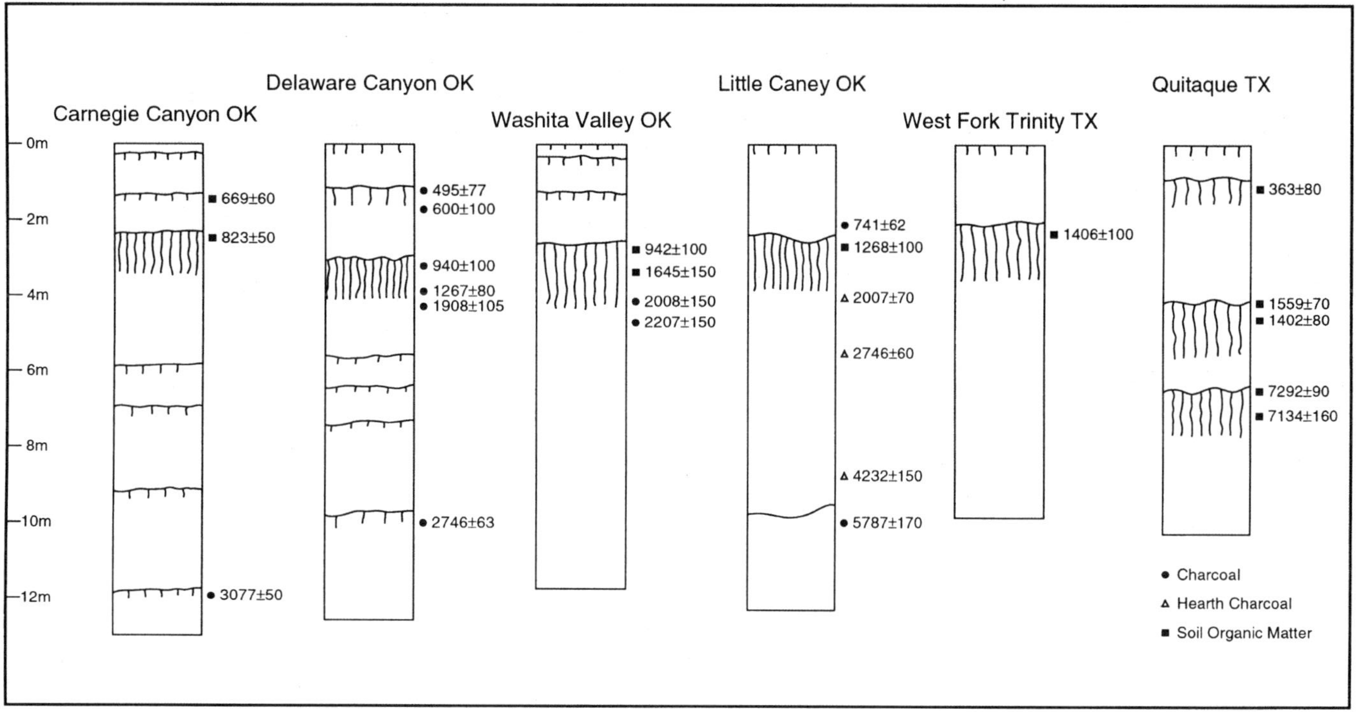

Figure 6.3 Stratigraphic correlation of the Copan soil and related soils at sites in Oklahoma and Texas (based on Ferring, 1990, fig. 5; 1992, fig. 1-3).

of these soils are of only local stratigraphic significance because of variability in their morphology and their numbers. A morphologically distinctive soil formed during floodplain stability ~2000 to ~1000 yr B.P., however, and it is a geoarchaeologically important marker bed throughout the region. In buried positions it has an A-C profile with cumulic A horizon. The soil has been used to identify the stratigraphic position of late Archaic, Plains Woodland, and late Prehistoric archaeological sites (Ferring, 1992, p. 13) in a variety of geoarchaeological investigations in the region. The A horizon typically has higher densities of stratified archaeological sites than the alluvium above or below. In the various geoarchaeological studies in the region, the cumulic soil is sufficiently distinctive to be given informal soil stratigraphic names (e.g., the Copan soil of Artz, 1985; the Caddo soil of Ferring, 1990, 1992; fig. 6.3). Moreover, in landscape positions in which it was not subjected to cumulization or burial, the soil exhibits A-Bw and A-Bt morphology (Artz, 1985; Ferring, 1992); that is, this soil stratigraphic unit exhibits facies variations depending on its age (figs. 5.3 and 10.13). Understanding this facies variation has proved important in understanding and predicting the stratigraphic location of sites of different ages. For example, in northeastern Oklahoma, the cumulized facies of the Copan soil contained Plains Woodland occupation zones at depths of 40–90 cm, whereas the unburied facies produced late Archaic artifacts at those depths and Plains Woodland material at or near the surface (Artz, 1985).

In the Arkansas and Kansas River basins of the central Great Plains and Central Lowlands, stratified alluvium with multiple buried soils is ubiquitous but somewhat more complicated to work with than the sequence in western Oklahoma and northwestern Texas, because of cutting and filling cycles and the formation of terraces (Johnson and Logan, 1990; Mandel, 1992, 1994, 1995; Mandel and Bettis, 2001a,b). Terraces were mapped along the master streams and tributaries of the region. Correlations have proven difficult because of the usual problems of terrace correlation, but soil morphology has been a significant aid (e.g., Sorenson et al., 1987; Johnson and Logan, 1990). The Holocene fills below the younger terrace surfaces largely are the result of episodic floodplain aggradation, and as a result, buried soils are common in these deposits (figs. 6.4 and 6.5; Johnson and Logan, 1990; Mandel, 1992). Most of these buried soils are of similar morphology: typically A-C and A-Bw horizonation with either thin A horizons or more distinctive cumulic A horizons (fig. 6.4). This minimizes their stratigraphic utility because they are so similar regardless of the age of the fill they are in. Most of the correlations of buried soils are based on radiocarbon dating, which has shown a general basinwide synchroneity in cycles of deposition and stability/soil formation. This indicates that the buried soils are useful for correlation once their association with fill of a particular age is confirmed. For example, soils buried in alluvium below a surface Alfisol or Argiudoll (i.e., soils with an argillic Bt horizon) probably are not correlative with soils buried in alluvium with a surface Entisol or simple cumulic Mollisol. In the alluvium farther south in Oklahoma and Texas, as noted above, the distinctive cumulic A horizon of the Copan and related soils can be correlated because of its stratigraphic position below one or a few very weakly developed soils. In the Kansas River basin, Johnson and Logan (1990, pp. 288–289) note that the location of Paleoindian sites might be predicted on

Figure 6.4 Multiple late Holocene soils buried in alluvium of Buckner Creek (locality BC-1), a tributary of the Pawnee River, Kansas (from Mandel, 1994, fig. 54B). Photo provided by R. D. Mandel; reproduced with permission of R. D. Mandel and the Kansas Geological Survey. BC is the Buckner Creek soil (cumulic Ab-Akb-ABkb-Bkb morphology); HC is the Hackberry Creek soil (Akb-Bkb-BCkb morphology). Note hearth associated with the BC soil.

the basis of the soil stratigraphy exposed at the famous Lime Creek site (Davis, 1953, 1962) in the Republican River drainage to the north, in Nebraska. Stratified Paleoindian occupations were found in association with multiple A horizons buried deep below the surface of an early Holocene terrace.

The advantages and difficulties of applying soil stratigraphy in geoarchaeological research along active alluvial settings are well illustrated by Mandel and Bettis (2001a,b). At the Farwell Locality (Mandel and Bettis, 2001a, pp. 13–25) along the Big Nemaha River in northeastern Kansas and southeastern Nebraska, multiple, discrete occupation zones are well preserved in association with multiple, moderately expressed buried soils. The soils and archaeology are in two packages of late Holocene alluvium: the Honey Creek member of the DeForest Formation inset against the Late Gunder Member (fig. 6.5). The sediments and soils in each set are generally similar, consisting of stratified silty sediment and moderately expressed buried soils (A-Bw-C and A-Btw-C horizonation). The two members can be distinguished only because of the clear contact separating them. The stratified late Holocene sediment is inset against somewhat more homogeneous late Pleistocene alluvium (fig. 6.5). On the surface, the different sets of deposits are distinguished only by low (<50 cm) scarps. Draped across the entire landscape is a layer of silty alluvium with a mollic A horizon. Thus, the stratigraphic relationships of the alluvium, the soils, and the archaeology cannot be distinguished from examination of the surface soil. Likewise, Mandel and Bettis (2001a,b) provide illustrations and discussion of multiple buried soils in alluvium that are facies of cumulic or welded soils on higher terraces (fig. 6.6). Archaeological material was found in discrete, well-stratified contexts in the buried soils

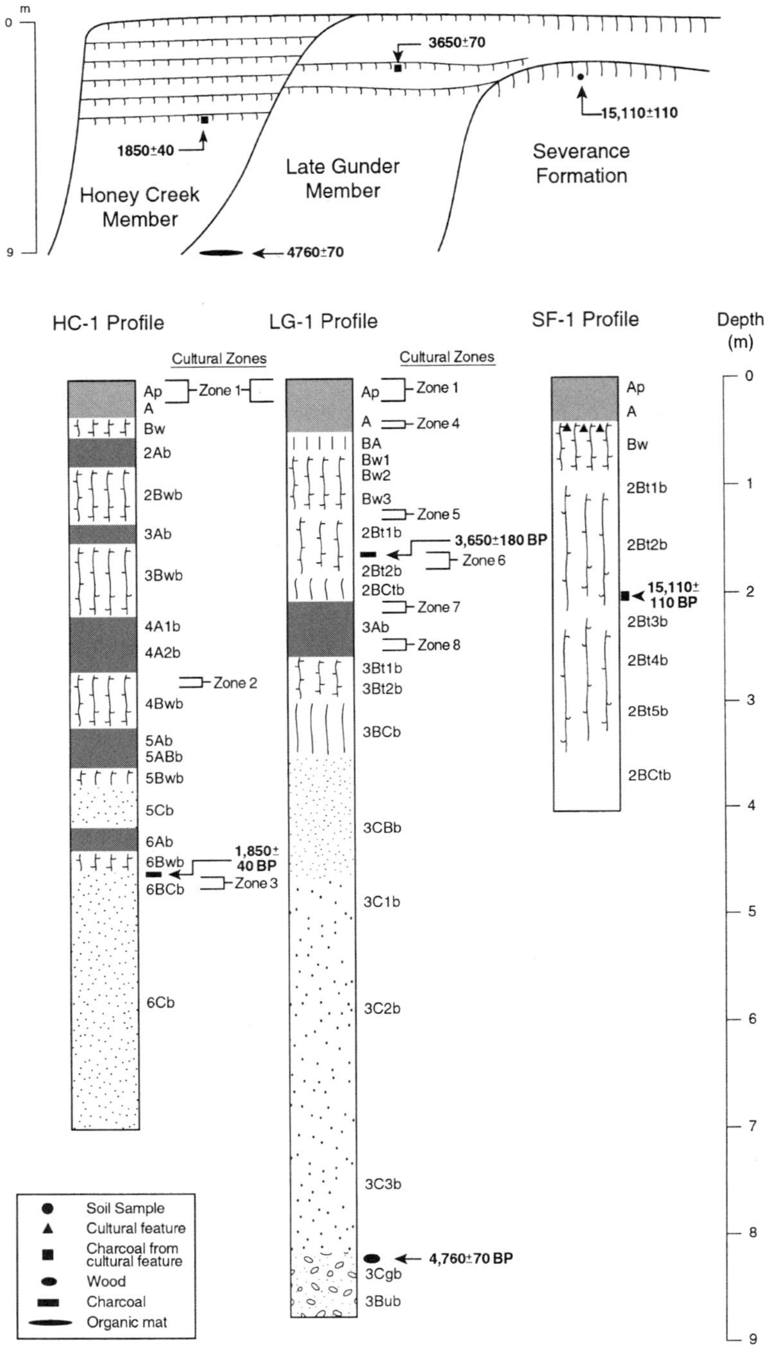

Figure 6.5 Lithostratigraphy, soil stratigraphy, and geochronology of Holocene alluvium at the Farwell locality along the South Fork of the Big Nemaha River, Kansas (based on Mandel and Bettis, 2001a, figs. 11, 12, and 14). This section shows how the simple soil geomorphology of the surface (flat with A-Bw profile) belies the complex subsurface soil stratigraphic and lithostratigraphic record. The general morphological similarities of soils throughout the Late Gunder and Honey Creek members further illustrates the difficulties of using the individual soils for stratigraphic correlation.

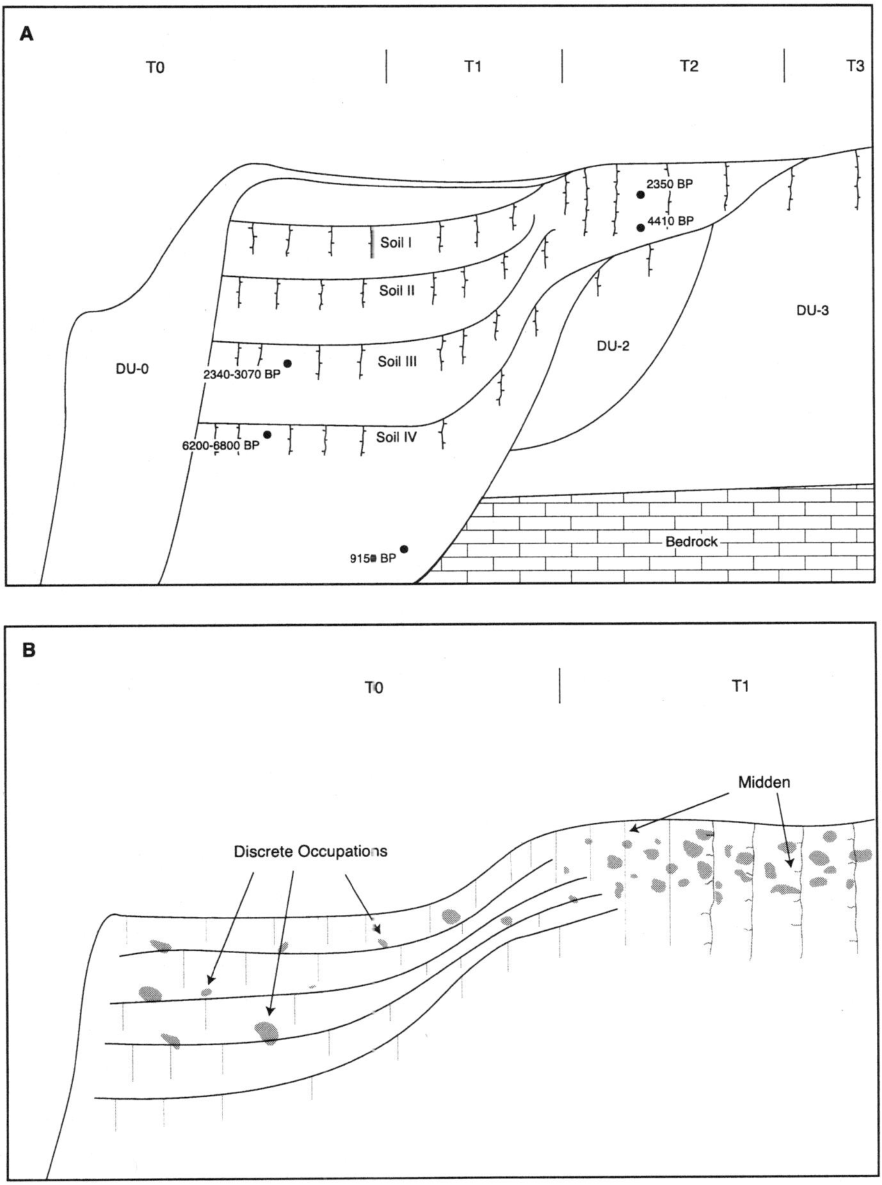

Figure 6.6 Illustrations of multiple floodplain soils merging to form a cumulic upland soil (modified from Mandel and Bettis, 2001b, figs. 7.7 and 7.8). (A) Soil stratigraphy of the Main site in the Cumberland River valley, eastern Kentucky. Several weakly expressed buried soils beneath T1 merge to produce a cumulic surface soil on T2. Late Woodland archaeology was found in alluvium above Soil II in the T1 fill. Early and Middle Woodland occupation zones were found in and below Soil III. In the T2 fill, Early Archaic debris was found below Soil IV, but Middle and Late Archaic as well as Woodland horizons were compressed into the well-developed soil of the T2 surface. (B) Soil stratigraphy of 13CF102 in the Boyer River valley, western Iowa. Discrete Woodland occupations were associated with weakly expressed buried soils beneath the floodplain (T0). The Woodland people also occupied the adjacent terrace (T1), where sedimentation rates were lower than those on T0. As a result, the discrete occupations in the weakly expressed soils in the T0 fill are represented by a single midden in a strongly expressed surface soil on T1.

but was significantly more difficult to correlate and trace in the cumulic or welded settings.

Buried soils and archaeological materials in terrace fill can be subjected to postburial alterations, particularly beneath older terraces. The Memorial Park site is in alluvium of a terrace along a tributary of the Susquehanna River in central Pennsylvania (Cremeens et al., 1998). Seven buried soils were identified in the alluvium, indicative of episodic aggradation when the terrace was the floodplain (fig. 7.7). The soils are A-C or A-Bw profiles or multistory A horizons. Fifteen archaeological components from Middle Archaic through Early Woodland were found in the alluvium, most associated with the buried soils. What is unusual about this site is the presence of a fragipan (Bx horizon) superimposed over the third through fifth buried soils (from the top down) and their associated archaeology. A fragipan is a somewhat enigmatic subsurface horizon usually found in the forested midlatitudes. It is seemingly cemented when dry but brittle when moist. It tends to restrict entry of water and roots into the soil matrix (Soil Survey Staff, 1999, pp. 39–41; Witty and Knox, 1989). A significant aspect of the welding of the Bx horizon to multiple buried soils and the overprinting of the archaeology is that this postburial soil genesis did not disturb the pedostratigraphy or archaeological integrity of the site.

The effects of soil cumulization and burial in an archaeological context are well illustrated at the Big Eddy site in Missouri (Lopinot et al., 1998). The site is well stratified with multiple alluvial layers and buried soils. Late Paleoindian components are in the upper part of "Buried Soil 1" in a horizon identified as "3Ab(2Btb6)" (see appendix 1), formed in late Pleistocene floodplain deposits (Hajic et al., 1998; Ray, 1998). The horizon nomenclature refers to a buried A horizon (3Ab) with a superimposed Bt horizon (2Btb6) resulting from postburial welding (fig. 6.7). Lithologic stratification was not observed in exposures of the buried A horizon, but cumulization was indicated by several characteristics. Dense artifact accumulations were present in a stratified sequence of occupation levels (Hajic et al., 1998, p. 90; Ray, 1998, pp. 199–207). In terms of artifact styles, occupation zones producing Dalton-style artifacts were found throughout the buried, cumulic A horizon, but San Patrice artifacts were found only in the middle and upper A horizon, which is the correct stratigraphic sequence for these types. Laboratory data are also indicative of some episodic sedimentation and stability during formation of the A horizon. Micromorphology and organic carbon content show that the lower portion of the A horizon probably represents the original A horizon based on an increase in organic carbon content in that position (fig. 6.7) and a discrete spike in silt content (fig. 6.7). In thin section, the lower third of the buried A horizon exhibited illuvial clay in the matrix and lining voids, characteristic of a Bt horizon (Hajic et al., 1998, pp. 86, 89). This indicates that the surface was aggrading and that the position of clay accumulation in the subsoil moved up (into the original A horizon) as the surface aggraded. The homogenous character of the A horizon, and some radiocarbon reversals, also indicate some mixing, probably caused by bioturbation.

Dry valleys or "draws" and arroyos on the Great Plains and in the desert West of the United States have long been the focus of archaeological and geoarchaeological research because of localized occurrences of stratified archaeological

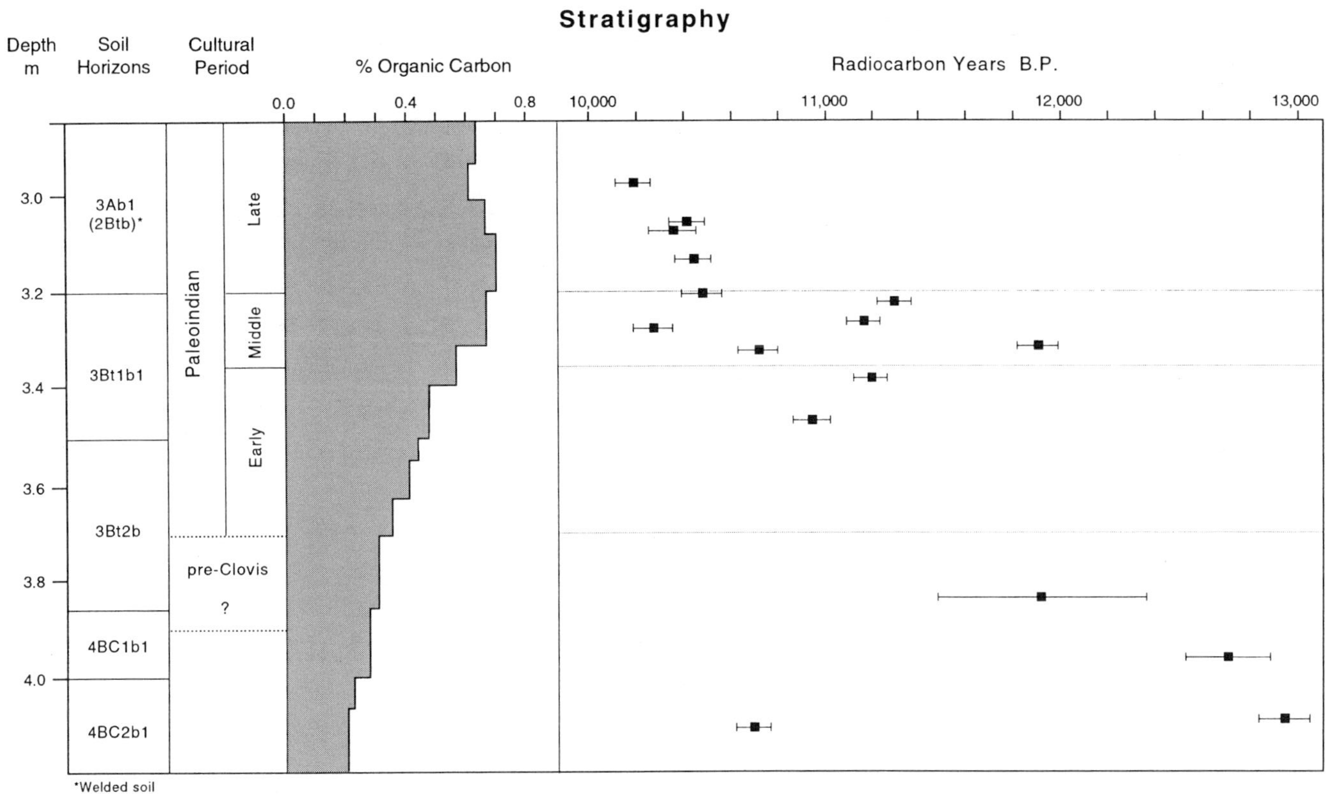

Figure 6.7 Soil stratigraphy of Paleoindian levels at the Big Eddy site, Missouri (based on Hajic et al., 1998, figs. 7.8 and 7.12). Organic carbon data illustrate the effects of cumulization. The radiocarbon dating is indicative of minimal mixing of the soil. Horizon designations are from the original publication.

sites. Indeed, some of the roots of geoarchaeology in North America can be found in these studies (e.g., Bryan, 1941b; Bryan and Albritton, 1943; see also historical perspectives in Haynes, 1990, and Mandel, 2000). The alluvial chronology of Haynes (1968) is based on many of these early studies of draws and arroyos. The complexities of stratigraphic correlations along and between these small drainages have become increasingly apparent (e.g., Bull, 1991; Holliday, 1987a; Waters, 1985, 1992; Waters and Haynes, 2001), but at the same time they have produced some good examples of soil stratigraphic research. In some settings, buried soils in these depositional environments are useful for regional stratigraphic correlation. Draws on the Southern High Plains have aggraded episodically since the end of the Pleistocene (Haynes, 1975, 1995; Holliday, 1985a, 1995; Stafford, 1981), resulting in both stratigraphy traceable over long distances and multiple buried soils in the valley fill. The best expressed and most widely traceable soil (identifiable at the Lubbock Lake, Clovis, Plainview, Marks Beach, Midland, and Mustang Springs archaeological sites, among others) is one developed in middle Holocene eolian sediments, forming through most of the late Holocene (since ~4500 yr B.P.; figs. 2.1 and 9.3; the "Lubbock Lake soil" of Holliday, 1985a,d). The most common facies of the soil is found along the valley axis, formed in loamy or finer sediments, and displaying an A-Bt-Bk profile (fig. 2.1A). In sandier facies or along sloping valley margins the soil is A-Bw or A-Bw-Bkw. It is buried in some settings but not in others. As a result, most or all of the late Archaic, late Prehistoric, and Historic period archaeological record is compressed into the A horizon (see also discussion in chapters 7 and 9).

On the northern Great Plains and especially in the Wyoming basin, arroyo fills typically include latest Pleistocene to early Holocene paludal muds in discontinuous deposits ("black mats," discussed in chapter 8). These layers were modified pedogenically and then buried, and today they form a distinctive marker bed (see Reider, 1990, pp. 335–339, for a comprehensive description and discussion). The soil has a dark gray (e.g., 10 YR 3/2 moist) A horizon over a Btg-Cg or Bwg-Cg horizon sequence. The soil is present at most of the Paleoindian sites in the region (Carter/Kerr-McGee, Agate Basin, Sheaman, Sister's Hill, and Horner). In settings in which the soil formed during the Clovis (11,500–11,000 yr B.P.) or Folsom (11,000–10,000 yr B.P.) occupations it tends to be darker and more heavily gleyed than in situations in which it formed during later Paleoindian time (~10,000–8000 yr B.P.; fig. 8.6). This contrast is attributed to wetter, cooler conditions promoting increased production and preservation of biomass during Clovis and Folsom time (Reider, 1990, p. 337). In the valley fill of draws on the Southern High Plains, Paleoindian-age marsh sediments and soils are also reported from a few archaeological sites though the sediments and soils are highly localized (Haynes, 1975, 1995; Holliday, 1995).

In the arroyo fills of the Basin and Range region of the southwestern United States, organic-rich layers also known as "black mats" are associated with Paleoindian sites (Haynes, 1968, 1991; Quade et al., 1998; Liu et al., 2000). However, they are not as useful as regional stratigraphic markers as the organic-rich layers on the Great Plains are, because black mats are reported from a variety of localities in the Basin and Range dating to the late Pleistocene, early Holocene, and late Holocene (Quade et al., 1998; see discussion in chapter 8).

Table 6.1. General soil-stratigraphic characteristics of alluvium on the Argolid and Argive plain of southern Greece

Soil name	Horizon sequence	Diagnostic features	Carb stage[1]	MS[2]	Soil classification	Age of alluvium
Kranidhi	A-C	Thin to nonexistent A horizon; stratified parent material.	None	1	Xerofluvent	
Upper Mbr Flambouro	A-C	Thin A horizon; some mixing of parent material stratification.	None	1	Xerofluvent	
Lower Mbr Flambouro	A-Bw/ Bk-C	Well-developed A horizon; 10YR 4/3 B horizon	I	1	Xerochrept & Calcixerollic Xerochrept	
Pikrodafni	A-Bt/ Btk-C	Well-developed A horizon; 5YR 4/4 Bt horizon with few, thin clay films.	I–II	2	Haploxeralf & Calcic Haploxeralf	2700–1400 B.C.
Upper Mbr Loutro	A-Btk-C	Eroded A horizon; 7.5YR 4/4 Bt horizon with common, thin clay films.	II	3/4	Calcic Haploxeralf	45,000–32,000 years
Middle Mbr Loutro	A-Btk-C	Eroded A horizon; thick 5YR 4/6 Bt horizon with thin, common clay films; no stratification in upper 2 m	II–III	5	Calcic Palexeralf	>60,000 years
Lower Mbr Loutro	A-Btk-C	Eroded A horizon; thick 2.5YR 3/6 Bt horizon; no stratification in upper 3 m	III	6	Calcic Palexeralf	>250,000 years
Upland soil	A-Bt-C	Thick, dark A horizon; thick Bt horizon with thick, continuous clay films	N/A		Palexeralf	

Modified from Pope and Van Andel (1984, tables 1 and 2) and Van Andel (1998, fig. 6).

[1] Carbonate stage terminology from Gile et al. (1996). See figure 3.3 for an illustration of the stages of carbonate accumulation.

[2] Maturation stage.

In the Old World, soils are crucial to correlation of upper Pleistocene alluvial deposits and associated archaeology throughout southern Greece (e.g., Pope and Van Andel, 1984; Runnels and Van Andel, 1993b; Zangger, 1993; Van Andel, 1998). Stratigraphic correlations are based on soil morphology, stratigraphic relationships of buried soils, and geomorphic position of alluvium containing buried soils. In the Argolid region, for example, alluvium containing Paleolithic sites and scattered artifacts is subdivided into three "members" (of the informally designated "Loutro alluvium"), identifiable and correlatable on the basis of degree of Bt, Bk, and Btk horizon development (table 6.1; Pope and Van Andel, 1984; Van Andel, 1998). The oldest unit (Lower Loutro) contains the most strongly expressed soil (Bt horizon with 2.5 YR hues and thick, continuous clay films; Bk with stage III calcic horizon) and predates human occupation. The Middle Loutro

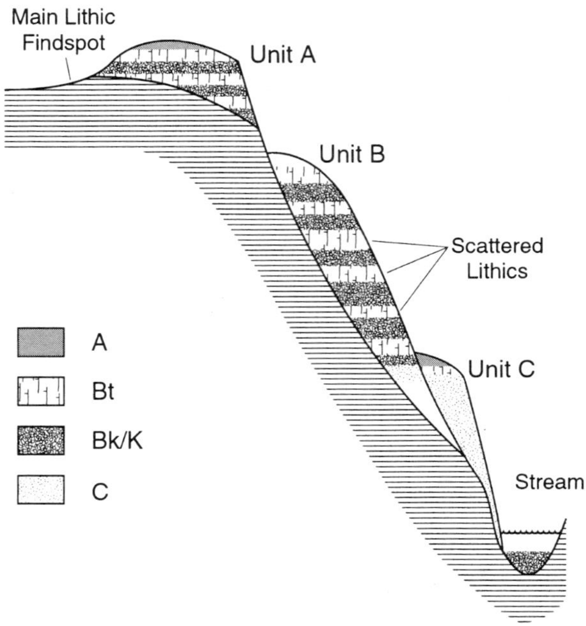

Figure 6.8 Soil stratigraphy of stream terraces in southern Greece (modified from *Geoarchaeology* v. 13, pp. 361–390, fig. 7, by T. H. Van Andel; ©1998 John Wiley & Sons, used by permission of John Wiley & Sons, Inc.). The term "lithics" refers to pre-Mousterian quartz pebbles and flake tools.

likewise predates human activity, but on the surface of the associated soil (Bt horizon with 5 YR hues and thin, continuous clay films; Bk with stage II-III calcic horizon), middle Paleolithic artifacts are sometimes found. The Upper Loutro alluvium contains both Middle and Upper Paleolithic finds. Its associated soil, the weakest of the three soils associated with the Loutro (Bt horizon with 7.5 YR hues and thin, discontinuous clay films; Bk with stage II calcic horizon), sometimes contains Early Helladic archaeology on the surface. In Thessaly, buried soils are found in sediments comprising three alluvial terraces (fig. 6.8; Runnels and Van Andel, 1993b; Van Andel, 1998) and are likewise important in correlating Paleolithic archaeology. All three terrace fills contain buried soils, which is indicative of episodic aggradation between major phases of valley incision. The highest terrace is found only on hilltops and represents sedimentation in an alluvial fan. Soils in the associated deposits are similar in morphology to the Lower Loutro soils. Lower Paleolithic (pre-Mousterian) artifacts are found below the fan sediments, resting on an older surface. Alluvium of the middle terrace, found along deeply incised valley walls, includes buried soils morphologically similar to those of the Middle Loutro. This soil correlation, scattered artifacts from these soils, and some numerical age control all indicate that middle Paleolithic peoples occupied the now buried landscapes of the Middle Loutro (Runnels and Van Andel, 1993b; Van Andel, 1998).

Alluvial Fans

Fans are considerably less common than most of the above-described alluvial landforms and depositional environments, but they are nevertheless geoarchae-

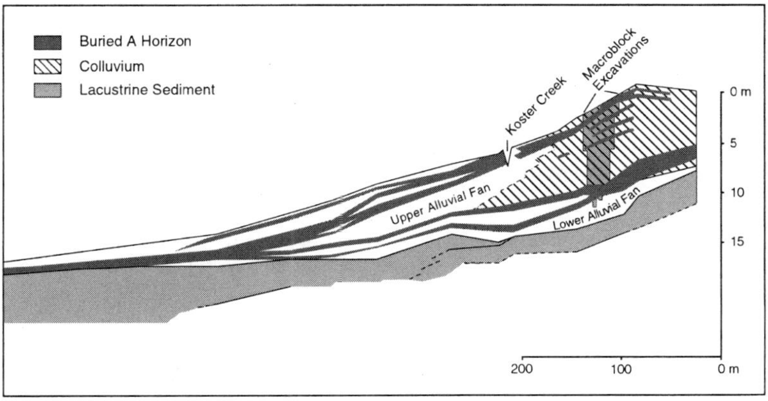

Figure 6.9 General cross section of the fan at the Koster site illustrating the soil stratigraphic relationships (see fig. 7.11; modified from Hajic, 1990, fig. 7).

ologically significant. They tend to be situated between upland and lowland settings (e.g., along the margins of alluvial valleys) and are therefore typically located between macroenvironments. This made alluvial fans attractive sites for habitation, based on the archaeological record (e.g., Anderson, 1970, 1988; Butzer, 1977; Hoyer, 1980; Van Andel and Zangger, 1990; Bettis and Hajic, 1995). Alluvial fans also are aggradational landforms and tend to be well stratified. They therefore lend themselves to geoarchaeological investigation.

Alluvial fans formed along the margins of many stream valleys in the midwestern United States throughout the late Quaternary. The fans were key sites of human activity. Geoarchaeological investigations of these sites illustrate a variety of soil stratigraphic relationships (Hoyer, 1980; Wiant et al., 1983; Styles, 1985; Hajic, 1990; Stafford et al., 1992; Bettis and Hajic, 1995; Running, 1995). Rapid, episodic sedimentation in the early and middle Holocene produced weakly expressed multistory A-C soils, but slower sedimentation rates in the late Holocene resulted in cumulic A horizons. Laterally, moving from the proximal facies to the distal facies, the multistory soils merge into cumulic profiles as the sediments thin (fig. 6.9).

The soil stratigraphic relationships displayed by the fans have a significant effect on the archaeological record. Wiant et al. (1983, p. 151) summarize the soil and archaeological stratigraphy at the Napoleon Hollow site in Illinois: "Of particular interest is the association of each major cultural stratum with an organic A horizon. In an overall sense, deposition of the fan was slow enough that nearly the entire thickness shows characteristics of A and B soil horizons. However, in each case the best-defined cultural strata are associated with periods of fan stability sufficient to produce well-defined organic A horizons that stand in good contrast to the general low-grade pedologic alteration of surrounding sediments. The Middle Archaic occupations are particularly good with each occurring in a discrete dark A horizon. The Late Archaic and Woodland occupations are stacked within the very thick [cumulic] mollisol-type A horizon of the modern surface soil."

Eolian Soil Stratigraphy

Eolian sediments cover between 10% and 20% of the Earth's surface (Pye, 1987, p. 200; Lancaster, 1995, p. 1). Most of these deposits are sands in sand sheets, sand seas, and sand dunes, and in silts in primary and redeposited loess, but they also include mixtures of sand and silt in "loess loams" and "cover loams." Evidence of prehistoric and historic occupations has been found in and around these eolian deposits in both the New and Old World. Many eolian deposits, in general, are stratified, and buried soils are common, particularly in loess. Thus, lithostratigraphy and soil stratigraphy are key components of the geoarchaeology of these deposits. In regions with thick and extensive eolian sediments, buried soils are uniquely suited for stratigraphic correlation and landscape reconstruction.

Dunes and Sand Sheets

Sand dunes and sand sheets commonly contain archaeological debris or bury archaeological sites (e.g., Farrand and Ronen, 1974; Ronen, 1975, 1977; Blair et al., 1990; Goring-Morris and Goldberg, 1991; Holliday, 1997, 2001b; Smith and McFaul, 1997; Gvirtzman et al., 1999; Hall, 2002; LaBelle et al., 2003). The easily erodible nature of many sand deposits and the dynamic nature of many dune systems, however, often mixes archaeological contexts and destroys stratigraphic relationships (e.g., Wandsnider, 1988; Buck et al., 2002). In addition, eolian sands typically are homogeneous and are often difficult to correlate. Buried soils can be preserved in some eolian sands and sometimes serve as important stratigraphic markers. Soil development in Holocene dunes usually is minimal because of the lack of weatherable minerals, the limited biomass input, and, in dry environments, the limited through-flow of water. Dusts containing carbonate and silicate clay are important additions to soils in eolian sand in some environments, however (dust inputs in soils are well summarized by Simonson, 1995). Soils in Holocene sands typically exhibit only A-C or A-Bw profiles (fig. 6.10; e.g., Valentine et al., 1980; Haynes et al., 1993; Smith and McFaul, 1997; Haynes, 2001; Holliday, 2001b; Hall, 2002), which can minimize their utility as stratigraphic markers. Soils formed through most of the Holocene and with significant dust inputs can exhibit Bt and Bk horizons, however (e.g., Holliday, 1997; Smith and McFaul, 1997). In Pleistocene sands, soils with more time to form can exhibit a wider range of morphologies, including argillic Bt horizons, calcic horizons, and even petrocalcic horizons (e.g., Farrand and Ronen, 1974; Ronen, 1975, 1977; Smith and McFaul, 1997; Gvirtzman et al., 1999; Hall, 2002).

A pedogenic feature characteristic of soils formed in sand are "clay bands" or "clay lamellae" (fig. 6.11; e.g., Dijkerman et al., 1967; Larsen and Schuldenrein, 1990; Rawling, 2000). The origins of these thin, relatively clay-rich zones within a sandy parent material have been debated for decades on the basis of both field and experimental observations (Rawling, 2000). Many are clearly pedogenic, and abundant descriptive and quantitative data are available for the bands and the soils containing them. Few studies have incorporated clay bands as stratigraphic markers, largely because data on their rates of formation and on their regional spatial variability are lacking, but their utility in this regard is being recognized,

Figure 6.10 Exposure of a late Holocene sand dune on the Sheyene Delta (late Pleistocene) in southeastern North Dakota. At the top of the dune (at and above the figure's cap) is a layer of historic sand, which buries a very weak and thin A horizon. The shovel is leaning against a weakly stratified, overthickened (cumulic) buried soil with A-C horizonation. To the left, that buried soil merges with the thin, weak A-C soil buried by the historic sand.

particularly in geoarchaeological contexts (Foss and Segovia, 1984; Larsen and Schuldenrein, 1990; Stewart et al., 1991; Jorgensen, 1992; Prusinkiewicz et al., 1998; Hall, 2002). Clay bands are ubiquitous in sand dunes on and near the Southern High Plains of Texas and New Mexico (Gile, 1979, 1985; Holliday, 2001b; Hall, 2002) and have proven to be excellent stratigraphic markers (Holliday, 2001b; Hall, 2002). In particular, Paleoindian occupations in the Andrews Dunes are associated with very well expressed (thick and numerous) clay bands (fig. 6.11; Holliday, 1997). The late Pleistocene and early Holocene sand containing the early sites was subjected to pedogenesis (in the form of silicate clay additions from dust) throughout most of the rest of the Holocene (Holliday, 2001b).

In the Chaco dune field of northwestern New Mexico, buried soils are common and stratigraphically significant in geoarchaeological research (fig. 6.12; Hall, 1990; Wells et al., 1990; Smith and McFaul, 1997). Smith and McFaul (1997) identified six episodes of late Quaternary landscape stability and soil formation, but only a few of the soils seem to be morphologically distinct enough to serve as regional stratigraphic markers. The two oldest buried soils, which were also recognized by Hall (1990) and Wells et al. (1990), are the most distinctive. The oldest (Tohatchi I of Smith and McFaul, 1997), formed on a landscape available to Paleoindians, is the best developed, with a reddish-brown (7.5 YR to 5 YR hues) argillic Bt horizon and stage I to II calcic horizons (Smith and McFaul, 1997; Wells et al., 1990). In overlying eolian sands is a slightly weaker soil (7.5 YR hues, cambic B horizon, stage I+ calcic horizon; Tohatchi II of Smith and McFaul, 1997)

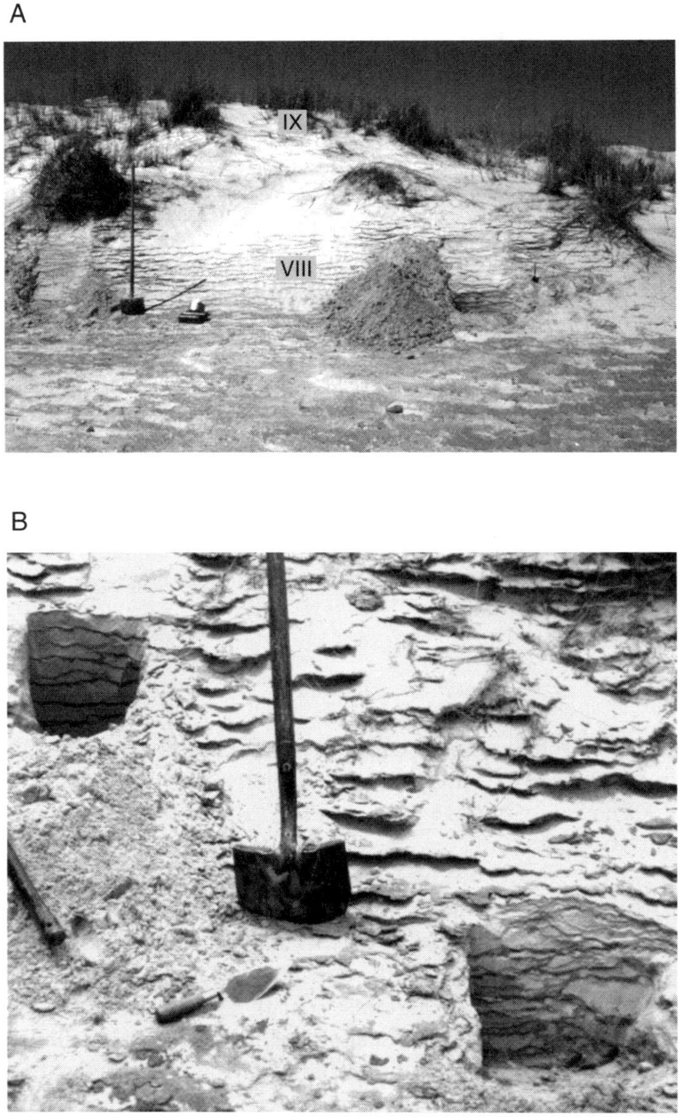

Figure 6.11 An outstanding example of clay bands exposed in eolian sand (Unit VIII) at the Bedford Ranch (Late Paleoindian) site, Texas. (A) Shows how the bands can appear in outcrop, especially if they are relatively well expressed. (B) shows the bands in detail, illustrating their complex microstratigraphy as they merge and bifurcate. The sands between the bands contain ~1% clay. The bands themselves contain only 2%–4% clay but hold enough extra moisture and iron oxide to readily stand out. These discrete zones of illuvial clay probably began forming by ~8000 yr B.P. and continued to develop until Unit VIII was buried by Unit IX, ~2000 yr B.P. (modified from Holliday, 1997, fig. 3.55B).

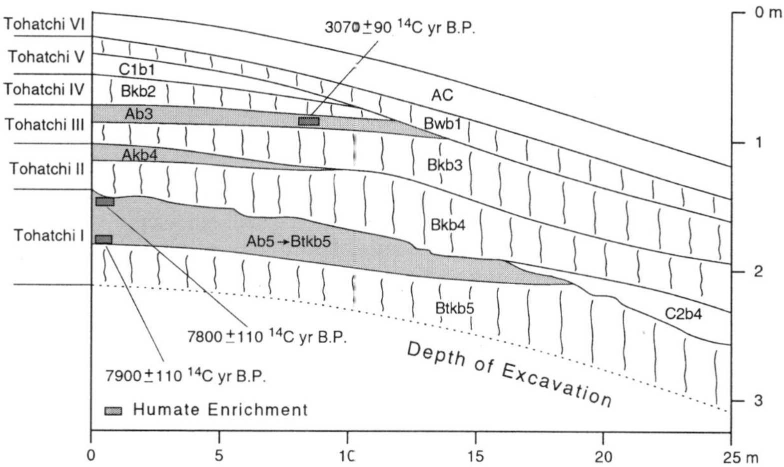

Figure 6.12 Soil stratigraphy and cultural chronology in the Chaco Dunes, New Mexico (modified from Smith and McFaul, 1997, fig. 3).

formed in the middle Holocene and containing Archaic occupations (Smith and McFaul, 1997; Wells et al., 1990). Soils with Bw and weak Bk horizons and associated with Late Archaic and Late Prehistoric archaeological sites are common in late Holocene sands (Tohatchi III-VI of Smith and McFaul, 1997) but are more discontinuous and, in any case, are difficult to differentiate unless stacked one on another (Smith and McFaul, 1997).

Degree of pedogenic development has proven to be useful for stratigraphic correlation and dating of eolian and alluvial deposits associated with the Selima sand sheet in the Darb el Arba'in Desert of southern Egypt and northern Sudan (Haynes, 1982, 1989, 2001; McHugh et al., 1988; Haynes et al., 1993; Maxwell and Haynes, 2001), though relatively few details are available. Deposits of alluvial sand and gravel (strata A and C) are differentiated on the basis of soil development, with stratum A having the best-developed soils (Haynes, 2001, p. 131) and containing Acheulian artifacts (≥300 ky). Stratum A is firm-to-hard, red- to reddish-brown (2.5 to 5 YR hues) gravelly sand (Haynes, 1989, p. 159). Strata B and D are sand sheets and are also differentiated on the basis of soil development. Stratum B is a firm, reddish-brown (5 YR hues) sand with coarse prismatic structure (Haynes, 1989, p. 159). It is older than 300 ky, has much stronger pedogenic development, and is more indurated than the light-brown, Holocene stratum D (Haynes, 2001, p. 132). Stratum D is further subdivided on the basis of degree or stages (0–4; table 6.2) of pedogenic development (Haynes et al., 1993, pp. 621–622; Haynes, 2001, pp. 132–134). Stages 0, 1, and 2 make up one of the more unusual weathering sequences, based on very subtle changes to the sand sheet in the form of varying degrees of cohesion and prismatic structure development. This incipient pedogenesis is the result of Holocene hyperaridity: Wetting of stage 0 sand by very infrequent rain produces the cohesion of stage 1; repeated wetting and drying produces stage 2 pedogenesis (Haynes et al., 1993,

Table 6.2. Stages of pedogenic development in Stratum D of the Selima Sand Sheet, Egypt and Sudan

Stage	Definition
Stage 0	No cohesion and will not stand as a vertical wall when excavated. Walls flow until the angle of repose is reached; that is, has undergone zero degree of pedogenesis. Bimodal grain-size distribution. Sedimentary bedding is distinct. Much of the eastern Sahara is covered by a 1-cm-thick layer of sand with stage 0 pedogenesis.[1]
Stage 1	Soft and unconsolidated, but has adequate cohesion to stand as a vertical wall when trenched, even though it may continually ablate with the impact of wind and sand. Bimodal grain-size distribution. No soil structure, and sedimentary bedding is distinct.[1]
Stage 2	Soft and unconsolidated, but has adequate cohesion to stand as a vertical wall and also has weak medium-prismatic structure. Cracks between peds are so fine that little or no cracking pattern emerges on scraping away the overlying sand with stage 0 pedogenesis. Bimodal grain-size distribution. Laminae ae distinct within each ped.[1]
Stage 3	No primary sedimentary features resulting from bioturbation by plant root activity, animal burrowing, and trampling by people and animals. Retains the bimodal grain-size distribution, with the addition of some clay that, if present in significant quantity, may cause a trimodal grain-size distribution. Soil colors are redder and browner than more youthful stages.
Stage 4	Redder than Stage 3, has stronger structure, and is apparently older. Stage 3 soils developed in stratum D often contain Neolithic cultural materials.

Based on Haynes and Johnson (1984), Haynes (1989, p. 159), Haynes et al. (1993, pp. 621–622), and Haynes (2001, pp. 132–134).

[1] Laminar bedding remains intact in incipient stages 0–2.

p. 622; Haynes, 2001, p. 134). Stages 0, 1, and 2 apparently have no chronological significance. They are simply a function of "the frequency and/or intensity of individual rainfall events" (Haynes, 2001, p. 134). Stage 3 and 4 soils in stratum D, however, are a good stratigraphic marker for Neolithic archaeology; that is, the soils were formed under relatively moist conditions during the Neolithic.

Buried soils are important stratigraphic markers in eolian deposits and archaeological sites along the coastal plain of Israel. The deposits include both loess and sand, but the best-known and most widely studied are the eolian sands and eolianites. The carbonate-cemented eolianites are known as "kurkar." Kurkar is expressed geomorphically as sandstone ridges or "kurkar ridges" that run parallel to the coast (fig. 6.13; Farrand and Ronen, 1974; Horowitz, 1979, figs. 5.20, 5.35, and 5.58; Gvirtzman et al., 1998). The eolianites are stratified and were deposited episodically throughout much of the Pleistocene. Intercalated with the kurkar are buried soils. Red, sandy, noncalcareous varieties are the well-known "hamra" soils. "Hamra" is a common soil association in coastal Israel, and the term is applied to morphologically similar buried soils (fig. 6.14; e.g., Dan et al., 1968; Dan and Yaalon, 1971; Gvirtzman et al., 1998, 1999; Tsatskin and Ronen, 1999; Gvirtzman and Wieder, 2001). Most of the hamra apparently are characterized by Bt horizons (fig. 6.14B) and are classified as Rhodoxeralfs on the basis of present morphology and climate. Buried red, sandy soils with carbonate are the "husmas" variant of the hamra (Dan and Yaalon, 1971; Wieder and Gvirtzman, 1999). They are Btk horizons (Calcic Rhodoxeralfs) reflecting recalcification

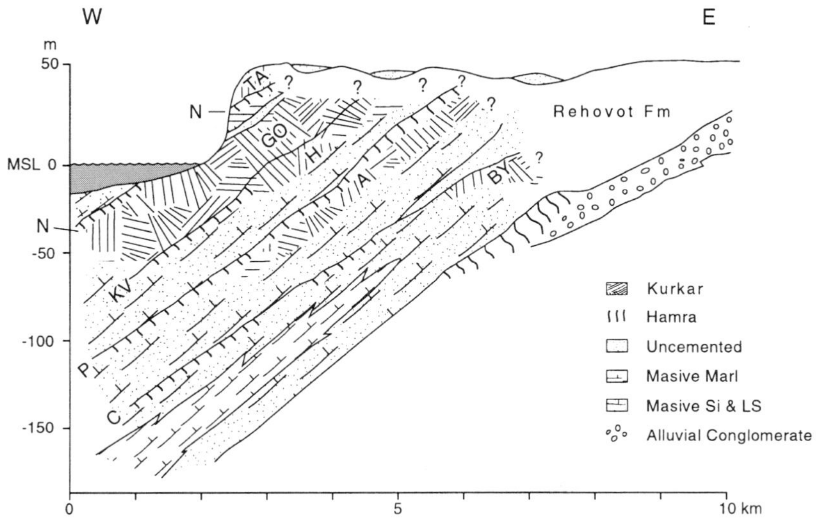

Figure 6.13 Quaternary stratigraphy in the Coastal Plain of Israel, illustrating the stratigraphic relationships of the principal Hamra and Kurkar units to transgressive marine units. The labeled Hamra and Kurkar comprise the Hefer Formation (for Hamra, N = Netanya Hamra, KV = Kefar Vitkin Member, P = Poleg Member, and C = Caesarea Member; for Kurkar, TA = Tel Aviv, GO = Giv'at Olga, H = Herzliyya, A = Ashod, and BY = Bat Yam; modified from Gvirtzman et al., 1999, fig. 3). The Tel Aviv, Giv'at Olga, and Herzliyya Kurkar form distinct kurkar ridges. The Rehovot Formation consists of alternating beds of unconsolidated dune sands and Hamra. The Kefar Vitkin hamra is associated with Mousterian artifacts. The younger Netanya hamra contains Epi-Paleolithic archaeology and locally consists of multiple buried soils (table 8.3). The "T & H" refers to the uncemented Ta'arukha and Hedera eolian sand units (table 8.3). See also table 6.3.

following burial; that is, the husmas is welded to overlying hamra (fig. 6.14B; Gvirtzman et al., 1999; Wieder and Gvirtzman, 1999). Individual layers of kurkar can be tens of meters thick or thin to a feather edge, allowing individual hamra and other buried soils to merge and form welded pedocomplexes (e.g., Gvirtzman et al., 1984, 1998, 1999; Tsatskin and Ronen, 1999; Wieder and Gvirtzman, 1999; Gvirtzman and Wieder, 2001). The hamra tend to be in the range of 1–5 m.

Cemented marine and eolian sands, hamra, and other deposits are known formally as the Kurkar Group (Gvirtzman et al., 1984). Along the western coastal plain of Israel, nonmarine continental facies of the kurkar and hamra are intercalated with marine sandstones and make up the Hefer Formation, whereas inland, along the eastern coastal plain, nonmarine continental facies of the kurkar and hamra are the dominant units and make up much of the Rehovot Formation (fig. 6.14; Gvirtzman et al., 1984, 1999; Sivan et al., 1999).

In detail, the soils buried in the eolian units, including the hamra, exhibit considerable variability in morphology. The thickness, clay, and carbonate content of kurkar, hamra, and husmas vary, though pedologic characteristics are reported from only a few sites (e.g., tables 6.3, 8.3, and 9.2). For example, along the northern coast, the Nes Amim hamra (middle Pleistocene?) is a discontinuous soil

Figure 6.14 Hamra and kurkar exposed along the coastal plain of Israel near Hadera (photos provided by P. Goldberg and W. R. Farrand; reproduced with permission of P. Goldberg and W. R. Farrand). (A) Section showing a full Kurkar-Hamra sequence. The calcareous Kurkar contrasts sharply with the deep red Hamra. (B) The Hamra section, with excavations for Mousterian artifacts. Note nodular carbonate in the lower half of the Hamra (Btk horizon), probably the Husmas variant of Hamra.

Table 6.3. Stratigraphic correlation of soils, geomorphic processes, human activity, and marine oxygen isotope stages in the area of the Revadim Quarry, Israel

Revadim unit	Unit name (soil type)[1]	Regional unit	Event	Marine oxygen isotope stage	Age (ka)
1	Dark Brown Grumusol (Vertisol)	"Grumusol" Netanya Hamra	Pedogenesis	?2 and ?3	
—	Missing	Missing	Accumulation of sand dunes, parent materials of the next soil	3?	
—	Missing	Paleosol 4 Kefar Vitkin Mbr	Pedogenesis	4	74–59
—	Missing	Sand body between paleosols 3 and 4	Accumulation of sand dunes, parent materials of the next soil	5	128–74
2 and 3	Quartzic Gray Brown Soil (Haploxeralf)	Paleosol 3 Poleg Mbr	Pedogenesis, development of the Quartzic Gray Brown Soil	6	185–128
—	Missing	Sand body between paleosols 2 and 3	Accumulation of sand dunes, parent materials of the Quartzic Gray Brown Soil	7	245–185
4	Red Hamra Soil (Rhodoxeralf)	Paleosol 2 Caesarea Mbr	Human living floor; pedogenesis, development of the Red Hamra Soil	8	300–245
5	Loose dune sand	Sand body between paleosols 1 and 2	Accumulation of sand dunes, parent materials of the Red Hamra Soil	9	340–300
6	Reddish-Brown Hamra (Rhodoxeralf)	Paleosol 1 Yad Mordekhai Mbr	Pedogenesis	?	?

Modified from Gvirtzman et al. (1999, table 1).

[1] Israel Soil Classification (= FAO-UNESCO) and U.S. soil taxonomy.

expressed as yellowish-to-reddish sand with varying amounts of clay and iron oxides (Sivan et al., 1999, p. 283). Above is the Evron hamra (oxygen isotope stage 6), described as the thickest Quaternary soil in the area, which is dark red and argillaceous, with ferruginous nodules at the top, grading downward into orange, fine-grained, sandy soil (Sivan et al., 1999, p. 283). Along the central coast, the Kefar Vitkin hamra (fig. 6.13; associated with Mousterian artifacts) is 1–5 m thick and dark-red to brown (Gvirtzman et al., 1998, p. 35). The younger (Epi-Paleolithic) Netanya hamra (Fig. 6.13) is composed of 2 m of yellow, clean sand below (C horizon) and 2 m of dark-reddish-brown, loamy hamra above (Bt horizon). In other sections the Netanya hamra is composed of a series of buried soils and sands, totaling up to 8 m in thickness (table 8.3; Gvirtzman et al., 1998, p. 37; Gvirtzman and Wieder, 2001); that is, it can have a multistory facies. Other buried soils include "dark brown soils" with more weakly expressed Bt and Bk horizons (Haploxeralfs; Gvirtzman et al., 1999; Wieder and Gvirtzman, 1999) and "brown sandy soils" (Ronen, 1975, pp. 231, 233–235; Engelmann et al., 2001, pp. 800, 801) or "sandy regosols" (Entisols, probably Psamments; the Nahsholim Sands; Gvirtzman et al., 1998, pp. 35, 37).

Paleolithic sites and artifacts are common components of the continental facies of the Kurkar Group (e.g., Farrand and Ronen, 1974; Ronen, 1975, 1977; Tchernov et al., 1994; Gvirtzman et al., 1999; Porat et al., 1999; Tsatskin and Ronen, 1999) and include the famous sites on Mt. Carmel (Tsatskin et al., 1995). The stratigraphy and geochronology of the coastal plain of Israel, as part of the "Levantine corridor," are crucial to understanding the migration of early Hominids "out of Africa" and, thus, the soils buried in the sands are important components of geoarchaeological research. Individual buried soils and eolian sand layers are difficult to trace over long distances because some of the soils and most of the kurkar units are physically similar, the kurkar ridges are discontinuous, exposures are widely scattered, and systematic attempts to trace individual soils and sands are few. Indeed, the presence of Paleolithic occupations and their association with buried soils in the kurkar was not even recognized until the 1970s, with the opening of quarries and highways cuts (Ronen, 1977). Correlation of some hamra units is based on contained archaeology (tables 6.3, 8.3, and 9.2; figs. 6.13 and 6.14; e.g., Farrand and Ronen, 1974; Ronen, 1975). Some workers have gone so far as to refer to certain soils as "Mousterian red loam," "Mousterian soils," "Mousterian paleosols," and "Epi-Paleolithic soils" (Ronen, 1975, p. 233; 1977, pp. 183, 186; Ronen et al., 1999).

Dating and sedimentological characteristics at several localities on the coastal plain indicate that the buried soils formed during low stands of sea level (Gvirtzman et al., 1999, p. 120; Sivan et al., 1999). In addition, marine kurkar must have formed during high stands of sea level (Gvirtzman et al., 1998, p. 42; Sivan et al, 1999; see also chapter 9). The continental kurkar deposits are facies of the marine kurkar, and therefore, the subaerially deposited eolian sands must have accumulated during both high stands and regressions. Those considerations plus fossil content, some numerical age control, and the correlation of the few dated sections using archaeological materials allowed correlations of sequences of soil formation and sedimentation with the marine oxygen isotope record (table 6.3).

The sequences of kurkar/hamra span much of the Pleistocene. One of the oldest hamra and one of the oldest archaeological sites in the region is at Evron Quarry (Ronen, 1991; Tchnerov et al., 1994). The site produced lower-middle Acheulian artifacts and a lower Pleistocene vertebrate fauna associated with the Nes Amim Hamra (Sivan et al., 1999; Dorot Hamra in Tchernov et al., 1994) and has an estimated age of ~1 Ma based on lithic and faunal composition (Tchernov et al., 1994). Middle-late Acheulian materials were recovered from the upper part of an eroded red hamra (welded to an overlying soil) at the Revadim site (Gvirtzman et al., 1999; Wieder and Gvirtzman, 1999). The age for pedogenesis, which would also bracket the human occupation, is estimated at 300–345 ka using lithostratigraphic correlation (table 6.3; Gvirtzman et al., 1999). The late Acheulian assemblage from the Holon site is sandwiched between two hamra and dated to ~200 ka from optical methods (Porat et al., 1999). The Evron hamra, one of the most strongly expressed buried soils on the northern Israeli coastal plain, is correlated with oxygen isotope stage 6, 186–128 ka using faunal and lithostratigraphic correlation (Sivan et al., 1999).

Mousterian and later occupation zones are relatively common in late Quaternary hamra. A hamra yielding Mousterian artifacts is reported from the northern (Carmel) coastal plain (fig. 6.14; Farrand and Ronen, 1974; Ronen, 1975; Ronen et al., 1999; Tsatskin and Ronen, 1999). Age control is very poor, but it is <120 ka (stage 5e) and probably younger than 112 ka (stage 5c; Farrand, 1994). Along the central (Sharon) coastal plain, Gvirtzman et al. (1998) identified the Kefar Vitkin Hamra and correlated it with stage 4 (~74–59 ka) and with the "Mousterian hamra" of the Carmel coastal plain. Sand and kurkar above the Mousterian hamra contain weakly expressed "brown sandy soils" (Ronen, 1975, pp. 231, 233–235; Engelmann et al., 2001, pp. 800, 801) or "sandy regosols" (Entisols, probably Psamments; the Nahsholim Sands; Gvirtzman et al., 1998, pp. 35, 37), but they have yielded no archaeology. The youngest well-studied, traceable, and archaeologically significant soil in the kurkar is the "Epi-Paleolithic hamra" or "Netanya Hamra" of the central and northern coastal plain (table 8.3; fig. 6.13; Ronen, 1975, pp. 231, 233; Gvirtzman et al., 1998, p. 37; Engelmann et al., 2001; Frechen et al., 2001). The soil, locally a pedocomplex of welded soils, probably dates to stage 2 and early stage 1 (~60–10 ka) based on optical dating and associated artifacts (Engelmann et al., 2001; Frechen et al., 2001).

Loess

The vast loess deposits of North America, Europe, Asia, and South America have long provided some of the classic opportunities for soil stratigraphic correlation of Pleistocene sediments (figs. 6.15 and 6.16; e.g., Schultz and Frye, 1968; Kukla, 1975, 1977; Wasson, 1982; Pécsi, 1984a, 1987, 1991; Liu, 1987, 1988; Pye, 1987, pp. 198–265; Cremaschi, 1990; Liu et al., 1991; Zárate et al., 1993; Pécsi and Richter, 1996; Pécsi and Velichko, 1995; Iriondo, 1999; Derbyshire, 2001). Indeed, the record of buried soils in loess deposits probably provided much of the impetus for the development of soil stratigraphy and confirmed the effect of the Earth's orbital (or "Milankovich") cycles on the Quaternary terrestrial stratigraphic record (Kukla, 1975, 1977; Imbrie and Imbrie, 1979; Smalley et al., 2001;

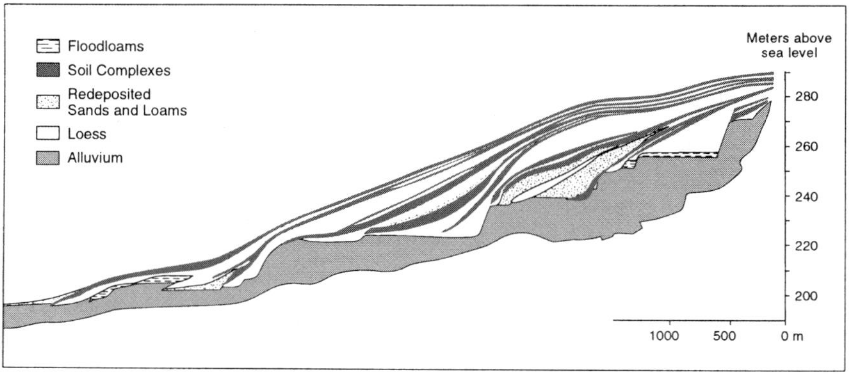

Figure 6.15 General cross section of loess stratigraphy near Brno (Czech Republic) illustrating the soil stratigraphic relationships (modified from Kukla, 1977, fig. 4).

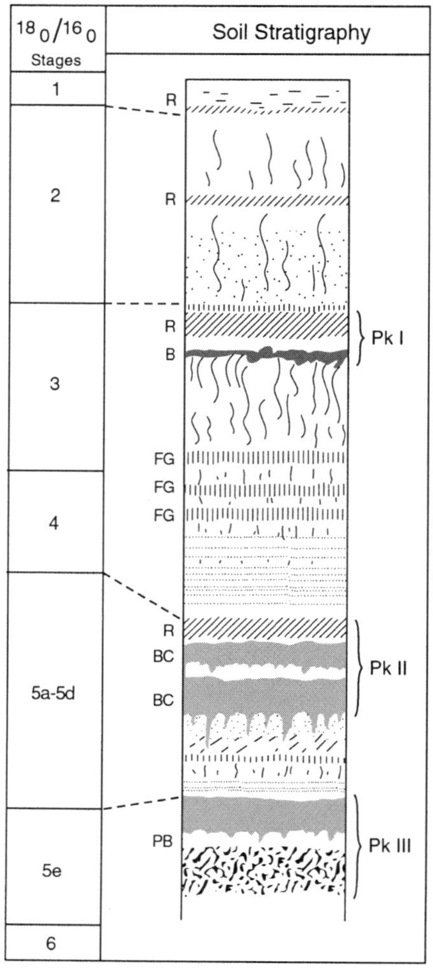

Figure 6.16 Generalized loess and soil stratigraphy of eastern Europe illustrating the basic sequence of pedocomplexes (Pk) and inferred soil types (R = Rendzina, B = Braunerde, FG = Frostgley, BC = Braunerede/Chernozem, PB = Parabraunerde; modified from Hoffecker, 1987, fig. 3; Kukla, 1977, fig. 6).

see chapter 8). Human remains and the evidence of human activities have also long been found in association with loess, particularly in the Old World (e.g., Movius, 1944; Aigner, 1981, pp. 174–244; see chapter 8), where the record of human occupation spans much if not all of the time of loess deposition (late Tertiary and Quaternary).

A variety of nomenclatures were developed to identify the loess sediments and soils of North America, Europe, and Asia. In the Old World there are various systems of terminology, depending on the region and the worker. These systems are generally similar, however, consisting of letters or numbers from the top down. The soils are usually identified as "soil complexes" or "pedocomplexes" (e.g., Dodonov, 1984; Lazarenko, 1984; Bronger and Heinkele, 1989b; Bronger et al., 1995; Dodonov and Baiguzina, 1995), and in some well-known localities the soils are identified and numbered as "PC" (PC 1, PC 2, etc.; Dodonov and Baiguzina, 1995) or as PK (PK I, PK II, etc.; fig. 6.16; Kukla, 1975; Bronger et al., 1995). The long, thick, complete, and spectacular loess soil sequence in the Loess Plateau of China has a more or less standardized and simple system of identifying regionally significant buried soils as S1, S2, and so forth, from the top down (a system also employed on the Russian Plain by Yakimenko, 1995). A few workers employed names for loesses and soils (e.g., Pécsi, 1984b; Velichko, 1990), though these terms are sometimes based on a genetic classification and are abbreviated to a letter nomenclature (e.g., in the Hungarian loess sequence the MF is the Mende upper forest–steppe soil complex, and PDK is the Paks-Dunakömlöd brownish-red soil; Pécsi, 1984b).

Several different approaches have been taken in making soil stratigraphic correlations in Old World loess. Local and long-distance correlations of individual loess strata and soils in Europe and Asia are usually based on soil morphology. In the drier regions of thick loess exposures and deep incision such as Central Asia or central China, soil strata can be physically, visually traced for tens to hundreds of kilometers. In some cases paleoenvironmental markers such as pollen zones or gastropods are also used for correlation. Luminescence techniques hold considerable promise for dating and correlating later Pleistocene deposits (e.g., Stremme, 1989; Wintle, 1990; Forman, 1991; Frechen and Dodonov, 1998), and magnetic susceptibility appears to be a powerful technique for correlation (e.g., Kukla and An, 1989; An et al., 1991; Maher and Thompson, 1991; Derbyshire et al., 1995a,b). In addition, magnetostratigraphy provides a first approximation of the ages of sets of sediments and soils (e.g., Burbank, 1985; Kusumgar et al., 1985; Kukla and An, 1989; Dodonov and Baiguzina, 1995; Bronger et al., 1998; Frechen and Dodonov, 1998; Davis and Ranov, 1999).

The morphologies (both macro- and micro-) of the soils buried in loess can vary dramatically over significant distances because of variation in loess thickness and texture, and variation in the environment (Kemp, 1999); that is, facies changes are apparent in the soils. This is well illustrated by the S1 Pedocomplex (last Interglacial) on the Loess Plateau of China (fig. 6.17; Derbyshire et al., 1995b; Kemp et al., 1996, 1997). In the drier regions near the source of the loess on the western and northwestern margins of the Plateau, S1 consists of several discrete and weakly expressed soils that are separated by minimally weathered loess. To the east and southeast, in the more humid regions of the Plateau, far

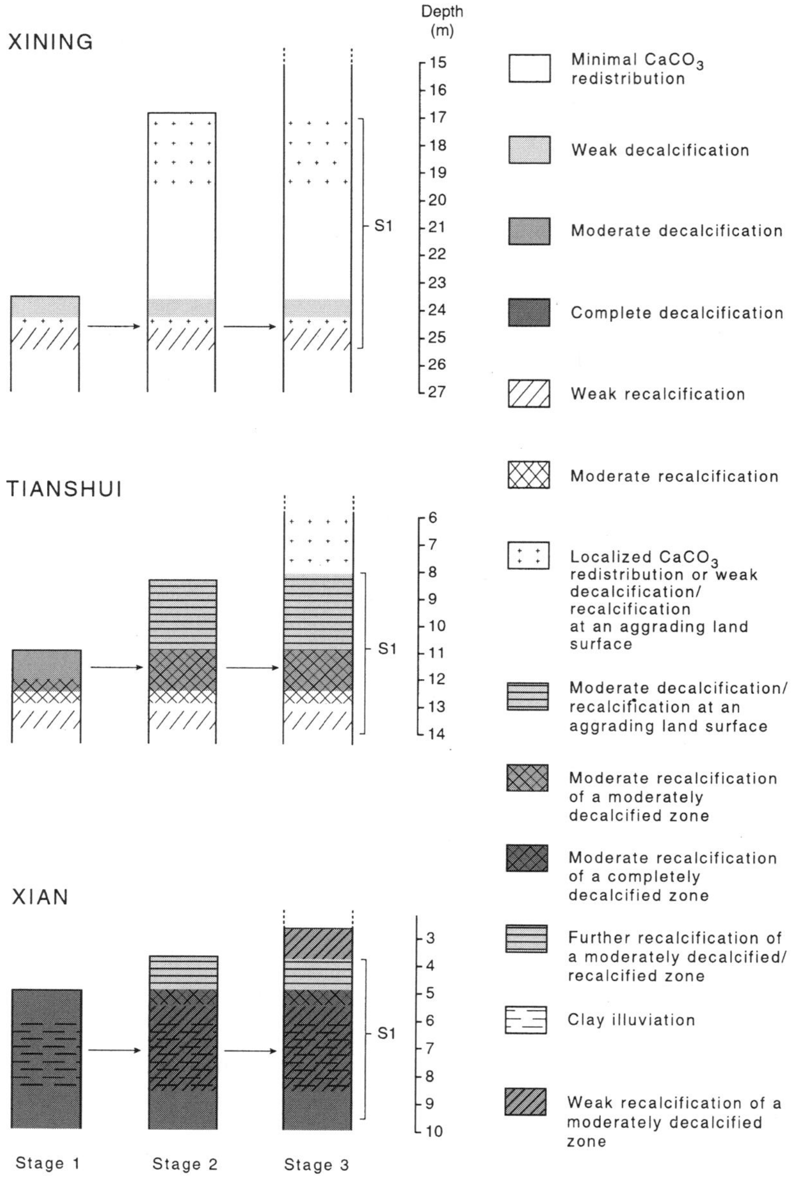

Figure 6.17 Schematic reconstruction of the pedosedimentary stages responsible for development of the buried S1 paleosol at three sites along a regional climatic and depositional gradient on the Loess Plateau of China (reprinted from *Catena*, v. 35, R. A. Kemp, "Micromorphology of loess-paleosol sequences: A record of paleoenvironmental change," pp. 181–198, fig. 7, © 1999, with permission from Elsevier Science). The Xining area (top) is the driest of the study areas and also nearest the source of the loess. Sedimentation rates are highest and pedogenic processes are slowest, resulting in multiple, relatively thin, discrete buried soils. The Xian area (bottom) is the most humid part of the Loess Plateau and most distant from the loess source. The rates of deposition are lowest and pedogenic processes most intense, resulting in thick, welded soils.

from the loess source, the soils making up S1 are more strongly expressed and welded. The interpretations assumed that through the Quaternary, the overall climate trends across the region were similar (i.e., more humid in the east, drier to the west), which seems reasonable. At each location studied, the soils of the S1 Pedocomplex exhibited evidence for two stages of development followed by a third stage subsequent to pedogenesis (fig. 6.17; Kemp et al., 1997). The processes and resulting features of each stage varied, however, depending on local depositional and climatic conditions. Stage 1 was characterized by leaching of carbonate and, in the more humid settings, translocation of clay. Stage 2 saw loess aggradation and overthickening of the soil, accompanied by carbonate translocation. In areas downwind of the source, where loess was thinner, this stage of carbonate was welded to the stage 1 soil. Stage 3 included more rapid loess deposition and, in arid near-source settings, burial of the overthickened S1. Where loess is thinner and the climate more humid, weak carbonate translocation welded the stage 3 loess to the overthickened stage 3 soil.

Recognition of the facies changes among soils buried in loess such as those identified for S1 on the Loess Plateau could be of considerable significance to correlation and interpretation of associated archaeological assemblages or remains, especially those in welded or overthickened soils. For example, hominid remains were recovered from a pedocomplex along the southeastern Loess Plateau near Lantian (chapter 8). Understanding the facies relationships of the soil complex across the region would be an important means of reconstructing hominid landscapes.

Archaeological materials in loess and in associated buried soils are used for correlation, but more often the soils are used to correlate occupation zones. Examples of pedo-archaeological correlations of loess soils, especially for geoarchaeological purposes, are unfortunately rare, at least in the English-language literature. Ranov (1995) summarized the data that are available in a discussion of the "Loessic Palaeolithic," referring to the various Paleolithic industries found in the loess of Europe and Asia. The most extensive record of archaeology in association with loess sediments and soils is from Tajikistan in Central Asia, which is the focus of the review by Ranov (1995; see also Davis and Ranov, 1999). The loess is thick (100–200 m) and contains multiple and well-expressed buried soils, many containing archaeology (fig. 6.18; Dodonov, 1991). Correlation of buried soils between archaeological sites is a key component of the cultural chronology (table 6.4). The significance of the soils in this context is well illustrated by the confusion that ensued following the apparent miscorrelation of buried soils at one of the archaeological sites (Ranov, 1995, p. 743): "Currently, two new chronological models may be considered. The first possibility is that the palaeosol representing the last (Riss-Würm) interglacial may not be palaeosol 5 (palaeosol 4 in A.A. Lazarenko's scheme), but palaeosol 3 (Lazarenko's 2). Another possibility is that palaeosol 2 (Lazarenko's 1) may represent the last interglacial. The first model implies that the ages of palaeosols 5 and 6 should be increased by some 200 ka, . . . while the second model implies that they should be increased much more—to around 300 ka older." The investigators also made assumptions about the relationship of the soils and loess to glacial–interglacial cycles as a means of dating, but their assumptions for Central Asia have yet to be confirmed.

Table 6.4. Stratigraphy of the "Loess Paleolithic" in southern Tajikistan

	Soil number	Soil type[1]	Pebble tradition: Sites	Pebble tradition: Industries	Blade tradition Mousterian and U. Paleolithic sites in caves and open air		China Paleolithic
	1	Serozem (Orthent? Cambid?)			Tutkaul III Shougnou I	Upper Palaeolithic ↑	
	2	Lt. Cinnamon (Ustochrept)			Shougnou II Shougnou III Shougnou IV		Shuidonggou (32–40 ka)
	3	Lt. Cinnamon (Ustochrept)	Khonako II (?)	Flake debris pebble tools technology			
	4	Brown (Inceptisol? Alfisol?)	Lakhuti III	Non-clear with crude flakes, debris, pebble-tools technology	Khudji Kara-Bura Ogzi-Kichik	Mousterian	
	5	Brown (Inceptisol? Alfisol?)	Lakhuti I	Evolved local pebble tools technology with a few levallois elements			Xuijiayao (100 ka)
	6	Brown (Inceptisol? Alfisol?)	Karatau I	Local pebble tools technology without systematic flaking			Zhoukoudian 1–3 (230 ka)
	7	Brown (Inceptisol? Alfisol?)					
	8	Brown (Inceptisol? Alfisol?)					
B→M	9						
	10						
	11		Kuldara	Short flakes, pebble-techniques without choppers			Donguttuo (1 Ma)

Modified from Ranov (1995, table 1). BM indicates position of the Brunhes-Matuyama magnetostratigraphic boundary.

[1] Soil type listed by Ranov (1995, table 1) is indicated, plus likely soil taxonomy equivalent in parentheses.

A

Figure 6.18 Views of the archaeologically significant loess in Tajikistan (provided by and published with permission of R. S. Davis). (A) The scale of the loess sequence in the region is illustrated at Karatau. Excavations in the sixth pedocomplex can be seen at right center. The arrow indicates the soil zone, which is 1–2 m thick. (B) The fifth pedocomplex in loess at the Lakhuti site. This photo provides a good example of a dark, well-developed (Btk morphology) buried soil in loess with compound prismatic and angular blocky structure (in contrast to the massive, light-colored loess). The flat surface at lower right is the base of the archaeological excavations, representing the lowermost zone of Paleolithic archaeology in this soil.

Another problematic aspect of the archaeological record of the soils in Tajikistan is that the occupation zones within the B horizons usually are not in discrete layers or in association with obvious living surfaces (fig. 6.19; Lomov and Ranov, 1984, 1985; Dodonov, 1991; Ranov 1995). There are probably several reasons for this situation. One is turbation from prolonged pedogenesis (chapter 9). Davis and Ranov (1999, p. 188) provide a good illustration of the lack of assemblage discreteness at Kul'dara, in southern Tajikistan. A small artifact assemblage, dated to ~800,000 yr by paleomagnetic correlation, was recovered from the 11th and 12th buried soils below the surface. "No features or faunal remains were found at Kul'dara. The density of the lithic artifacts there averaged less than 3 per cubic meter. They were distributed vertically over a span of 1.8 meters with no sign of patterning or concentration. It is evident that the lithics were subject to some postdepositional disturbance. . . . This lack of patterning is true of all of the Paleolithic loess sites in Tajikistan, as is the general lack of features and fauna."

Elsewhere in southern Tajikistan, somewhat more discrete concentrations of artifacts are reported from the 6th buried soil below the surface at Karatau I (~200,000 yr based on thermoluminescence) and in the 5th buried soil at Lakhuti I (~130,000 yr based on thermoluminescence; see fig. 6.18; Lomov and Ranov, 1984, 1985; Dodonov, 1991; Ranov, 1995; the TL ages may be underestimates, based on Frechen and Dodonov, 1998). These soils are pedocomplexes 2–4 m thick. The zones of maximum artifact concentration are in association with the zones of maximum pedogenic expression, but clearly, no discrete occupation levels are apparent. Lomov and Ranov (1984, 1985) argue for repeated occupation of the site during slow aggradation of the soil. This could well be the case on a loess landscape, as described above by Derbyshire et al. (1995b) and Kemp et al. (1997) for the Loess Plateau (fig. 6.17), and in the model of soil upbuilding by Kemp (2001; fig. 5.10). Turbation cannot be discounted, however. In comparing the vertical distribution of artifacts at Lakhuti, Karatau, and Kuldara, the youngest site (Lakhuti) has the most discrete concentration, whereas the oldest site (Kuldara) has the least discrete concentration (fig. 6.19).

Hoffecker (1987) provides a good case study in the use of buried soils in loess for long-distance correlation and dating of archaeological sites and complexes. The issue was establishment of a better chronological record for the poorly dated Paleolithic occupation of the Russian Plain. The key to stratigraphic dating was the intensely studied and relatively well-dated loess–soil sequence of the Pannonian Basin of Central Europe. The Central European soil sequence (described and discussed further in chapter 8) was dated using numerical techniques and by correlation to the oxygen isotope record of the deep ocean (Kukla, 1970, 1975, 1977). The sequence was correlated with the loess and soils of the Russian Plain using soil morphology and some radiocarbon ages from the latter area. The loess stratigraphy of Central Europe spans the Quaternary (chapter 8), but the archaeological record of concern was the Middle and Upper Paleolithic record, which only began during the last interglacial (oxygen isotope stage 5e).

The loess-soil record of the Russian Plain exhibits a pedologic record very similar to that of Central Europe in that the thickest loess layers are capped by the best-developed soils, representing full glacial conditions (and loess

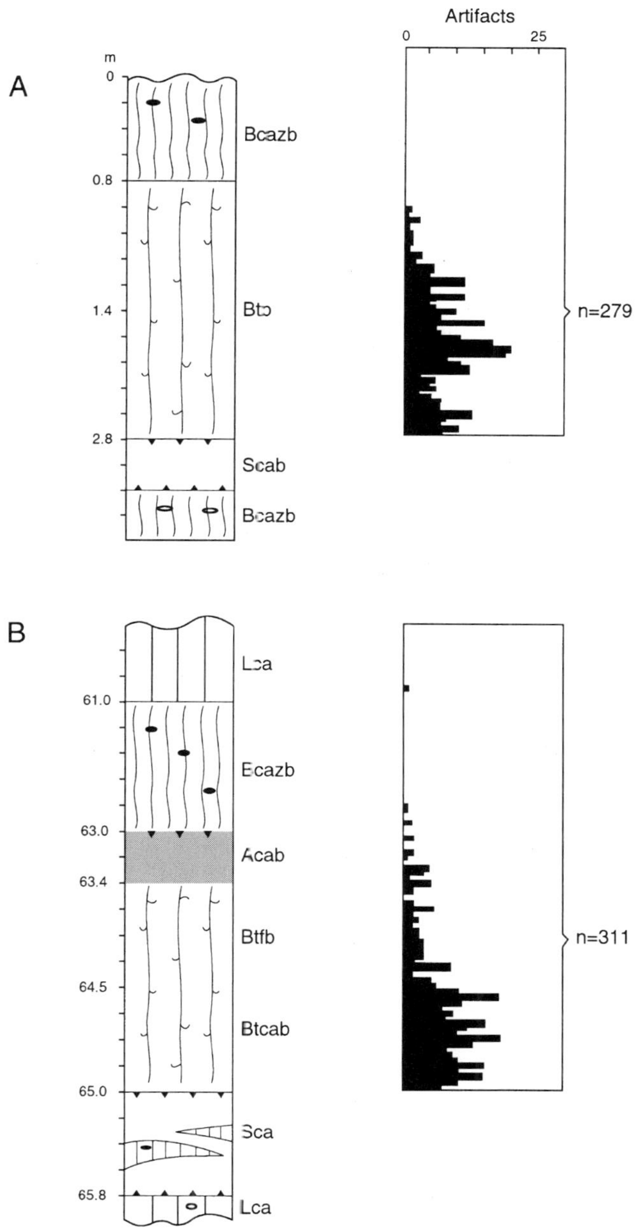

Figure 6.19 Distribution of Paleolithic artifacts in buried soils in Tajikistan loess (based on Lomov and Ranov, 1985, fig. 30-77). (A) Fifth pedocomplex at Lakhuti; (B) sixth pedocomplex at Karatau. Horizon nomenclature was not translated in the article. The Btca is a Btk horizon. The Bcaz and Btf horizons are unknown types of Bk and Bt horizons. The Lca is probably carbonate accumulation in otherwise unweathered loess, a Ck horizon. The Sca is probably carbonate accumulation in sand, also a Ck horizon.

deposition) followed quickly by full interglacial conditions (and soil formation; fig. 6.16; Hoffecker, 2002, pp. 28, 30–34). The interglacial soil was buried first below thin loess and eolian sediment, then increasingly thicker layers of loess, with soils that become progressively weaker. This loess–soil sequence represents the onset of the next glacial cycle, with a progressive increase in loess deposition and decrease in pedogenesis because of the loess and the colder environment. The sequence culminates in thick loess deposition during the subsequent full glacial period and then another well-expressed soil in the subsequent interglacial. The last interglacial soil in Central Europe, Pedokomplex III (PK III), is the most strongly expressed in the sequence, described as "podzolic" by Hoffecker (1987, p. 262), but clearly expressing Alfisol characteristics according to Kukla (1970, pp. 150–154; 1975, pp. 109–113; 1977, pp. 326–329). The equivalent soil in the Russian Plain sequence is the well-expressed lower portion of a buried pedocomplex (the Mezin Soil), which exhibits latitudinal gradation from podzolic (forested) north to chernozemic (grassland) south. The loesses and soils of early glacial Central Europe and the Russian Plain probably are related to the cycles of cool and temperate climate of oxygen isotope stages 5a–5d. The soils formed under cold steppe grasslands. Throughout both regions, the few soils associated with full glacial conditions (oxygen isotope stages 4, 3, and 2) are "frost-gleys" or permafrost soils.

The Paleolithic archaeological record of the Russian Plain is contained largely in river valleys of the region and, therefore, is associated with redeposited loess (Klein, 1969, pp. 36–49; 1973, pp. 18–27; Hoffecker, 1987). This, of course, raises questions about the validity of cross-continental soil stratigraphic correlations for archaeological dating. Many of the archaeological horizons on the Russian Plain are associated with the soils buried in the redeposited loess. Radiometric dating of the soils, the reworked loess, and archaeological features indicate that cycles of sedimentation and stability in the valleys are synchronous with the same cycles on the uplands and, therefore, that regional soil stratigraphic correlations based on the loess record are applicable to the redeposited loess (Klein, 1969, pp. 45–49; 1973, pp. 18–33; Hoffecker, 1987, 1988; Sinitsyn, 1996).

The soil stratigraphic associations of the Paleolithic archaeological remains on the Russian Plain provide insights into the timing and nature of human activity. The loess–soil sequence of the Russian Plain provides little evidence of human occupation before the last interglacial, indicating the probability that the Russian Plain was the last major region of temperate Europe to be colonized by human population (Hoffecker, 1987, pp. 277–278). Occupation density remained low in the last interglacial; few sites are associated with the well-developed soil of that period (indicative of lengthy landscape stability under attractive environmental conditions). The early glacial loesses and soils, in contrast, provide abundant evidence of Middle Paleolithic occupation. The transition from the Middle to Upper Paleolithic is difficult to investigate because sites correlatable with oxygen isotope stage 4 are rare. The Upper Paleolithic record, however, is relatively rich and found in soils and redeposited loess correlated to stages 3 and 2.

Also working on the Russian Plain, Kurenkova et al. (1995) used the stratigraphic relationship of archaeological sites in loess to reconstruct settlement patterns for the Upper Paleolithic. Their discussion is brief, but buried soils were key markers, indicative of landscape stability and, therefore, likely to attract inhabitants. They also discovered occupations in unweathered loess, indicating short-term stability of the landscape not manifested pedologically. In addition, they supported Hoffecker (1987) in noting the absence of sites older than the last interglacial stage.

Loess stratigraphy has played a much less prominent record in New World geoarchaeology because most loess deposition ended before significant human occupation of either North or South America (one notable exception is Alaska). The most extensive loess deposits in North America are on the Central Lowlands and Great Plains. In contrast to the Old World terminologies, the loess and associated buried soils are named, generally following formal lithostratigraphic and pedostratigraphic nomenclature. This system avoids the problems inherent in fitting newly discovered soils into existing numbering schemes. The archaeologically significant components of the Midwest loess sequence are only the latest Pleistocene and Holocene sediments and soils—typically the upper Peoria Loess and the Brady soil (or paleosol or geosol)—formed in the loess and recognized only where buried by the overlying Bignell loess (fig. 6.20). Deposition of Peoria loess ended at the close of the Pleistocene, probably by ~10,500 yr B.P. Pedogenesis in the Peoria loess (i.e., formation of the Brady soil) occurred during some part of the Paleoindian occupation of the region, although the dating is unclear. The best estimate seems to be 10,500–8800 yr B.P. (Mason and Kuzilla, 2000). In any case, stability of upland landscapes probably was the norm throughout Kansas and Nebraska during much if not most of Paleoindian times. Depending on the validity of the dating, late Paleoindian occupations may have been

Figure 6.20 An exposure of the Brady Soil near North Plate, Nebraska. This prominent, buried, dark-gray Ab-Btb profile formed in late Pleistocene Peoria Loess and was buried by the Bignell Loess in the very late Pleistocene or early Holocene. The Brady Soil represents the buried regional landscape during the Paleoindian occupation of the area.

coincident with subsequent, localized accumulation of the Bignell loess (Martin 1993; Mason and Kuzilla, 2000). Paleoindian sites on the uplands may occur, therefore, in the upper Peoria loess, in the Brady soil, or in the Bignell loess, though few are reported (Johnson and Logan, 1990). Archaic sites on the uplands should be found within the Bignell loess, which is locally stratified and contains buried soils (Mason and Kuzilla, 2000), but they are likewise rare (Johnson and Logan, 1990; Mandel, 1995, table 1). In areas in which the Peoria loess is not covered by the Bignell loess, the entire archaeological record of the region is compressed or mixed into the regional surface soil formed in the Peoria loess.

Silt and sand deposited by wind during the Holocene across North Dakota records cyclic landscape instability and soil burial, producing the sediments and buried soils of the Oahe Formation (Clayton et al., 1976, 1980; Artz, 1995). The deposits, derived from late Pleistocene outwash and loess, are draped across glaciated uplands and on stream terraces. Archaeological sites and artifact scatters are widely reported from the Oahe, and soils have proven to be useful for some stratigraphic correlation (Jorstad et al., 1986; Artz, 1995; Waters and Kuehn, 1996). The formation is composed of three members (Aggie Brown, Pick City, and Riverdale, bottom to top) and two informally named buried soils: the Leonard paleosol formed in the Aggie Brown Member and the Thompson paleosol formed in the lower Riverdale member. The Leonard is the more prominent of the two buried soils, characterized by a dark-gray-to-dark-brown (10 YR 4/1, 3/1, 3/0 dry) buried A horizon with conspicuous worm or insect burrows (Clayton et al., 1976, pp. 5, 7). The soil was formed in the terminal Pleistocene to early Holocene and produces Paleoindian sites and artifacts (Artz, 1995). In both stratigraphic position and age, it appears to be a correlate of the Brady soil to the south. The Riverdale member consists of multiple layers of eolian sediment, locally containing multiple buried soils, including or in addition to the Thompson paleosols. These multiple buried soils can contain stratified occupation zones from the Archaic up through the late Prehistoric (Jorstad et al., 1986).

Volcanic Deposits

Archaeological sites have been found buried in volcanic deposits, particularly in tephra layers, throughout the volcanically active regions of the world (e.g., Bishop, 1978; Sheets and Grayson, 1979; Olson, 1983; Sheets, 1983; Steen-McIntyre, 1985; Davis, 1986; Lippi, 1988; Wilson, 1990; Sheets and McKee, 1994; Scudder et al., 1996; Panfil et al., 1999; James et al., 2000). Human activity in and near volcanoes spans the archaeological record, from early Hominid evolution (e.g., Laury and Albritton, 1975; Bishop et al., 1975, 1978; Hay, 1976; Bonadonna et al., 1998) to the advent of agriculture (e.g., Sheets, 1983; Scudder et al., 1996; Panfil et al., 1999). Buried soils also are common components of volcanic deposits, again particularly tephras (fig. 6.21; e.g., Dalrymple, 1967; Kato and Matsui, 1979; Campbell, 1986; Shoji et al., 1993, pp. 8–9, 12–16, 42–44, 62–63; Panfil et al., 1999; Baümler and Zech, 2000; James et al., 2000; papers in Lowe, 1996, and Cavarretta et al., 1998). The frequency of buried soils in volcanic layers is largely the result of the repeated, episodic nature of volcanic eruptions. The very common association of buried soils with volcanic ash also is because of the rapid

Figure 6.21 Buried soils in volcanic ash near Tronadora, Costa Rica (from *Archaeology, Volcanism, and Remote Sensing in the Arenal Region, Costa Rica*, fig. 1-7, P. D. Sheets and B. R. McKee, Eds., © 1994. University of Texas Press, Austin; reproduced with permission of P. D. Sheets and the University of Texas Press; photo provided by P. D. Sheets). The white and light-gray zones (e.g., units 10, 40, 55, and 61) are tephra layers. The dark-gray zones are buried soils formed in the ash. The cyclical nature of the eruptions is illustrated by the dating of the ashes: 61 = 1800 BC; 55 = 800 BC; 40 = AD 800–900; 10 = AD 1968.

rates of tephra weathering resulting from the instability of volcanic glass (Dahlgren et al., 1993), which, when combined with the cyclic nature of eruptions, produces an ideal environment for creating and burying soils. Indeed, the unusual weathering characteristics of tephra resulted in development of the soil order Andisol (Soil Survey Staff, 1999). Soil stratigraphic research in association with volcanic deposits is rare, however, in or out of archaeological contexts. Most of the data that are available are from tephra layers.

The morphologies of soils formed in volcanigenic parent material can vary significantly depending on the parent material, local environment, and duration of weathering. Very generally, the soils buried in tephra tend to be relatively weakly expressed (A-C or A-Bw-C profiles) because of the relatively frequent eruptions of the world's silicic volcanoes (e.g., table 6.5; fig. 6.21). Many regions that witnessed repeated tephra eruptions through the Quaternary also are regions of humid, temperate climates with high biological productivity (e.g., Japan, islands of the southwest Pacific, Central America, and South America; Ugolini and Zasoski, 1979; Shoji et al., 1993, p. 3). The Holocene soils formed under these conditions, therefore, typically have black, relatively thick A horizons high in organic matter that provide a strong contrast with the light-colored tephra layers (table 6.5; fig. 6.21; Ugolini and Zasoski, 1979; Campbell, 1986; Shoji

Table 6.5. Description of soils and tephrastratigraphy at the Nambillo site, Ecuador

Stratum	Thickness, cm	Description
I	15	*A horizon:* very dark brown (10YR 2/2) sandy silt with decaying organic matter
II	20–30	*Volcanic sediment:* ash, coarse pumice sand & fine pumice gravel; light olive brown (2.5YR 5/4)
II	15	Transition between strata IV & II; dark brown (10YR 3/3) sand and clay
IV	70–80	*Ab1 horizon* (anthropic epipedon): black (10YR 2/1) clay; ~1665 to ~820 C14 yr BP
V	0–20	*Volcanic sediment:* ash, coarse pumice sand and pumice gravel
VI	0–50	*Ab2 horizon* (anthropic epipedon): very dark grayish brown (10YR 3/2) clayey silt; ~2515 to ~1665 C14 yr BP.
VII	0–140	*Volcanic sediment:* several indistinct bands of ash and pumice sand of varying coarseness and color.
VIII	25–55	*Ab3* (anthropic epipedon): black (10YR 2/2) clay; ~5325 to 2315 C14 yr BP
IX		*Weathered tephra:* yellowish-red (5YR 5/8) cemented clay
Total thickness	145–405	

Modified from Lippi (1988, table 1).

et al., 1993, pp. 7–10). This particular variety of A horizon gave rise to the melanic epipedon of soil taxonomy (Soil Survey Staff, 1999). Soils formed in tephra in other regions and in older soils are more variable in morphology, however (James et al., 2000). For example, soils in tephra are reported with calcretes in drier environments (Bishop et al., 1975, 1978), and with Bt horizons (or otherwise well-developed profiles) when eruptions are less frequent (Pillans, 1988; Frezzotti and Narcisi, 1996; Palmer and Pillans, 1996). The weathering of tephra also sometimes results in formation of highly expandible clays and development of shrinking–swelling soils (Vertisols or soils with vertic properties; see chapter 10), most notably characterized by slickensides (Wendorf and Schild, 1974b; Laury and Albritton, 1975; Ahmad, 1983; Brown and Feibel, 1986).

Geochemical characteristics of tephra provide one of the best means of stratigraphic correlation in the geologic record, but soils can be useful for correlating tephra in the field, especially if the tephra are similar in physical appearance. The utility of soils buried in tephra for stratigraphic correlations probably will be uniquely related to the characteristics of the individual eruptions and tephra layers, however (Shoji et al., 1993, pp. 42–44). The thickness and distribution of tephra are functions of the amount of ash produced, the duration or each eruption, and the direction and strength of the wind that carried the tephra across the landscape (fig. 6.22). Furthermore, the particle size and mineralogy of individual layers of tephra can vary significantly both laterally and vertically. Though data are rare, the many variables that may affect soil genesis in tephra indicate that the variation in the morphology of individual soil strata may be significant and make for difficulties in correlation (e.g., Melson, 1994, pp. 38–39). In a study of late Quaternary, tephra-derived buried soils from Italy, however, Frezzotti and

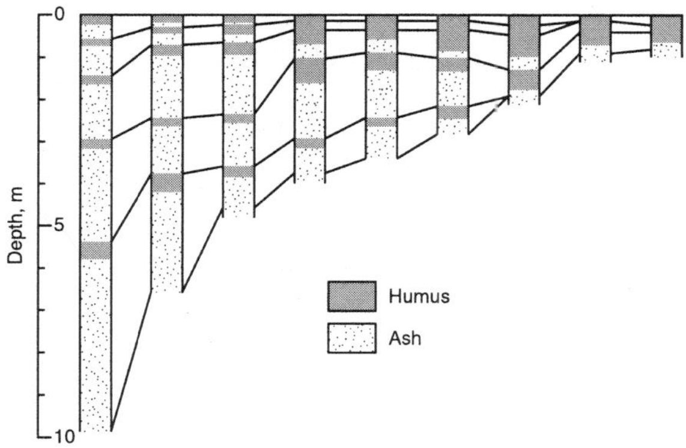

Figure 6.22 The Towanda Andisols, Japan, showing variability in lithostratigraphy and soil stratigraphy over a distance of ~60 km (based on Shoji et al., 1993, fig. 3.3).

Narcisi (1996) report two soil stratigraphic units ("Pedomarkers") that can be traced over a significant portion of the central Apennines. These two soils are very similar in morphology (black Ab-brown Bw profiles), raising the possibility that in the field, rather than being difficult to trace because of high variability, they are difficult to distinguish from one another. The correlations of these soils are also based on radiometric dating and parent material lithologies. Dating of tephra and the geochemistry of the sediments has proven to be more effective in stratigraphic correlation as a result of the often weak (i.e., nondistinctive) nature of soils buried in volcanic ash (e.g., Steen-McIntyre, 1985; Melson, 1994).

Several examples of tephra geoarchaeology illustrate the wide range of environments and archaeological contexts associated with volcanic ash deposits and also illustrate the very limited soils data available. Some of the most intensive and extensive work involving the geoarchaeology of tephra was conducted by P. Sheets and his colleagues in Central America (fig. 6.21; Sheets, 1983; Sheets and McKee, 1994). Soil studies were limited and tended to focus on chemical weathering (Melson, 1994) or soil fertility (Olsen, 1983), discussed in chapter 11. The weathering in the upper part of each tephra layer is a soil phenomena, and therefore, weathering studies were undertaken to aid in the difficult task of correlation of multiple, grossly similar tephra layers produced by a single volcano (Melson, 1994, pp. 38–42). Depletion of potassium and phosphorus, and the relative proportion of residual minerals through each layer proved to be useful indicators in correlation of individual strata. In central Mexico, buried soils in tephra were significant components of geoarchaeological investigations around the Popocatépetl volcano (Panfil et al., 1999). The soils were used to trace tephra and correlate archaeological sites of the Tetimpa Occupation (2300–1300 yr B.P.). Of particular interest was a posteruptive reoccupation of a volcanic landscape denoted by weak soil development.

In Alaska, along the Middle Susitna River in the Alaska Range, tephra stratigraphy was a key in correlating archaeological sites as part of a regionally extensive survey and testing program (Dixon and Smith, 1990). The volcanic ash is interbedded with eolian and lacustrine silt. Following each ash fall, the landscapes were apparently somewhat stable, and soils formed in the upper Oshetna tephra, Watana tephra, and Devil tephra. Localized, weakly expressed soils were noted within the Watana tephra, indicative of multiple ash falls from a series of eruptions. Data are not available on the soils themselves, but they were important markers for establishing an archaeological stratigraphy.

At an intersite scale, Lippi (1988) used soils buried in tephra to map and correlate occupation zones at the Nambillo site in Ecuador. The field data were recovered by systematic coring of the site (deep in a remote rainforest setting),

Table 6.6. Soil stratigraphy and archaeology at Gademotta, Ethiopia

Unit no.	Unit no.[1]	Soil horizon	Archaeology	Description[2] (soil thickness)
19				Sand/sandstone
18	34	A Btk	MSA artifacts in upper half (site 72-5)	Brown soil (90 cm) with slickensides
17	33	Bw		Sand
16	32	Bk		Colluvium with Fe patches
15	31	Ck		Ash
14				Sand/sandstone
13	30	A Bk	MSA artifacts in upper half (site 72-6)	Brown soil (70 cm) in fine colluvium
12	27, 28, 29		MSA artifacts at base of 27 (site 72-7B)	Fine colluvium and ash
11	26	A AB Bk	MSA artifacts at top (site 72-7B)	Brown soil (100 cm) in fine colluvium; with dessication cracks
10	25			Ash (regional marker bed)
	24[3]	A C		Weak brown soil
	23[3]			Fine colluvium
9	22	A Btk	"Final Acheulian" artifacts at base; MSA artifacts at top (both are site 72-8B)	Thick brown soil (220 cm) in fine colluvium; with Fe, Mn mottles
8A–D	21a–c			Wadi fills
7	20	A Btk		Brown soil (70 cm) in sand; with Fe, Mn mottles
6[4]				Ash[5]
5[4]		Bt		Sand[5]
4[4]				Mudstone[5]
3	19			Muddy colluvium with gravel
2	18	A Btk		Brown soil (90 cm) with slickensides
1	17	Bw		Silty colluvium with pebbles

Laury and Albritton (1975). Below Unit 1 is interbedded rhyolitic lava and volcaniclastic deposits of the Kulkuletti Volcanics; identified as the lower and middle Gademotta Formation (units 1–16) by Wendorf and Schild (1974b).

[1] Wendorf and Schild (1974b).

[2] Description based on Wendorf and Schild (1974b), except as noted.

[3] Units 23 and 24 of Wendorf and Schild (1974b) are localized and apparently not recognized in the area investigated by Laury and Albritton (1975).

[4] Stratigraphic units 4, 5, and 6 reported by Laury and Albritton (1975) appear to have no equivalent in the area investigated by Wendorf and Schild (1974b); the soil of unit 5 of Laury and Albritton (1975) is nowhere described except to note that it exhibits "strong development" (Laury and Albritton, 1975, fig. 9).

[5] Description based on Laury and Albritton (1975, fig. 9).

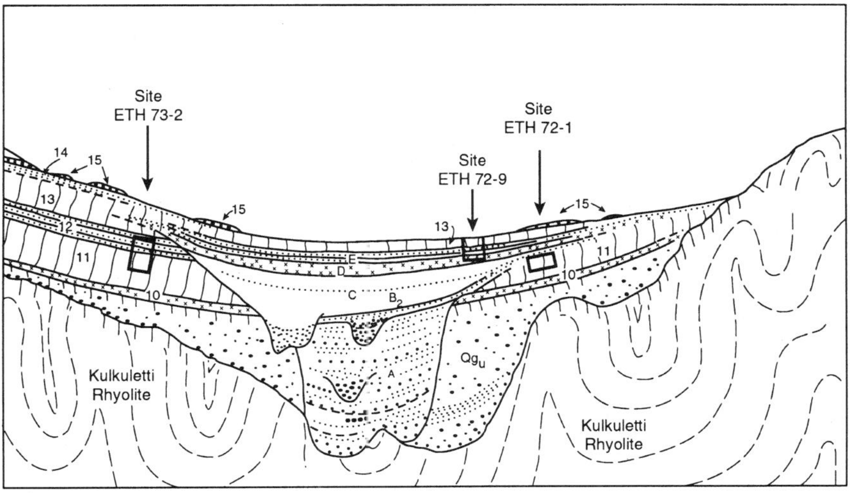

Figure 6.23 Cross section of the Gademotta Formation, illustrating the relationship of the soil stratigraphy and archaeological sites ("ETH"; modified from Laury and Albritton, 1975, fig. 6). The numbers refer to stratigraphic units listed in table 6.6.

using a portable hand auger. Three distinct buried soils, each with occupation debris, were recognized (table 6.5) within the accumulation of tephra (with a total thickness ranging from 1.5 to 4.0m). Mapping of each buried soil, based on the coring data, allowed construction of a series of paleotopographic maps. Analyses of soil samples for organic matter and phosphate content provided data for hypothesizing the location of ancient activity areas across the buried landscapes.

In applications very different in setting and age from those summarized above, soils buried in volcanic sediment and rock in the East African Rift Valley have proven useful for correlation of hominid and Paleolithic localities. Details on the soils are rare, in general, but some information is available from a few sites. Buried soils were identified in upper Pliocene and lower and middle Pleistocene volcanic deposits at the Chesowanja hominid locality in the Northern Rift Valley of Kenya (Bishop et al., 1975, 1978). The sediments and rock include pumiceous tuff, basalt, and trachyte, as well as layers of alluvial and lacustrine deposits. Archaeological materials recovered from the sequence include australopithecine remains and Developed Oldowan and Acheulian stone tools. In the upper basalt of the Chesowanja Formation is a prominent "complex palaeosol" 3–4m thick, identified as the "palaeosol member" of the formation and used as a key stratigraphic marker. The upper part of the soil also produced a significant Acheulian assemblage.

Buried soils in the middle and late Pleistocene tephras of the Gademotta Formation in the Ethiopian Rift Valley were used to correlate Middle Paleolithic sites throughout the area (Wendorf and Schild, 1974a; Laury and Albritton, 1975). Soils are prominent and widespread stratigraphic markers, formed in multiple layers of primary and reworked tuff, rhyolite, and laharic mudstone. The reworked sediments occur in both lacustrine and alluvial clastic rock. The

Gademotta Formation was subdivided into "units" numbered from oldest to youngest (table 6.6, fig. 6.23). The units included both sediments and soils. Some pedologic details are provided by Wendorf and Schild (1974b): The buried soils exhibit organic matter, secondary (amorphous) clay and secondary carbonates, and staining and cementing by iron- and manganese-oxides. Profiles, therefore, are typically A-Bw, ABt or Bt, and A-Bt-Bk or Bt-Bk. Most of the Middle and Late Paleolithic sites are associated with these soils, indicative of occupations on stable landscapes. The oldest Paleolithic site in the area includes "Final Acheulian" artifacts and a Middle Paleolithic occupation in the soil in unit 9. Above are multiple, welded buried soils (associated with units 11, 13, and 17–18), each associated with Middle Paleolithic occupation.

Differentiating buried soils, occupation layers, and unweathered tephra can be difficult, however, because of the complex chemical makeup of volcanic deposits. For example, at the Yagi site, Japan, six occupation layers were identified within four buried soils formed in tephra layers (Davis, 1986). Levels of organic matter were high throughout the section because of complexes of humus and amorphous materials. Phosphorus levels, often indicative of human activity (see chapter 11), were variable through the section and were lowest in some archaeological zones, probably because of natural levels of P in the tephra.

7

Soils and Time

The influence of time on soil formation is a unique characteristic of pedogenesis among geomorphic processes that, like lateral variability, serves to distinguish soils and soil-forming processes from other geomorphic phenomena. Another unique aspect of time is that, among the five factors of soil formation, it does not contribute directly to soil formation. However, the passage of time allows the various pedogenic processes operating at a given location to alter the parent material and produce a soil. The physical, chemical, and biological processes of soil formation (table 3.1) generally are much slower than many, if not most, processes of sedimentation and erosion (table 5.1). Moreover, most soil-forming processes are so slow that their effect on the soil is markedly time dependent (Birkeland, 1999, p. 144).

Time as a factor of soil formation is a key concept in soil geomorphology and has driven much soil geomorphic research (Yaalon, 1975, 1983; Knuepfer and McFadden, 1990; Birkeland, 1999). Because time is also a key consideration of much archaeological research, the time-factor concept of soil genesis can likewise play a significant role in geoarchaeological research (Holliday, 1990a, 1992a). The concept that some time must elapse before a soil can form is arguably one of the most significant aspects of soil development in an archaeological context. This chapter is a discussion of some approaches to the issue of time in archaeology, using soils. The first section is a look at the archaeological implications of soils as indicators of stable landscapes and stratigraphic discontinuities. A number of case histories are presented. The validity of intersite and intrasite archaeological correlations using soils and interpretations of archaeological assemblages associated with soils are profoundly dependent on recognition of soils as depositional hiatuses. The subsequent section reviews the concept of the soil chronosequence and its use in archaeological dating. This is one of the most widely applied aspects of Jenny's state factor approach to soil geomorphology, and it has considerable potential in archaeology. The last part of the chapter is a discussion of the radiometric dating of pedogenic features. A number of components of soils are dateable by several different techniques, and all of them have archaeological applications. There are also some pitfalls in this sort of dating, and they are outlined.

Vreeken (1975) presents a valuable discussion of the concept of time as a factor of soil formation. An important point he makes is that referring to "the age of a soil" can mean several very different things. One is simply the duration of soil development; that is, how many years it took to form. For a soil at the surface, this is the time from the beginning of soil development to the present. For a buried soil, the duration of soil formation is the time from the beginning of formation to its cessation. For example, we can think of a soil that took 10,000 yr to form. This could be the past 10,000 yr for a surface soil or an interval such as 25,000–15,000 yr ago for a buried soil. Another aspect of time and soil formation is when on the geologic time scale did the soil form; that is, the historical context within which pedogenic and geomorphic processes took place. For example, knowing whether a soil formed during the 10,000 yr of the Holocene or for 10,000 yr during the last full glacial period is of the utmost importance in interpreting and understanding the evolution of a soil, because the environmental conditions of the Holocene were very different from those of the last glacial maximum. When applied to buried soils, the term "soil age" sometimes refers to the length of time since burial. In this sense the meaning can be ambiguous. Is a "10,000-yr-old soil" one that formed for 10,000 yr, or one that was buried 10,000 yr ago? The difference in usage is quite significant. The former is a pedogenic question, and the latter is a chronostratigraphic issue. As with terms discussed in chapters 1 and 5, "soil age" should be used very carefully and very clearly.

Soils, Time, and Stratigraphic Interpretation

The association of archaeological materials with a soil or the presence of a soil or soils in a stratigraphic sequence at an archaeological site is of direct and fundamental significance to the interpretation of the archaeological record. Soils require some amount of time to form, and soil development requires a relatively stable landscape, one that is neither aggrading nor eroding at rates that exceed pedogenesis. This aspect of pedogenesis was formalized conceptually as the "time factor" by Jenny (1941, 1980; Birkeland, 1999) and expressed stratigraphically in the "K-cycle" concept of Butler (1959, 1982) (chapter 3). The presence of a soil, therefore, denotes the passage of some amount of time under conditions of relative landscape stability. Furthermore, a buried soil in a stratigraphic sequence denotes a hiatus in deposition, a kind of unconformity. In such a sequence, the sediment, which is the parent material for the soil, may have accumulated rapidly or slowly, but a significant period of nondeposition had to occur for the soil to form. Under some circumstances deposition can occur relatively instantaneously;

possibly in a matter of days, years, or decades. Soil formation, in contrast, almost always takes longer—usually at least a century or several centuries, and commonly millennia (table 5.1, fig. 7.1; though see the exceptions given, e.g., Birkeland, 1999, and as discussed below under "Chronosequences"). Put another way, in a stratigraphic sequence, the accumulation of deposits represents the passage of time sedimentologically, and the alteration or weathering of those deposits represents the passage of time pedologically. In the author's experience, however, it seems that many investigators assume that in an archaeological site of some thickness, especially a stratified site, sedimentation occurred more or less continuously. Certainly this approach was the case in much of the earlier geoarchaeological literature, particularly before rates of soil formation were well established (e.g., Green, 1962; Haynes, 1968). In settings in which sedimentation is episodic and buried soils are common, soil formation may take up a significant amount of the time represented in a stratigraphic sequence (fig. 7.1; e.g., Kraus and Bown, 1986; Pawluk, 1978; Holliday, 1992a).

There are several notable, archaeologically significant exceptions to the concept of brief, relatively rapid intervals of sedimentation or erosion separated by relatively lengthy intervals of stability and pedogenesis. Bogs or marshes (Histosols) are characterized by prolonged accumulation and accretion of organic matter and do not necessarily require surface stability to form. Other cumulizing or accretionary settings (chapters 5 and 6), locally found on many landscapes, also are exceptions. In some floodplain or toeslope settings, multiple, very weakly developed soils occur, formed during brief intervals (a few years) of stability between episodes of deposition (chapter 6). These soils are characterized

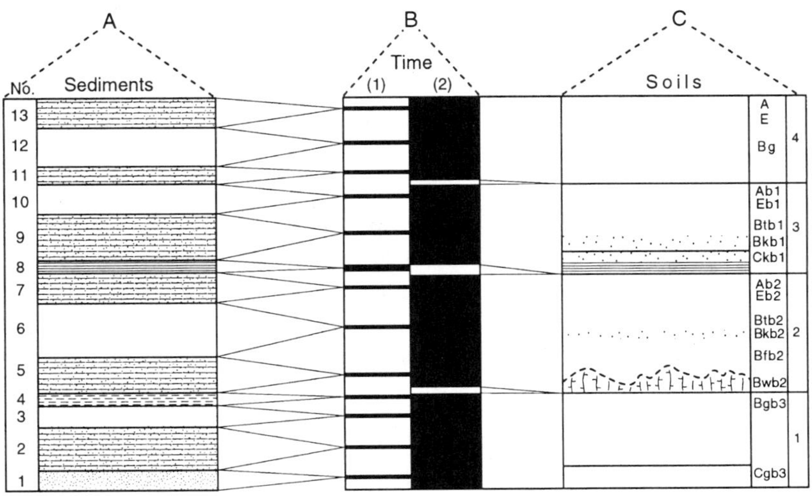

Figure 7.1 Diagram illustrating hypothetical alluvial sequence consisting of layers 1–13 (column A) and associated soils 1–4 (column C; modified from Kraus and Bown, 1986, fig. 1). Column B indicates the relative proportion of time occupied by deposition (B1) and soil formation (B2).

simply by accumulation of organic matter (from both *in situ* growth and redeposition) and by some bioturbation and depositional mixing.

"In archaeological investigations, consideration of landscape stability, as indicated by soil development, is important in locating cultural deposits, interpreting artifact associations and contexts, defining site stratigraphy, reconstructing the depositional and landscape history, and establishing cultural chronologies" (Mandel and Bettis, 2001b, p. 175). In searching for archaeological sites or other evidence of human activity in stratigraphic sequences with buried soils, the surfaces of the soils (i.e., the A horizons) will be more likely than other parts of the soil or parent material to contain artifacts and features because they represent surfaces that were stable for some amount of time and because human occupation is more likely on a stable surface than it is on a landscape undergoing rapid sedimentation. This aspect of A horizons has influenced development of strategies in surveying and prospecting for archaeological sites, especially in buried settings (Michlovic et al., 1988; Ferring, 1992; Pincha and Gregg, 1993; Stafford, 1995).

Furthermore, the degree of soil development can be of considerable help in deciphering the archaeological record associated with buried soils and buried surfaces. For the purposes of this discussion, buried soils can be viewed as representing time in two basic stratigraphic forms: as weakly developed soils, each representing a relatively short time (i.e., a relatively brief interval between periods of deposition or between periods of deposition and erosion), or as better-developed soils representing longer intervals of nondeposition and landscape stability (fig. 7.2; though, obviously, soil development is expressed in a wide array or continuum of degrees of development). Multiple, weakly developed soils are, therefore, indicative of rapid, multiple episodes of sedimentation with relatively brief intervals of stability and pedogenesis. The relative amount of time taken up by pedogenesis in these settings will still greatly exceed that taken up by deposition (Kraus and Bown, 1986; Kraus and Aslan, 1999; Ferring, 1992; Gladfelter, 2001), but the time will be parceled out among the individual deposits and soils (fig. 7.2). Archaeological sites in these settings will be well stratified, with "superposition of artifacts and features that resulted from serial occupation of sites" (Ferring, 1986, p. 264; fig. 7.2).

In contrast to weakly developed soils, more strongly expressed soils represent long intervals between periods of instability. In a stratigraphic sequence with several well-expressed buried soils, most of the time is probably represented by the soils. Archaeologically this is particularly significant because slow deposition during multiple episodes of occupation results in accumulation of archaeological debris as mixed assemblages on paleosurfaces (figs. 7.1 and 7.2; Ferring, 1986, p. 264). That is, the archaeological evidence for repeated occupations is compressed into the A horizon of a single surface. Put another way, assuming "that the probability of cultural utilization of a particular landscape position is equal for each year, it follows that the surfaces which remain exposed for the longest time [i.e., the surfaces of the better-developed soils] would represent those with the highest probability of containing cultural remains" (Hoyer, 1980, p. 61). The formation of a soil at an archaeological site, therefore, can profoundly influence the nature of the archaeological record and interpretations of the cultural history of the

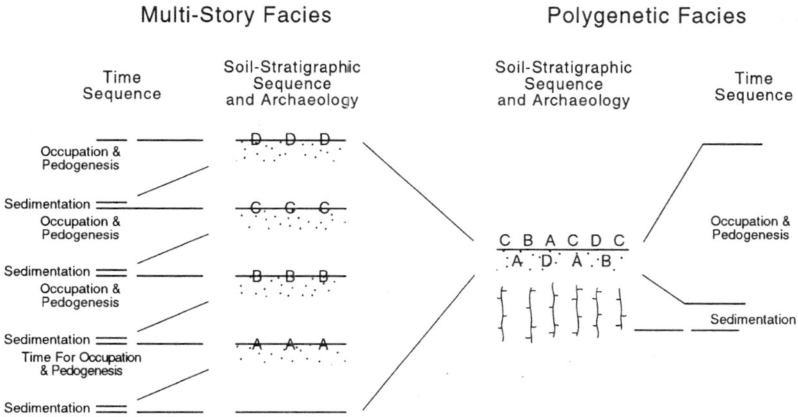

Figure 7.2 Diagram illustrating the effect of varying rates of deposition on the archaeological record (see also figs 5.3, 5.9, and 6.6). Frequent episodes of flooding and sedimentation produce multistory facies with multiple, weakly expressed soils containing relatively discrete occupation zones, whereas prolonged stability results in a few or a single polygenetic soil with a palimpsest of occupation debris.

locality. Long spans of time are represented by the surface of a well-developed soil, and thus long records of habitation can be compressed into zones only a few centimeters thick. In searching for archaeological sites, the A horizons of well-expressed buried soils are more likely than weakly developed ones in the same area to contain occupation debris, but that debris is more likely to be a mixture of artifacts from multiple occupations.

Alluvial and eolian depositional environments are settings in which sedimentation tends to be episodic and in which buried soils are common. In particular, settings characterized by episodic deposition with longer intervening periods of nondeposition and pedogenesis include floodplains, alluvial fans, and dune fields and some other eolian environments (chapter 6). Multiple, weakly expressed buried soils are characteristic of some alluvial fans (e.g., Hajic, 1990; Hoyer, 1980; Styles, 1985), and many floodplain settings and other aggradational alluvial environments (e.g., Ferring, 1986, 1990, 1992, 2001; Kraus and Bown, 1986; May, 1989, 1992; Johnson and Logan, 1990; Gerrard, 1992, pp. 106–107; Mandel, 1992, 1994; Kraus and Aslan, 1999; Gladfelter, 2001; Huckleberry, 2001), and are a fundamental trait of the "Fluvent" suborder of the Entisol soil order (chapter 6). This sort of soil stratigraphic record can also be found in dunes and some lowland settings that accumulate eolian sediment from air fall (e.g., Valentine at al., 1980; Holliday, 1985a, 1995, 2001b; Muhs, 1985; Hall, 1990; Wells et al., 1990; Jorgensen, 1992; Muhs et al., 1996, 1997a,b) or via redeposition (e.g., Klein, 1969, pp. 36–49; Hoffecker, 1987; Sinitsyn, 1996). Accretionary eolian environments such as loess landscapes can contain more strongly expressed buried soils representing longer intervals between sedimentation (e.g., Catt, 1991; Bronger et al., 1995; Holliday, 1995; Kemp et al., 1996; Kemp and Derbyshire, 1998; Kemp, 2001), though these settings may also be characterized by somewhat slower rates of sedimentation.

Some case histories illustrate different kinds of archaeological records that can be produced by varying rates and periodicities of sedimentation and soil formation. The discussion is organized around different kinds of depositional environments. The focus is on alluvial settings, which have yielded more information than the other kinds of settings on rates of sedimentation, rates of soil formation, and the nature of associated human occupation. The Wilson-Leonard site is in a low-order tributary of a major river. Lubbock Lake is in the headwaters of a master stream, but the stratigraphy was created by paludal and eolian processes with some slopewash additions. The Shangqui area is on a large floodplain. The Cherokee, Koster, Napoleon Hollow, and Onion Portage sites are alluvial fans along the margins of large valleys. Most of the other localities are found on floodplains. The one exception is site HaRk1, in eolian sand atop an alluvial terrace in British Columbia, Canada.

Alluvial Valleys, Floodplains, and Fans

The broad array of alluvial settings produces a significant range in the rate and cyclicities of sedimentation and complex soil stratigraphic relationships within and among archaeological sites (chapters 5 and 6). The following case histories provide examples from a range of settings, including small tributaries; large, active floodplains; and alluvial fans.

The Wilson-Leonard site is on Spanish Oak Creek, a low-order tributary of the Brazos River in central Texas. The site contains a stratified record of late Paleoindian through late Prehistoric occupation (Collins, 1998a). The archaeological material at Wilson-Leonard is contained within an alluvial and colluvial sedimentary sequence about 6.5 m thick (fig. 7.3; Holliday, 1992a; Goldberg and Holliday, 1998). Different stratigraphic nomenclatures are used by Holliday (1992a) and by Goldberg and Holliday (1998), and there are also somewhat varying interpretations of the soil stratigraphic sequence. Except for a greatly refined radiocarbon chronology (Collins, 1998a), however, this discussion essentially follows Holliday (1992a) because the basic geoarchaeological conclusions have not changed. The sediments accumulated episodically and are numbered 1 through 6, oldest to youngest. Soils formed in several of the strata, and each soil was named (fig. 7.3).

Strata 5 and 6 and associated soils are of primary concern here. Alluviation at the site began >11,200 yr B.P., and by ~9500 yr B.P. about 4.5 m of sediment had accumulated (strata 1–4). Stratum 5, averaging 50 cm thick, rests on an erosional surface on stratum 4 and consists of silty overbank deposits. The Stiba soil is a moderately well-developed soil with an A-Bw or A-Btw profile formed in stratum 5. The A horizon is dark and particularly prominent and was a useful stratigraphic marker. Deposition of stratum 5 occurred ~8700 yr B.P. Pedogenesis then began and continued until about 7000 yr B.P., when stratum 5 and the associated Stiba soil were buried.

Stratum 6 is the surficial deposit at the site (fig. 7.3). The unit lies unconformably on stratum 5, is up to 1.5 m thick, and is also composed of silty overbank sediments. The Wilson-Leonard soil formed in stratum 6 and is moderately well developed, with an A-Bw, A-Bk, or A-Bt profile. The lower half of stratum

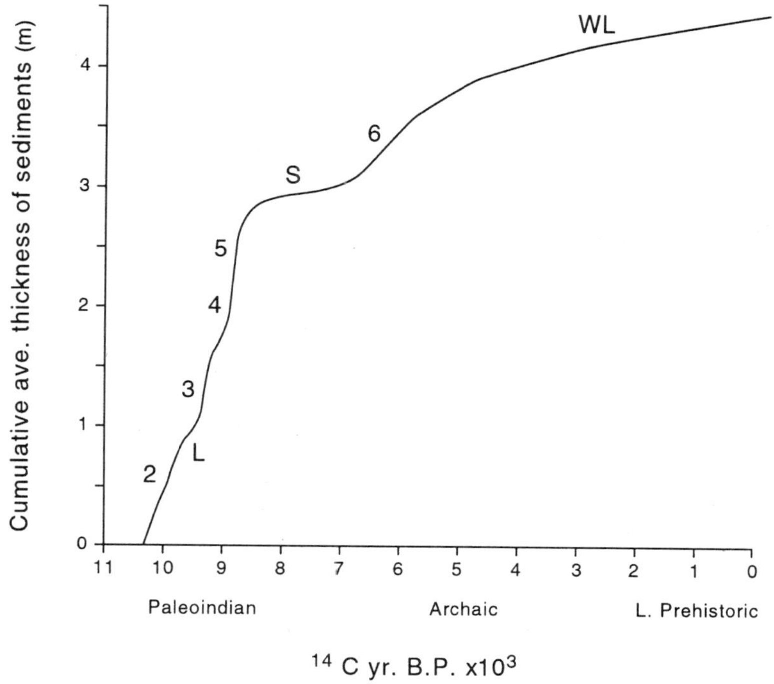

Figure 7.3 Sedimentation curve and soil formation at the Wilson-Leonard site, Texas (modified from Holliday, 1992b, fig. 3–2). Numbers refer to stratigraphic units and letters refer to soils (L = Leanne soil, not discussed; S = Stiba soil; WL = Wilson-Leonard soil).

6 was deposited between about 7000 and 4000 yr B.P., and the upper half slowly aggraded over the last 4000 radiocarbon years. Formation of the Wilson-Leonard soil probably began about 4000 yr B.P. and kept pace with the slow aggradation.

Archaeological material is common throughout the alluvium at Wilson-Leonard and is especially dense in stratum 5 (the A horizon of the Stiba soil) and stratum 6 (the A horizon of the Wilson-Leonard soil). The association of various artifact styles recovered from the upper portions of strata 5 and 6 were initially confusing in the field until the pedological relationships were established. Stratum 5 produced a wide range of late Paleoindian and Early Archaic artifact styles along with considerable occupation debris (Bousman, 1998; Collins et al., 1998). These types (e.g., Angostura, Baker, Gower, Thrall) appear to be chronologically distinct and span the late Paleoindian and early Archaic periods (Prewitt, 1981, 1983; see also Turner and Hester, 1993). The surface represented by the Stiba soil was clearly stable for almost 2000 yr, however, spanning the late-Paleoindian to early-Archaic transition (fig. 7.3). Stone tools, lithic manufacturing debris, and refuse from successive occupations throughout this time must have been deposited on the surface of the Stiba soil and were probably mixed by subsequent turbation.

Similarly, archaeological material from the late Archaic and late Prehistoric periods (Collins, 1998b; Collins et al., 1998) was found concentrated in the A horizon of the Wilson-Leonard soil, which began forming by 4000 yr B.P. The surface associated with this soil has been relatively stable for the last several thousand years, and therefore, archaeological materials from occupations spanning the late Archaic through late Prehistoric stages were concentrated and mixed in the relatively thin zone of the A horizon (fig. 7.3).

The North China Plain is an area famous for rapid aggradation over a large region in a relatively short period of time. The sedimentation and flooding have been significantly influenced by human activity, but the area nevertheless provides a good example of a stratified archaeological record within sediments that accumulated at varying rates. Jing et al. (1995) investigated the geoarchaeology of Neolithic and Bronze Age occupations at the Laonanguan site in the Shangqui area (see also discussion in chapter 9).

The alluvial deposits at the site are 10–12 m thick and are subdivided into six lithostratigraphic units (1–6) and five pedostratigraphic units (PS-1–PS-5; table 7.1; figs. 7.4 and 9.10). The soils formed in the top of each lithostratigraphic unit. Four occupation zones ("anthropogenic units" A-1–A-4) were found in the deposits, each associated with a buried soil (table 7.1, fig. 7.4). The oldest occupation zone (A-1) spans the Neolithic to the Bronze-age Zhou dynasty (from ≥7000 cal yr B.P. to 2000 cal yr B.P.). The occupation debris is essentially welded to upper PS-1, which exhibits a moderately well-developed profile (A-Bw-Bk). The strength of soil development does not seem to reflect the amount of time (>5000 yr) for genesis. Perhaps the soil was eroded. In any case, the compression of multiple occupations into a surface marked by the soil may account for the difficulties in the identification, differentiation, and interpretation of different periods of early archaeological remains (Jing et al., 1995, p. 507).

Above PS-1 and A-1 is archaeological zone A-2 (the Han through the Tang-Song Dynasties, ~2000 cal yr B.P. through the early 12th century). Zone A-2 represents successive occupations that occurred as the floodplain slowly aggraded. Soil-forming processes kept pace with the aggradation, resulting in a cumulic A-Bw profile. PS-1 and A-2 are 1.5–2.5 m thick. The occupation debris is found therefore more or less in stratigraphic sequence. Individual occupation surfaces, however, were destroyed by pedogenic (probably mostly bioturbation) processes (Jing et al., 1995, p. 496).

After the early 12th century the flood frequency and sedimentation rate increased at the Laonanguan site. The oldest sediments from this sequence, Unit 3, include at least two or three buried soils that constitute a cumulative or composite soil profile (PS-3) depending on its distance from the river channel (Jing et al., 1995, p. 493). Anthropogenic unit A-3 (Song-Yuan Dynasties, middle 12th through the 15th century) is associated with Unit 3 and PS-3. Relatively little archaeological material came from A-3, possibly because of the "dilution" effect of the rapid sedimentation.

The youngest occupation zone, A-4, is represented by the present city of Shangqui, which started in A.D. 1511. Continuous occupation debris from A-4 is associated with Units 4, 5, and 6 and PS-4 and -5. Because of the relatively high occupation density and frequent flooding, occupation zones are found through these deposits and soils, the latter being weakly expressed A-C or A-Bw profiles.

Table 7.1. Correlation of stratigraphic units, episodes of sedimentation and pedogenesis, and hydrologic regimes at Laonanguan, North China Plain (see figure 9.10)

Period	Stage	Stratigraphic unit[1] Litho	Soil	Anthro	Flooding pattern	Sedimentation/pedogenesis	Soil horizonation
Middle 19th century –present		Unit 6	Ap PS-5	A-4 A-4	Small	Very low sedimentation rate; moderate soil development	A-Bw-C
Middle 16th– middle 19th century	III	Unit 5 Unit 4	PS-4	A-4 A-4	Frequent large and catastrophic flooding	High sedimentation rate; weak soil development	A-Bw-C
Early 12th– middle 16th century		Unit 3	PS-3	A-3	Very frequent large and catastrophic flooding	Very high sedimentation rate; very weak soil development	Multiple, cumulic A-C
2000 B.P.– early 12th century	II	Unit 2	PS-2	A-2	Small (and large flooding)	Moderate sedimentation rate; week/moderate soil development	Cumulic A-Bw
Late Pleistocene– ca. 2000 B.P.	I	Unit 1	PS-1	A-1	Rare	Strong soil development; very low or negligible sedimentation	A-Bk

From Jing et al. (1995, table 1).

[1] Stratigraphic units: PS = pedostratigraphic unit; A = anthropogenic unit.

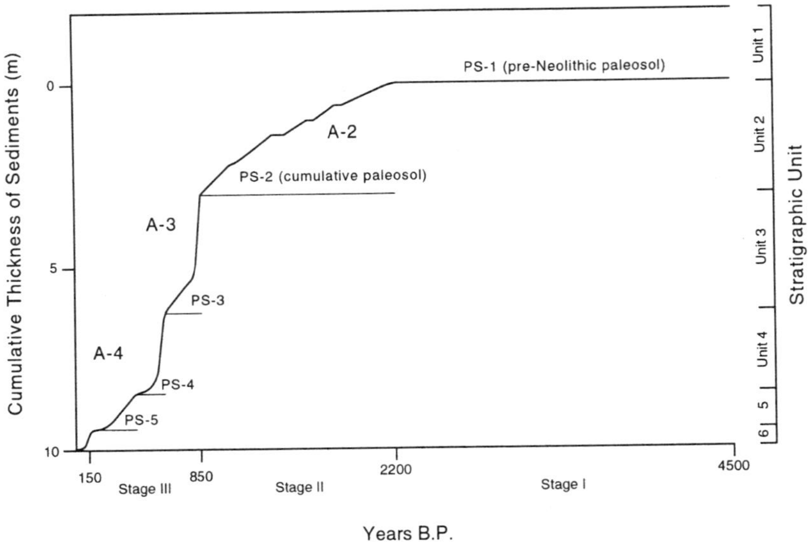

Figure 7.4 Sedimentation curve and soil formation at the Laonanguan site on the North China Plain. "PS" refers to soil stratigraphic units (see table 7.1; see also fig. 9.10) (modified from *Geoarchaeology* v. 10, pp. 481–513, fig. 9, by Z. Jing, G. Rapp, and T. Gao, © 1995, John Wiley & Sons, used by permission of John Wiley & Sons, Inc.).

The stratigraphic separation, and hence age resolution, of occupation zones at the Laonanguan site is related to both intensity of occupation and rates of sedimentation (table 7.1). When people lived on the surface during prolonged stability (A-1), the archaeological record was compressed and mixed. Occupation during rapid aggradation (A-3) produced more discrete, stratigraphically separate occupation zones.

Active, aggrading floodplains provide perhaps the best setting to find discrete, stacked occupation zones, each representing relatively brief intervals of human activity (Ferring, 1986, 2001). The archaeological layers are usually separated by layers of sterile sediment that are often delimited by buried soils. Because of the brief periods of landscape stability, the soils are usually weakly expressed and typically composed of multiple A-C soil profiles (the classic Fluvent of soil taxonomy; see chapter 6). Several sites illustrate the "dilution" effect of repeated, rapid sedimentation.

Site 450K197 along the Columbia River, Washington, provides a spectacular example of such a situation (Chatters and Hoover, 1986; fig. 7.5). Multiple occupation zones were found in association with 11–12 buried A-C soils. The alluvium accumulated over the past 2000 yr. Radiocarbon dating and flood frequency analysis showed, however, that most of the sedimentation and formation of eight or nine A horizons took place between 1020 and 1390 A.D., allowing an average of no more than 40 yr duration for each exposed floodplain surface.

A geoarchaeological situation very similar to 450K197 is reported by Stevenson (1985) for the Peace Point site along the Peace River in northeastern Alberta,

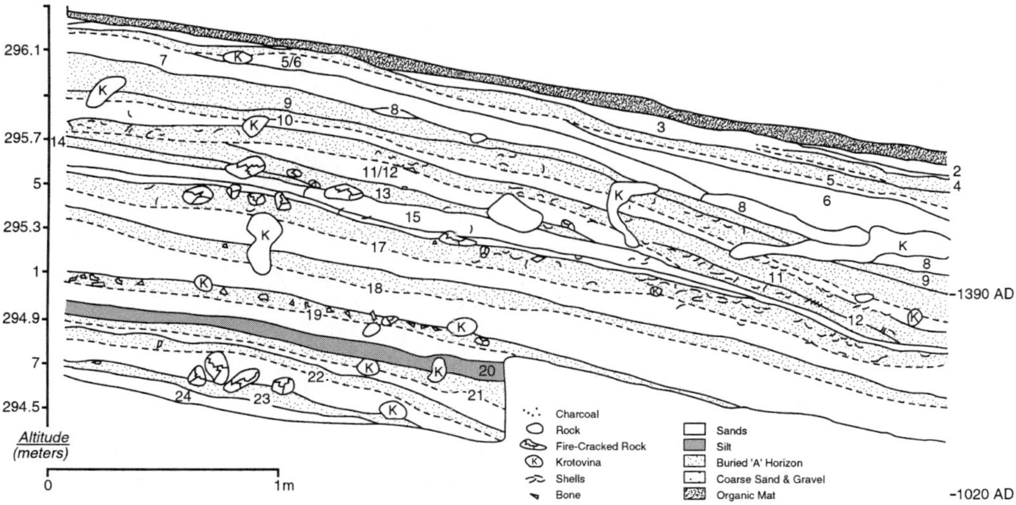

Figure 7.5 Soil stratigraphy of alluvium at site 45OK197, along the Columbia River of Washington (reprinted from *Quaternary Research*, v. 26, J. C. Chatters and K. A. Hoover, "Changing late Holocene flooding frequencies on the Columbia River, Washington," pp. 309–320, fig. 4, © 1986, with permission from Elsevier Science). The sediments dip toward the channel off to the right. The soils and sediments are indicative of rapid, episodic deposition with brief intervals of landscape stability during which occupation debris accumulated.

Canada. Floodplain deposits ~2 m thick contained over 20 buried A horizons; 18 of which yielded cultural debris (fig. 7.6). The sediments accumulated between ~2000 and ~1000 yr B.P. The dating and the weak soil development indicate that cycles of flooding and stability recurred approximately every 50 yr. This provides an unusually high-resolution record of occupation. The discreteness of each occupation zone, resulting from the rapid but low-energy burial, allowed some inferences to be made regarding the formation of each artifact assemblage.

Farther up the Peace River, at the Peace River Pulp Mill site in central Alberta, multiple archaeological horizons are reported from multiple buried soils formed in alluvium (Bobrowsky et al., 1990). Nine major and a number of minor soils were identified in the alluvium; the oldest dated to ~7300 yr B.P. (Bobrowsky et al., 1990, pp. 107–108). All of the soils produced debris from human occupation. Unlike the Peace River or 45OK197 sites, however, periods of stability were longer. The nine major buried soils each represented ~600 yr of stability, and therefore, each soil contained a palimpsest of archaeological features.

The Memorial Park site in central Pennsylvania contains 15 archaeological components in stratified alluvium (Cremeens et al., 1998). The site is in a terrace of the West Branch of the Susquehanna River. The terrace fill contains seven buried soils (fig. 7.7). The preburial horizonation of the buried soils is not clear because of postburial overprinting by a fragipan (chapter 6), but the soils appear to be weakly expressed with A-C or A-Bw profiles. The occupation zones range in age from Middle Archaic (ca. 7100 yr B.P.) to early Late Prehistoric (ca.

Figure 7.6 Multiple thin, weakly expressed buried soils (A-C horizonation) in alluvium at the Peace Point site, Alberta, Canada (from Stevenson, 1985, fig. 5; reproduced by permission of the Society for American Archaeology from *American Antiquity*, v. 50, no. 1, 1985; photo provided by and reproduced with permission of M. G. Stevenson). Discrete occupation zones were associated with most of the buried surfaces represented by the A horizons.

1000 yr B.P.). The soil morphologies and radiocarbon ages show that each cycle of sedimentation and soil formation was ~1000-yr duration or a little less (fig. 7.7). Archaeological debris was found throughout the alluvium, but the highest artifact densities were associated with the buried A horizons (fig. 7.7). Given that each landscape represented by a buried soil was stable for roughly the same amount of time, the artifact densities may well be an indication of relative occupation intensity.

Three alluvial fans in the midwestern United States provide excellent examples of the effect on soils and archaeology of varying rates of sedimentation. The Cherokee site is along the eastern margin of the Great Plains (Anderson and Semken, 1980). The site is in an alluvial fan formed along the valley wall of the Little Sioux River. The fan developed throughout the Holocene, and within its alluvial deposits are two late Paleoindian bison kills and two Archaic bison kills.

Hoyer (1980) describes and discusses the soils and stratigraphy of the Cherokee site. The stratigraphic sequence is 12–15 m thick, with 20 sedimentary units and 19 soils (figs. 7.8 and 7.9). Of the four cultural horizons identified within the deposits, three of the occupation zones were found in buried A horizons (fig. 7.8). Most of the sediments were deposited and most of the soils formed in the early to middle Holocene (10,000–4600 yr B.P.; fig. 7.9). There were no long periods of nondeposition during this time, and therefore, most of the buried soils are weakly developed (with A-C or A-Bw profiles). The strongest pedogenic expression is

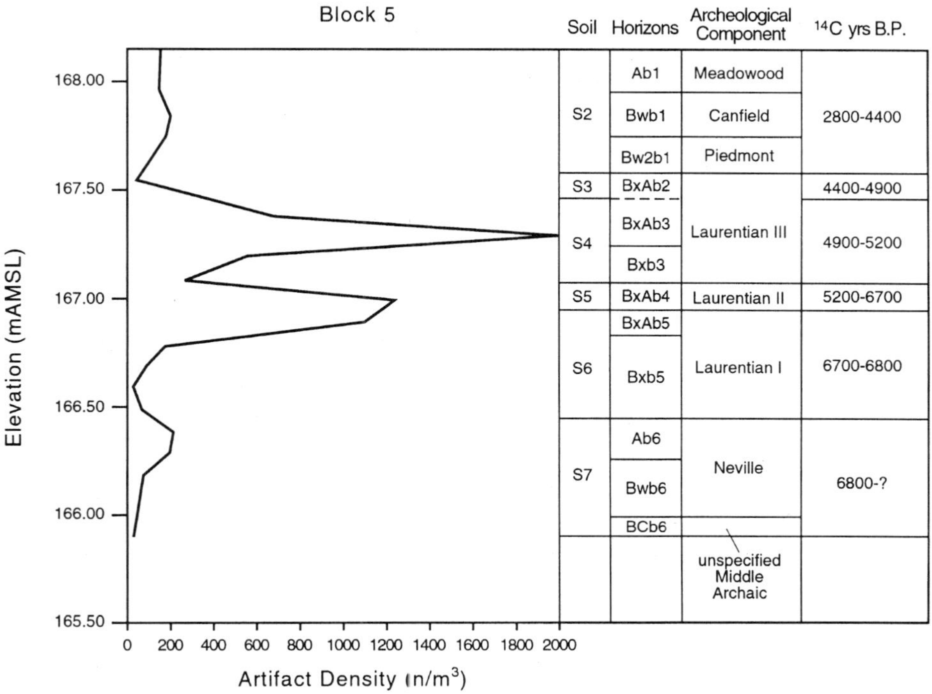

Figure 7.7 Soil stratigraphy and artifact distribution with depth at the Memorial Park site, Pennsylvania (modified from *Geoarchaeology* v. 15, pp. 339–359, fig. 8, by D. L. Cremeens, L., J. P. Hart, and R. G. Darmody, © 1998 John Wiley & Sons, used by permission of John Wiley & Sons, Inc.). Peaks in artifact density coincide with buried A horizons. The horizon nomenclature reflects the superimposition of a fragipan (Bx horizon) over several buried soils, thus welding them. Radiocarbon age ranges apply to both sedimentation and soil formation. Horizon designations are modified from the original publication following the conventions presented in appendix 1.

seen in the surface soil and in the three buried soils that formed over the last 4600 yr, resulting in a "complex, polygenetic" soil (Hoyer 1980, p. 34) with, very generally, an A-Bt profile. The cultural horizons were in the early to middle Holocene sediments and soils, below the upper soil complex, and therefore, the geologic/pedologic situation (i.e., nearly continuous sedimentation with only brief periods of stability; fig. 7.9) allowed for the preservation and recognition of the remains from several short-term occupations.

The Koster and Napoleon Hollow sites, in Illinois, are in stratified alluvial fans formed along the bluffs of the Illinois River in the Central Lowlands of the United States (fig. 6.9). They are on opposite sides of the river from one another and are about 60 km apart. They contain broadly similar stratigraphic and archaeological records (fig. 7.10; Wiant et al., 1983; Styles, 1985; Hajic, 1990). Both fans are composed of silty sediment derived from the adjacent uplands, and through much of the Holocene, both localities apparently were attractive to people for habitation. The following discussion of the soil stratigraphy, geoarchaeology, and

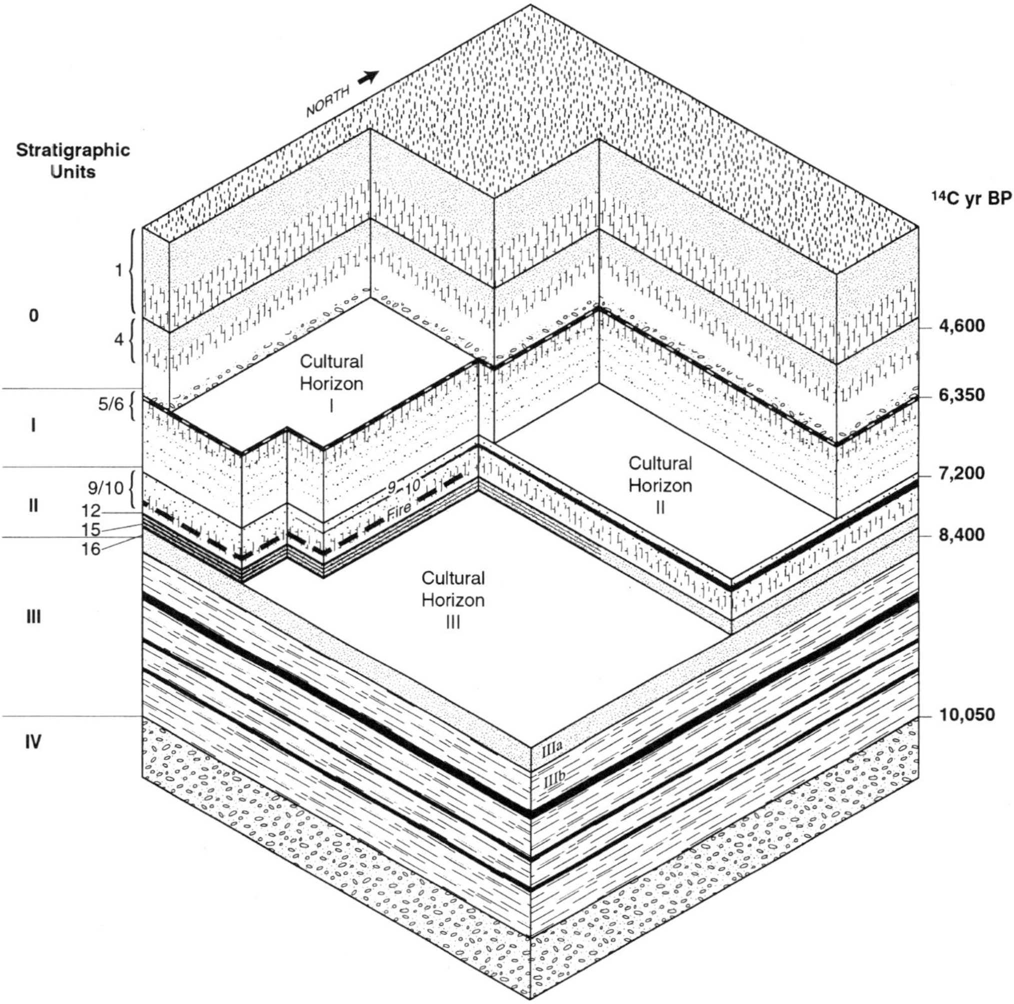

Figure 7.8 Block diagram of the Cherokee fan illustrating the geoarchaeological relationships of the soils and human occupation ("Cultural Horizons"; modified from Hoyer, 1980, fig. 2.5; reprinted from *The Cherokee Excavations*, D. C. Anderson and H. A. Semken, Eds., pp. 21–66, © 1980, with permission from Elsevier Science).

chronology of Koster is based on the monograph by Hajic (1990). The geoarchaeology of Napoleon Hollow is presented by Styles (1985). The archaeological record at both sites spans the last ~9000 radiocarbon years of the Holocene. At Koster, nine archaeologically significant buried soils (lettered a–i, oldest to youngest) were identified in the fan sediments (figs. 6.9, 7.11, and 7.12). The chronology is well constrained by 38 radiocarbon ages determined on archaeological charcoal or carbonized plant remains (Hajic, 1990, pp. 16–19, table 2). At Napoleon Hollow, buried soils were identified as components of eight geomorphic surfaces in the fan or adjacent terraces and floodplain, labeled "GS-a" to

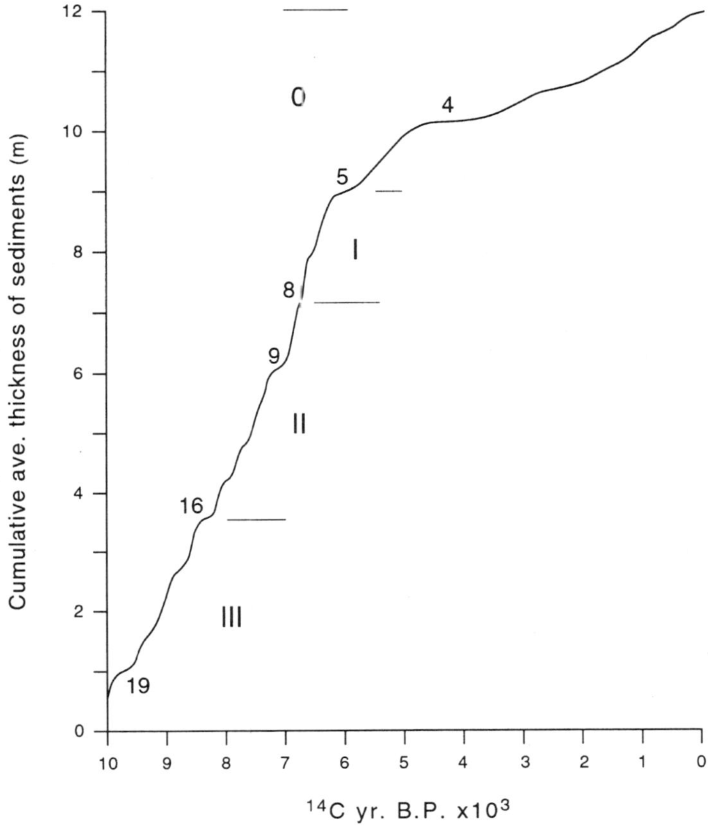

Figure 7.9 Sedimentation curve and soil formation at the Cherokee site, Iowa (modified from Holliday, 1992b, fig. 3–4). Roman numerals refer to principal stratigraphic units and Arabic numerals refer to soils (see fig. 7.8).

"GS-h" (fig. 7.10). All but one of the surfaces (GS-f) were identified in the fan, and all surfaces contained evidence of human activity. The chronostratigraphy is based on 25 radiocarbon ages determined largely on charcoal from cultural features (Styles, 1985, p. 81, table 9). Most of the aggradation that built the two fans took place in the early and middle Holocene (figs. 7.10 and 7.12), similar to the record at the Cherokee site and in other fans from the Central Lowlands (e.g., Running, 1995; Bettis, 2000).

Though the lithostratigraphy and cultural stratigraphy of both sites are similar, there are some significant differences that illustrate the effect of variable rates of sedimentation and stability/soil formation on the archaeological record. Koster has a thicker sequence of deposits (as much as 10 m) than does Napoleon Hollow (<6 m; fig. 7.10). Higher rates of sedimentation at Koster, therefore, result in more discrete buried soils and more discrete occupation zones, though most probably represent a palimpsest of activities. All buried soils at Koster are either A-C or A-Bw profile. The soil morphology and dating indicate that pedogenesis was relatively brief: the A-C soils formed for ≤500 radiocarbon yr and the A-Bw soils

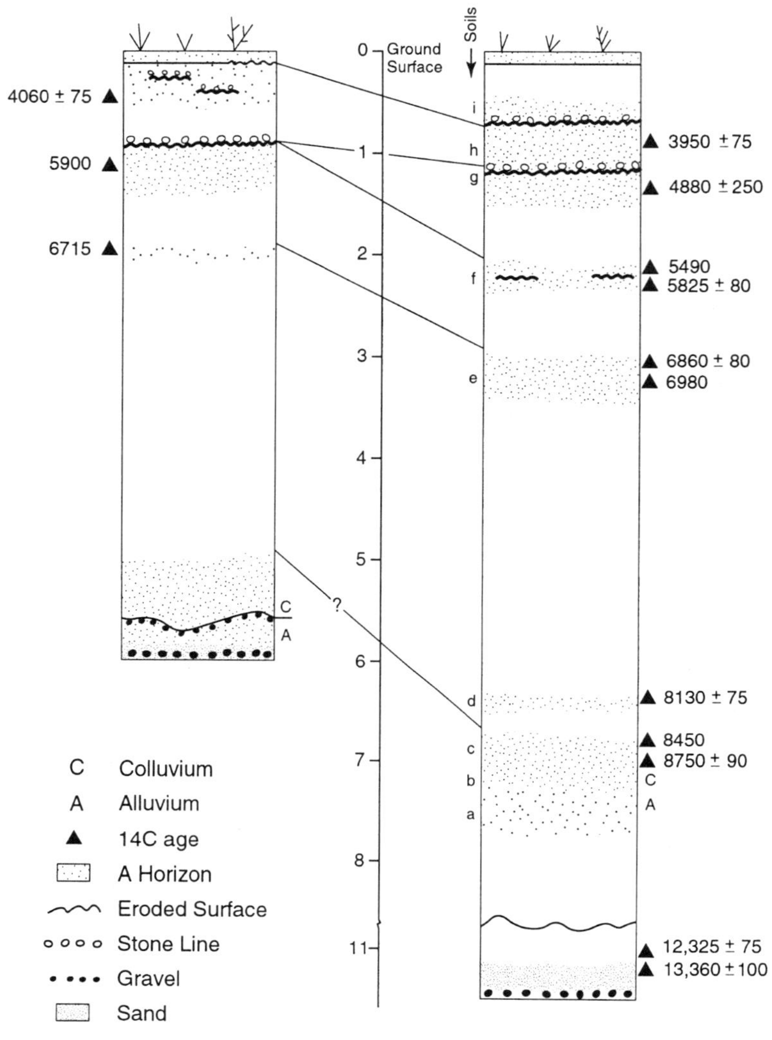

Figure 7.10 Correlation of fan stratigraphy between the Napoleon Hollow and Koster sites, Illinois, (modified from Wiant et al., 1983, fig. 8–6).

formed for 1000–1500 radiocarbon yr. The result is an archaeological record found throughout the fan; that is, not confined to the buried A horizons, but sandwiched within the fill and in the soils. In contrast, Napoleon Hollow formed under slower but more continuous deposition. The archaeologically significant soils (on GS-c, -d, -e, -g, and -h) are mostly cumulic A-Bw profiles. Relative stability and pedogenesis varied from as little as 100 yr (GS-e, A-C soil) to at most 1000 yr (GS-d; A-Bw soil). The result is that the cultural levels at Napoleon Hollow tend to be more compressed within a set of cumulic soils.

The patterns of site distribution through the alluvial fans at Koster and Napoleon Hollow are representative of sites in fans throughout much of the mid-

Figure 7.11 Buried soils (e–i) exposed on the north wall of the Koster site, Illinois ("macroblock" or main excavation block; see fig. 6.9; from Hajic, 1990, fig. 17; photo provided by M. Wiant; reproduced courtesy of the Center for American Archeology, Kampsville, Illinois). Occupation debris is apparent in soils g and f.

western United States and are well summarized by Bettis and Hajic (1995, pp. 93–94): "Archaeological deposits can be preserved throughout the sediments comprising . . . alluvial fans. Given an equal probability of occupation from one year to the next, the density of archaeological materials is likely to be greater in the upper, pedogenically altered deposits of each sediment-soil cycle because the sedimentation rate is slowest there. In lower deposits of each sediment-soil cycle the sedimentation rate [was] faster and archaeological materials are more likely to be diluted by culturally sterile sediment. Although potentially not as concentrated, the record of human activity in the rapidly aggrading parts of a sediment-soil cycle can be very well preserved and temporally discrete."

Onion Portage is a stratified archaeological site in an alluvial fan along the Kobuk River, in northwestern Alaska (Giddings, 1962; Anderson, 1970, 1988; Hamilton, 1970; Schweger, 1985). The fan formed by slow, episodic redeposition of sediment derived from bluffs along the river (fig. 7.13), and the result is a sequence of weakly expressed buried soils identical to Fluvents from floodplains. The buried soils were identified on the basis of their association with archaeological horizons or "Cultural Bands." The soils were associated with Cultural Bands 3, 5, 6, 7, and 8 (top to bottom; figs. 7.13 and 8.1). Schweger (1985, p. 130) presents the only description of the soils at Onion Portage, and only in a general

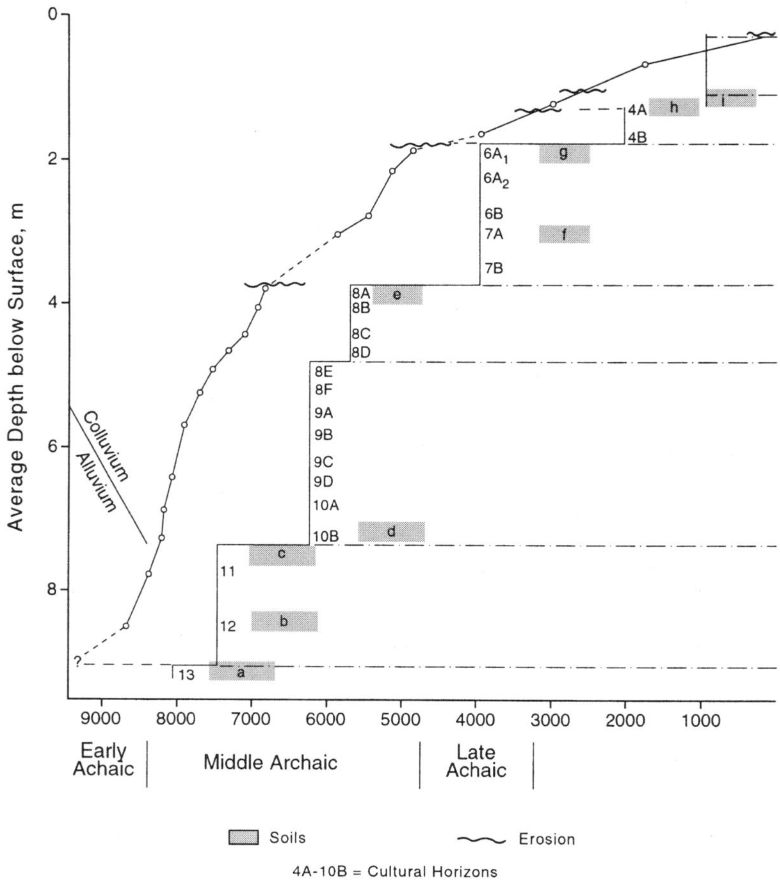

Figure 7.12 Sedimentation curve and soil formation at the Koster site (modified from Hajic, 1990, fig. 20). Buried soils are indicated by a–i; cultural horizons by 4A–13.

manner. Based on the descriptions and photos presented by Schweger (1985), the "bands" are thin (up to a few centimeters, at most), and most are composed of individual lenses of organic matter associated with archaeological features called "levels" (figs. 7.13, 8.1, and 10.10B). The levels in Band 8, associated with the Kobuk occupation, are only gleyed or mottled horizons. Band 7 has a weakly developed A-E-Bw (or Bs?) soil associated with Northern Archaic occupations. Band 3 (Choris Complex), and Bands 5 and 6 (Northern Archaic) have somewhat better expressed "podzolic" A-E-Bw (Bs?) profiles. The organic lenses associated with the occupations apparently are in the A horizons. As described by Hamilton (1970, p. 72), the main occupation area at Onion Portage is marked by a vertical succession of buried surfaces, occupied for intervals varying from about 100 to 1000 yr. The brief episodes of stability on the fan apparently provided an attractive setting for repeated occupation throughout the Holocene. The relatively rapid aggradation, however, resulted in preservation of these occupation zones as relatively discrete archaeological levels.

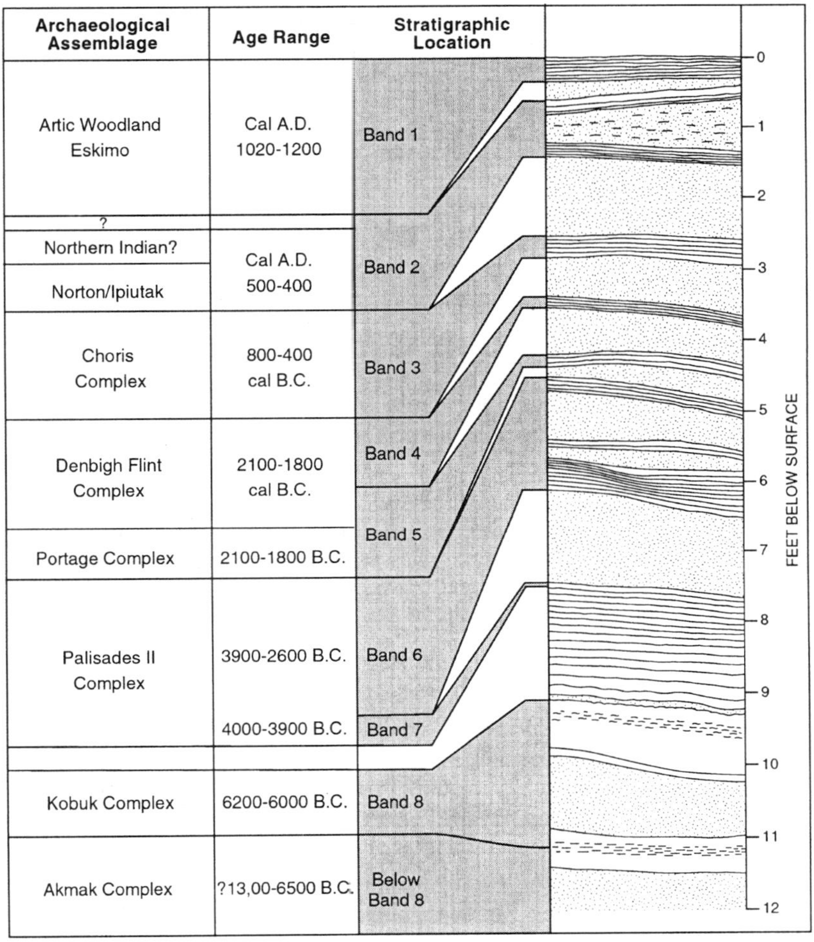

Figure 7.13 Generalized cultural chronology and soil stratigraphy at Onion Portage, Alaska (modified from Anderson, 1968, p. 28; see also figs. 8.1 and 10.10B). Individual "bands" are composed of thin lenses of organic matter that represent very weak A horizons. Age ranges for the cultural complexes are calibrated back to the Denbigh Flint Complex following Mason and Gerlach (1995).

During initial investigation of Onion Portage, artifacts of a previously unknown lithic assemblage were recognized and named Akmak. The artifacts were spread throughout Bands 5, 6, 7, and 8, indicating that they were redeposited from a source up-slope and that the source area was of some antiquity. Subsequent radiocarbon analyses showed that the lowest buried soil dated to ~8500 yr B.P.; thus, the Akmak assemblage, if redeposited, was older than that date (fig. 7.13). Soils were used to trace the archaeological stratigraphy up-slope to a narrow ridge-top setting overlooking the river. On the ridge, Akmak materials were found mixed throughout a soil exposed at the surface (except where buried by spoil from a 300-yr-old house pit). This soil exhibits a relatively well-developed podzolic profile (A-E-Bs). The original Akmak occupation was likely on the ridge on top of the soil. The artifact assemblage, therefore, represents an accumulation of occupation debris that was subsequently mixed throughout the soil by various turbation processes (chapter 10), probably cryo- and floralturbation.

Eolian Landscapes

Eolian landscapes are well known for preserving buried soils, but these settings also can be affected by wind deflation (e.g., Buck et al., 2002). Stratigraphic sequences, especially in eolian sands, therefore can be incomplete. Likewise, multiple, well-preserved archaeological sequences are not widely reported from eolian deposits, except for some loess records in the Old World. The following examples contrast the time significance of buried soils associated with archaeological records preserved in a range of eolian settings.

Site HaRk1 is in a cliff-top dune on a terrace of the Peace River, British Columbia, Canada. The geoarchaeology is described and discussed by Valentine et al. (1980). This dune is somewhat unusual in that a number of buried soils are preserved—most associated with archaeology—and there seem to be no erosional unconformities. The soil stratigraphic record is very similar to a floodplain soil sequence: Seven relatively weakly expressed buried soils (A-C, A-Bw, and A-Bk profiles) were identified in eolian sediment that accumulated through the middle and late Holocene (table 10.1, fig. 7.14). The soil development is indicative of relatively brief periods of pedogenesis, similar to a floodplain. Radiocarbon dating supports this interpretation, indicating that the cycles of sedimentation and stability each lasted as little as ~500 radiocarbon yr, and at most ~1500 radiocarbon yr (fig. 7.14). The result is a site with a sequence of occupation zones, each associated with discrete buried soils. The amount of time and occupational "compression" on each buried surface is relatively limited.

The Lubbock Lake site on the Southern High Plains of northwestern Texas and eastern New Mexico provides an example of an archaeological site with buried soils in an unusual eolian setting. The archaeology is described and discussed by Johnson (1987b) and Johnson and Holliday (1986, 1989). The site is along Yellowhouse Draw, a now-dry headwaters tributary of the Brazos River. Most of the buried soils at the site formed in eolian sediment that filled the draw through much of the Holocene (Stafford, 1981; Holliday, 1985a,b,d; Holliday and Allen, 1987; see also chapters 5, 6, and 9). The fill totals up to 8 m thick. These sediments are divided into five principal strata, numbered oldest to youngest (figs. 7.15 and 7.16; strata 2–5 are pertinent to this discussion). Soils formed in most of these strata, and they are named: The Firstview Soil formed in paludal and eolian sediment of upper stratum 2 from about 8500 to about 6400 yr B.P. The soil is weakly developed and includes a lowland, palustrine facies with an A-Cg profile and a well-drained, eolian valley-margin facies with an A-Bw profile.

Strata 3 and 4 combined are up to 4 m thick, and both include eolian and marsh sediments. These deposits accumulated between about 6400 and 4500 yr B.P., with a period of nondeposition and formation of the Yellowhouse Soil separating the two and occurring sometime between about 6000 and 5000 yr B.P. (fig. 7.16). The Lubbock Lake soil (fig. 2.1) developed in the eolian facies of stratum 4 and, where

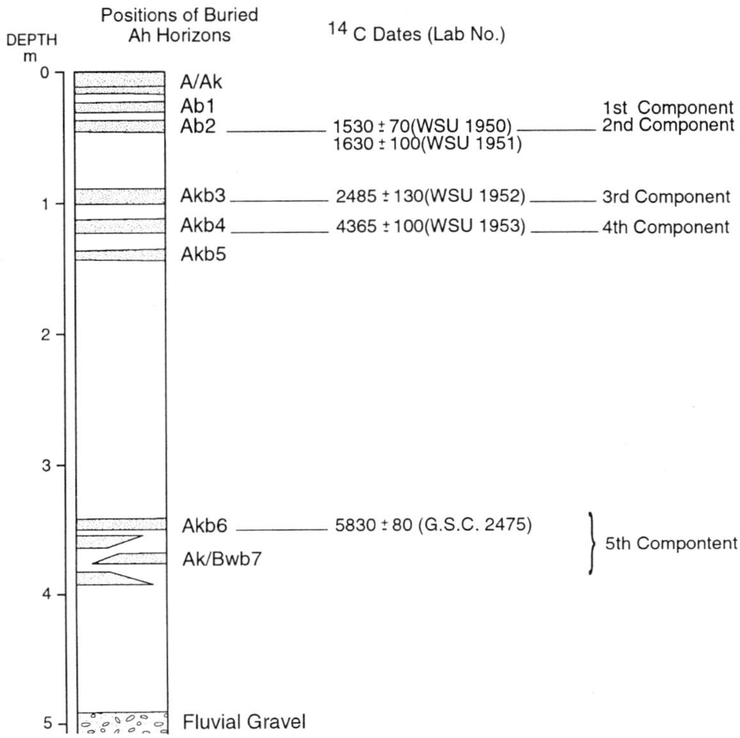

Figure 7.14 Chronology of buried Ah horizons at the Peace River eolian section and correlation with archaeological zones ("components"; based on Valentine et al., 1980, fig. 3).

not buried, is the present surface soil at the site. Within the last 1000 yr localized deposits of eolian, slopewash, and marsh sediment (stratum 5) accumulated on top of stratum 4 (figs. 2.1A and 7.16).

Nondeposition, marked by soil formation, occurred during several significant cultural periods at Lubbock Lake: the late Paleoindian to early Archaic transition during formation of the Firstview Soil, and the late Archaic, Ceramic, Protohistoric, and Historic occupations on the developing Lubbock Lake soil (fig. 7.16). Cultural debris from these occupations is compressed into the relatively thin A horizons of the respective soils (e.g., Johnson and Holliday, 1986). In some areas late Ceramic through Historic material is well stratified within stratum 5 (Johnson, 1987a), but otherwise, establishment of a cultural chronology for the latest Paleoindian to early Archaic occupations and for the late Archaic to early Ceramic occupations is virtually impossible. The stratigraphic and pedologic record at Lubbock Lake is essentially representative of that in draws throughout much of the Southern High Plains (Holliday, 1995). There was little deposition during the early and late Archaic, and therefore, conditions were not suitable for preservation of remains from discrete occupations. This is probably one of the reasons that the Archaic record of the Southern High Plains is poorly known (Johnson and Holliday, 1986).

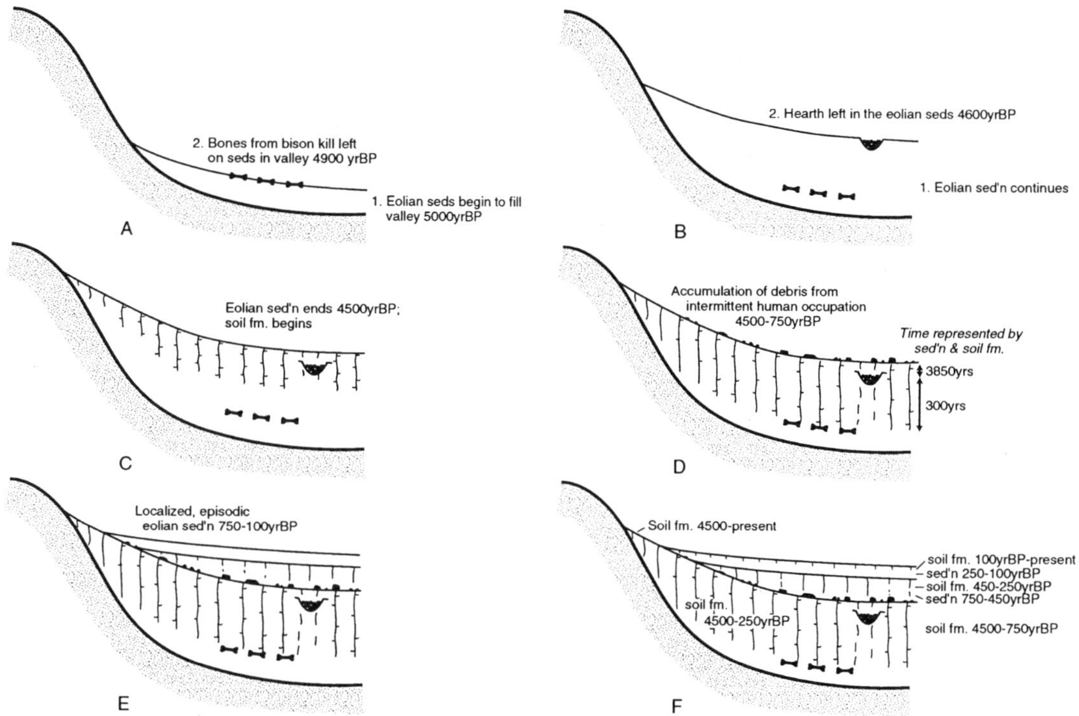

Figure 7.15 Generalized sequence of late Holocene occupation, sedimentation, and soil formation at the Lubbock Lake site, Texas (modified from Holliday, 1990b, fig. 5; reproduced with permission of the Geological Society of America). Numbers 1 and 2 in (A) and (B) represent specific events; (F) is a schematic section of the present-day soil stratigraphic and geochronologic relationships (see also figs. 2.1A and 9.3).

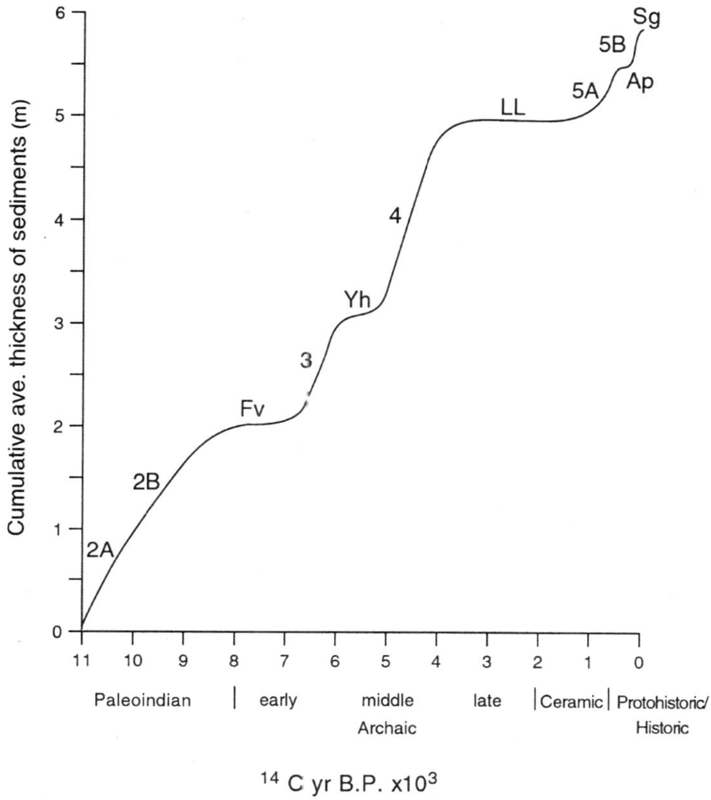

Figure 7.16 Sedimentation curve and soil formation at the Lubbock Lake site, Texas (modified from Holliday, 1992b, fig. 3–3). Numbers refer to stratigraphic units and letters refer to soils (Fv = Firstview soil, not discussed; Yh = Yellowhouse soil; LL = Lubbock Lake soil; Ap = Apache soil; Sg = Singer soil). See also figures 2.1 and 9.3.

The thick, stratified loess deposits of the Old World contain archaeological features and sites usually associated with buried soils, but the duration of pedogenesis in loess (usually at interstadial and interglacial time scales) tends to exemplify the "cumulative" or mixed character of archaeological assemblages associated with well-expressed soils. This characteristic is well shown in the loess of Tajikistan (Lazarenko, 1984; Dodonov, 1991; Bronger et al., 1995). As noted in chapter 6, many, if not most, of the buried soils are pedocomplexes, some including welded soils, usually exhibiting A-Bt-Bk, A-Bt, Bt, or Bt-Bk horizons (A horizons are not well preserved). The occupation zones in the soils usually are not in discrete layers or in association with obvious living surfaces (fig. 6.19; Lomov and Ranov, 1984, 1985; Dodonov, 1991; Ranov, 1995), the result of either occupation on a slowly aggrading surface (chapter 6) or the result of turbation from prolonged pedogenesis (chapter 10).

Soil Chronosequences and Archaeology

A "soil chronosequence" is a group of soils whose properties vary primarily as a function of age variability. The soils in a chronosequence formed in similar parent materials in similar landscape positions under a similar climate and vegetation. The one difference among the factors of soil formation is that the age of the soil is different from landscape to landscape. For example, many chronosequences are defined for the surfaces of a set of alluvial terraces whose ages vary but for which the parent materials and landscape settings are the same and the climate and biota either have changed little or the changes can be controlled for. In such a terrace chronosequence, the soils on the higher terraces typically will have more strongly developed profiles than the soils on the lower terraces because the higher terraces are older and therefore had more time for pedogenesis (fig. 7.17; e.g., Chartres, 1980; Dethier 1988; Rockwell et al., 1985; Scully and Arnold, 1981; Blum and Valastro, 1992; Blum et al., 1992; and other papers listed in table 7.2). As a corollary to this definition of a chronosequence, in an area in which there are a number of soils and in which the influence of parent material, landscape position, climate, and flora and fauna was similar among the soils and did not vary through time, the soils with stronger profile development probably formed over longer intervals of time than those that are less developed.

Several kinds of chronosequences were identified by Vreeken (1975; see also Birkeland, 1999, p. 178; fig. 7.18). The terminology is awkward, but Vreeken's categorization is conceptually useful in discussing both the applications and the pitfalls of chronosequence studies in archaeology. A "postincisive" (PtI) chronosequence is an array of soils that began forming at different times in the past and now are all exposed or were all buried at the same time. The PtI chronosequence is a set of soils on successively younger landscapes such as terraces, moraines, beach ridges, and alluvial fans. A "preincisive" (PrI) chronosequence is a group of soils that began forming at the same time, but were buried at successively more recent times, and can include an unburied soil. This kind of chronosequence would have to be produced by localized events such as human activity and some mass wasting processes (Vreeken, 1975, p. 386). These two sets of soils (PtI and PrI) coexisted at the surface during different intervals of geologic time and therefore have partial historical overlap. Because these soils either began at different times or were buried at different times, they are "partially time-transgressive."

In evaluating PtI and PrI chronosequences, the latter probably can tell us more about the evolution of soils, but the former are much more common. A PtI chronosequence implies that the soils in the group represent an evolutionary sequence, with each soil following the same genetic pathway. This is probably not the case with most chronosequences and clearly is not the case in chronosequences that began in the early Holocene or earlier because the older soils likely were subjected to significant environmental changes. Preincisive sequences probably are the best for tracing soil development because they provide time-lapse information on soil development. In particular, if environmental changes took place during such a sequence, older polygenetic soils should be explainable based on the characteristics of the soils buried earlier (Vreeken, 1975, p. 386). The PrI

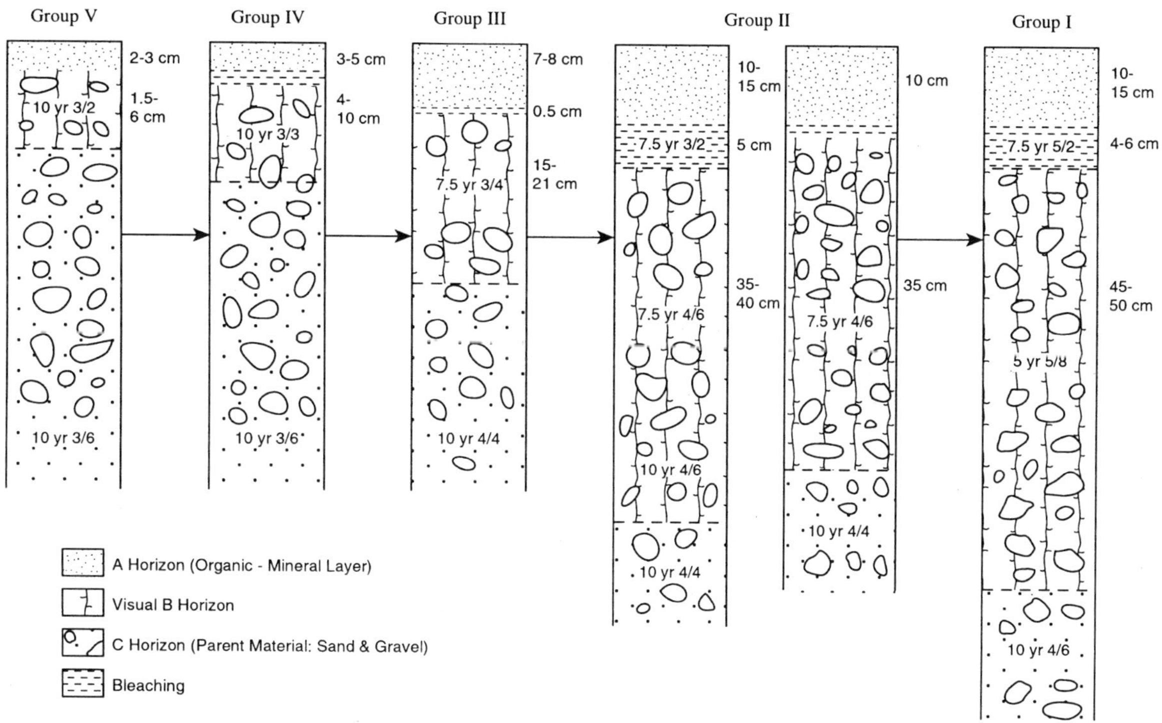

Figure 7.17 An example of soil development through time as shown by pedogenesis in a terrace chronosequence in Scotland (modified from Robertson-Rintoul, 1986, fig. 5).

Table 7.2. Examples of chronosequences

Location[1]	Lithology/setting	Rate characteristics[2]	Reference
Coastal Oregon	Sediments in marine terraces	Overall profile morphology, Fe and Al, podzolization, clay illuviation	Langley-Turnbaugh and Bockheim, 1997
Metolius River, OR	Glacial and volcanic sediments	Alteration and translocation of Fe, Al, and P	W. Scott, 1977
Rocky Mountains, CO, Sierra Nevada, CA[3]	Glacial and periglacial sediments, loess	Cambic and argillic horizon formation; overall profile morphology and thickness; rubification; alteration and neoformation of clay minerals	Shroba and Birkeland, 1983
Rio Grande; Las Cruces, NM area	Alluvium in terraces and fans; dunes	Argillic and calcic horizon formation, rubification and overall morphology in parent materials of different lithologies and in different landscape positions	Gile et al., 1981
South Platte River, CO	Alluvium in terraces	Argillic and calcic horizon formation; overall profile morphology and thickness	G. Scott, 1963; Machette, 1975; Holliday, 1987a;[4] McFaul et al., 1994[4]
Mojave Desert, CA and AZ	Alluvium in fans and terraces	Cambic, argillic and calcic horizon formation	Shlemon, 1978; Bischoff et al., 1981; Shlemon and Budinger, 1990
Eastern Mojave Desert, CA	Loess	Argillic and calcic horizon formation; overall profile morphology; rubification; alteration or translocation of Fe	McFadden et al., 1986
Ventura Basin, CA	Alluvium in fans and terraces	Argillic horizon formation; overall profile morphology and thickness; rubification	Rockwell et al., 1985
Transverse Ranges, CA	Alluvium in terraces	Alteration of Fe and clay minerals	McFadden and Hendricks, 1985
Transverse Ranges, CA	Alluvium in terraces	Argillic horizon formation; rubification; overall profile morphology; alteration or translocation of Fe	McFadden and Weldon, 1987
Sacramento Valley, CA	Alluvium in terraces	Cambic and argillic horizon formation; overall profile thickness and morphology; rubification	Busacca, 1987
Silver Lake Playa, CA	Alluvium in fans	Calcic horizon formation; overall profile morphology; rubification	Reheis et al., 1989
Black Mesa, AZ	Alluvium in terraces	Cambic, argillic, natric, and calcic horizon formation, argillic horizon thickness; overall profile morphology and thickness; rubification	Karlstrom, 1988[4]

Table 7.2. (*cont.*)

Location[1]	Lithology/setting	Rate characteristics[2]	Reference
South-central AZ	Alluvial arroyo fill	Cambic and argillic horizon formation, rubification, leaching of salts	Huckleberry, 1997
Laramie Basin, WY	Alluvium in terraces	Argillic and calcic horizon formation; overall profile morphology	Reider et al., 1974[4]
Prairie Divide, CO	Glacial deposits	Cambic and argillic horizon formation; overall profile morphology and thickness; clay mineralogy; rubification	Reider, 1975
Lubbock Lake site, TX	Eolian valley fill	Cambic, argillic and calcic horizon formation; rubification	Holliday, 1985a, 1988[4]
Dune fields, northwest TX	Eolian sand in dunes	Clay band formation	Holliday, 2001b
Ridge and Valley area, PA	Alluvium in terraces	Argillic and cambic horizon and fragipan formation; overall profile morphology; neoformation of clay minerals	Bilzi and Ciolkosz, 1977a
Susquehanna River, NY	Alluvium in terraces	Cambic horizon formation; overall profile morphology; rubification	Scully and Arnold, 1981
Susquehanna River, PA	Alluvium in terraces	Clay illuviation and argillic horizon thickness; overall profile morphology and thickness; rubification; alteration or translocation of Fe	Engel et al., 1996
Des Moines River, IA	Alluvium in terraces	Cambic and argillic horizon formation; rubification; overall profile morphology	Bettis, 1992[4]
Northern Michigan	Lacustrine sediment in lake terraces	Overall profile morphology; podzolization	Barrett and Schaetzl, 1992
Lower Ohio River Valley, OH and KY	Alluvium in terraces	Cambic and argillic horizon formation; rubification; overall profile morphology	Stafford and Creasman, 2002[4]
Ohio River Valley, IN	Alluvium in terraces	Cambic and argillic horizon formation and podzolic features; overall profile morphology and thickness	Cantin and Stafford, 1997[4]
Blue Ridge Mountains, GA	Alluvium in terraces	Clay illuviation and argillic horizon thickness; overall profile morphology and thickness; rubification; alteration or translocation of Fe	Leigh, 1996;[4] Leigh and Cable, 1997[4]

Table 7.2. (*cont.*)

Location[1]	Lithology/setting	Rate characteristics[2]	Reference
Southeast United States[5]	Marine and alluvial deposits	Argillic horizon formation; alteration and translocation of Fe	Markewich and Pavich, 1991
Rio General Valley, Costa Rica	Alluvium in fans	B horizon thickness; rubification, alteration or translocation of Al and Fe, overall profile morphology and thickness	Kesel and Spicer, 1985
Colca Valley, Andes Peru	Alluvium in terraces	Argillic and calcic horizon and duripan formation; rubification, overall morphology; mineral weathering	Sandor, 1987; Eash and Sandor, 1995
Kennet Valley, United Kingdom	Alluvium in terraces	Argillic horizon formation; mineral weathering; rubification	Chartres, 1980
Glen Feshie, United Kingdom	Alluvium in terraces	B horizon thickness; podzolization, overall profile morphology and thickness; rubification	Robertson-Rintoul, 1986
Almar River, Spain	Alluvium in terraces	Argillic horizon formation; overall profile morphology and thickness; Fe transformation	Dorronsoro and Alonso, 1994
Opatowice, Poland	Eolian sand	Clay band formation	Prusinkiewicz et al., 1998[4]
Sava River Valley, Slovenia, Yugoslavia	Alluvium in terraces	Argillic horizon formation; overall profile thickness; Fe transformation	Vidic et al., 1991
Voidomatis River, Pindus Mountains, Greece	Alluvium in terraces	Argillic horizon formation; overall profile thickness; carbonate leaching, Fe alteration, magnetic susceptibility	Woodward et al., 1994
Southern Greece[6]	Alluvium in terraces; colluvium	Argillic and calcic horizon formation, rubification, overall morphology	Van Andel, 1998[4]
Nichoria, Greece	Alluvium in terrace	Argillic and cambic horizon formation; alteration of Fe	Haidouti and Yassoglou, 1982[4]
North Caucasus, Russia	Loess and burial mounds	E and Bt horizon formation	Alexandrovskiy, 2000[4]
Harappa site, Pakistan	Alluvium in terraces	Argillic and calcic horizon formation, rubification; alteration of Fe	Pendall and Amundson, 1990[4]
Western Gangetic Plains, India	Alluvium in terraces	Argillic and calcic horizon formation; overall profile thickness	Kumar et al., 1996

Table 7.2. (*cont.*)

Location[1]	Lithology/setting	Rate characteristics[2]	Reference
Selima sand sheet, Egypt[4,7]	Sand dunes and sheets	Destruction of bedding structure, rubification	Haynes, 1982, 2001; Haynes and Johnson, 1984; McHugh et al., 1988; Haynes et al., 1993; Maxwell and Haynes, 2001
Southern Israel	Alluvium in fans and terraces	Gypsum, other salts, and argillic horizon formation; overall profile morphology	Birkeland and Gerson, 1991
New South Wales, Australia	Alluvium in terraces	Argillic horizon formation, overall profile morphology; Fe accumulation	Walker and Green, 1976

[1] Study areas are in the United States unless otherwise noted. In addition to these studies, the U.S. Geological Survey published a series of chronosequence studies from throughout the United States: Gulf and Atlantic Coastal Plain (Bulletins 1589-A and 1589-D); Horry County, SC (Bulletin 1589-B); east-central Alabama (Bulletin 1589-C); Merced, CA (Bulletin 1590-A); Ventura, CA (Bulletin 1590-B); Big Horn County, WY (Bulletin 1590-C); Carbon County, MT (Bulletin 1590-D); Colorado Front Range (Bulletin 1590-E); and Cowlitz River, WA (Bulletin 1590-F).

[2] Rate of argillic horizon formation can include data on rates of clay illuviation; Fe alteration can refer to Fe oxidation or alteration of Fe oxides.

[3] Summary of data from a number of localities.

[4] Archaeological context.

[5] Summary of data from United States Geological Survey Bulletins 1589 A, B, C, D.

[6] Data from Southern Macedonia, Larisa Basin, coastal Epiros, and the Argive Plain.

[7] See table 6.2.

sequence, however, may include properties that are the result of postburial alterations or slope position (Birkeland, 1999, p. 178). In any case, the PtI sequences have produced almost all of the chronosequence literature (as shown below and in table 7.2). This is because these kinds of landscapes are common and accessible and have long provided a means of assessing landscape evolution and reconstructing geologic history. The PrI chronosequences are much rarer in the literature, probably because they are mostly buried and because the sequence of events required to produce them is not common.

Two other kinds of chronosequences are "fully time-transgressive," in which both the beginning and the end of pedogenesis were at different times. These chronosequences are differentiated depending on whether pedogenesis has partial overlap or was completely separated on the geologic timescale (fig. 7.18): the "time-transgressive chronosequence without historical overlap" (TTwo), in which soil-parent material packages are stacked on top of the other (the "vertical chronosequence" of Stevens and Walker, 1970), and the "time-transgressive chronosequence with historical overlap" (TTw), which has characteristics of the other three. The stacked sequence of sediment and soils in the TTwo is the most familiar package of chronosequences for archaeologists and many geoarchaeologists. As described throughout this volume, these sequences are important for

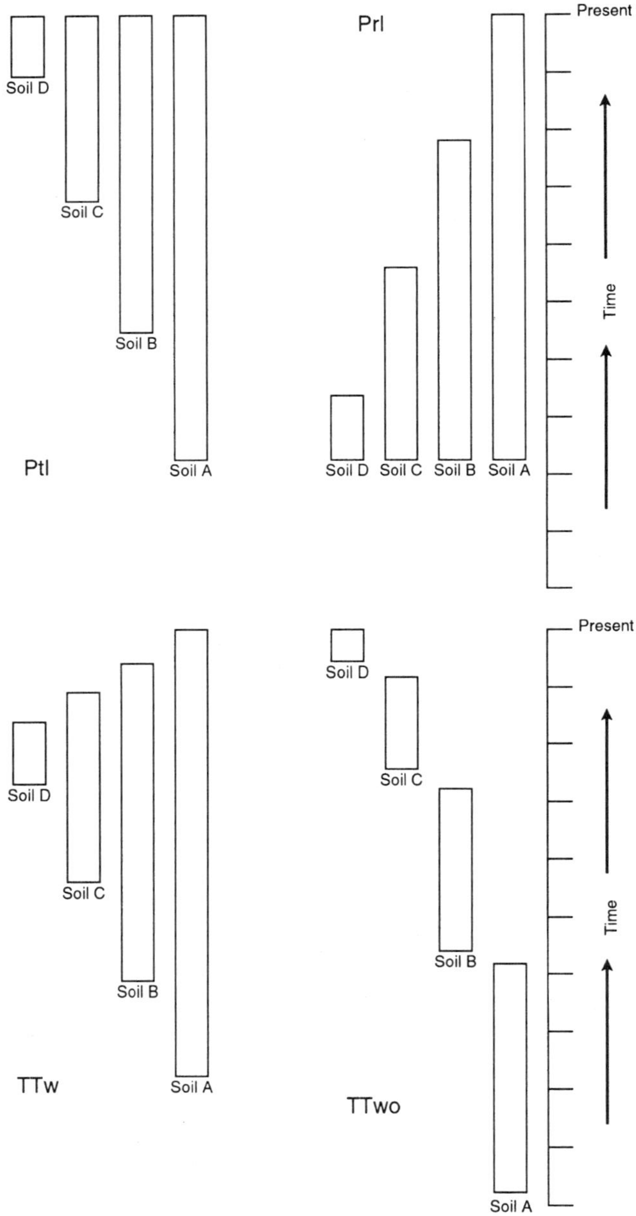

Figure 7.18 Diagrammatic illustration of the four principal kinds of chronosequences (modified from Vreeken, 1975, figs. 1, 2, 3, and 4). PtI is the postincisive sequence; PrI is the preincisive sequence; Ttw is the time-transgressive sequence with historical overlap; TTwo is the time-transgressive sequence without historical overlap.

stratigraphic correlation and differentiating sequential human activities, for deciphering landscape evolution and some site formation processes, and, to some extent, for reconstructing environmental changes. The TTwo, however, provides little insight into soil development because there is no continuity between the successive soils and their temporally separated land surfaces (Vreeken, 1975, p. 387). The TTwo, therefore, cannot be used to assess rates of soil development for dating purposes or to understand the relationship between pedogenesis and paleoenvironments. The TTw may be the most common kind of buried soil-landscape because most geomorphic processes are space- and time-transgressive, but for the same reasons these are the most problematic kinds of chronosequences (Vreeken, 1975, p. 387). Use of these sequences depends on solid data regarding the age of parent material deposition, degree of subsequent geomorphic stability, and duration of soil development.

The application of chronosequence studies in geomorphic and geoarchaeological research poses a number of interpretive problems, resulting in cautionary discussions (Vreeken, 1975; Schaetzl et al., 1994; Huggett, 1998) and wholesale rejection of their utility (Daniels and Hammer, 1992, pp. 195–202; see also Paton et al., 1995, discussed in chapter 3). Clearly there are many potential pitfalls in chronosequence applications, particularly in using them to assess rates of soil profile development (Huggett, 1998). Pedogenic thresholds (Muhs, 1984; chapter 3) can shift the direction of pedogenesis and significantly affect rates of soil-forming processes. Environmental changes (particularly climate and vegetation) are a significant factor in the development of most soils more than a few thousand years old, and thus, they likely affect rates of soil development (Huggett, 1998). There are several ways to handle or avoid this problem in soil chronosequence studies. One is to look at short-range sequences; that is, those that are no more than a few thousand years old. The environment has changed at that timescale, but in many regions the changes appear to have relatively minimal effect on chronosequences. Another approach is to work with very long chronosequences, that is, on a scale of hundreds of thousands of years or longer, whereby environmental changes are simply "averaged" as the environmental component of the factor equation. These kinds of chronosequences are relatively rare, however, and are of limited utility in archaeology except for general age estimates in Old World studies.

More generally, a steady rate of soil development (i.e., soil development in the absence of environmental fluctuations) does not necessarily have to be assumed in using soils as age indicators. Relative age estimates require only a qualitative assessment or estimate of soil age (e.g., is the soil "young" or "old"?). A soil with a well-expressed profile probably had more time to form than one with a poorly expressed profile if both formed in the same area in the same parent material and in a similar landscape setting. If degree of soil development is used to make numerical age estimates, then one must start with independent age control such as radiocarbon determinations. Then the rate of soil development is calculated relative to either a given number of years or a specified period in the geologic past (e.g., the early Holocene). With sufficient comparative data, the general effect of climate changes on soil formation, if any, should be detectable. These data can then be applied accordingly when using the soil information to determine ages for other sites in the region. Ultimately, the best argument in support of the utility of chronosequence studies is that they are shown to work in a wide variety of environments and settings (Yaalon, 1983; Birkeland, 1984, 1992, 1999; Knuepfer and McFadden, 1990).

Once a chronosequence is identified and investigated, it can be used as a dating tool in several ways. One application is simply for stratigraphic correlation of modern surfaces; that is, using the soils in a PtI chronosequence. Knowing that a soil of a given age in a particular landscape setting will exhibit a certain morphology is very useful for estimating the age of soils and, hence, landscapes throughout a region. The effect of environmental changes on the soil is not an issue here. All the investigator is doing is assessing the degree of soil development given a particular age of landscape (an environmental history). This is a common application of chronosequence studies in Quaternary stratigraphy and was at the heart of much early chronosequence research (e.g., papers in Morrison and Wright, 1967).

Several examples of the geoarchaeological application of unburied PtI chronosequences are available—most from North America. In one of the first applications of soil geomorphology in archaeological research in North America, Leighton (1934) used relative degree of soil-profile development to estimate the ages of Indian mounds in Illinois. The soils formed in the regional glacial drift in Illinois were deeply leached and had strongly developed A-Bt profiles. The drift was estimated to be 30,000–35,000 yr old (Leighton, 1934, p. 83). Soils in the mounds were weakly expressed and minimally leached. On the basis of tree-ring analysis, Leighton knew that the mounds were in excess of 400 yr old and reasoned that the mounds were probably between 2000 and 5000 yr old. This is a remarkably accurate date given the broad parameters Leighton worked within.

Bettis (1992) presents a much more recent but classic case study in the use of unburied PtI chronosequences for archaeological applications. He developed a model (expanded by Bettis and Hajic, 1995) for predicting the ages of alluvial landscapes, alluvial deposits, and associated archaeological remains for Holocene streams in the midwestern United States (table 7.3). The model is intended to aid archaeologists in adequately sampling the archaeological record, predicting site locations, and estimating the ages of sites. The distinguishing criteria for differentiating Historic, late Holocene, and early and middle Holocene alluvial deposits are easily observed properties of deposits and soils that can be recorded by archaeologists with only modest knowledge of soils and geomorphology (Bettis, 1992, p. 120). The model is based on soil morphology and on color and mottling characteristics in the subsoil alluvium. The resulting model is a good example of increased soil development as well as mottling and subsoil iron oxidation with time. A similar approach to the use of soil geomorphology for archaeological site prediction and dating has been applied successfully elsewhere in the midwestern United States (e.g., Artz, 1985; Stafford and Creasman, 2002; both discussed in chapter 4).

Along the Kansas River, soil geomorphic and Quaternary stratigraphic studies of alluvium and associated terraces (Johnson and Logan, 1990; Johnson and Martin, 1987; Mandel, 1995; Sorenson et al., 1987) provide data on soil

Table 7.3. Criteria used to group Upper Midwestern alluvial deposits into age-morphologic groups

Age-morphologic group	Bedding	Weathering zone[1]	Mottles	Surface soil (horizon sequence; B horizon color)
Early to middle Holocene	Restricted to lower part of section	O; MO; R; or U in part of some sections	Common; brown, reddish brown, or gray	A-E-Bt A-Bt; brown B horizon
Late Holocene	Usually restricted to lower part of section	Color usually 10YR hue, values 4 or less, chroma 3 or less; disseminated organic carbon imparts dark colors; may be oxidized or unoxidized but matrix colors are dark because of organic carbon content	Rare—usually not present	A-Bw; dark-colored B horizon
Historic	Present throughout section if >50 cm in thickness	O; MO; R; some sections dark colored because of high organic carbon content	Can be present or absent; brown, reddish brown, or gray	A-C; no B horizon

From Bettis (1992, table 4-3).

[1] O = oxidized, M = mottled, R = reduced, U = unoxidized.

morphology as a function of soil age (see also chapter 4 and table 4.2). The landforms most likely to contain *in situ* archaeological sites are the Newman and Holliday terraces (Johnson and Logan, 1990, pp. 275–277; also discussed in chapter 4). The surface of the Newman terrace typically exhibits a well-drained Mollisol with a cumulic A horizon (caused by occasional flooding). The associated alluvium aggraded from ~11,000 to 5000 yr B.P. The Holliday terrace usually has Entisols and some poorly drained Mollisols. The underlying alluvium was deposited from 5000 to <1000 yr B.P. The fills of both terraces contain multiple buried soils, indicative of episodic floodplain stability and soil formation during aggradation. During archaeological surveys the soils on the terrace surface provide a good first approximation of the age of the landform. Furthermore, the buried soils have the potential for containing archaeological sites.

Soil development was used to provide age estimates for the highly controversial Calico site in the Mojave Desert of California. Until the late 1970s the age of the deposits bearing putative artifacts was largely unknown, but estimates varied from 30,000 to >500,000 yr B.P. (Shlemon and Budinger, 1990, p. 306). Where in this age spectrum the true age fell would have a significant effect on the acceptability of the site as an early archaeological locality. That is, many archaeologists might be willing to accept an age of perhaps 30,000 yr B.P. given that fully modern humans were in Asia long before that time. Dates in the hundreds of thousands of years, however, would render the site much less likely to be archaeological because of the implications for prehuman hominid ancestors in the Western Hemisphere. The key feature at Calico in terms of soil dating is a well-developed soil (A-Bt-Bk profile) in alluvial fan deposits that bury the layer in question. The degree of soil development was compared to soil chronosequence data from the region (Shlemon, 1978) as well as to data from throughout the desert Southwest (e.g., Machette, 1985). The resulting age estimate of ~200,000 yr B.P. (Bischoff et al., 1981; Shlemon and Budinger, 1990) agreed with the results of uranium-series (U-series) dating of the Bk horizon (discussed later; Bischoff et al., 1981).

In the Old World, the chronosequence approach has been successfully applied in geoarchaeological contexts in Greece. Research on late Quaternary stratigraphy, geomorphology, and geoarchaeology, resulted in definition of soil "maturation stages" (MS 1–6, youngest to oldest) based on degree of color (rubification) and structural expression, and clay film and calcic horizon development (table 6.1; Van Andel, 1998). The sequence is calibrated on the basis of numerical dating methods. One geoarchaeological example of the use of the MS index is in assessing the association of Mid-Paleolithic artifacts with buried soils throughout Greece. Many artifacts are found in areas that were once internally drained basins with ponds, associated with sediments expressing little or no pedogenesis, implying brief occupation of an active landscape (Van Andel, 1998, p. 383). Other artifacts are found within "very mature Bt horizons," indicating occupation of a slowly aggrading but otherwise stable landscape lacking the attractions offered by the wetlands (Van Andel, 1998, p. 383).

In New Zealand, Jones (1990) used soil morphology to provide approximate maximum ages for pre-European settlements on river terraces along the east

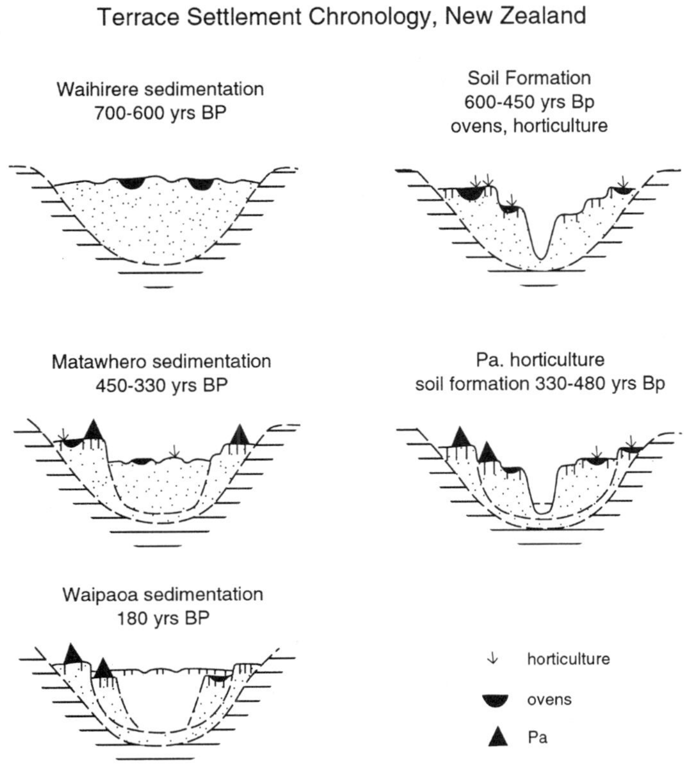

Figure 7.19 Schematic illustration of human settlement and development of a late Holocene chronosequence in valleys of the North Island of New Zealand (modified from *Geoarchaeology* v. 5, pp. 255–273, fig. 2, by K. L. Jones, © 1990 John Wiley & Sons, used by permission of John Wiley & Sons, Inc.).

coast of the North Island (fig. 7.19). New Zealand was first populated by Polynesians about 1000 yr ago. The first European contact took place in 1769. Thus, the precontact archaeological record is relatively brief. It has also proven difficult to date because of the lack of radiocarbon age-control and absence of time-diagnostic artifacts. During the precontact occupation, however, rivers in the study area underwent several cycle of erosion and aggradation, producing two alluvial terraces and the modern floodplain. Soils formed on each terrace as aggradation waned, resulting in a chronosequence. The degree of soil development on each terrace provided a means of correlating surfaces and, therefore, provided relative ages for archaeological sites on the terraces. Moreover, the radiocarbon dating of the alluvium and some archaeological sites resting on soils formed in the alluvium provided a means of assigning age estimates to the soils. Archaeologists can now look at the soil under sites on terraces and use the soil morphology plus some stratigraphic characteristics to provide a maximum age for the occupation. The Waihirere surface, formed by ~600 yr B.P., has a soil with a thin (<18 cm thick) A horizon over a Bw horizon. The Matawhero surface, formed by 350 yr B.P., is lower and subject to some flooding, producing a soil with a thick (>18 cm) A horizon or multiple buried A horizons. The formation of either a cumulic or multistory profile on the Matawhero surface depends on proximity to the stream channel. The modern floodplain is characterized by multiple, thin (<18 cm) buried A horizons.

Another approach to using chronosequences is to assess rates of soil profile development or development of specific pedologic characteristics to understand soil genesis and to use soil characteristics as relative, and in some cases numerical, age indicators. This aspect of chronosequence studies is a more direct application of the state factor approach in soil geomorphology and is the basis for much contemporary soil geomorphic research (Knuepfer and McFadden, 1990; Birkeland, 1992, 1999). The development of a variety of pedologic features is known to be time-dependent (table 7.4). If rates of soil formation can be determined, this information can be carried to other sites in similar situations in the region to provide age estimates for both natural and cultural deposits.

Archaeological investigations at both site-specific and regional scales are especially useful settings for studying soil chronosequences and for establishing rates of soil development because of the age control often available (e.g., Foss and Segovia, 1984; Foss and Collins, 1987; Bettis, 1992; Foss et al., 1995). The archaeological sites themselves have been the focus of some chronosequence studies; several investigators examined soil development in (and postburial alteration under) Late Prehistoric mounds constructed throughout much of the eastern United States (Parsons et al., 1962; Bettis, 1988; Kolb et al., 1990; Ammons et al., 1992; Cremeens, 1995) and in the Old World (Wilkinson, 1990; Alexandrovskiy, 2000). Historical records have also proven very useful in documenting rates of soil development over the last few hundred to a thousand years, particularly in Europe. Such records include construction activity (Hallberg et al., 1978; Gubin, 1984; Alexandrovskaya and Alexandrovskiy, 2000), moraine recessions (Jenny, 1941, pp. 37–38; Stevenson, 1969; James, 1988), and land drainage (Jenny, 1941, pp. 39–44). The ability to date deposits and soils and to produce high-resolution stratigraphic chronologies using time-diagnostic artifacts or numerous radiocarbon ages or some other form of numerical age control is another unique aspect of archaeological research that often is the envy of geoscientists.

There are only a few examples of geoarchaeological applications of chronosequences based on archaeological or historical data. The Lubbock Lake site provided an exceptional opportunity to date rates of soil development and then apply the data in other geoarchaeological studies. The late Holocene stratigraphic record includes examples of PrI, PtI, and TTwo chronosequences (Holliday, 1985d, 1988). Rates and characteristic features of soil development were established by combining field and laboratory data with the well-dated geochronology (Holliday, 1988). A variety of pedogenic characteristics are time-dependent at a Holocene time scale. Mollic epipedons form in about 100 yr, and steady-state conditions of organic carbon content are attained in ~1000 yr. Calcic horizons form in 200 yr. Minimally developed argillic horizons (20 cm thick with a 3% increase in illuvial clay content between the A and Bt horizon) form in 450 yr, the result of rapid infiltration of carbonate and clay derived from aerosolic dust, but argillic

Table 7.4. Pedogenic features dated in soil chronosequence studies

Soil characteristic	References
Soil indices	Walker and Green, 1976; Bilzi and Ciolkosz, 1977a,[1] b; Harden, 1982; Harden and Taylor, 1983;[1] Birkeland, 1984,[2] 1999;[2] McFadden et al., 1986;[1] Schaetzl and Mokma, 1988; Reheis et al., 1989; Dorronsoro and Alonso, 1994; Leigh and Cable, 1997[1,3] (*See also* chapter 2)
Profile thickness	Machette, 1975; Reider, 1975; Foss and Segovia, 1984;[2,3] Birkeland, 1984,[2] 1999;[2] Kesel and Spicer, 1985; Rockwell et al., 1985; Robertson-Rintoul, 1986; Karlstrom, 1988; Dorronsoro and Alonso, 1994; Eash and Sandor, 1995; Leigh, 1996;[3] Engel et al., 1996; Kumar et al., 1996; Cantin and Stafford, 1997; Leigh and Cable, 1997[3]
Cambic horizon	Reider, 1975; Scully and Arnold, 1981; Haidouti and Yassoglou, 1982;[3] Foss and Segovia, 1984;[2,3] Holliday, 1988;[3] Bettis, 1992;[3] Foss et al., 1995, table 1-2;[3] Kumar et al., 1996; Cantin and Stafford, 1997; Stafford and Creasman, 2001
Illuvial clay content, argillic horizon and rubification	Reider, 1975; Gile et al., 1981; Haidouti and Yassoglou, 1982;[3] Foss and Segovia, 1984;[2,3] Birkeland, 1984,[2] 1999;[2] Rockwell et al., 1985; McFadden et al., 1986; Holliday, 1988,[3] 2001b;[3] Karlstrom, 1988; Reheis et al., 1989; Bettis, 1992;[3] Dorronsoro and Alonso, 1994; Woodward et al., 1994; Eash and Sandor, 1995; Foss et al., 1995 table 1–2;[2,3] Engel et al., 1996; Kumar et al., 1996; Cantin and Stafford, 1997; Langley-Turnbaugh and Bockheim, 1997; Leigh and Cable, 1997;[3] Alexandrovskiy, 2000[3]
Clay bands (lamellae)	Foss and Segovia, 1984;[2,3] Gile, 1979, 1985; Prusinkiewicz et al., 1998;[3] Holliday, 2001b[3]
$CaCO_3$ content, calcic horizon	Machette, 1975, 1985;[2] Gile et al., 1981; Birkeland, 1984,[2] 1999;[2] McFadden et al., 1986; Holliday, 1988;[3] Reheis et al., 1989; Wilkinson, 1990;[3] Eash and Sandor, 1995; Kumar et al., 1996; Van Andel, 1998; Tandon and Kumar, 1999[2]
Podzolic characteristics (E, Bs, Bh)	Barrett and Schaetzl, 1992; Callum, 1995; Langley-Turnbaugh and Bockheim, 1997; Anderton, 1999; Alexandrovskiy, 2000[3]
Fragipan	Bilzi and Ciolkosz, 1977a; Foss et al., 1995;[3] Segovia, 1997;[3] Cremeens et al., 1998[3]
Pedogenesis in tephra	Ugolini and Zasoski, 1979;[2] Lowe, 1986; Neal and Paintin, 1986; Shoji et al., 1993, pp. 62–63[2]
Alteration or neoformation of clay minerals	Reider, 1975; Shroba and Birkeland, 1983; Birkeland, 1984,[2] 1999;[2] McFadden and Hendricks, 1985
Alteration or translocation of Fe, Al, and P	Walker and Syers, 1976; W. Scott, 1977; Birkeland, 1984,[2] 1999;[2] Kesel and Spicer, 1985; McFadden et al., 1986; Dorronsoro and Alonso, 1994; Engel et al., 1996; Leigh, 1996; Langley-Turnbaugh and Bockheim, 1997; Leigh and Cable, 1997[3]
Magnetic susceptibility	An et al., 1990, 1991; Maher and Thompson, 1991, 1992, 1995, 1999; Woodward et al., 1994; Liu et al., 1995; Evans and Heller, 2001; Singer et al., 1992

[1] Application.
[2] Summary/review.
[3] Archaeological context.

horizons 100 cm thick with a 4%–6% increase in illuvial clay can form in 3500 yr. Overall horizonation is also better expressed with time, and B horizon hues can redden from 10YR to 5YR in 3500 yr, though most soils have 7.5YR B horizons. The development of some of these specific characteristics of B horizons also varies with slope position of the soil (see below). The strongest development of argillic and calcic horizons is along the valley axis of Yellowhouse Draw where the soils were on a flat landscape. The weakest expression is along the valley margins where the soils slope toward the valley axis. These data compare favorably with stratigraphic and pedologic information from other draw localities on the Southern High Plains (Holliday 1985c,e, 1995). These results indicate, therefore, that the degree of development of various pedologic features such as argillic and calcic horizons can be used to correlate and date deposits at other draw localities in the region.

In the Upper Peninsula of Michigan, Anderton (1999) used the development of the Spodic horizon along with other pedogenetic and weathering characteristics to define a "soil-artifact context model" for assessing the relative age of archaeological sites on paleoshorelines of Lake Superior and Lake Michigan (fig. 7.20). Numerical age control is very difficult to come by because dateable carbon is rarely preserved and few time-diagnostic artifacts are present. In addition, the shorelines contain Archaic sites, which were occupied as the shorelines formed (~5000–2000 yr P), and Woodland sites, which postdate shoreline formation (~2000–1000 yr B.P.). Archaeologists, therefore, had a very difficult time sorting out the palimpsest of occupations on the raised shorelines of the region.

Anderton conducted fieldwork at eight archaeological sites on shorelines plus soil geomorphic research along these landforms to develop his model for relative dating of sites (fig. 7.20). The model was based on the principle that sites contemporaneous with or slightly younger than shoreline formation ("correlative sites") were affected by the same pedogenic and other weathering processes that modified the littoral deposits. Significantly younger sites on these landforms ("non-correlative sites") probably were not significantly modified by these processes. Sites that were correlative with shoreline development (Archaic) have artifacts that are deeper within the soil profile, soil horizon boundaries that cut across midden concentrations, and some artifacts that are iron stained from Spodic horizon development. Sites that are not correlative with the time of shoreline development (Woodland) have artifacts that are at or very near the ground surface, and archaeological features, if present, will cut across soil horizons and will tend not to be iron stained.

In Greece the calibration of rates of Bt and Bk formation using numerical dating methods allows archaeologists to use the soils and soil stratigraphic relationships to estimate the ages of Paleolithic artifacts found in and between buried soils (Runnels, 1988; Runnels and Van Andel, 1993a; Van Andel, 1998; Zangger, 1993). At the Kokkinopilos site in the Epirus region, for example, soil development was used to date a significant artifact find (Runnels and Van Andel, 1993b). A hand ax was discovered embedded in red sediments, ~15 m below a well-expressed Bt horizon, found just below uneroded surfaces in the region. The Bt horizon contains Middle Paleolithic artifacts, and on the basis of the archaeo-

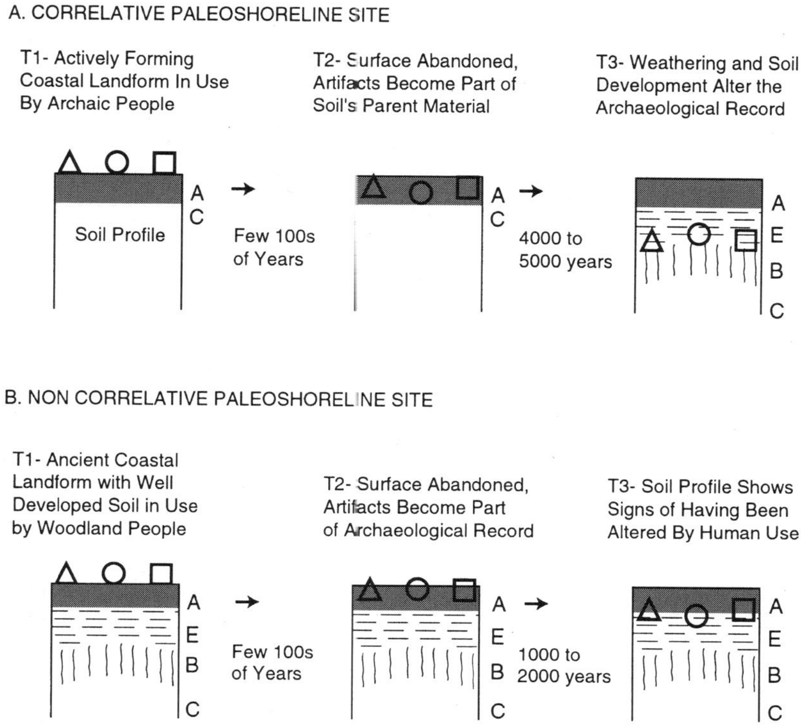

Figure 7.20 The soil-artifact context model in "correlative" and "noncorrelative" sites, Upper Peninsula of Michigan (modified from *Geoarchaeology* v. 14, pp. 265–288, fig. 3, by J. B. Anderton, © 1999 John Wiley & Sons, used by permission of John Wiley & Sons, Inc.).

logical inclusions, morphology, and stratigraphic position, it was correlated with Middle Paleolithic soils elsewhere in Greece. The archaeology, numerical age control from other sites, and sea level correlations provide a minimum age for the beginning of soil formation at approximately 60,000 yr B.P. These data, combined with the thickness of the sediments between the soil and the artifact and estimates of sedimentation rates, indicate that the hand ax is a Lower Paleolithic specimen. Lower Paleolithic sites are very rare in Greece, but subsequent research shows that the age estimate based in part on soil morphology was roughly correct (Zhou et al., 2000).

In making comparisons of soils from site to site for dating purposes, the soils being compared must be in similar landscape positions and of similar parent material. These factors can exert a strong influence on soil morphology even in very young soils (see chapters 5, 6, and 9). At Lubbock Lake, for example, the best-developed soil in the sequence, the Lubbock Lake soil, varies morphologically depending on landscape position and parent material (i.e., it exhibits soil facies, fig. 2.1; chapter 5; see also chapter 9). A-Bw or A-Btw soils are common in the sandier, valley margin settings where the soil slopes toward the valley axis; more strongly expressed A-Bt-Bk soils are found in finer-textured valley axis settings where the landscape was flat (fig. 9.3; Holliday, 1985d, 1988). Regional studies of valley fill (Holliday, 1995) show that this relationship is common. Therefore, in making comparisons from site to site, soils in valley axis settings must be compared only to other soils in valley axis settings. Furthermore, other considerations, such as stratigraphic relationships and archaeology, must be taken in account. Soils similar in morphology can form at different periods in time.

Radiometric Dating of Soils

The dating of chronosequences and calibration of rates of soil formation are ultimately dependent on numerical age estimates of soils. A variety of analytical methods have been applied to dating soils (e.g., Evans, 1995; Birkeland, 1999, pp. 137–140). Some of this dating is based on determining the age of deposition of soil parent material; that is, by dating the sediments themselves. For example, luminescence techniques can be applied to dating sand grains of an eolian deposit, or the radiocarbon method can be used to date wood fragments in alluvium. Another approach is the dating of pedogenic features themselves. Several soil materials, principally soil organic matter and calcium carbonate, are directly dateable by isotopic methods (following the terminology of Colman et al., 1987, and Colman and Pierce, 2000). In archaeology, pedology, soil geomorphology, and Quaternary stratigraphy, by far the best known and most widely used radioisotopic dating method is radiocarbon dating. Charcoal, wood, and plant macrofossils tend to be preferred materials for radiocarbon dating, but pedogenic organic carbon and calcium carbonate are dateable and with some success. Pedogenic $CaCO_3$ is also dateable by the Uranium-series disequilibrium method. The genesis of pedologic organic matter and calcium carbonate impose some important considerations on the interpretation of the resulting ages. The final two subsections of this chapter, therefore, are summary discussions of radiocarbon dating of soil organic matter and calcium carbonate, followed by comments on the U-series dating of soil calcium carbonate. The emphasis is on cautionary notes concerning the interpretations of the results. For in-depth discussion of radiocarbon, a vast literature is available, particularly in archaeological contexts. Good summaries are found in Taylor (1987) and Taylor et al. (1992). Reviews of the U-series method are presented by Blackwell and Schwarcz (1995) and Ku (2000), and a succinct summary is provided by Bradley (1999, pp. 76–80).

Radiocarbon Dating

Organic matter and calcium carbonate accumulate in soils during pedogenesis. Except for air and water, soil organic matter (SOM) is the most common constituent added to soils. SOM initially accumulates at the surface of a soil, forming an A horizon and, under some conditions, an O horizon. SOM is also translocated lower in the soil in some soil-forming environments, producing the Bh horizon, for example. Calcium carbonate is a common soil constituent, but the pedogenic accumulation of carbonate tends to occur in nonleaching environments such as arid and semiarid, and some subhumid, settings. There, Bk and calcic horizons are nearly ubiquitous soil features. Many investigators have applied the radiocarbon method to both SOM and inorganic soil carbonate since the early days of method, but with mixed results. Under the right conditions and with appropriate caution in sampling and interpretation, however, radiocarbon dating of SOM and calcium carbonate can help in establishing the broad outlines of a soil chronology and a site chronology.

The organic matter in an A horizon comes from plant and animal debris left on and just below the soil surface throughout pedogenesis. For a surface soil this includes current and past carbon inputs. For a buried soil the SOM includes the carbon added throughout pedogenesis until burial. Organic matter in soils includes (from Matthews, 1985, p. 271) living flora and fauna, dead and partly decomposed biomass, relatively simple organic compounds (e.g., cellulose and lignin), and humus substances, which have a complex chemical structure and form during humification by biochemical synthesis from the products of decomposition. Over time during pedogenesis, the SOM undergoes "turnover": processes such as decomposition, formation of humus, and translocation that operate at varying rates in surface soils (Costin and Polach, 1969; Paul, 1969; Stevenson, 1969; Scharpenseel, 1971a; Goh, 1980; Matthews, 1985; Stein, 1992c; Wang et al., 1996). The dead biomass and simple compounds decompose rapidly, usually within a few years. The humus substances, however, can persist for hundreds to thousands of years. Humus, therefore, is the compound that is dated by radiocarbon.

A radiocarbon determination on SOM of a surface soil therefore dates a mixture of young and old carbon and produces an "apparent mean residence time" (AMRT; Campbell et al., 1967b). In general, and barring contamination by older carbon, the AMRT is always younger than the time that pedogenesis began (Matthews, 1985). In dating a buried soil, if the A horizon was completely cut off from carbon inputs, then a radiocarbon age on that A horizon will reflect the AMRT of the soil prior to burial and the time elapsed since burial (Scharpenseel, 1971b). As discussed below, a number of factors will influence the specific AMRT produced by a soil. Most radiocarbon dating of SOM from soils in geoarchaeological contexts has been applied to buried soils. In any case, when dating SOM in an A horizon, the resulting radiocarbon determinations will not provide an age for the beginning, the duration, or the end of pedogenesis (discussed below). Barring contamination, the SOM in a buried A horizon will provide a maximum age for burial of the soil and a minimum age for the end of deposition of that soil's parent material.

The applicability and interpretation of radiocarbon ages from buried A horizons depends on a number of factors. An important consideration is what fraction or fractions of the SOM are dated. Most radiocarbon labs date the NaOH-soluble (humic acid), NaOH-insoluble (humin or residue) fractions, or the total ("bulk") organic carbon (i.e., humus) after an acid treatment to decalcify the sample (Campbell et al., 1967a; Costin and Polach, 1969; Martin and Johnson, 1995). Many investigators have shown that these various components can produce different ages for the same sample. Humic acids tend to be more mobile than humin and traditionally are thought to produce ages younger than those produced by humin (Campbell et al., 1967a; Matthews, 1985), but a large literature now shows that this relationship is not consistent (see the discussion and summary by Martin and Johnson, 1995). In a study of dates from seven buried soils in the central Great Plains, Martin and Johnson (1995) suggest that in late Holocene soils the total decalcified bulk sample and the humin fraction tend to be older, but in late Pleistocene soils no fraction was consistently the oldest. They attribute the variability to contamination (discussed later).

A larger data set for comparing radiocarbon ages on SOM is now available from the Southern High Plains. One hundred pairs of dates on bulk decalcified SOM and humic acids were determined on organic-rich muds from fills in playa basins (fig. 7.21, left panel) and twenty-six pairs of dates on humic acid and humin were determined on soils from eolian sand (fig. 7.21, right panel). Some of the comparative data seem to be at odds with the conclusions of other investigators, though no consistent trends are apparent. The ages determined on the humic acid fraction tend to be older than the bulk sediment (fig. 7.21, left panel). This trend is strongest for Holocene dates from playas. For the late Pleistocene samples, the trend shifts somewhat, with the bulk sediments producing older ages. In contrast, the humin or residue ages are generally older than the humic acid ages (fig. 7.21, right panel). In both cases, the humic acid ages tend to become younger in the older, especially late Pleistocene samples.

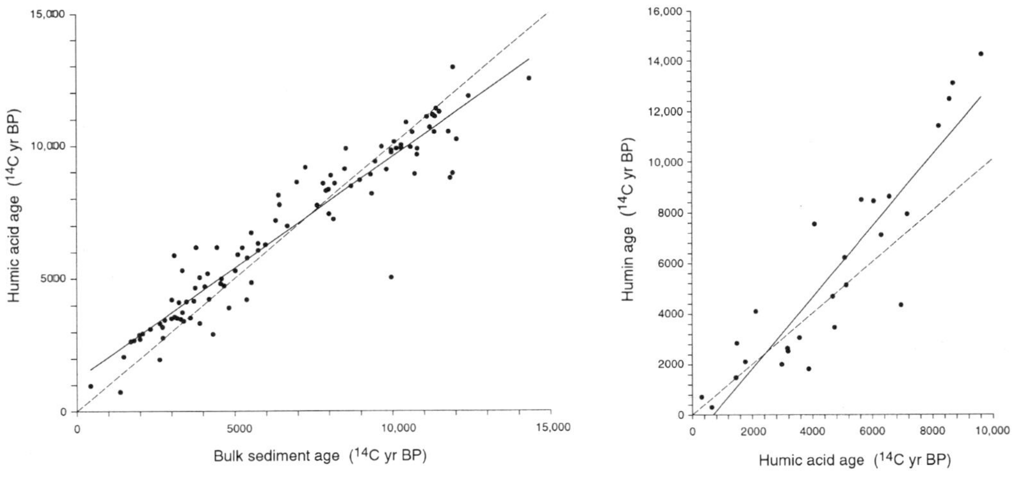

Figure 7.21 Comparisons of radiocarbon results for organic-rich sediments and buried soils on the Southern High Plains (the dashed line has a slope of 1; that is, it represents a 1:1 correspondence). (Left panel) Results of dating bulk sediment and humic acid from playa muds. (Right panel) Results of dating humic acids and humin from buried A horizons in dunes.

Radiocarbon ages from soils generally are youngest at the surface and increase in age with depth (Matthews, 1985; Wang et al., 1996). Organic carbon in surface horizons is dominated by younger carbon, stored as both partially decomposed organic tissues and stabilized carbon held in cell walls and microbial structures. Floral and faunal mixing of this young organic carbon with organic carbon from earlier stages of pedogenesis yields a radiocarbon age that is older than the age of the most recent accumulation of organic matter. The age of the organic carbon in surface horizons reflects the mean annual rate of turnover of the easily biodegradable organic matter, also described as the rate of rejuvenation of the soil organic matter. In many settings the radiocarbon age of the uppermost A horizon may be only a few decades old, and therefore, dating of the surface of a buried A horizon can provide a close approximation of the age of burial (barring contamination; Holliday et al., 1983; Matthews, 1985; Martin and Johnson, 1995). The oldest carbon in the soil, stored in the more persistent humin and in some humic acid, often is located in subsurface horizons, where it was transported through years of illuviation (Stein, 1992c, pp. 202–203). Stored with it is younger illuviated carbon and carbon from rootlets, both of which dilute the isotopic age of the older carbon. Older carbon persists with depth, and some carbon from the earliest stages of pedogenesis should be present in the lower A horizon or as translocated carbon in the B horizon. However, more abundant and more active younger radiocarbon tends to overwhelm the oldest radiocarbon. The result of dating the lower A horizon, therefore, is likely to be an age significantly younger than the initiation of pedogenesis. Wang et al. (1996) show, for example, that the oldest radiocarbon ages from a buried A horizon may only be one-tenth the age of initial pedogenesis.

The age–depth relationship resulting from the radiocarbon dating of a soil is strongly conditioned by the soil-forming environment (Matthews, 1985). Older carbon will persist even at the surface of a soil under saturated or freezing conditions. Well-drained and warm settings produce the more "typical" age–depth relationship reported by Wang et al. (1996) and others (Matthews, 1985, pp. 273–276), with the top of the A horizon dominated by carbon from young SOM and carbon from older, more persistent SOM becoming more dominant with depth.

Dating of cumulic A horizons and associated archaeological features can be somewhat more informative regarding the history of landscape evolution and human occupation. The process of overthickening or cumulization (chapter 5) tends to minimize (though not always overcoming) the effects of bioturbation or translocation on radiocarbon ages. Radiocarbon ages from the oldest buried soil at the Big Eddy site, Missouri (Hajic et al., 1998; discussed in chapter 6), provide a good example. The upper part of the oldest buried soil is a cumulic A horizon ~35 cm thick (3Ab in fig. 6.7). The horizon also contains stratified, late-Paleoindian cultural features yielding Dalton and San Patrice artifacts. The intact nature of the archaeology is indicative of minimal mixing. Six radiocarbon samples were dated: two wood charcoal, two undifferentiated charcoal, and two bulk soils (Hajic et al., 1998, p. 95). The resulting radiocarbon ages generally are in correct stratigraphic order (fig. 6.7), indicating only minimal mixing or translocation of particulate carbon or SOM.

Another important consideration in assessing radiocarbon ages from buried A horizons is whether the soil surface was eroded before burial (summarized by Matthews, 1985, and Martin and Johnson, 1995). Because of the age–depth relationship common in many soils, erosion of the upper part of the A horizon removes much of the most recently incorporated organic carbon (some will have been mixed or translocated more deeply). Some burial processes such as glaciation, mass wasting, and solifluction can mix and even invert soil horizons and thus further confuse a radiocarbon chronology. Archaeological sites in these settings are relatively uncommon, however.

Contamination is a big problem in dating buried soils. The sources of older or younger carbon are many (summarized from Matthews, 1985). Old carbon from soils or charcoal on surrounding uplands can be transported by alluvial, colluvial, or eolian processes onto soil parent material as it accumulates, or these older allochthonous materials can accumulate on a soil surface during pedogenesis. The introduction of old carbon can be a particular problem on archaeological sites in arid or cold settings in which old wood or old charcoal can persist on the landscape for hundreds of years. After burial, younger carbon can be introduced mechanically by burrowing animals or roots and rootlets. Carbon in the mobile humic acids from an overlying soil can also be translocated into a buried soil. The likelihood and degree of this sort of contamination is in part related to the soil-forming environment; in particular, to settings with abundant water percolation, usually wetter environments. These situations typically are forested ones, and therefore, acids from forest litter can be carried to depth, bringing abundant younger carbon into the buried soil. Podzolizing environments are particularly notorious (Matthews, 1980, 1985). The problem is significantly less common in drier environments (Martin and Johnson, 1995). However, even in dry environments, the problem can be acute if the sediments are sandy, thus allowing deep and rapid infiltration of water. Approximately half of the radiocarbon ages on buried soils from Holocene dunes in the Southern High Plains likely were contaminated by this mechanism (Holliday, 2001b). Shallow burial also increases the likelihood of contamination from overlying soils.

Sampling poses some problems, too. Haas et al. (1986) showed that resampling from the walls of excavation units and trenches over a period of years produced progressively younger radiocarbon ages. Why this happened was not determined, but it could be the result of the movement of water carrying younger humic acids down the walls of the exposures, or perhaps of some sort of microbial activity moving into the soil laterally from the exposure. Samples, therefore, should be collected from fresh exposures. Ideally walls should be dug back at least 50 cm before sampling. Obviously, if a wall is sloping, and especially if cracks are present, a sampling surface might have to be farther back.

To summarize, the radiocarbon analysis of buried A horizons can yield reasonable results and aid in soil stratigraphic and geoarchaeological chronologies under some conditions. The method appears to work best in soils formed and preserved in nonleaching conditions. For best results, samples should be collected from the top of the buried A horizons. This should produce ages that approximate the age of burial. Samples from deeper in the A horizon soil tend to be progressively older, but rarely do they approach the initiation of pedogenesis.

The accumulation of inorganic carbon in pedogenic calcium carbonate is somewhat similar to the accumulation of SOM in that the carbonate slowly builds up in a Bk horizon as the soil forms (Machette, 1985), but the similarities end there. Inorganic carbon in a Bk horizon is not subjected to turnover. In theory, the carbon from the initial through-terminal stages of carbonate buildup will remain in the soil and a radiocarbon determination on the Bk horizon should provide a sort of average age (akin to the AMRT of an A horizon). A host of variables can affect the resulting age, however, and render it problematic if not outright inaccurate (see reviews by Amundson et al., 1994; Head, 1999, pp. 314–316; Tandon and Kumar, 1999, p. 141). For this reason the dating of inorganic carbon in calcium carbonate in soils has received significantly less attention than the dating of SOM. In part this is because soils with pedogenic carbonate are found only in drier climates, limiting widespread applicability. The bigger problem, however, is succinctly summarized by Birkeland (1999, p. 138): "$CaCO_3$ is readily soluble, and solution and reprecipitation can take place; every time this happens, new carbon is added to the system because of the CO_2 in the soil atmosphere." Other problems include welding of younger carbonate to the zone in question and old carbonate in the Bk horizon, from clasts in the parent material (e.g., particles derived from limestone bedrock), brought in as dust, or brought in by groundwater (the "hard water effect").

Amundson et al. (1994) summarize the conditions necessary for reliable dating of inorganic carbon from Bk horizons: First, ^{14}C content of the atmosphere is constant or its variation with time is known; second, ^{14}C content of the sample is the same as that of the atmosphere, or the relationship between the two is known; and third, there is no exchange of C with an external source, following pedogenic carbonate formation. Because climate change and the "accompanying dissolution/precipitation will always be a problem," Birkeland (1999, p. 140) concludes that radiocarbon dating of Bk horizons will be more successful for Holocene soils than for late Pleistocene ones. An additional caveat is that the method probably works best for soils in settings that have remained dry and well drained. Several examples illustrate these conditions.

The Lubbock Lake site provides an example of successfully dating a Bk horizon in a buried Holocene soil. The Lubbock Lake soil, described above, formed ~4500–1000 yr B.P. in loamy eolian parent material deposited ~5500–4500 yr B.P. The Bk horizon in the soil yielded dates averaging 2400 yr B.P., which seems very reasonable for carbon inputs spanning most of the late Holocene. The method probably worked because the situation was relatively straightforward: The parent material contained no old carbonate clasts, the Bk horizon was not affected by hard water, there was no dissolution or precipitation following burial, and there was no welding of younger carbonates to the buried Bk. Some of the carbonate probably was derived from dust (Holliday, 1988), but it apparently was dissolved and then reprecipitated in equilibrium with the soil-forming environment.

Several investigators have used radiocarbon ages from Bk horizons for relative and broad numerical age control. Williams and Polach (1971) report successful dating of calcic horizons from arid South Australia. Their study focused on soils formed in eolian and alluvial deposits just after ~30,000, between ~16,000 and 12,000, and between ~6000 and 1500 yr B.P., based on radiocarbon dating of organic carbon. The respective mean radiocarbon ages for pedogenic carbonate formed in the sediments are 27,450 ± 7000, 13,900 ± 2000, and 7750 ± 1000 yr B.P. The dates on pedogenic carbonate in the two upper Pleistocene units are reasonable approximations, but in this study the date on carbonate from the Holocene unit is somewhat older than expected.

Blum and Valastro (1992) and Blum et al. (1992) dated calcic horizons as part of geoarchaeological research in northwestern Texas. The soils formed in alluvial terraces, and the carbonate dates were used to provide minimum ages for underlying deposits and maximum ages for sediment overlying the soils. Along the Colorado River, early Holocene alluvium, dated by artifact typology and a suite of internally consistent radiocarbon ages on organic-rich sediment, are inset against older, higher terraces. This relationship indicated that the youngest of the higher terraces contained late Pleistocene alluvium. This terrace and alluvium exhibits a soil with a stage II-II+ calcic horizon. Radiocarbon ages on the Bk horizon are 14,300 ± 1190, 11,430 ± 540, and 10,360 ± 150 yr B.P. (Blum and Valastro, 1992, pp. 428–429, 431), supporting the interpretation that the alluvium of the younger high terrace is late Pleistocene. The deposits are likely to be at least somewhat older than ~14,000 yr B.P. when additions of Holocene carbonate are taken into account. Along reaches of the upper Brazos River, soils in high terraces buried by alluvial fans yielded a roughly similar set of dates that conformed to the regional geochronology dated by artifact styles and radiocarbon ages from organic-rich sediment (Blum et al., 1992, p. 349). The fans began forming by ~10,900 yr B.P. Carbonate from a truncated soil in late Pleistocene gravel buried by the fan was dated to 13,500 ± 280 and 12,555 ± 940 yr B.P.

Under the right circumstances pedogenic carbonates apparently can be reliably used for radiocarbon dating. These circumstances include Holocene soils, soils not subjected to groundwater infiltration, and soils formed in parent material with no detrital carbonate (Birkeland, 1999, pp. 138–140). Late Pleistocene soils also seem adequate if they remained well-drained and dry, although Goring-Morris and Goldberg (1991, p. 117) report falsely young ages on carbonate nodules from late Pleistocene soils. In addition, more useful radiocarbon control probably will be obtained for soils formed over relatively brief periods. That is, a radiocarbon age of 2500 yr B.P. on carbonate from a soil formed 5000–500 yr B.P. is more archaeologically useful than a date of 20,000 yr B.P. on a soil formed 50,000–10,000 yr B.P. Another approach might be dating of the interiors of carbonate pebbles to estimate the initial time of pedogenesis (as done in U-series dating, noted later in the text). Dating the exteriors of carbonate pebbles might produce ages that approximate the final stages of carbonate precipitation, but only if the exterior were not subjected to dissolution.

U-Series Disequilibrium

U-series disequilibrium dating refers to a range of methods that are based on decay products of ^{238}U and ^{235}U. In deposits containing uranium that remain undisturbed for at least one million years, a dynamic equilibrium is established between uranium and its daughter products such that the decay is at the same rate as formation by the parent isotope. When the system is disturbed, the balance between production and loss will no longer prevail and the different proportions of isotopes will change. Measuring the degree to which a disturbed system of decay products has returned to a new equilibrium provides an indication of the time elapsed since the disturbance. In practice, U-series dating focuses on ^{230}Th (a decay product of ^{238}U with a half-life of 7.52×10^4 yr) and ^{231}Pa (a decay product of ^{235}U with a half-life of 3.24×10^4 yr). Disturbance of the decay series in natural systems is common because of the different physical properties of the decay products. Most significant is that ^{230}Th and ^{231}Pa are insoluble in water. Uranium, in contrast, is soluble.

Of particular interest in soils research is the fact that uranium is coprecipitated with calcite or aragonite from natural waters that are free of thorium and protactinium. Thus, as carbonate minerals are formed they will contain uranium but no daughter products. Over time, however, daughter products will accumulate. The method has been applied to a wide range of carbonate materials, including calcic horizons in soils. Some success is reported in applying U-series for dating pedogenic carbonate. Ku et al. (1979) discuss the application of U-series to calcareous soils developed on upper Quaternary alluvial fans in the Mojave Desert of southern California. The pedogenic carbonate yielded ages that are internally consistent and that agree with the geomorphic and stratigraphic relative age estimates. The dating provided minimum ages for each surface, and one suite of 14 samples yielded an average age of 83,000 ± 10,000 yr B.P., which was in agreement with a variety of independent age controls.

U-series dating has been applied with some success in several more expressly geoarchaeological situations. As mentioned above, the method was applied at the controversial Calico site in the Mojave Desert (Bischoff et al., 1981). Groundwater carbonates coating cobbles and putative artifacts were dated to ~200,000 yr. This was in agreement with an estimate of ~200,000 yr for development of a soil in deposits that rested on top of the zone with carbonate cement. In the Southern Argolid of Greece, the U-series method was applied to pedogenic carbonates above and below Pleistocene soils that yielded middle Paleolithic (Mousterian) artifacts and sites (Pope and Van Andel, 1984; Pope et al., 1984; Van Andel, 1998; see discussion in chapter 6). The dates were from the Lower and Upper Loutro alluvium; the artifacts from the Middle Loutro alluvium. Groundwater carbonate on the artifacts dated to ~52,000 yr, which was in reasonable agreement with dates on Mousterian materials from other regions. The pedogenic carbonates from below and above the archaeological zone were dated to ~272,000 and ~33,000 yr, respectively. These dates were in agreement with the artifact ages and in broad agreement with the degree of associated soil maturity and trends in soil development dated by other means (fig. 7.22; Van Andel, 1998). In the al-Jubah region of Yemen, pedogenic carbonate from upper Pleistocene fanglomerates were dated by U-series (Overstreet and Grolier, 1988, pp. 185–186). The resulting age of ~44,000 yr was interpreted as a minimum age for the deposits.

A number of problems are inherent in U-series dating of soil carbonates (e.g., Ku and Liang, 1984; Slate et al., 1991). Dating of bulk samples will include the

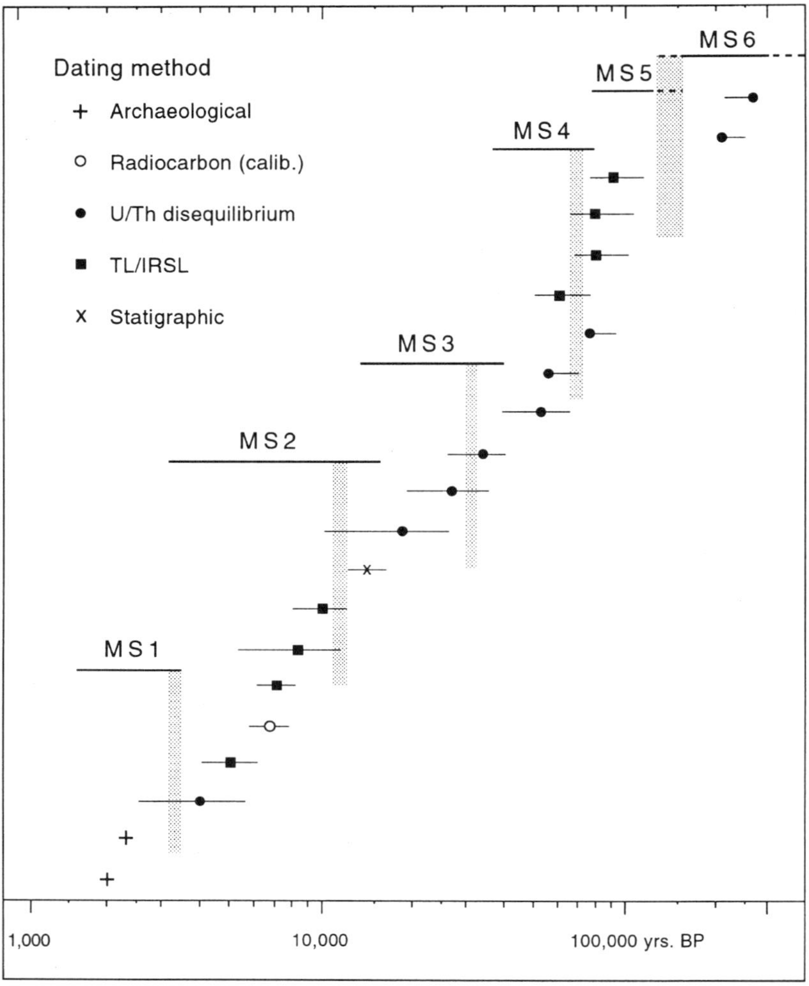

Figure 7.22 Radiometric dating of the soil chronosequence from Greece illustrating the general agreement of U-series and other dating methods (modified from *Geoarchaeology* v. 13, pp. 361–390, fig. 5, by T. H. Van Andel, © 1998 John Wiley & Sons, used by permission of John Wiley & Sons, Inc.). MS-1 through MS-5 refer to soil "Maturity Stages" (see table 6.1).

most recently precipitated uranium-bearing carbonate as well as the oldest material, thus producing a mean residence time. To get around this problem, the interiors of carbonate concretions can be dated to estimate the age of initial carbonate precipitation and, therefore, initial soil genesis. More recent studies (Sowers et al., 1988; Slate et al., 1991) show, however, that U-series dating of pedogenic carbonates involves the same pitfalls of dissolution/reprecipitation and old carbonate encountered in radiocarbon dating of Bk horizons. In addition, contaminates in the form of Th, Pa, and Ra also can be leached from clays, organic matter, and other noncarbonate components of the parent material by groundwater and during laboratory pretreatment. Experienced U-series investigators remarked that all dates for pedogenic carbonates must be regarded with skepticism until confirmed by more reliable methods (Blackwell and Schwarcz, 1995, p. 191).

8

Soils and Paleoenvironmental Reconstructions

One of the earliest uses of soils in archaeological research, in addition to stratigraphic markers, was as paleoenvironmental indicators. Similar to soil stratigraphy, the use of soils as environmental indicators in archaeological research probably has its roots in Quaternary geology (e.g., Leighton, 1937; Bryan, 1941a, 1948; Bryan and Albritton, 1943; Movius, 1944; Ruhe, 1965; Haynes, 1968; Valentine and Dalrymple, 1976). Quaternary geologists and geomorphologists working with archaeologists were quick to use soils as clues to past environments (e.g., Leighton, 1936; Antevs, 1941; Bryan, 1941a; Hopkins and Giddings, 1953; Haynes, 1968). Likewise, the nature of prehistoric environments has long been a fundamental question in archaeology.

Recognition of the relationship of soil development and morphology to environmental conditions goes back to the beginning of modern pedology, in the later 19th century in Russia and in the early 20th century in the United States (Thorp, 1941, 1949; Tandarich and Sprecher, 1994; Johnson and Hole, 1994). Climate and vegetation in particular were understood as important soil-forming factors long before Jenny produced his landmark volume on *Factors of Soil Formation* (1941). What Jenny (1941, 1980) brought to the discussion was a theoretical means, using the state factor approach, of assessing the effect of vegetation and climate on soils. By understanding these relationships via biosequences or climosequences, we are theoretically able to pick out the morphological and chemical characteristics of soils that are linked to climate or to vegetation. Climate most directly influences pedogenesis through precipitation and temperature and influences pedogenesis indirectly through vegetation. The most direct effects of biota probably come from the addition of a wide range of chemical compounds, from bioturbation, and from rooting.

This chapter is a discussion of those characteristics of soils that have some utility for environmental reconstructions, including climate and vegetation estimates. The chapter also includes some discussion of the potential pitfalls in using soils as paleoenvironmental indicators. Longer and more in-depth discussions of soil–environment relationships in the context of soil geomorphology or environmental reconstruction are presented by Birkeland (1999, pp. 268–306) and chapters in Wilding et al. (1983b) and Martini and Chesworth (1992, pp. 155–306).

The discussion is divided into five sections. The first is a discussion of some issues that greatly complicate if not completely stymie the use of soils as environmental indicators. These fundamental issues are critical to the subsequent discussions. The next three sections deal with the use of soil morphology (both specific pedogenic features and overall profile morphology) and of soil classification (following soil taxonomy) for environmental reconstructions and include many cautionary notes. Soils have been misunderstood and misused in environmental reconstructions, so both the uses as well as the misuses are presented. Understanding the potential problems in the misapplication of soils is as important as understanding the positive applications. The fifth section of the chapter deals with the use of stable isotopes as indicators of past plant communities and atmospheric conditions, a rapidly expanding area of paleoenvironmental research. A considerable amount of this work focuses on stable isotopes that accumulated in soil horizons. The preservation of these isotopic signatures of past vegetation and climate is intimately related to pedogenic properties and processes, so this method is appropriate to a discussion of soils as environmental indicators. However, there is no discussion of the use of plant fossils such as pollen or phytoliths recovered from soils. That is the realm of paleobotany, and the reader is directed to the large archaeologically oriented literature in that field (Dimbleby, 1985; Piperno, 1988; Rovner, 1988; Pearsall, 1989; Pearsall and Piperno, 1993; Gremillion, 1997; and the comprehensive summary by Dincauze, 2000, pp. 343–365). Magnetic properties of soils, especially susceptibility, have been investigated for paleoenvironmental clues, particularly paleoprecipitation (e.g., An et al., 1990, 1991; Maher and Thompson, 1991, 1992, 1995, 1999; Singer et al., 1992; Liu et al., 1995; Evans and Heller, 2001; Maher et al., 2002). Some data indicate that the magnetic signal of buried soils is directly linked to the duration and climatic conditions of pedogenesis (Verosub et al., 1993). Others argue, however, that the origin of the magnetic properties, whether related to primary loess deposition or to pedogenesis or to some combination, is unclear (Meng et al., 1997; Kemp et al., 1996; Kemp and Derbyshire, 1998). This technique, therefore, is not included in the chapter.

Several definitions must be presented before proceeding with the discussion of soils as environmental indicators in archaeological contexts. The terms "environment" and "climate" (and especially "paleoenvironment" and "paleoclimate") often seem to be used interchangeably, but one is relatively specific and the other more general. Environment "encompasses all the physical and biological elements and relationships that impinge upon a living being. Specification of an organism's environment emphasizes those variable relevant to the life of that organism—ideally, almost every aspect of its surroundings" (Dincauze, 2000, p. 3). Climate, in contrast, refers to the regional, long-term characteristics of weather patterns such as average precipitation, average temperatures, and seasonal fluctuations; "the atmospheric conditions typical of the location" (Aguado and Burt, 1999, p. 381). In looking at proxy indicators of climate the intent is usually to get at relatively specific kinds of information about temperature or precipitation. The reconstruction of environments, however, can be much more general but also very localized, such as the reconstruction of poorly drained, bog-like conditions. Soils are better suited for environmental reconstructions, though as discussed below, under some circumstances a degree of climatic or even vegetation specificity can be involved.

Caveats and Complications

Before proceeding with a discussion of how soils can be employed as paleoenvironmental indicators, some fundamental problems must be presented. Using soils to reconstruct past climatic and biotic conditions has proven to be very difficult for a number of reasons. Vegetation is closely linked to climate, and the two are difficult to sort out in the factorial approach (Jenny, 1941, pp. 197–199; Birkeland, 1984, pp. 165, 260, and 1999, p. 268). Climosequences in which vegetation has not changed along with climate, and vice versa, are difficult to find. Furthermore, soils are not sensitive to discrete climate changes that may be culturally significant. Such changes can be more easily detected from plant or animal remains, particularly in a high-resolution, microstratigraphic context. However, as is the case with many environmental proxies, the microclimate and local vegetation can dominate local soil-forming processes, thus obscuring the signal of regional climatic and vegetation conditions that may be of interest. In addition, climate changes in the Holocene, the time period with which most North American archaeologists deal, were often of insufficient magnitude to be detectable in the pedologic record.

Equifinality is also an issue in the paleoenvironmental interpretation of soils. In particular, long periods of soil formation apparently can produce some pedologic characteristics similar to those produced under particular climatic conditions. For example, red, clay-rich Bt horizons are typical of warm, humid conditions, but they can also form on old landscapes in drier environments (Boardman, 1985a; Catt, 1988). As a corollary to equifinality, relatively few specific soil features or types of surface or buried soils are related to unique or easily circumscribed environments of formation. To further the example of equifinality and the Bt horizon, argillic horizons occur in modern surface soils in a wide variety of environments throughout North America.

Another characteristic of soils that has received little attention is that soils are the products of immediate, localized conditions. The physical, chemical, and biological characteristics of a soil profile were determined by the parent material at that site, by the flora and fauna living at that site, and by the meteorological conditions (such as precipitation and temperature) that operated on the site over time. This is in contrast to other proxy environmental indicators that are affected by both regional and local conditions. Pollen accumulation, for example, is dependent on vegetation both far and near. Sedimentological characteristics are also influenced by a number of regional factors that influence the local depositional environment. The characteristics of alluvium at a given site, for instance, are influenced by a variety of factors that may have been far removed from the local setting. The soil-forming factor that can provide a nonlocal effect on a soil is landscape or catenary setting. The slope position of a soil will influence its evolution by means of water movement over or through the soil and through erosion and deposition; these processes will largely be influenced by factors operating upslope of the soil.

Local pedogenic conditions may not necessarily reflect regional environmental conditions, and therefore, the reconstruction of regional environments using soils should not be based on one or a few soils. Indeed, the reasons an archaeological site is located where it is may be unusual characteristics such as access to water or other resources—characteristics that may have a profound effect on local pedogenesis but that may not be expressed in the regional soils of the same age and stratigraphic position. This point is well made by Sanborn and Pawluk (1980), working near Calgary, Alberta. They investigated a well-expressed soil (A-E-Bt profile with strong prismatic and blocky structure and abundant illuvial clay) buried below the Mazama tephra (~6900 yr B.P.) and formed in alluvium. This example is not from a geoarchaeological context, but it easily could be, given the age and stratigraphic context of the section. The degree of soil development could be indicative of lengthy landscape stability or perhaps genesis under podzolizing conditions. The latter interpretation could have implications for shifting vegetation in the region, also documented nearby in a soil stratigraphic record at an archaeological site (Reeves and Dormaar, 1972), discussed below. Well-expressed early Holocene soils are rare in the region because the landscape is quite young (i.e., postglacial), so the locality was of considerable interest.

Sanborn and Pawluk (1980) showed that the buried soil is regionally extensive, but the well-expressed variant (or facies) is highly localized. The soil apparently was affected by groundwater discharge high in sodium. The sodium dispersed the clays and produced the strongly developed Bt horizon. The soil was initially a Solonetz in the Canadian classification system (probably a Natrustalf or Natrudalf in soil taxonomy). The sodium was subsequently leached, producing a Solod (Ustalf or Udalf). Thus, the morphology was related to localized conditions and not to landscape age or regional environment.

In archaeological contexts, the Clovis site in New Mexico and the Lubbock Lake site in Texas are good examples of sites in microenvironments that were inviting to human occupation, but that are not necessarily representative of the region—in this case the Southern High Plains. The Clovis site, in eastern New Mexico, is in a spring-fed basin that drained into Blackwater Draw through much of the late Quaternary (Haynes and Agogino, 1966; Haynes, 1975, 1995). The site is well stratified, with late Pleistocene alluvial, lacustrine, and muddy paludal deposits (Units B, C, D, and lower E; following Haynes, 1975, 1995) and Holocene paludal mud and eolian deposits (Units upper E, F, and G). Unit E, containing late Paleoindian occupation zones, was a muddy aggrading marsh and locally exhibits a cumulic A horizon that resulted from slow eolian additions to the marsh (Holliday, 1995, pp. 36–37; 1997, p. 67). In the draws elsewhere on the Southern Plains, late Pleistocene and early Holocene paludal muds are not common, and an early Holocene paludal soil is rare (Holliday, 1995). The Unit E soil at the Clovis site, therefore, is useful for reconstructing site-specific environments, but it is not a regional environmental indicator.

A somewhat similar situation can be found at Lubbock Lake, further described below and in chapters 6, 7, and 9. The site is directly in Yellowhouse Draw, in a reach fed by springs (Stafford, 1981; Holliday, 1985a, 1995). The oldest buried soil (the "Firstview soil" of Holliday, 1985b) was a marsh associated with late Paleoindian occupations and formed in early Holocene paludal muds (stratum 2m; fig. 9.3). This soil is not directly correlative with the cumulic soil at Clovis. The Firstview soil is younger than the soil at the Clovis site, and it also appears to reflect local fluctuations in spring activity (Holliday, 1985b, 1995). Overlying eolian sediments contain the moderately to well-expressed Lubbock Lake soil (A-Bt-Bk; fig. 9.3), which formed throughout the late Holocene (Holliday, 1985a,d). Locally, the soil exhibits gleying. This characteristic is important in understanding the water table record at Lubbock Lake, which influenced the availability of resources for animals and human, but regionally, the gleying is rare in the correlative soil (Holliday, 1995). The soil morphology at Lubbock Lake is thus important for understanding the environmental history of the site but is not necessarily an indicator of regional climate history.

A further complication in using soils for environmental reconstructions is that the polygenetic nature of soils can confound their paleoenvironmental signal. The very characteristics that make soils useful as age indicators work against their utility as paleoenvironmental proxies. As soils form through time they may be subjected to a succession of environments. The longer the duration of pedogenesis, the more changes a soil is likely to have experienced; that is, the more polygenetic it will be. At best, therefore, soils represent some sort of "averaging" or mixing of whatever morphological and chemical characteristics may be linked to the environment. Broadly speaking, soils that formed over relatively short periods of time and did not experience many environmental changes are more useful for reconstructing environmental conditions of pedogenesis than are soils that formed for a longer time and were subjected to a variety of environments (or at least much more effort is required to reconstruct the environmental history of polygenetic soils). The problem here, however, is that pedologic features that develop relatively quickly tend not to persist in buried soils (table 5.1; Yaalon, 1971c).

Finally, a stratigraphic sequence with a buried soil, though suggestive of changes in landscape stability, is not necessarily indicative of environmental changes. Certainly this is the case with multiple weakly expressed A-C soils buried in floodplain deposits (i.e., Fluvents; chapter 5). The cycles of sedimentation and stability are simply part of the natural evolution of a floodplain, relating to variability in precipitation and runoff from year to year (Ferring, 1986, 2001; Brown, 1997, pp. 96–103). Soils can also be buried as a result of human activity. As Wilkinson (1997, pp. 855–856) notes, burial of the middle Holocene Jahran soil, in northern Yemen (chapters 9 and 10) coincides with human activity that destabilized the landscape, resulting in widespread erosion and sedimentation. Purposeful burial of the landscape by construction of check dams and terrace walls (e.g., Overstreet and Grolier, 1988; Sandor, 1992) also produces buried soils in which sediment that might otherwise be transported across the landscape is impounded.

In general, therefore, the comments of Valentine and Dalrymple (1976, p. 218) regarding the interpretation of past climate and vegetation from soils are well taken and are still valid: "Although soil science is under great pressure to furnish environmental evidence, it is debatable whether we understand the interaction of the soil-forming processes with the site and environmental factors well enough yet to make confident extrapolations." In some situations, however, soils have proven useful in providing paleoenvironmental clues, if not data. This rest of the chapter is a discussion of those situations and related methods and approaches.

Pedogenic Features

Many, if not most, physical and chemical characteristics of soils are directly or indirectly related to environmental conditions. In using soils to reconstruct environmental conditions the key is selecting those pedogenic characteristics that are indicative of relatively specific conditions such as drainage, topographic setting, rainfall, or plant communities. The best results in using soils (in archaeological or nonarchaeological contexts) as paleoenvironmental indicators seem to be for local environmental reconstructions. Soils are particularly useful for assessing local drainage conditions and paleotopographic setting because pedogenesis is sensitive to both surface and subsurface water movement and because topographic setting affects water movement. The properties of soils that seem to be indicative of the climatic conditions of soil formation include organic-matter content, the depth of leaching—which determines the presence or absence of $CaCO_3$ and more soluble salts—depth to the top of the zone of accumulation of the carbonate or salts, overall profile morphology, and, as discussed in a separate section, stable isotope content of organic carbon and calcium carbonate. There is only a limited amount of information on which pedologic features can be related to past plant and animal communities. Furthermore, as noted above, plant and animal distribution is so closely linked to climate that sorting out the effects of each is often difficult. The soil characteristics that seem to be most directly indicative of vegetation and also persist in buried soils are the E horizon and related podzolic characteristics and some overall profile morphologies.

Soil Color

Dark colors (low chromas and low values in the Munsell system) in buried A or O horizons have been used as indicators of relatively high SOM production under relatively wet conditions. In general, this approach seems valid for reconstructing local environmental conditions and perhaps indirectly getting at regional climate. Dark colors are produced when SOM production is high relative to SOM decay, which typically occurs under relatively moist conditions (high water table, poor drainage, or climates that are subhumid or wetter). Dark colors seem to persist in buried soils that were once high in SOM, long after most of the SOM oxidized or was destroyed by microbial activity (chapter 10; e.g., figs. 2.1 and 2.2). A buried A horizon with a low chroma and a low value (e.g., 10YR 3/1) probably formed under relatively lush vegetation. Dark coloration from SOM does not persist forever in buried soils (chapter 10), so the absence of dark colors in older soils (e.g., soils buried before the late Pleistocene) has little significance.

Organic soils and peats (Histosols in soil taxonomy) tend to be black, dark gray, dark brown, or dark reddish-brown and composed of organic matter in various states of decomposition (Everett, 1983; Shotyk, 1992). They form in wetlands and, because of their association with permanent water, they often contain evidence of human occupation. Indeed, they are famous for containing exceptionally well-preserved archaeological remains (chapter 10; fig. 10.1; Coles and Coles, 1986, 1989; Coles and Lawson, 1987; Purdy, 1988, 1991). They also contain a wealth of paleobotanical remains that, in general, provide much more diverse and specific paleoenvironmental data than can be gained from pedologic study. The paleoenvironmental significance of these soils will not be further discussed, therefore, except to note that peats can be stratified (e.g., Doran and Milanich, 2002; papers in Purdy, 1988) and that this characteristic can be important in reconstructing chronologies of environmental change.

Organic matter and humus can be translocated in soils under some conditions, so their presence is not always an indicator of a former surface. Podzolization (discussed later), common under acidic coniferous forests in sandy parent material, results in translocation of iron and humus, producing "coffee-brown" Bh horizons. Thorson (1990, p. 406) also describes the formation in cold climates of "pseudo-paleosols," which are translocated organic materials and other ions and are characterized by dark colors that can be mistaken for buried soils (discussed in chapter 10).

Iron in various states imparts many of the distinctive color characteristics of soils and also is indicative of local environmental conditions. Iron oxides produce a range of reddish-brown colors characteristic of many B horizons (table 8.1; Schwertmann et al., 1982; Kemp, 1985a; Schwertmann, 1993; Birkeland, 1999, pp. 288–289). Uniformly red or reddish-brown zones in soils are indicative of well-drained settings. This information is useful in reconstructing site settings and in assessing preservation conditions for perishable artifacts (see chapter 10), though these are very broad generalizations. One relatively specific relationship of iron oxides, B horizons, and local environments is the formation of a "podzolic B" horizon under coniferous forests or lichen-heath tundra (discussed below).

Degree of redness also is generally related to climate—mostly temperature, it seems. Very broadly, increasing temperatures produce redder soils because temperature helps drive chemical reactions. For example, soils of the southwestern and southeastern United States and around the Mediterranean are redder than those of the northern United States, Canada, and northern Europe. Time is also an important component of redness, however. Other factors being equal, older soils will be redder than younger ones because more time was available for oxidation of free Fe. Many of the coastal and alluvial landscapes in the southeastern United States, for example, are much older than the glaciated ones of the

Table 8.1. Iron oxide minerals and their pedoenvironments

Iron oxide (formula)	Munsell color	Pedoenvironments	Soils
Goethite ($FeOOH$)	7.5YR–2.5Y	Wherever weathering takes place	All soils with Fe released
Hematite (Fe_2O_3)	7.5R–5YR	High soil temperature, low water activity, rapid biomass turnover, high Fe release rate from rocks	Aerobic soils of the tropics and subtropics with dry seasons (?)
Lepidocrocite ($FeOOH$)	5YR–7.5YR Value ≥6	Anaerobic/aerobic systems, noncalcareous	Aquic subgroups in temperate regions (Pseudogleys)
Ferrihydrite ($Fe_5HO_8 \cdot 4H_2O$)	5YR–7.5YR Value ≤6	Rapid oxidation in humic environments	Gleys Podzolic B horizons
Maghemite (Fe_2O_3)	2.5YR–5YR	Usually a product of fires	Mainly tropical and subtropical soils

From Schwertmann (1993, table 4-1).

northeastern and midwestern United States. Thus, considerable care must be taken in using redness as a climate indicator. Red interglacial soils in North America (e.g., the Sangamon soil) and in Europe, for example, were used to infer warmer-than-present, temperate interglacial conditions (Zeuner, 1959, pp. 35–41, 93, 112, 164–165; Butzer, 1971, p. 95; Ruhe, 1970; Kemp, 1985b; Catt, 1979, 1988). Additional stratigraphic work and better age control in those regions now show that these soils formed on landscapes that persisted for many tens of thousands of years—landscapes that may have had as much as 10 times the time to form than did postglacial soils of the late Quaternary (e.g., Boardman, 1985a; Catt, 1988; Curry and Pavich, 1996).

A specifically geoarchaeological misapplication of red color as an environmental indicator can be found in the work at the Paleolithic site of Swanscombe on the Lower Thames in England (see also chapter 5). The Lower Loam, which rested on the Lower Gravels with Clactonian artifacts, had been interpreted as an interglacial soil based on its red coloration. The issue is summarized by Kemp (1985b), who used laboratory data, primarily micromorphology, to show that the reddening of the soil was caused by iron concretions formed by a fluctuating water table. This study also showed that the soil was subjected to minimal pedogenic modification; that is, it was not subjected to prolonged or intense weathering and thus has none of the "supposed . . . characteristics" of interglacial soils (Kemp, 1985b, p. 354).

Gray, greenish-gray, bluish-gray, or gley colors (chroma typically <2) and some yellow colors are indicative of reducing conditions, promoting either reduction and reprecipitation of Fe or Mn, or leaching of Fe or Mn out of the soil (Vepraskas, 1992). These "redoximorphic" features occur either as mottles (or soft masses), where they are found mixed in a repetitive color pattern with oxidized colors, or as a horizon with a uniformly gray chroma. These colors tend to persist long after burial and are good indicators of high water tables (uniformly gray colors) or fluctuating water tables (mottled colors). Combined with stratigraphic and paleotopographic information, soils with gley colors or redoximorphic features are useful in reconstructing site settings in or near lakes, bogs, and marshes or other wetlands (fig. 9.9; e.g., Holliday, 1985b; Reider, 1990; Jacob, 1995a,b; Quade et al., 1995; Beach, 1998a,b; Ashley and Driese, 2000). Gleying is also described as "the most widespread weathering process in Beringia" by Thorson (1990, p. 405). The process is commonly associated with permafrost because the frozen ground impedes drainage. Typical permafrost soils (now grouped as Gelisols in the 1999 edition of *Soil Taxonomy*) exhibit gray to bluish-gray (N5/) and often mottled Bw horizons below thick O horizons (Thorson, 1990, p. 405).

In using soil color for any sort of environmental interpretations, considerable care must be taken to ensure that the color characteristics of interest are related genetically and chronologically to the soil under scrutiny. Color can be inherited from the parent material of the soil (i.e., red soils or rocks that undergo erosion can produce red sediment that then imparts its color to soils that form in it). On the Southern High Plains, Holocene eolian sediments in draws and dune fields were derived from the regional Pleistocene soils with 5YR colors. The resulting sediment varies from 10YR to 5YR (Gile, 1979; Holliday, 1988, 1995, 2001b). The color of soils formed in those sediments, therefore, cannot be used to make environmental inferences.

Postburial alteration of color (discussed in chapter 10) is another potential problem, particularly in settings with fluctuating water tables. A soil that was poorly drained during one particular phase of human occupation could undergo a significant change in morphology if the water table subsequently lowered and the soil became well drained and subjected to subaerial weathering. This was the situation at the Lubbock Lake site in the late Holocene. In the initial stages of pedogenesis in the Lubbock Lake soil a Bg horizon formed, probably because of a high water table. Through time, however, drainage conditions changed, the soil was oxidized, and a Btk horizon formed (Holliday, 1985d, 1988). The reverse situation could also affect a site. As floodplains aggrade and soils are buried, the soils are submerged beneath the low seasonal water table. In the presence of sufficient organic matter and reducing conditions, ground waters can produce gley features in the buried soils (Retallack, 1991b).

Establishing the time of oxidation or reduction is very difficult to do, however. One clue is the presence of secondary soluble materials such as calcium carbonate or salts in a gleyed or mottled zone. The soluble materials likely postdate the wet conditions; otherwise, they probably would be dissolved. Furthermore, under some conditions secondary calcium carbonate can be radiocarbon dated (see chapter 7). The ^{14}C age can then provide a minimum-limiting date for the reduction. In the Lubbock Lake soil, mentioned above, a calcic horizon was superimposed over the gleyed zone (producing a Btgk; Holliday, 1985d). The bulk carbonate in the calcic horizon yielded radiocarbon ages averaging 2400 yr B.P. (chapter 7). Deposition of the parent material ceased by ~4500 yr B.P. The iron reduction, therefore, must have taken place between 4500 and 2400 yr B.P., and probably much closer to 4500, given that the carbonate date is a kind of average (chapter 7).

Zones of $CaCO_3$ Accumulation

The accumulation of $CaCO_3$ in soils forms distinctive horizons (Bk, Bkm, K, calcic, or petrocalcic; figs. 2.1, 2.3, 6.4, and 8.6). These zones form when calcium and bicarbonate ions are available in the soil solution and evaporation exceeds precipitation. They are thus ubiquitous pedogenic characteristics in arid and semiarid environments and, therefore, have some climatic significance. These soil horizons are also easily confused with carbonates formed under a wide variety of depositional environments (e.g., Tandon and Kumar, 1999, tables 3 and 4). A considerable amount of research has gone into understanding the genesis and morphology of these horizons because they are so common in soils and because of their effect on soil water movement and on geomorphology (e.g., Brown, 1956; Gile et al., 1965, 1966, 1981; Goudie, 1983; Klappa, 1983; Machette, 1985; McFadden and Tinsley, 1985; Tandon and Kumar, 1999). A wide variety of factors in addition to precipitation and evapotranspiration have a significant effect on the morphology of calcic horizons: the texture, porosity, permeability, uniformity, and carbonate content of the soil parent material, clay content of any Bt horizon, and slope position.

Of particular paleoclimatic interest to many workers is the depth to the top of the Bk horizon. A variety of investigators have shown that this parameter is directly related to climate (e.g., Jenny and Leonard, 1934; Arkley, 1963). Among a set of soils in which all other factors are the same, the top of the Bk will be deeper among the soils receiving higher precipitation. In theory, therefore, depth to the top of the Bk in a buried soil is a good indicator of paleoprecipitation. Retallack (1994) compiled data on depth-to-Bk in modern soils from a variety of environments and produced a regression equation that he applied to buried soils. That is, by measuring the depth from the top of a buried soil to the top of an associated Bk and then plugging the measurement into the equation, the mean annual precipitation (MAP) during pedogenesis could be estimated with "reasonable accuracy" (Retallack, 1997b, p. 381). This is an attractive approach because of its simplicity. There are serious problems with it, however. The calculation provides only the grossest estimate of MAP. The standard deviation (at 2 δ) determined in the calculation is ±282 mm. The full range of the standard deviation (564 mm) is half of the range of the MAP plotted (Retallack, 1994, fig. 3-2) and roughly equivalent to the range in MAP between Denver, Colorado, and Indianapolis, Indiana. For any given MAP the depth variation can be up to several meters; moreover, the curve is only meaningful if all the other factors of carbonate accumulation are held constant and if compaction and erosion can be accounted for. The general relationship calculated by Retallack (1994) is instructive, but to select a soil (buried or otherwise) with a Bk horizon and simply "fit it" to the curve to predict a past MAP without taking into account other factors yields ambiguous if not meaningless results.

E Horizon and Podzolic Characteristics

The formation of an E horizon represents leaching (eluviation) of clay, iron, aluminum, organic matter, or a combination of these constituents, leaving a lighter-colored zone with a concentration of sand-sized or silt-sized quartz just below the A horizon and above the B (figs. 1.1B, 1.1C). This process usually takes place under humid, forested settings. "Podzolization" is a particularly intense form of leaching and translocation of iron, aluminum, or organic matter out of the E horizon and precipitation in the B horizon. The result is a strongly expressed E or albic horizon and Bh, Bs, or Bhs spodic horizon (fig. 1.1B). Key environmental ingredients in podzolization are coarse parent material (which enhances rapid, deep leaching), abundant water, and vegetation that produces a thick, acidic litter such as coniferous forests, heather, or lichen-heath in the tundra (McKeague et al., 1983; Schaetzl and Isard, 1996). The presence of an E horizon, therefore, is broadly indicative of the pedogenic environment; in particular, humid, forested conditions.

The presence or absence of an E horizon in a soil has been particularly useful in tracking shifts of the prairie/forest boundary in the midwestern United States, the Rocky Mountains, the Canadian prairies, the Caucasus, and the tundra-forest boundary in the Arctic (e.g., Ruhe and Cady, 1967; Ruhe, 1970; Reeves and Dormaar, 1972; Sorenson, 1977; Al-Barrak and Lewis, 1978; Reider et al., 1988; Alexandrovskiy, 2000), discussed in more detail later. As forests expanded at the expense of prairies during full glacial conditions, the prairie soils were converted to forest soils, resulting in formation of an E horizon in the prairie soil profile. As prairie expanded into formerly forested regions in postglacial time, the formation of prairie soils only partially obscured the E horizon. Today an E horizon (or E horizon characteristics such as platy structure) in prairie soils is out of equilibrium with the prairie setting and is indicative of shifting vegetation (Jungerius, 1969; Ruhe and Cady, 1967; Ruhe, 1970; Al-Barrak and Lewis, 1978). Likewise, in a sequence of buried soils near the forest/tundra boundary, presence or absence of E horizons can be used to denote shifts in the treeline (Sorenson et al., 1971; Reeves and Dormaar, 1972; Sorenson and Knox, 1973; Sorenson, 1977).

Similar to the E horizon, a podzolic B horizon (Bh, Bs, or Bhs) is a good indicator of a well-drained soil formed under humid, acidic, usually forested conditions. Thorson (1990, p. 405), however, cautions archaeologists and geoarchaeologists against assuming that reddish-brown (or coffee-brown) B horizons are necessarily podzolic. In Alaskan Beringia, most archaeological sites are in well-drained settings because site discovery is skewed toward them. These settings are characterized by "Arctic Brown" soils (Ochrepts), which have reddened B horizons (5YR 4/4) resulting from minimal leaching, but also bonding of iron, clay, and humus. These soils can be confused with podzolic soils.

Cryogenic Characteristics

Cryogenic characteristics are those formed by "cryoturbation," which refers to soil mixing caused by freezing and thawing of the ground (also discussed as a site-formation process in chapter 9). The process is pervasive in Arctic, mountain, and many middle latitude settings (Wood and Johnson, 1978, p. 333). Frost is an important pedogenic agent in these settings (Van Vliet-Lanoë, 1985, 1998; Valentine et al., 1987; Tarnocai and Valentine, 1989), and frost features in soils are described as "among the most reliable climatic indicators" (Catt, 1991, p. 176).

Reconstructing the presence or absence of frost in soils in the high and middle latitudes, and how it is affected by alternating colder and warmer climates at timescales of millennia and centuries, is important in understanding the conditions humans adapted to. The effects of freezing and thawing are apparent at both macroscopic and microscopic scales.

Macroscopic features produced by cryogenic processes include convolutions from gelifluction, ice-wedge casts, patterned ground, silt caps on stones, and in the Bw or cambic horizon, subhorizontal play structure, if the parent material was loamy, and gray to bluish-gray mottles (i.e., formation of a Bg horizon). "Gelifluction" refers to the slow down-slope movement of water-saturated soil above permafrost (Benedict, 1976). The process is a type of solifluction, which refers to the more general process of slow, down-slope movement of water-saturated material above an impervious substrate (Wood and Johnson, 1978, pp. 346–347). Gelifluction features have been useful in environmental reconstruction. One of the earliest examples is the work of Hopkins and Giddings (1953) at the Iyatayet site on Cape Denbigh, Alaska, which is also an early example of soils being used in geoarchaeological research. Archaeological debris from the Denbigh Flint Complex (poorly dated; probably 3700–3300 yr B.P.; Mason and Gerlach, 1995) is found just above a buried "podzolic" soil (a zone of organic matter accumulation over an oxidized B horizon). The soil and the archaeology were heavily disturbed by cryoturbation before burial (fig. 10.11), but gelifluction is not an active process in the region today. The field data were taken to indicate, first, genesis and occupation of the buried soil under relatively warm conditions free of cryogenic activity; second, the onset of cold conditions that resulted in cryoturbation of the soil; and third, burial of the soil and subsequent human occupation and soil formation under conditions similar to today.

Schweger (1985) used a similar approach in studying multiple buried soils with associated archaeological occupation zones at the Onion Portage site, Alaska. The site is in a stratified Holocene alluvial fan along the Kobuk River. Five buried soils were identified (in Cultural Bands 3, 5, 6, 7, and 8, top to bottom; figs. 7.13 and 8.1). The soils in Cultural Bands 7 and 8 exhibited weakly developed soils characteristic of permafrost (A-Cg or A-Bg profile) and also were disturbed by a variety of cryogenic processes (fig. 10.10B). The overlying soils, however, exhibited better expressed "podzolic" soils (A-E-Bh/Bs?), with little evidence for cryogenesis beyond soil creep. Schweger (1985) used these data (fig. 8.2), supported by examination of plant macrofossils, to argue that the climate during the earlier occupations 10,000–6000 yr B.P. was significantly colder than today. After ~6000 yr B.P. conditions warmed and allowed tree growth and associated pedogenesis on successive landscapes created by periodic alluviation.

Contorted beds and soils are widely reported from archaeological and nonarchaeological sites and identified as "solifluction" features; for example, in Paleolithic horizons at several of the Kostenki localities in Russia (fig. 10.10A; Klein, 1969, pp. 41–49; Sinitsyn, 1996) and at sites in the Ukraine (Klein, 1973, pp. 18–32). The evidence for solifluction is often used as an indicator of past periglacial environments (i.e., as evidence for gelifluction). However, the process is not unique to regions with permanently or seasonally frozen ground (Wood

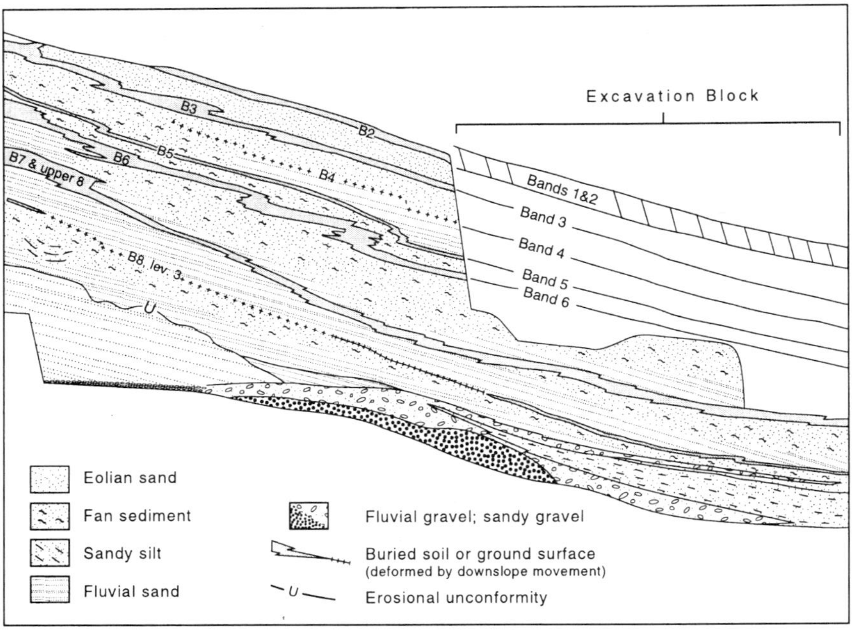

Figure 8.1 Stratigraphic section of the Onion Portage site, Alaska, showing the buried soils (Bands or B 1–8; modified from Hamilton, 1970, fig. 62). See also fig. 7.13. Bands 5 and 6 were modified by creep, but 7 and 8 were subjected to creep, along with a variety of cryogenic processes (see fig. 10.10B).

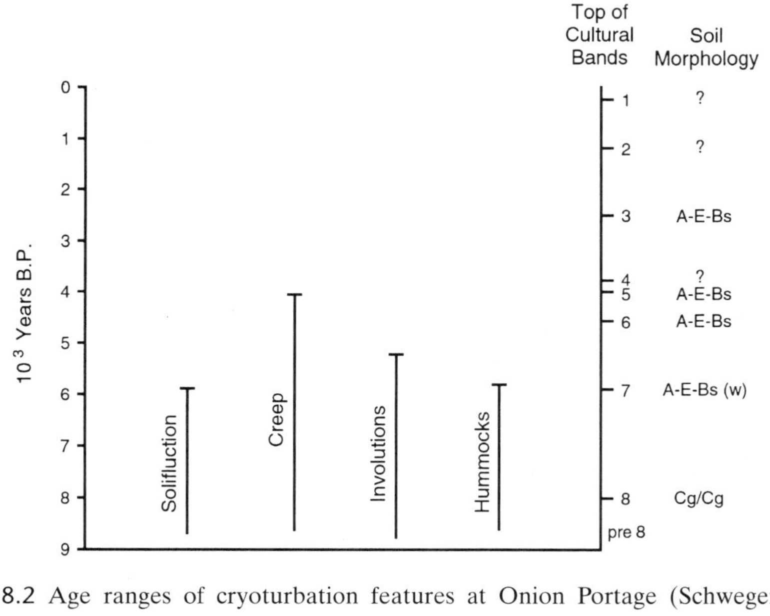

Figure 8.2 Age ranges of cryoturbation features at Onion Portage (Schweger, 1985, fig. 8).

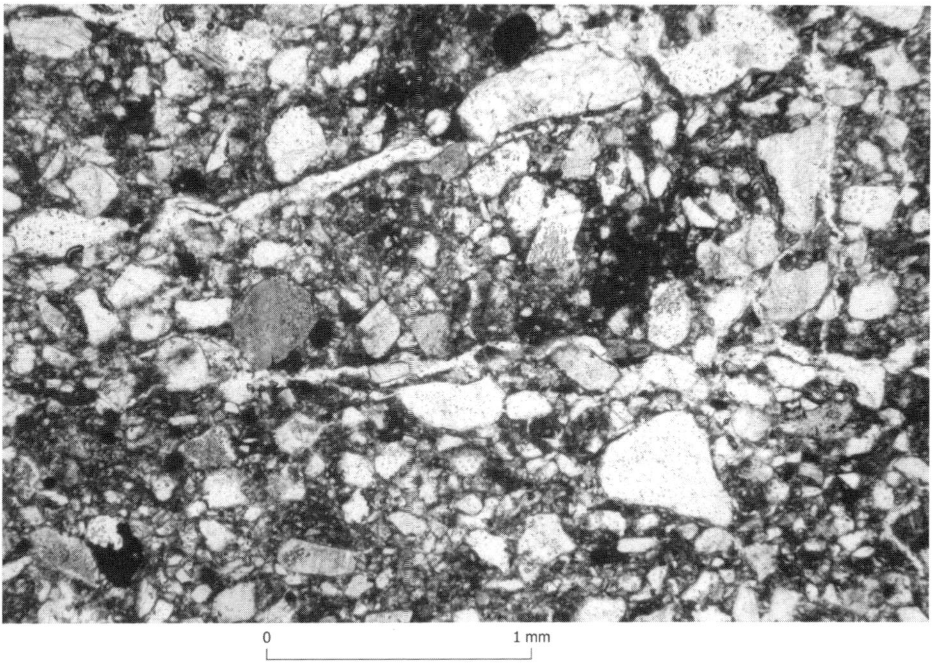

Figure 8.3 Thin-section photomicrograph (under plain polarized light) cut vertically through a soil at Malmö, Scania, Sweden (provided by and published with permission of R. I. Macphail). It shows the effects of ice lensing in the form of horizontal, planar voids. Traces of "dusty clay" coatings (thin, gray films), from rapid infiltration of fines, line the voids.

and Johnson, 1978, pp. 346–347). For an interpretation of frozen ground in regions far removed from cryic conditions today, other kinds of evidence (e.g., other kinds of cryogenic features or plant remains) are necessary.

A variety of microscopic features are also indicative of frost activity (fig. 8.3; Van Vliet-Lanoë, 1985, 1998; Tarnocai and Valentine, 1989; Fedoroff et al., 1990), and under the proper circumstances, inferences can be made regarding the timing of cryogenesis relative to pedogenesis; that is, whether the cryogenic features were acquired by the parent material before soil genesis, during pedogenesis, or after. Micromorphological frost features include disruption of clay films lining voids, formed in earlier temperate periods and incorporated into the soil matrix (as "papules"); silty cappings on sand particles, platy structure; rounded granules of illuvial silt; accumulation of soil or sand in fissures or coarse pores; and large rounded pores or vesicles produced by air bubbles during thawing of ground ice (modified from Catt, 1991, pp. 175–177). These soil characteristics provide valuable clues that a soil was affected by frost during its formation. Cryogenic processes are also important postpedogenic factors that alter both soils and archaeological materials and sites (see chapter 9).

Overall Profile Morphology

The overall morphology of a soil (both macro- and micromorphology)—that is, the combination of horizon characteristics—probably has more utility in paleoenvironmental reconstruction than the presence or absence of individual horizons. This is because a particular combination of horizons may require a more narrow range of environmental conditions and because sets of horizons may provide a sharper contrast with modern conditions.

Prairie/Forest and Forest/Tundra Boundary

The morphologies of surface and buried soils have been used very effectively to track shifts in vegetation and climate along the prairie/forest and forest/tundra boundaries. Along the prairie/forest border, this work is based on the following general relationship: Alfisols with E horizons and Spodosols will typically form under forested conditions, whereas Mollisols, Alfisols without E horizons, and Inceptisols are common in the nonforested areas. Archaeological contexts provided two of the few studies of these shifts based on soil stratigraphy. In the Absaroka Mountains of Wyoming, Reider et al. (1988) report a buried soil sequence of O-E-Bw-2Bkb-2Ckb. The surface O-E-Bw soil, associated with the modern coniferous forest cover, probably formed over the last ~4000 yr. The older soil with the Bk horizon developed in underlying deposits probably formed during the middle Holocene "Altithermal" (~8000–4500 yr B.P.). The Bk horizon is a clear indicator of drier and probably grassland conditions. A Bk horizon typically will not form with an E horizon or under forest vegetation because of the intense leaching. Phytoliths confirmed this interpretation. The soils therefore seem to be a record of a downward shift in elevation of the forest/grassland boundary coincident with a shift from warmer drier conditions of the middle Holocene to the somewhat wetter, cooler conditions of the Neoglacial.

A somewhat similar study was carried out in an archaeological context at the foot of the Canadian Rockies in southern Alberta (Reeves and Dormaar, 1972). At the Gap site, archaeological materials associated with buried soils were found in a sequence of colluvial deposits just below the lower tree line. Soil morphology and the infrared spectra of the humic acids in the soils (Dormaar and Lutwick, 1969) indicated that the tree line shifted down and up across the site several times during the Holocene, probably in response to climate changes. The deepest soil, formed in the early Holocene, has an A-Bt-C profile with chemical characteristics indicative of open canopy forest. A middle Holocene soil in the overlying deposits exhibited a simple A-C profile (with the Mazama Ash, ~6900 yr B.P., bifurcating the A horizon) but contained relatively high SOM and N along with humic acids indicative of grassland soils. The early Holocene tree line, therefore, was lower than present, probably because of cooler, more humid conditions, and the middle Holocene tree line was likely higher than present, because of warmer, drier conditions. In a regional study of buried Holocene soils in southern Alberta, Waters and Rutter (1984) identified an essentially similar soil sequence. Combining soils and phytolith data, they also concluded that the middle Holocene in the foothills of the southern Canadian Rockies was warmer and drier, with grasslands replacing forest.

In the North Caucasus of Russia, the construction of artificial mounds (kurgan) allows some unusual insights into shifting vegetation patterns (Alexandrovskiy, 2000). The sites are in the foothills of the northern slope of the West Caucasus. A number of investigators noted that the mounds bury soils characteristic of steppe vegetation (Chernozems or unleached Mollisols), but the soils on the mounds formed under forests (Luvisols or leached Alfisols), thus preserving a record of environmental change. The timing of these shifts was unclear, however, until Alexandrovskiy (2000) provided numerical age control for the mounds. The kurgan are composed of varying mixtures of unweathered loess, redeposited Chernozem material, and sandy alluvium. The mounds were constructed in several phases between 5500 and 1000 yr B.P. The oldest are 5500–4500 yr old. The next youngest mounds date to ~3500 yr B.P. Some mounds fall in the range of 3500–2000 yr B.P. The youngest mounds are 2000–1000 yr B.P.

The regional surface soils today are the "Grey Forest" soil variety of Luvisol (probably Udalfs), which formed in loess and related silty parent material but that appear to represent "degraded Chernozems"; that is, grassland soils modified by the podzolizing effects of forest vegetation. This interpretation is based on the morphology and other physical characteristics of the lower Luvisols. Just below the E horizon is a humus zone and a deeper calcic horizon and krotovinas, which are identical to the lower A and deeper subsoil portions of Chernozems preserved under the oldest mounds. The buried soils are similar to grassland steppe soils found 30–50 km away. These data plus radiocarbon ages on buried Chernozems indicate that through the first half of the Holocene the surface soils were Mollisols, probably forming under grassland. Soils buried under mounds dated to ~3500 yr B.P. are also described as "degraded Chernozems" and represent the initial stages of alteration of grassland soils under tree cover. The data from the buried soils therefore indicate that forest replaced grassland between 5500 and 350 yr B.P. The soils formed in mounds support the interpretations based on the buried soils. All mound soils are Luvisols, with the older soils exhibiting stronger expression of the A-E-Bt morphology.

An excellent example of the use of soils for both vegetation and climatic reconstructions along the forest/tundra boundary comes from the Keewatin District of northern Canada (Sorenson et al., 1971; Sorenson and Knox, 1973; Sorenson, 1977). The forest soils are podzolic Inceptisols with Oe-E-podzolic B horizons; the tundra soils are Entisols ("Arctic Brown" soils) with A-C profiles (fig. 8.4). Both kinds of soils also are found in buried contexts in eolian deposits on either side of the ecotone (i.e., within the forests and on the tundra). The dating and mapping of these buried soils allowed for a reconstruction of shifts in the forest/tundra boundary throughout the Holocene. Furthermore, by mapping the distribution of modern forest and tundra soils and comparing the data to the position of present-day air masses, dated shifts in the biome boundary were used to reconstruct paleo–air mass frequencies and to correlate these shifts to climatic changes during the Holocene.

Figure 8.4 Buried podzolic soil (PP or "paleopodzol" of Sorenson, 1977) under an "Arctic Brown soil" (AB) in the Northwest Territories of Canada (photo provided by and reproduced with permission of J. C. Knox). The O and E horizons of the buried soil are readily apparent.

Forest or Prairie?

Soil data can also effectively document the absence of forest on a landscape. Pollen studies on the Southern High Plains of Texas and New Mexico in the late 1950s and early 1960s resulted in an interpretation of "boreal forest" covering the region in the terminal Pleistocene (Wendorf, 1961, 1970; Wendorf and Hester, 1975). This interpretation was archaeologically significant because it suggested that the region was cool and wet during the Folsom occupation, an interpretation substantially different from previous ones. Soil morphology and chemistry indicated otherwise, however (Holliday, 1987b). The hypothesized boreal forest would have grown in the Blackwater Draw Formation (Pleistocene) and, therefore, in the well-expressed surface soils of the region (Paleustolls and Paleustalfs; chapter 4). A boreal forest would impart distinctive pedologic characteristics to the soils. Pine and spruce cover would produce podzolized soil profiles characterized by both an eluvial (E) horizon, which is depleted of organic matter, clay, and free iron and aluminum, and an illuvial (B) horizon with accumulations of clay, organic matter, or iron and aluminum (McKeague et al., 1983; Fanning and

Table 8.2. Physical and chemical data for selected soils of the eastern United States and High Plains of Texas

A. Podzolized soils of the eastern United States[1]

Becket Fine Sandy Loam; Haplorthod, Massachusetts			Matapeake Fine Sandy Loam; Hapludult, Virginia			Cahaba Fine Sandy Loam; Hapludult, Virginia		
Horizon	% Clay	% free Fe_2O_3	Horizon	% Clay	% free Fe_2O_3	Horizon	% Clay	% free Fe_2O_3
A	4	1.08	A	6	0.56	Ap	5	0.30
E	7	1.69	E	13	0.78	E	7	0.47
BEt	11	3.99	BE	20	1.06	BE	19	0.99
Bt	10	3.58	Bt	39	2.08	Bt	26	1.29
C	9	3.15	BC	20	—	BC	20	0.90
...	...	...	2C	7	0.64	C	23	0.58

B. Common regional soils of the Southern High Plains, Texas[2]

Amarillo fine sandy loam; Paleustalf			Acuff sandy clay loam; Paleustoll			Pullman silty clay; Paleustoll		
Horizon	% Clay	% free Fe_2O_3	Horizon	% Clay	% free Fe_2O_3	Horizon	% Clay	% free Fe_2O_3
A	20	0.30	Ap	22	0.7	Ap	35	0.8
Bt1	26	0.39	Bt1	24	0.7	Bt1	44	0.9
Bt2	21	0.31	Bt2	27	0.7	Bt2	42	0.8
Bt3	22	0.34	Bt3	21	0.5	Bt3	39	0.90
...	...	...	Bt4	23	0.3	Bt4	38	1.0
...	...	...	Bt5	20	0.1	Bt5	42	0.4
...	...	...	...	...	...	Bt6	39	0.5

[1] From Holliday (1987b, table 2A).
[2] From Holliday (1987b, table 2B).

Fanning, 1989, pp. 98–102; Schaetzl and Isard, 1996). These characteristics are well expressed in examples of podzolized profiles from pine forests of the humid eastern United States (table 8.2A). However, even Pinyon, which occurs just west of the Southern High Plains but prefers to grow in dry climates and thin, rocky sediments (notably different from the surficial cover of the High Plains; Brown, 1982), will produce soils that exhibit profiles with eluvial horizons (Folks, 1975).

Physical and chemical data from High Plains soils (table 8.2B) show that there is no evidence for podzolization, however. Holocene pedogenesis could obscure or mask some podzolic features, but the E horizon position should still be considerably lower in organic matter, clay, iron, and aluminum relative to the B horizon. In addition, the content of illuvial clay in the argillic horizons and carbonate content in the calcic horizons of the soils of the Blackwater Draw Formation are much greater than that found or expected in Holocene soils in the region (Holliday, 1985d, 1988, 1995). Additional research in the region, focusing on paleontology (Johnson, 1986, 1987c, 1991), stratigraphy and soils (Holliday,

1995, 1997, 2001b), pollen (Hall and Valastro, 1995), and stable isotopes (Holliday, 2001b) support the interpretation that the region had no forest cover and further indicate that during Folsom time the area was in a drying phase, possibly experiencing drought (Holliday, 2000).

Black Mats, Peats, and Related Organic-Rich Soils

Buried organic-rich or formerly organic-rich soils have been important in environmental reconstructions for archaeological sites in the middle latitudes of North America. A common association of soil colors and horizons in many buried situations is a black or very dark gray A horizon, sometimes cumulic and relatively high in SOM content, over a fully reduced or mottled C or Bw horizon (Cgb or Bgb). This relationship is indicative of a high water table, producing the reduced zone and also promoting high rates of SOM production in the A horizon. This sort of soil profile has been reported from a wide variety of paleotopographic lows in now well-drained settings at sites throughout the Central and Western United States and dating to the late Pleistocene or Holocene. These zones are most widely known as "black mats" but have gone by a variety of names (see Quade et al., 1998, p. 129, for a review of the terminology). They are not always described as soils (exceptions to this include Haynes, 1975; Holliday, 1985a,b, 1995, 1997; and Reider, 1990) but clearly represent some form of pedogenesis (accumulation of SOM on a stable land surface along with postdepositional alteration of parent material below the surface; figs. 8.5 and 8.6). These soils are also clearly indicative of local environmental conditions (perhaps a "wet meadow" or bog) and may be tied to climate if a direct relationship can be shown between the local groundwater conditions and climate.

On the Great Plains and in neighboring areas these "black mats" date to the late Pleistocene and early Holocene, are associated with spring-fed marshes and streams, and often contain Paleoindian archaeological debris (figs. 8.5A, 8.6, and 9.4; Haynes and Agogino, 1966; Haynes, 1968, 1975; Reider, 1980, 1982a,b, 1990; Stafford, 1981; Holliday, 1985b). At the Lubbock Lake site, for example, the oldest buried soil (the Firstview Soil of Holliday, 1985b) is a black mat (fig. 8.5A). It has dark colors, indicating that it contained abundant organic matter; a zone of reduction or gleying immediately below the surface horizon, indicating a water table just below the surface; and locally abundant silicified plant remains in the surface horizon. This marsh soil is associated with late Paleoindian features and formed in early Holocene paludal muds. Many of the black mats reported from valleys on the Great Plains are buried by younger sediments and soils. The soils formed in the later Holocene deposits have oxidized B horizons and zones of carbonate accumulation (A-Bw-Bk or A-Bt-Bk horizonation; e.g., the "Lubbock Lake soil" at the Lubbock Lake site, discussed above; figs. 2.1 and 9.3) and sometimes more soluble salts. These soil characteristics are indicative of drier, better-drained settings. The sequence of soils shows that groundwater levels declined in the valleys from early to middle Holocene time. As noted above, specific morphological characteristics of the soils may be related to local conditions, but collectively the soils suggest a shift from more humid to semiarid or arid climates (Reider, 1990; Holliday, 1995, 1997).

A

B

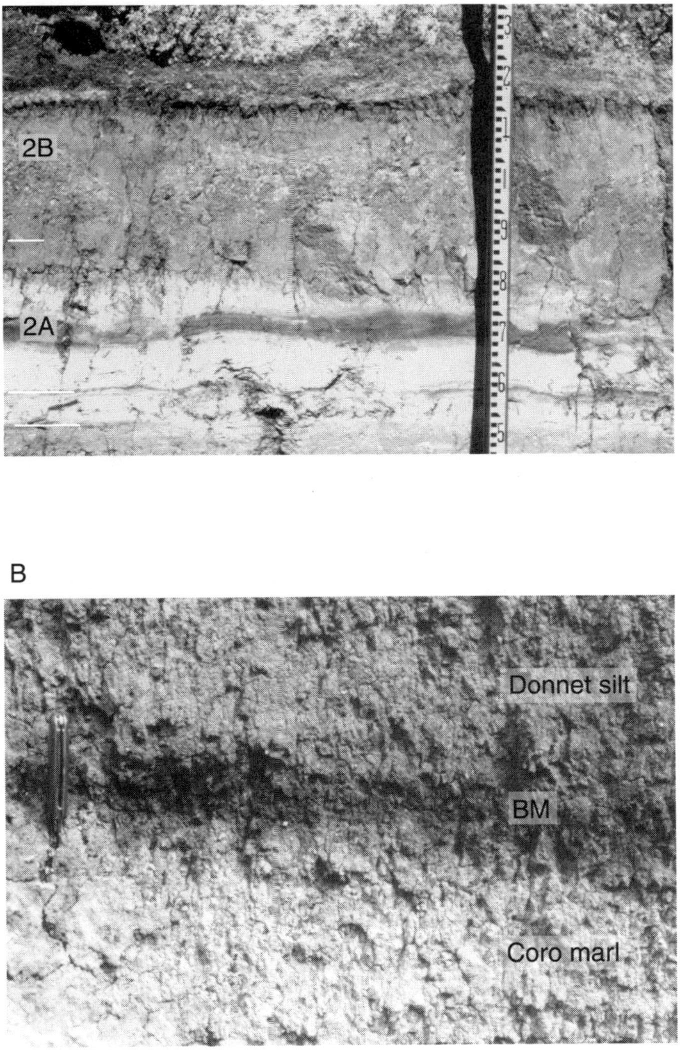

Figure 8.5 "Black mats" at archaeological sites in the western United States. (A) Stratum 2 at the Lubbock Lake site, Texas. Stratum 2A is mostly diatomite (white) with interbeds of organic-rich mud (dark gray bands), which locally contain Folsom occupation features. Stratum 2B is an organic-rich mud that aggraded slowly from ~10,000 to ~8500 yr B.P. The Firstview soil is apparent, formed in upper 2B, probably as a "wet meadow" or "cienega" (spring-fed marsh) soil. (B) The "black mat" (BM or "Clanton Clay") in Curry Draw, near the Murray Springs (Clovis) site, Arizona. This soil dates from ~10,600 to 9700 yr B.P. and is draped over the Clovis features at Murray Springs (Haynes, 1987). The finer structure typical of A horizons is well expressed.

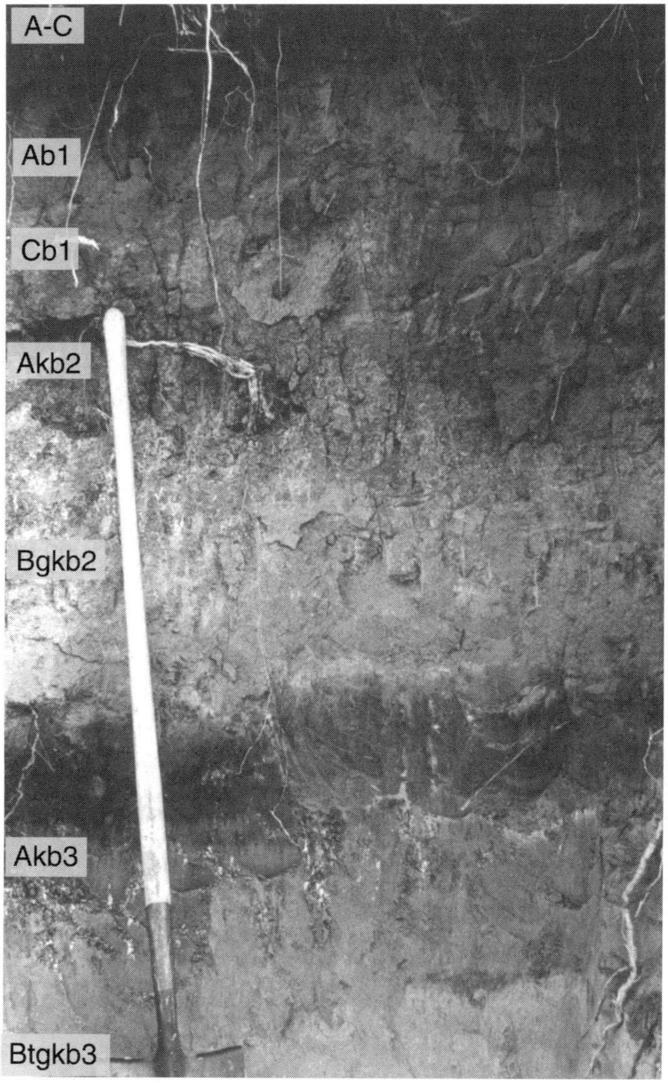

Figure 8.6 A "black mat" and younger buried soils at the Carter/Kerr-McGee site, Wyoming (modified from Reider, 1990, fig. 7; reproduced with permission of the Geological Society of America and R. G. Reider; photo provided by R. G. Reider). The Akb3 "black mat" is Clovis-Folsom age. It probably formed as a kind of "wet meadow soil" that also resulted in gleying of the B horizon below. The b2 soil contains Agate Basin, Hell Gap, and Cody Complex occupation debris. The carbonate in the b3 and b2 soils probably is from the overlying soils and represents welding of the soils. Horizon designations are modified from the original publication following the conventions presented in appendix 1.

At the Laddie Creek site, in the higher elevations of the Big Horn Mountains of Wyoming, the most striking buried soils are "black mats" associated with an intense Archaic occupation (Reider and Karlstrom, 1987). The soils are characterized by overthickened, dark-gray-to-dark-brown (values and chromas of 3/2, 4/2, 4/4 dry) A horizons (>1.25 m) with organic matter content of ~0.5%–1.0%, which is relatively high for a buried soil in an arid or semiarid setting. The soils formed in colluvium that slowly aggraded in the early to middle Holocene (~9000–6000 yr B.P.). The high production of organic carbon and the accumulation of colluvium is attributed to spring discharge, based on comparisons with modern springs in the area. In the lower elevations of the region the early to middle Holocene was characterized by arid Altithermal conditions. The presence of active springs likely accounts for the occupation intensity.

The distribution of peats provides an approach to using soils for environmental reconstruction in the high Arctic of Canada (Valentine et al., 1987; Tarnocai and Valentine, 1989; Van Vliet-Lanoë, 1998). With retreat of glacial ice and regional warming following the last glacial maximum, permafrost disappeared from much of the Canadian Arctic, and peat soils were allowed to form. These conditions were apparently obtained in the early to middle Holocene (during the Altithermal or "Hypsithermal"; Zoltai, 1995). With the return of colder conditions in the late Holocene, peat development in the Arctic ceased, shifting south to the subarctic and boreal regions. These organic-rich soils became relict and heavily modified by cryoturbation.

Loess—General

The most numerous, extensive, and widely investigated buried soils in the world are found in the vast loess deposits that blanket parts of the interiors of North America, Europe, Asia, and South America (see also chapter 6). They have long been used as stratigraphic markers (chapter 5) and, in the last decades of the 20th century, were undergoing considerable scrutiny as paleoenvironmental indicators (e.g., Schultz and Frye, 1968; Kukla, 1975, 1977; Wasson, 1982; Pécsi, 1984a, 1987, 1991; Liu, 1987, 1988; Pye, 1987, pp. 198–265; Cremaschi, 1990; Liu et al., 1991; Zarate et al., 1993; Pécsi and Richter, 1996; Pécsi and Velichko, 1995; Iriondo, 1999; Derbyshire, 2001, 2003). In the Old World, archaeological sites are widely reported from loess (e.g., Movius, 1944; Aigner, 1981, pp. 174–244; Kurenkova et al., 1995; Ranov, 1995; Davis and Ranov, 1999), though there is surprisingly little information on the specific relationship of the occupations with the soils or the environmental implications of the archaeologically associated soils.

Several approaches have been taken to interpreting the paleoenvironments of loess pedogenesis. A fairly straightforward approach is the recovery of plant and animal fossils such as molluscs, phytoliths, and pollen associated with the soils (e.g., Kukla, 1975, 1977; Lozek, 1990; Rousseau, 1990; Engel-DiMauro, 1995; Keen, 1995; Rousseau et al., 2001). Gastropods appear to be especially abundant in many loess sections. In the absence of these remains or, more commonly, in the absence of facilities or funds to carry out such studies, soil morphology (macro- and micromorphology) and loess characteristics are used to make paleoenvironmental interpretations. There are few examples of this approach in an expressly geoarchaeological context, however. A review of some case histories provides insights into the potential of using soils as environmental indicators in aggrading eolian environments, as well as some of the pitfalls of such an approach.

Broadly speaking, for glacially derived loess the episodes of loess deposition coincide with glacial or stadial periods, and loess stability and soil formation occur during interglacial or interstadial periods (e.g., Kukla, 1970, 1975, 1977; Bronger and Heinkele, 1989b; Semmel, 1989; Antoine et al., 1999; Kemp, 2001). For loess in China derived from the deserts of Asia, the loess deposition is correlated with a weak summer monsoon (during a glacial period) bringing cold dry conditions, whereas relative landscape stability and pedogenesis dominated under a stronger summer monsoon (during an interglacial), bringing relatively warmer and more humid conditions (e.g., An et al., 1990, 1991; Kemp et al., 1996). Specific episodes of loess deposition have proven difficult to date, however, and they probably vary from region to region, depending on the source and proximity of the loess and the rate of loess deposition, among other factors. Kemp (2001) also makes an important point regarding the geologic interpretation of soils buried in loess sequences (following McDonald and Busacca, 1998): The most useful way to think about them is as representing shifts in the balance between loess deposition and loess pedogenesis (fig. 5.10). Changes in this balance may be driven by regional environmental conditions, but the relationship between the soils and the environment may not be as direct as proposed by some (e.g., Morrison, 1978).

Loess in Central and Eastern Europe

The buried soils in the loess of Europe have long been the basis of paleoenvironmental reconstructions at glacial–interglacial time scales. Little of this work is directly geoarchaeological, but it provides excellent examples of the kinds of interpretations that can be made from soil morphology given proper checks based on independent sources of paleoenvironmental data.

The classic example of environmental interpretations of buried soils in loess is the work of Kukla (1970, 1975, 1977) in (then) Czechoslovakia and Austria. His soil stratigraphic correlations with the oxygen isotope record of the deep sea confirmed the effect of Milankovitch cycles on the Quaternary terrestrial stratigraphic record (Kukla, 1970, 1975, 1977; Imbrie and Imbrie, 1979; Smalley et al., 2001). Discussions and interpretations of the soils are presented in several key papers (Kukla, 1970, pp. 150–154; 1975, pp. 109–113; and 1977, pp. 326–329). The basic approach was to compare the morphology of the buried soils with the surface soils of Europe and then make inferences regarding the vegetation and climate conditions of pedogenesis. Simple comparisons of profile morphologies can produce misleading interpretations resulting from equifinality (chapter 3), but Kukla's case is a strong one because he was making comparisons with modern surface soils in the region in settings similar to the buried ones and because of the repetition of sequences of soil types through a section and from section to section. Further, abundant gastropod assemblages were recovered from some buried soils, providing paleoenvironmental data that confirmed the soil interpretations.

The loess soils of Central Europe are buried within loess deposits that span the Quaternary. The principal buried soils are pedocomplexes (PK; figs. 6.16, 8.7, and 8.8). Each pedocomplex represents a set of stacked soils sandwiched between unweathered loess. Apparently, between the intervals of major loess deposition (i.e., during interglacial and interstadial periods), the soils formed on surfaces that aggraded slowly and episodically because of additions of slope wash, colluvium, and perhaps minor amounts of loess. Some if not most of the soils in each pedocomplex probably are welded, though this is unclear from the discussions. Each cycle of full-glacial loess deposition was followed by stability and soil formation during the subsequent interglacial. These soils are the best expressed in the PK sequence and include (in German soil terminology) "parabraunerdes" (or "podzolic forest soils"; Udalfs and Aqualfs) and "braunlehms" (Xeralfs, Udalfs; e.g., PKIII formed during the last interglacial; figs. 6.16 and 8.7). The parabraunerdes have decalcified sola with Bt horizons, including minimally expressed argillic horizons, and are typical of soils under mixed hardwood forests in southern and Central Europe today. The braunlehms are more strongly developed soils with redder and more clay-rich argillic horizons and are typical in the Mediterranean environments of Europe under deciduous forests. These soils could be used to indicate warmer, more intense weathering conditions, but all other proxy indicators are indicative of conditions identical to those of braunerde formation, and the soil morphologies are polygenetic (though no descriptive details are provided).

Typically found above the soils with Bt horizons are "Braunerdes," "Chernozems," and "Rendzinas" (from German, Russian, and FAO-UNESCO terminology) formed in thin loesses (e.g., PKII; figs. 6.16 and 8.7). The lower, older soils in this part of the sequence begin as A-Bw soils (Braunerdes) with "Chernozemic" (humus-rich) A horizons welded to the top (Chernozems are the classic grassland soils). The upper, youngest soils in this sequence have simple low-humus A-C profiles (Rendzinas). The morphologies and faunas of these soils are indicative of cool to cold continental grasslands, similar to the steppe regions of central to northern Europe and central Asia. This sequence of loess and soils is correlated with the fluctuating cold/warm but progressively colder environments that appear to characterize the early stages of glaciation (e.g., oxygen isotope stages 5a–d; figs. 6.16 and 8.7).

Above the weakly to moderately expressed early glacial soils are thin, unweathered deposits of slopewash and eolian sand, and thick loess. Some of these deposits contain thin, weakly expressed Chernozems, Rendzinas, and frost-gley or frost-pseudogley soils (figs. 6.16 and 8.7). These soils and their faunas are indicative of cold conditions, including permafrost. A somewhat better-expressed Braunerde welded to a Rendzina is found just below massive full-glacial loess and is correlated to slightly warmer interstadial conditions (e.g., PKI and oxygen isotope stage I; figs. 6.16 and 8.7).

These sediment/soil cycles repeat through the section and closely match (in time and in climatic implications) the record of glacial–interglacial cycles suggested by the oxygen isotope record from the oceans (see also chapter 6). Furthermore, the terrestrial record mimics the isotope record by indicating rapid deglaciation followed by slow onset of the next glacial phase. A broadly similar

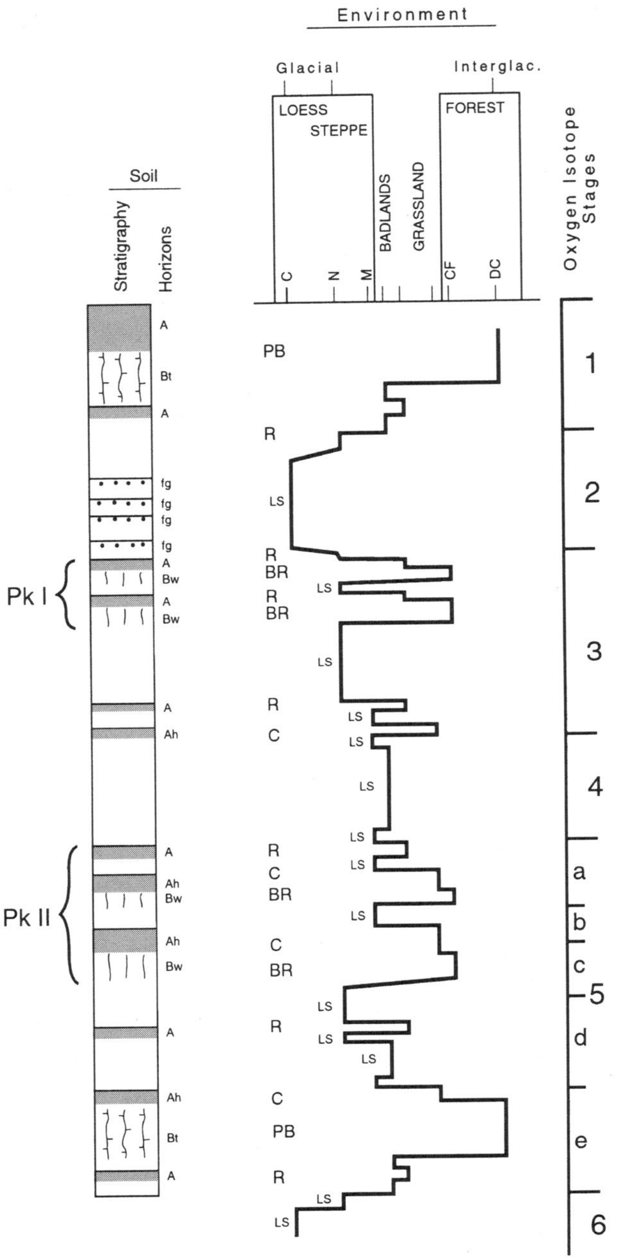

Figure 8.7 Interpretations of the late Quaternary paleoenvironments of Eastern Europe based on loess and soil stratigraphy (based on Kukla, 1977, fig. 6). For the environmental interpretation (Loess Steppe), C = cold, N = normal, M = mild, (Forest) CF = coniferous forest, DC = deciduous forest. Abbreviations to the left of the curve describe soils and sediments: PB = Parabraunerde, R = Rendzina, C = Chernozem, LS = Loess. European pedocomplex notation (Pk) added for comparison to fig. 6.16. Buried soil notation not included in horizonation.

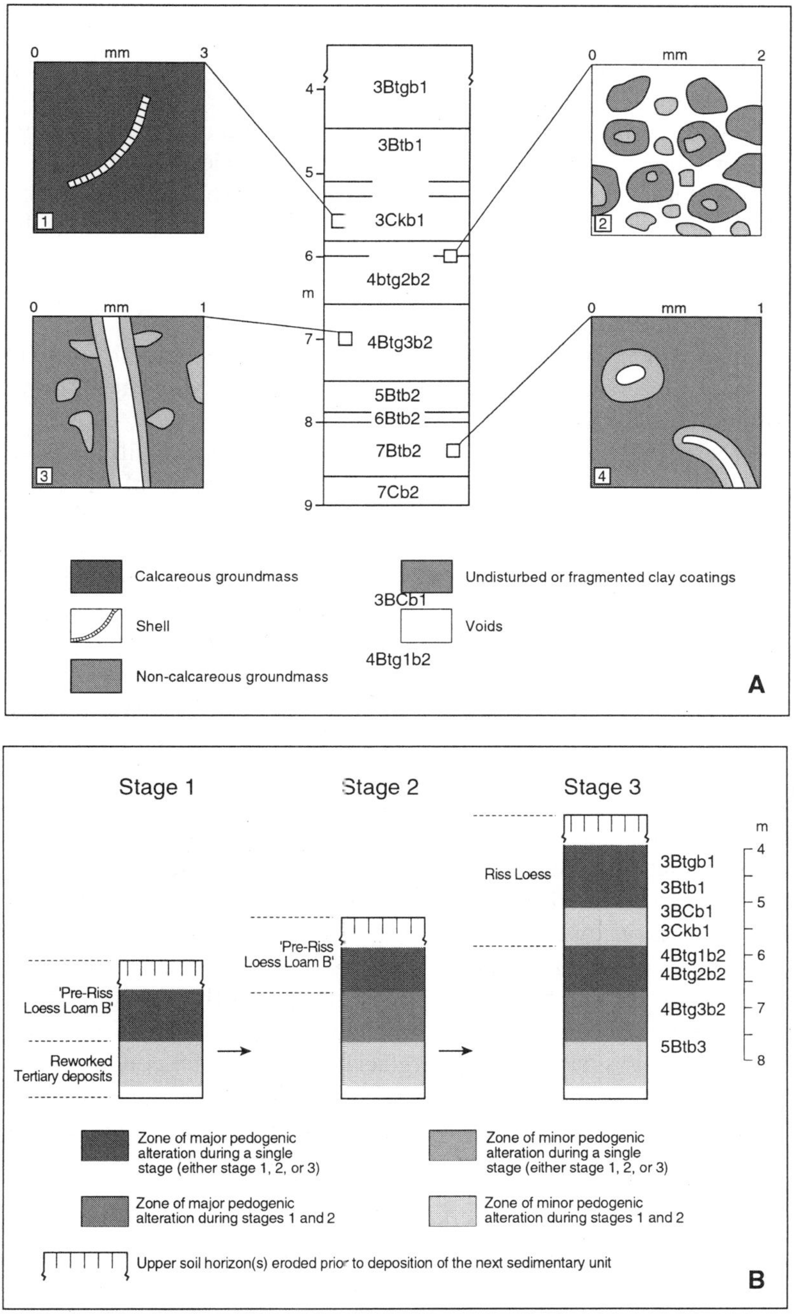

Figure 8.8 The result of thin-section and field studies of a complex, welded buried soil in loess near Attenfeld, southern Germany. (A) Schematic illustration of key micromorphological fabrics in the buried soil complex (reprinted from *Catena*, v. 35, R. A. Kemp, "Micromorphology of loess-paleosol sequences: A record of paleoenvironmental change," pp. 181–198, fig. 3, © 1999, with permission from Elsevier Science). (B) Schematic reconstruction of the stages of sedimentation and weathering leading to development of the welded pedocomplex (reprinted from *Catena*, v. 35, R. A. Kemp, "Micromorphology of loess-paleosol sequences: A record of paleoenvironmental change," pp. 181–198, fig. 4, © 1999, with permission from Elsevier Science).

interpretation of soils buried in loess, but described and discussed in more detail and from northwestern France, is provided by Antoine et al. (1999).

In a study in Germany, Kemp et al. (1994) showed that some pedocomplexes in European loess span several glacial–interglacial cycles, based on micromorphological features (fig. 8.8). A buried soil complex exhibited undisturbed clay coatings—indicative of clay illuviation during an interglacial period—but also fragmented clay coatings, some incorporated within rounded aggregates of silt, because of cryogenic disruption during onset of the subsequent glacial stage (fig. 8.8). Several such cycles were identified in a single pedocomplex, based on thin section studies. The upper part of each soil apparently was eroded before the subsequent phase of deposition. Pedogenic processes then penetrated the younger sediment and altered the soil buried beneath it (fig. 8.8).

On the Dnieper Plain in the Ukraine of eastern Europe, multiple buried soils in loess provided an opportunity to compare the local record of late Pleistocene paleoenvironments with the regional model of loess deposition, soil formation, and paleoenvironments developed in Central Europe (Rousseau et al., 2001). The sequence here differs from that in Central Europe by containing a stacked sequence of deposits and soils rather than welded soils. The interpretations are based on soil micro- and macromorphology as well as palynology and low-field magnetic susceptibility. Three pedocomplexes spanning the time from marine oxygen isotope stage 5 to stage 2 were identified (fig. 8.9). The Kaydaky and Pryluky pedocomplexes are correlated with interglacial conditions of oxygen isotope stage 5. The oldest soil, the Kaydaky, appears to represent the maximum interglacial warmth (~stage 5e), based on the recovery of deciduous pollen taxa and the "interglacial character" of the soil; that is, relatively strong degree of pedogenesis (leached A-Bt horizonation). Deposition of the Tyasmyn Loess and the parent materials for the multiple Pryluky soil complex are believed to be related to deposition by continental dust storm. Pollen is indicative of cooler, drier steppe-like conditions, and the Pruluky soils are interpreted as forming initially under forest, but with a trend toward grasslands through time; that is, the Pryluky complex may be reflecting, in a rough sense, the fluctuating transition of oxygen isotope stages 5a–d from interglacial to full glacial conditions. There is no way, however, to sort out the relative effects of duration of pedogenesis among these two soils and the underlying Kaydaky soil. The Uday Loess accumulated under dry conditions with one phase of stability and pedogenesis, perhaps under more moist conditions.

The Vytachiv pedocomplex and the overlying Bug loess reflect the glacial conditions of oxygen isotope stages 4, 3, and 2. The soils of the Vytachiv complex are leached but otherwise weakly expressed, interpreted as forming under cold, open forest conditions, with burial of the lower soil by loess under cold, dry conditions. The Bug Loess was deposited under the cold, dry full glacial conditions of oxygen isotope stage 2.

The sequence of buried soils and loesses of the Dnieper Plain appear to nicely reflect the glacial–interglacial, and interstadial cycles recorded in the marine oxygen isotope record. They are also well supported by pollen data. This record tends to support the environmental interpretations made for the pedocomplexes of Central Europe. In the absence of age control, stratigraphic provenience, and

Strats	Soil	Horizons*	Soil Types/Lithology	Micromorphology
Holocene	Hlb2	Ah AB	Brown forest soil (Luvisol or Alfisol)	
	Hlb1	Btf		
	Hla	Bw		
Bug Loess	bg		Loess	Microortsteins & manganese hydroxide segregations
Vytachiv Complex	vt_{b2}	A Bw	Boreal brown & dark brown soils (*Cambisol or Inceptisol*)	Microcrystalline calcite
	vt_{b2-1}	Ck	Loess	
	vt_{b1}	A Bw Bk	Boreal brown & dark brown soils (*Cambisol or Inceptisol*)	Microcrystalline calcite
Uday Loess	ud	Ck	Loess	
Pryluky Complex	plc	Ah Bt Bk	Boreal brown soil (*Alfisol?*)	Cutans of optically-oriented clay; thin carbonate films around pores; carbonate microconcretions; calcite in plasma
	plbx	Ah Bt	Brown/gray forest soil (*Luvisol or Alfisol*) with "leached Chernozem" (*Mollisol*)	Cutans of optically-oriented clay
	pla	A C	"Meadow soil" (*Luvisol or Alfisol*)	
Tyasmyn	ts		Loess	
Kaydaky Complex	kd_{b2}		Leached "meadow Chernozem" (*Mollisol*)	
	kd_{b1}	A-E-Bt-Bth-Btf-Cg	Parabraunerdes; gray and light-gray forest soil (*Luvisol or Alfisol*)	Cutans of optically-oriented clay; particles of clayey humus

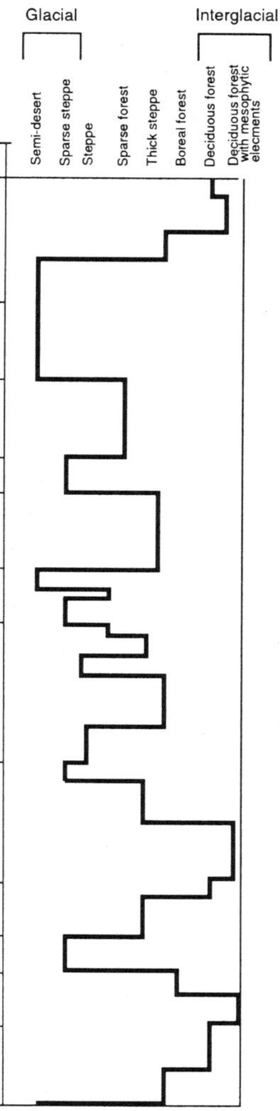

Figure 8.9 Stratigraphy and environmental interpretations of soils in loess on the Dnieper Plain (based on Rousseau, 2001, figs. 2 and 4, table 1).

the pollen, however, the individual soils would be difficult to interpret environmentally. For example, the weakly expressed soils of the Vytachiv Complex could, given time, evolve into the well-expressed soils of the Pryluky or Kaydaky complexes.

Loess in Central Asia

Probably the best-known archaeological record in a loess–soil stratigraphic context (in English-language literature) is in Central Asia, particularly Tajikistan (e.g., Dodonov, 1991; Ranov, 1995; Davis and Ranov, 1999). Paleolithic sites dating back to ~1 Ma are reported in association with soils buried in the loess. Descriptive data on the pedology or soil geomorphology are scarce, however, and paleoenvironmental interpretations based on the soils diverge considerably. Most of the stratigraphically significant soils are described as pedocomplexes (Lazarenko, 1984; Dodonov, 1991; Bronger et al., 1995); that is, they consist of several closely stacked and even welded soils. The typical morphology of each pedocomplex includes a relatively well-expressed, rubified Bw-Bk or Bt-Bk profile in the lower part of the complex, overlain by more weakly developed and less leached Bw-Bk soils with increasing amounts of soluble salts. Several investigators studied the soils, and each has somewhat different descriptions and interpretations, though all see the soil complexes as indicative of interglacial conditions, with a trend toward aridity accompanying the gradual onset of glacial conditions. Lazarenko (1984, pp. 127, 129) and Dodonov (1991, p. 188) provide similar genetic classifications of each pedocomplex (e.g., table 6.4) that appear to oversimplify the soil characteristics. Dodonov (1991) proposed that the main Bw-Bk or Bt-Bk complex represents soils forming first under drier conditions, producing the Bk horizon, that then evolved into conditions of "optimum . . . humidity and paleotemperature" (Dodonov, 1991, p. 188) and formation of the redder Bw or Bt horizon. A shift from drier to more humid conditions, however, should result in removal or at least partial destruction of the Bk. Lazarenko (1984) sees the main Bw-Bk or Bt-Bk part of each complex as genetically related and forming under semiarid to subhumid tallgrass steppe or steppe/open forest. This seems reasonable because it is a more straightforward interpretation supported by somewhat more complete descriptive data plus some proxy evidence from molluscs and pollen.

Bronger et al. (1995) criticized Lazarenko's and Dodonov's interpretations based on general descriptive data and applied micromorphology to the interpretation of the pedocomplexes. Micromorphology aided in documenting and quantifying clay illuviation and in differentiating primary from secondary (pedogenic) carbonate. Their approach to environmental interpretation of the buried soils is also problematic, however. Their conclusion is that the main soil in each complex formed under subhumid broadleafed forest. They assume that climate and vegetation are the two most important soil-forming factors in the region and that the Holocene soils in the region offer the key to past soil-forming processes (Bronger et al., 1995, pp. 69–70; Bronger et al., 1998). This approach ignores the role of time and equifinality. A wide range of environmental conditions could produce soils with Bw, Bt, and Bk horizons. For example, prolonged pedogenesis under drier conditions could produce reddish Bt horizons as well as Bk horizons. Bronger et al. (1995, p. 77) also believed that clay translocation occurs only under humid, forested conditions. Dust infiltration under more arid conditions is clearly demonstrated in a variety of soil geomorphic studies (e.g., Syers et al., 1969; Gile et al., 1981; Holliday, 1988; Reheis et al., 1995). Bronger and colleagues correctly point out the possibility that most of the pedocomplexes underwent recalcification following burial, thus weakening interpretations of the Bk horizons. However, the Bk horizons tend to be in the proper position (i.e., at the base of each soil in each complex) for horizons genetically related to the B (Bronger et al., 1995, figs 4 and 5).

Loess Plateau, China

The thickest and most extensive accumulation of loess in the world is the Loess Plateau of China (e.g., Kukla, 1987; Pye, 1987, pp. 200, 246–250; Zhang et al., 1991). These deposits contain many buried soils (e.g., Liu et al., 1985; Kukla, 1987; Zhang et al., 1991; Derbyshire et al., 1995a,b) and also produce evidence of human occupation (e.g., Movius, 1944; Jia, 1980; An et al., 1987; Madsen et al., 1998). Unfortunately, however, little of this archaeological record has been clearly linked to the soils, and the soils at the archaeological localities are not well described. Nevertheless, the soils hold considerable potential for paleoenvironmental reconstructions (e.g., Bronger and Heinkele, 1989a,b; Kemp et al., 1996; Han et al., 1998; Kemp and Derbyshire, 1998), and a review of several approaches is instructive for what may be learned regarding hominid environments.

Perhaps the best known evidence for ancient human activity associated with the loess soils are the fossils of "Lantian Man" from the southeastern Loess Plateau near Xian (Jia, 1980, pp. 13–18). These *Homo erectus* remains are from two different buried soils, widely separated in time, from two localities. The loess sequence is referred to on the basis of the "type section" or reference section at Luochuan, about 200 km north of Xian (Liu et al., 1985; Kukla and An, 1989). The older of the two discoveries is from the Gongwangling section, where the fossil material was found in or in association with a zone of carbonate nodules in loess layer L15 (An et al., 1987; An and Ho, 1989) and dated to ~1.15 Ma (An and Ho, 1989). This zone probably is a truncated soil (An and Ho, 1989). Above L15, stone artifacts were recovered from S14 (An et al., 1987). The vertebrate fauna found in the nodule zone is indicative of a warm, forested environment similar to South China.

Initial stratigraphic work at the Gongwangling locality resulted in some difficulties in reconciling the fauna with the implications of a cold, arid environment based on loess deposition (An et al., 1987, p. 201). Subsequent recognition of the carbonate zone as a truncated soil rectified the situation to some degree, but the faunal implications of a humid environment are still seemingly in contrast to the suggestion of drier conditions based on the genesis of carbonate nodules. However, (1) the exact relationship of the vertebrate fauna and the human remains to the carbonate nodules are not clear; thus, the carbonate could postdate the fossils; and (2) carbonate accumulation is associated with some soils formed under steppe woodlands (Eyre, 1968, pp. 253–254). Buried soil S14 probably is a "leached cinnamon soil" if the Luochuan section can serve as a proxy (Liu et al., 1985, fig. 7). This "moderately weathered polygenetic soil" (Kukla and An, 1989, p. 219) is equivalent to the parabraunerde of Europe (Kukla, 1987, p. 195, table 1; i.e., Udalfs and Aqualfs). On the basis of the author's visit to the section, the pedocomplex could be considered an overthickened (3–4 m) Alfisol of some kind. An Alfisol that thick but leached of carbonate probably formed under subhumid or wetter conditions.

The younger set of hominid remains from the loess soils was found in the Chenjiawo section. The fossils were from the uppermost of a three-soil pedocomplex (S6; An et al., 1987; An and Ho, 1989) dated to ~0.65 Ma (An and Ho,

1989). At the Luochuan section, S6 is a "leached cinnamon soil" similar to S14 (Liu et al., 1985, fig. 7). The S6 at Chenjiawo, however, is more strongly developed than at Luochuan (An et al., 1987, p. 194; An and Ho, 1989, p. 216) and is thus more likely to be a "luvic cinnamon soil" (Liu et al., 1985, fig. 7), which is still a parabraunerde (Kukla, 1987, p. 195, table 1), though perhaps either thicker or more clay rich. Again the implications are for a temperate climate. The fossil flora and fauna is indicative of forest steppe or savanna under warm, seasonal, semiarid to subhumid conditions (An et al., 1987, p. 201; An and Ho, 1989, p. 219).

The S14 and S6 soils apparently formed under somewhat similar conditions; the S14 climate was perhaps more humid. The S6 soil, however, is more strongly developed, indicating that it had more time to form. In any case, the conventional interpretation of the soils, along with the paleobiology of the Loess Plateau, show that *Homo erectus* lived in the area in the Early and Middle Pleistocene under warm, semiarid to subhumid conditions, in some kind of forest steppe setting. These interpretations may have to be reconsidered to some degree, however, based on study of S5 (Liu et al., 1985; Han et al., 1998), which is just above the S6. The S5 soil is the most strongly expressed of the buried soils of the Loess Plateau. It has a morphology generally similar to that of the other well-developed soils, but it is thicker and darker brown and contains more illuvial clay. Similar to the other soils, it was believed to represent forested conditions, but formed under some sort of "climatic optimum" warmer and wetter than today (Liu et al., 1985).

Bronger and Heinkele (1989a) come to similar conclusions, arguing that "stronger" development (5YR Bt horizon, 2m thick) must be the result of a warmer and more moist environment compared to the Holocene. Recent work completely revises that notion, showing that the soil probably formed under semiarid grasslands (Han et al., 1998). This is based on (1) particle-size data, showing a high content of silt in the A horizon, perhaps from dust; (2) presence of 2:1 clays, which are less stable in more humid environments; and (3) $\delta^{13}C$ values (discussed in a following section) indicative of grasslands. The strong soil development is attributed to time rather than intense pedogenesis. This study points out that morphologies conventionally believed to represent humid, forested conditions (decalcified reddish or reddish-brown Bt horizons) can also form under drier conditions given time. Interpretations of the other soils buried in the Loess Plateau may have to be reconsidered as well as those with similar morphologies in semiarid Tajikistan.

Hamra and Husmas Soils, Israel

The distinctive red soils formed in eolian sand along the coastal plain of Israel are referred to as hamra and husmas soils (chapters 6 and 9; e.g., Dan et al., 1968; Dan and Yaalon, 1971; Gvirtzman et al., 1998, 1999; Tsatskin and Ronen, 1999; Gvirtzman and Wieder, 2001; Frechen et al., 2002). They are common in stratigraphic successions and are important components of the Quaternary soil geomorphic and geoarchaeologic record of the region, particularly at Paleolithic sites (see also chapters 6 and 9). The hamra and husmas soils also are increasingly recognized for their potential to yield paleoenvironmental information (e.g.,

Gvirtzman et al., 1998; Wieder and Gvirtzman, 1999; Gvirtzman and Wieder, 2001), but relatively little of this work is in an archaeological context.

The study by Gvirtzman and Wieder (2001) provides a good example of the approaches taken in the paleoenvironmental interpretation of hamra/husmas sequences. Their study dealt with upper Quaternary buried soils along the central coastal plain. The approach to interpreting the soils is essentially the same as that taken in studies of the loesses of Central and Eastern Europe, discussed above. Eolian sediments are interpreted as resulting from arid environments, and pedogenesis is linked to stable, vegetated, wetter environments, varying in duration and intensity. Some proxy-climate indicators are available from the region to support the interpretations. The sequence consists of sandy eolian deposits, carbonate-cemented dune sand (kurkar), and buried soils (hamra/husmas and others) (table 8.3). The chronology is well dated by radiocarbon, providing age control for the duration of pedogenesis, which helps to constrain the time factor in assessing the effect of environment on the soils.

Six buried soils were identified in the sequence: (from bottom to top) the Mikhmoret Hamra, Ein Haqore Grumusol (a Vertisol), Ganot Hadar Hamra, Lower Raqit Regosol (an Entisol), and two Entisols within the Ta'arukha Sands (table 8.3). Each soil was buried by eolian sediment (typically, but not exclusively, dune sand), representing cyclic aridity. The four lower soils (Mikhmoret, Ein Haqore, Ganot Hadar, and Lower Raqit) make up the formally recognized "Netanya Hamra" or "Netanya Palaeosols" of the Hefer Formation; that is, in this area the Netanya Hamra consists of multiple buried soils (chapter 6). The Mikhmoret is typically a hamra soil (leached of carbonate), except in some sections in which carbonate from above was translocated into it (i.e., welding it), forming a husmas soil. The Mikhmoret soil is very well developed: thick (2.0–3.5 m), dark reddish-brown (5YR 3/3), with significant amounts of illuvial clay displayed as clay films on ped faces and in thin section. The soil formed under conditions significantly more moist than today. This is indicated by several characteristics. The soil parent material was highly calcareous kurkar, but it is fully leached of carbonate except in the above-noted husmas facies. Much of the illuvial clay is from dust (indicated micromorphologically by silt grains in the plasma and as coatings on large sand grains), which requires significant moisture for translocation. The Ein Haqore soil is a Grumusol or Vertisol, indicating seasonal contrasts between wet and dry. It is also mostly to completely leached of carbonate and contains illuvial clay and silt from dust.

The Ganot Hadar is another hamra, but with more yellow and brown (5YR 5/6, 4/4) in the Bt horizon and otherwise not as strongly expressed as the Mikhmoret soil. It is thinner (1 m) and has significantly less illuvial clay than the older hamra. This soil also likely formed under a climate wetter than today, but its weaker expression compared to the deeper hamra is the result of the significantly shorter duration of time and dust influx (2500 yr vs. >25,000 yr; table 8.3)

The Lower Raqit soil and the two soils within the Ta'arukha Sands are all weakly expressed (table 8.3). The Lower Raqit has a dark, organic-rich A horizon over sand; the Ta'arukha soils have less well-developed A horizons. The Lower Raqit soil had the same or less time to develop as the Ta'arukha soils (table 8.3), so the interpretation of more moist, heavily vegetated conditions for

Table 8.3. Soil-geomorphic evolution of Upper Quaternary hamra and kurkar along the coast of Israel

Stratigraphic unit	Soil Horizon	Duration (ka)	Description (soil type)[1]	Micromorphology
Hadera sand	C	1.3 to 0	Dune sand	
Ta'arukha 2 soil	A C	4.0 to 1.3 (2.7)	"Thin soil crust" (Xerorthent?)	
Ta'arukha sand 2		4.6 to 4.0 (0.6)	Dune sand	
Ta'arukha 1 soil	A C	~5.0 to 4.6 (~0.4)	"Thin soil crust" (Xerorthent?)	
Ta'arukha sand 1		~6.5 to ~5.0 (~1.5)[2]	Dune sand	
Tel Aviv Kurkar			Dune sand; partly cemented with carbonate	
Netanya Paleosols				
Raqit sand			Dune sand	
Lower Raqit soil	Ah C	~7.0 to ~6.5 (~0.5)	Dark, organic-matter rich; ~1.0 m (Regosol or Xerorthent)	
Sand		~7.5 to ~7.0 (~0.5)	Dune sand	
. . .				
Ganot Hadar soil	Bt C	~10.0 to ~7.5 (~2.5)	Reddish brown; 1.0–1.5 m (immature Hamra or Rhodoxeralf)	Illuvial clay as bridges, coats on grains and voids
Sand		~10.5 to ~10.0 (~0.5)	Dune sand	
Ein Haqore soil	Bss Bk[3]	~11.5 to ~10.5 (~1.0)	Gray-brown; 0.7–1.5 m (Grumusol or Xerert	Common pressure cutans; illuvial silty and clayey dust
Parent material of Ein Haqore soil[4]		~12.5 to ~11.5 (~1.0)	"Loessial" dust	
Mikhomoret soil	Btk[3] Bt C	~40.5 to ~12.5 (~27.5)	Dark reddish brown; clayey with abundant clay films; 2.0–3.5 m (Hamra/husmas or Rhodoxeralf)	Common pressure cutans; illuvial silty dust; disrupted illuvial clay coats
Dor Kurkar			Dune sand; cemented with carbonate	

From Gvirtzman and Wieder (2001, tables 2, 3).

[1] Israel Soil Classification (FAO-UNESCO) and U.S. Soil Taxonomy.

[2] Age of deposition of Ta'arukha sand 1, Tel Aviv Kurkar, and Raqit sand not differentiated.

[3] Bk is nodular carbonate and localized.

[4] No unweathered parent material reported.

development of its soil may be correct. Likewise, given that the Ta'arukha 2 soil and the Ganot Hadar soil developed for about the same amount of time (~2500 yr; table 8.3), but the former is significantly more poorly expressed than the latter (Entisol vs. Hamra), the Ta'arukha 2 soil probably developed under much drier conditions than the Ganot Hadar soil.

Soil-Forming Intervals

The concept of the "soil-forming interval" is also important in the use of soils for climate reconstructions. This idea was formulated and expanded on primarily by Morrison (1967, 1978) and commonly applied in the western United States. The soil-forming intervals were defined as discrete periods of accelerated soil formation related to particular climatic episodes. Classically, the soil-forming intervals were related to warmer and wetter climate. Little soil formation would take place during intervening cooler and drier episodes. This approach has been criticized by some (summarized by Birkeland and Shroba, 1974, and Birkeland, 1984, pp. 330–334; 1999, pp. 309–311), who argue that basically weathering and soil formation always occur when the landscape is stable, although specific aspects of pedogenesis may vary as a function of climate and landscape setting. Kemp (1999, 2001) also argues that some landscapes cannot be simply classified as either stable or unstable and are better understood in terms of a balance between rates of sedimentation and rates of pedogenic processes, which may or may not be directly in phase with particular environments.

The best example of the soil-forming interval in archaeological pedology in North America is the "Altithermal soil." This is a soil profile, typically buried, reported from many Holocene, primarily alluvial, stratigraphic sequences in the central and western United States (e.g., Leopold and Miller, 1954; Malde, 1964; Haynes, 1968; Reider, 1980, 1982a,b, 1990; Albanese, 1982). The soil is characterized by a moderately well-developed Bw or Bt horizon underlain by a distinct zone of calcium carbonate accumulation (Bk; fig. 8.6). The carbonate accumulation in particular is taken to represent pedogenesis under drier conditions related to the middle Holocene "thermal maximum" or Altithermal (Reider, 1990). Little data are available to show that pedogenesis is related to drier or warmer conditions; moreover, the duration of genesis of the Altithermal soils at most localities is not well established. In the absence of such data, an alternative hypothesis is that such a soil is not related to a short interval of drier climate, but formed over a longer period of time following the Altithermal. For example, at the Clovis site on the Southern High Plans, Haynes (1975) reported a well-developed soil (Bt-Bk profile) believed to have formed in the early to middle Holocene and considered to be related to a short period of warmer and drier climate. Data now show that at that site and across the Southern High Plains the middle Holocene or Altithermal was characterized by eolian deposition and landscape instability rather than soil formation (Holliday 1985a, 1989b, 1995). The well-developed soil observed at Clovis and reported at many other sites in the area in the same stratigraphic position formed in the late Holocene (post-5000 yr B.P.; Holliday 1985a,c,e, 1995). Furthermore, late Holocene soils on the Great Plains, including some formed in the last few hundred to a thousand years, show evidence of the same pedogenic characteristics as the "Altithermal soil" (i.e., development of both argillic and calcic horizons; Machette, 1975; Holliday 1985d, 1988, 1995; McFaul et al., 1994), indicating that these soil features can form under modern conditions and are not unique to arid "thermal maximum" conditions. On the Great Plains at least, the drier and probably warmer conditions of the middle (and perhaps early) Holocene resulted in erosion and deposition and general landscape instability. The genesis of the common Bt-Bk profiles on the middle Holocene deposits took place under more geomorphically stable conditions of the late Holocene.

Soil Classification and Environmental Reconstruction

In the 1980s and 1990s, a significant interest in pre-Quaternary soils emerged with a focus on the use of soil or soil-like features for paleoenvironmental interpretations (Retallack, 1983, 1990, 1997a, 2001; Wright, 1986; Reinhardt and Sigleo, 1988). One of the methods that evolved from this work is the use of soil classification based on the U.S. soil taxonomy as a direct indicator of past environments (e.g., Retallack, 1990, 1993, 2001), an approach referred to as "Taxonomic Uniformitarianism" (Retallack, 1994, pp. 51–53; 1998, pp. 205–208). Taxonomic Uniformitarianism likens soil taxonomy to biological taxonomy in the sense that "identification of a paleosol within a modern soil taxonomy may be taken to imply past conditions similar to those enjoyed by such soils today" (Retallack, 1994, p. 51).

There are a number of problems associated with the use of soil classification to infer past environments. The basic premise behind Taxonomic Uniformitarianism is flawed, as stated by Fastovsky (1991, p. 182):

> The idea that analogies can be drawn between soil classifications and biological classifications is misleading. Biological classifications are developed for their abilities to reflect genetic relationships among organisms. The assumption underlying this is that there exists one true phylogeny, the reflection of which is sought in the classification. Biological classifications are hierarchical because the character distributions in nature that they reflect are likewise hierarchical. Established soil taxa, however, have no single, inferred historical and genetic connection, such as is presumed to exist among organisms. Indeed, soil classifications have been developed for multiple purposes and impose an arbitrary typological system upon natural continua of overlapping processes. Because of this, soil taxonomies do not have the predictive power inherent in biological classifications.

Another difficulty with using soil taxonomy for ancient, usually buried, and typically lithified soils is that classification to the great group level (even to order level in some cases; e.g., Aridisols) requires initially that the nature of the soil-forming environments must be known (chapter 1). To then use the classification to infer the environment is circular reasoning. More to the point, ancient buried soils often do not preserve the information required to assign diagnostic terms used in soil taxonomy because of preburial erosion, polygenesis, and postburial diagenetic overprints in the sedimentary rocks. Erosion affects thickness, which is key to some taxonomic classification such as mollic epipedons. Polygenetic soils, of course, are difficult to associate with a specific soil-forming environment because they went through two (or more) different phases of pedogenesis under different conditions. Identification of diagnostic horizons in a buried soil further requires documentation that the zone in question is pedogenic and that its characteristics are not the result of postburial diagenesis. A number of processes (described in chapters 4 and 10) can mimic, block, or otherwise interfere with our ability to see through the effects of diagenesis to past environmental conditions. For example, the designation "Aquept" (e.g., Retallack, 1990, p. 85, fig. 4.11) requires detailed information about how long a soil remains saturated during the year—information that is simply not available for ancient soils. In the case of buried soils in floodplain environments, preburial aquic soil characteristics are difficult to sort from aquic soil features produced by diagenesis. The evolution of polygenetic soils or those affected by diagenesis can be worked out with careful investigation (examples above and in chapter 10), but not through classification. If diagnostic horizons cannot be identified because of erosion, polygenesis, or diagenesis, then further classification using the soil taxonomy is groundless.

Several examples illustrate the problems in using diagnostic horizons and taxonomic classification to make paleoenvironmental inferences. The identification of ancient grasslands has been the focus of much paleopedology, especially in the reconstruction of hominid habitats in Africa and south Asia (Retallack, 1991a, 1997b). Much of this work has revolved around the identification of Mollisols as clues to identifying grasslands (Retallack, 1990, pp. 109, 211; 1997a, pp. 18, 54–55, table 3.2) and the identification of mollic epipedons to recognize Mollisols, even though Mollisols and mollic epipedons are not exclusive to grasslands (Fenton, 1983; Bartelli, 1984, p. 10). Mollic epipedons were identified largely on the basis of granular ped structure, carbonate content, clayeyness (i.e., to be mollic they had to be clayey), and dark color (7.5YR–10YR hues; 2–5 values; Retallack, 1988, p. 9; 1990, pp. 109, 249, 369; 1993, fig. 1; 1997a, pp. 54, 102; 1997b, pp. 380–381). Of these characteristics, only color is part of the definition for mollic (Soil Survey Staff, 1975, pp. 15–16; 1999, p. 25), and the color criteria used by the U.S. Department of Agriculture are not the same as the color criteria presented by Retallack (1997b, p. 380). Granular structure and clayey texture are not unique to mollic epipedons, nor to A horizons (Soil Survey Division Staff, 1993; Buol et al., 1997). Retallack (1983, pp. 39, 43), for example, describes granular structure in C horizons of ancient soils in the South Dakota Badlands.

To distinguish nongrassland from grassland soils, Retallack (1997b, p. 381, fig. 2) defined "non-mollic" horizons. These zones characterize soils of deserts and woodlands and are based on platy to blocky structure and Bt horizons and Bk horizons. None of these soil structures or horizons is unique to either deserts or woodlands, however.

Argillic horizons and other Bt horizons as well as Bs horizons frequently are identified in the literature on ancient, buried soils on the basis of having high clay content or being iron rich (Retallack, 1990, pp. 110–111, 402). Such clayey zones then are used to infer the presence of Alfisols and to reconstruct forest vegetation (Retallack, 1986, pp. 25, 37; 1988, p. 8; 1991a, pp. 194–195). These interpretive leaps are misleading. Argillic horizons are not necessarily clay rich (they may also be silty, sandy, or even gravelly), but they must contain illuvial clay (Soil Survey Staff, 1975, pp. 19–27; 1999, pp. 29–34; Brasfield, 1984, p. 14). Moreover, argillic horizons are not associated with any particular vegetation or climate. They occur in modern surface soils throughout North America, from coast to coast, and from Mexico to Canada. Specific soil orders with argillic horizons range in environmental setting from Aridisols to Ultisols. The process of clay illuviation, which produces argillic horizons, can be caused by lessivage in response to moist, forested environments (under a wide array of tree species and climates) or the translocation of clay-rich dust in response to irregular precipitation in grasslands and deserts (Fanning and Fanning, 1989, pp. 93–97; Birkeland, 1999, pp. 112–120).

Another problem in applying soil taxonomy for environmental reconstructions is pointed out by Pawluk (1978, p. 64). Some of the arbitrary rules used for differentiating various kinds of diagnostic horizons in soil taxonomy result in the taxonomic separation of soils that developed along similar pedogenetic pathways. Essentially this is the reverse of the situation with equifinality and creates serious problems in using taxonomic classification for environmental reconstruction. An example, noted in chapter 2, can be found on the Southern High Plains, where many of the regional soils are well developed with thick (>1 m), red (5 YR hues) argillic horizons over thick (>1 m) calcic horizons. Thickness of the A horizon varies, however: Some qualify as mollic, but others do not. Otherwise the soils are identical morphologically and chemically, and they are the same stratigraphically, representing pedogenesis in an extensive "cover loam" (the Backwater Draw Formation; Holliday, 1989b, 1990b). The variation in thickness of the epipedon, likely caused by recent wind erosion, however, results in the soils being classified in two different orders: Paleustoll (thick epipedon) or Paleustalf (thin epipedon). Applying the principles of "Taxonomic Uniformitarianism" results in an interpretation of the Paleustalf (argillic horizon; no mollic) as formed under a forest and the Paleustoll (mollic present) as formed under a grassland.

Taxonomic Uniformitarianism also implies that much or most variability in soils is the result of environmental variations. In particular, it does not take into account variation in pedogenic expression caused by time. Thus a Mollisol with an A–C profile on a floodplain would be linked to a climate very different than an Alfisol on a nearby high terrace with strongly expressed Bt horizon, even though the major differences in soil morphology may simply be the result of different age landscapes.

A variation on this application of soil taxonomy for environmental reconstruction is also found in geoarchaeological work in North America. Buried soils that can be classified as Aquolls (thick, dark, organic-rich A horizon over a gleyed B or C horizon) are used rather broadly as paleoenvironmental indicators (e.g., Reider, 1982a,b, 1987, 1990; McFaul et al., 1994). For example, a buried soil at the Horner site, Wyoming, was classified as a Calciaquoll (Reider, 1987, pp. 352–353). Because Calciaquolls in particular and Aquolls in general are now common in eastern North Dakota, northern South Dakota, and western Minnesota, Reider (1987, p. 355) proposes that the environment at Horner 7000 yr B.P. was therefore similar to the present environment of the eastern Dakotas and Minnesota. Two problems exist with this interpretation. First, the entire section with the buried soil has secondary carbonate (used as a criteria to assign the Aquoll to the calci- great group), raising the possibility that the carbonate in the buried soil is a postburial phenomena. Second, Aquolls require only poor drainage with high production of organic matter, conditions that can occur in a wide variety of settings, regardless of climate, as long as the water table is high.

Stable Isotopes

The analysis of stable isotopes of carbon (C) and oxygen (O) is a rapidly expanding area of paleoenvironmental research (Bowen, 1991; Swart et al., 1993; Nordt et al., 1998). Stable isotopes of C (^{12}C and ^{13}C) and O (^{16}O and ^{18}O) are environmentally sensitive and can accumulate in soils through pedogenic processes. These isotopes, therefore, provide a means of reconstructing soil-forming environments. Nordt (2001) provides an excellent summary of the theory and some applications of stable isotope analysis of soils in geoarchaeological contexts. The following discussion of principals and potential pitfalls is based largely on his chapter, but also see Cerling and Quade (1993), Pendall et al. (1994), and Cerling (1999), and chapters in Swart et al. (1993), International Atomic Energy Agency (1998), and Nordt et al. (1998).

The fractionation between ^{13}C and ^{12}C is expressed as $\delta^{13}C$ and is indicative of the relative proportion of C_3 and C_4 biomass (fig. 8.10). Plants with C_3 photosynthetic pathways (trees, shrubs, forbs, and cool-season grasses) discriminate most against atmospheric $^{13}CO_2$ and produce $\delta^{13}C$ values around −27‰. They grow in a broad range of environments and are therefore difficult to use as paleoenvironmental indicators. The C_4 plants (warm-season grasses) discriminate less against $^{13}CO_2$ and yield $\delta^{13}C$ values near −13‰. The C_4 plants also are a powerful proxy for paleoclimatic reconstruction because C_4 species abundance is strongly and positively related to environmental temperature (Nordt, 2001, p. 423). Stable C isotopes accumulate both in SOM of A horizons and in inorganic pedogenic carbonate of Bk horizons. Isotopic analyses of these soil horizons can therefore yield clues to the vegetation under which the soil formed.

The ratio of the oxygen isotopes ^{18}O and ^{16}O, expressed as $\delta^{18}O$, is determined largely on the basis of the source of moisture for meteoric water. Very broadly, the $\delta^{18}O$ values for meteoric water are inversely proportional to temperature: The ^{16}O content of water vapor is enriched toward the poles as tropical and subtropical moisture is depleted of the heavier ^{18}O isotope (Nordt, 2001, p. 425). This relationship is considerably more complex than originally proposed, however (Cerling and Quade, 1993; Nordt, 2001). Isotopes of stable O can be found in the inorganic carbonate of Bk horizons.

The principal problems that can be encountered in looking at $\delta^{13}C$ and $\delta^{18}O$ values in A and Bk horizons for environmental reconstructions are similar to those that can arise in radiocarbon dating of carbon in SOM and inorganic carbonate (chapter 7). Detrital carbon derived from surrounding older landscapes and deposited in the parent material for Bk horizon or onto the A horizon can skew the fractionation in those horizons. In A horizons, the C in the SOM will reflect all plant communities that have grown on the surface. Thus, the more

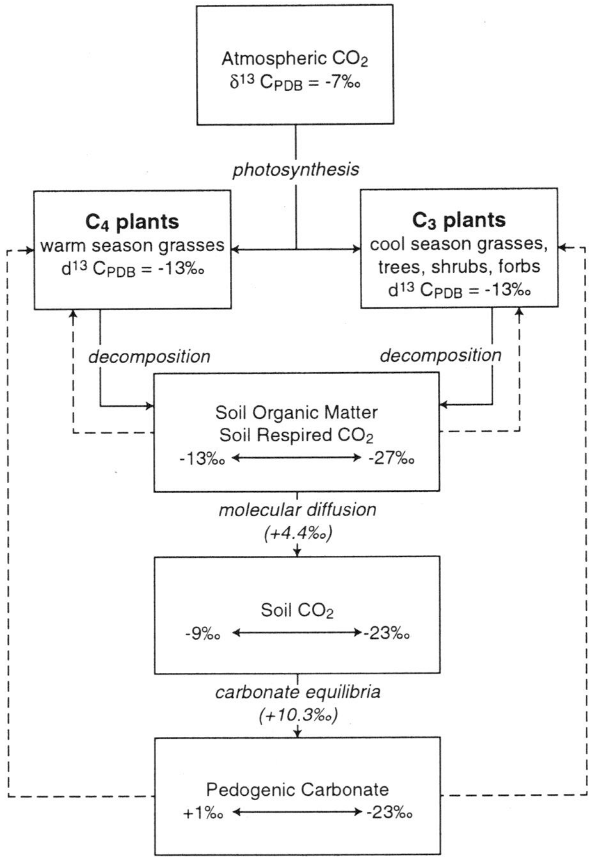

Figure 8.10 Illustration of the photosynthetic pathway for C_4 and C_3 plants and associated $\delta^{13}C$ values for soil organic matter, respired soil CO_2, and pedogenic carbonate (Nordt, 2001, fig. 15.2; from *Earth Sciences and Archaeology*, P. Goldberg, V. T. Holliday, and C. R. Ferring, Eds., © 2001, Kluwer Academic/Plenum Publishers. Reproduced with permission of Kluwer Academic/Plenum Publishers and L. C. Nordt).

environmental changes to which an A horizon is subjected, the more problematic are the interpretations of the $\delta^{13}C$ values resulting from the likelihood of mixing various types of plant biomass. In the investigation of O and inorganic C in carbonate of Bk horizons, the principal problems are analysis of samples that are lithogenic rather than pedogenic, lithogenic engulfed by pedogenic carbonate, pedogenic but detrital, or produced in association with groundwater (Nordt, 2001, p. 429). In addition, the more time that was involved in the formation of a Bk horizon, the more likely that the $\delta^{13}C$ and $\delta^{18}O$ values represent some mixing resulting from vegetation changes. This problem can be circumvented to some degree by selecting small carbonate nodules from soils that formed over relatively brief intervals or by microscopically sampling increments within larger nodules.

The best results in using $\delta^{13}C$ from soils for paleoenvironmental clues is in the analysis of A and Bk horizons of buried soils. A wide variety of examples are available, including some in geoarchaeological investigations. Several of these studies come from the south-central United States. Paleoenvironments and the nature of environmental changes during the Paleoindian occupation of North

America have been the subject of considerable discussion and research since the association of humans and extinct fauna were confirmed at the Folsom site (1926–1928). The nature of Clovis (~11,300–10,900 yr B.P.) and Folsom (~10,900–10,200 yr B. P.) environments on the southern Great Plains has been of particular interest and debate for decades (e.g., Wendorf, 1961, 1970; Haynes 1991; Holliday, 2000). Classical views of Paleoindian environments and some proxy data from the region indicated warming and drying from Clovis to Folsom and later Paleoindian times (Sellards, 1952; Sellards and Evans, 1960; Johnson, 1986, 1987a). Alternative views indicated that Clovis times were warmer and drier than Folsom and that Folsom environments were characterized by a wet, cool boreal forest (Wendorf, 1970; Haynes, 1991). Stable C isotopes from buried A horizons in dunes and from playa muds show a pronounced shift from cooler to warmer conditions from Clovis to Folsom times and, moreover, that the region was dominated by grasses throughout the Paleoindian occupation of the region (Holliday, 2000). The analysis of stable C isotopes in buried soils at the Big Eddy site in southwestern Missouri (Hajic et al., 1998) and along the Medina River of central Texas (Nordt et al., 2002) show a very similar trend (fig. 8.11)—a shift toward warm-season grasses 11,000–10,000 yr B.P., in particular. Furthermore, the isotopic data from all three regions are suggestive of a return to somewhat cooler conditions in early post-Folsom time.

At the Fort Bliss Military Reservation, south-central New Mexico and far western Texas, Monger (1995) reconstructed paleoenvironments for the last 10,000 yr of human occupation in alluvial fan settings. Three proxies for paleoclimate were used: erosion-sedimentation history, palynology, and stable C and O isotope analysis of pedogenic carbonate from buried soils. The carbon isotope analysis of paleosols showed that the $\delta^{13}C$ of pedogenic carbonate decreased from between −1‰ and −4‰ to between −10‰ and −6‰ at approximately 8000 yr B.P. This change was interpreted as a response to a major increase in C_3 biomass production that coincided with a significant increase in Cheno-am pollen. Together these data indicated a decline in grasslands and an increase in shrublands in the early Holocene. Rates of upland erosion were also increasing at this time, probably in response to reduced vegetative cover from a spreading C_3 shrub plant community. The $\delta^{18}O$ values of pedogenic carbonate varied little during this time, indicating that temperature or precipitation was not affecting vegetation changes. As a consequence, it was inferred that increasing atmospheric CO_2 concentrations beginning 8000 yr B.P. contributed to the decline in grasslands in fragile ecosystems of alluvial fan settings (Cole and Monger, 1994). Moreover, the vegetation shift may have lead to a change in prehistoric subsistence strategies, widespread erosional and depositional events, and differential preservation of the prehistoric archaeological record

Ferring (1995a) conducted geoarchaeological investigations at the Aubrey site on the Trinity River of north-central Texas and incorporated analyses of stable C and O isotopes from pedogenic carbonate for paleoenvironmental reconstructions (Humphrey and Ferring, 1994). The data indicated relatively high C_3 biomass production 11,000–7500 and 3500–2000 yr B.P. resulting from more humid conditions. C_4 biomass production increased in the middle Holocene, 7500–3000 yr B.P., and again in the late Holocene, 2000–1000 yr B.P., because of

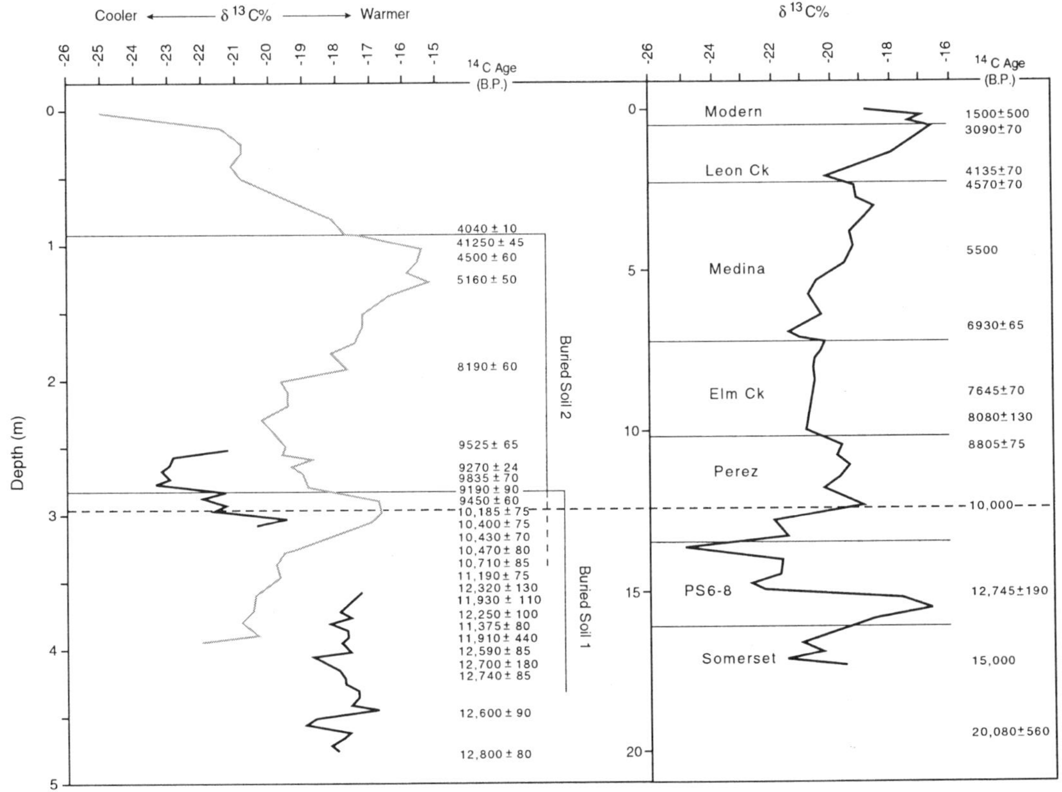

Figure 8.11 Stable carbon isotopic data for the Big Eddy site, Missouri (from Hajic et al., 1998), and along the Medina River, Texas (Nordt et al., 2002). Dashed line highlights 10,000 yr B.P.

drying. The $\delta^{18}O$ values from Holocene paleosols were stable, however. This indicates that shifts in C_3 and C_4 vegetation were responding to changes in seasonal precipitation and not to average precipitation or temperature. These data combined with the stratigraphic record led Ferring (1990, 1995a) to suggest that the middle Holocene dry interval led to a possible decrease in plant biomass production, a reduction of bison as a resource for human subsistence, and a decrease in human population densities.

In geoarchaeological work at Fort Hood Military Reservation in central Texas, Nordt et al. (1994) investigated the stable C isotope signatures of soils formed in a series of late Quaternary alluvial deposits. In a departure from most isotopic studies, the isotope signature throughout several soil profiles was analyzed, the theory being that over time the SOM is translocated down through the soil: Deeper C will be older. There was some mixing with detrital carbon, but the interpretations and trends could be checked by comparison with stable C values from dated buried A horizons. The late Pleistocene values for $\delta^{13}C$ are the lowest in the study (–21‰), indicative of greater contributions from cool-season grasses or trees. Early Holocene values indicate a gradual increase in C_4 biomass production during this time. The highest $\delta^{13}C$ values (–13‰ to –14‰) are interpreted as an increase in C_4 biomass production because of middle Holocene warming, along with pedogenic superimposition of SOM from C_4 plants onto the late Pleistocene detrital values. The shift back to depleted $\delta^{13}C$ values and less C_4 biomass production in the upper profiles probably is caused by cooler conditions in the late Holocene.

Prehistoric settlement patterns in the Rift Valley of Kenya were investigated by Ambrose and Sikes (1991) by looking at the stable C isotopes from surface soils. The purpose of the study was to document shifts in the boundary between montane forest and lowland savanna grassland during the Holocene and to assess the effect of these changes on archaeological settlement patterns. Samples were collected from the upper 50 cm of surface soils along an altitudinal transect. A number of sites in the forest up to 300 m in elevation above the modern forest/savanna boundary displayed a significant increase in $\delta^{13}C$ values with depth (e.g., from –24‰ to –15‰), indicating the presence of a C_4 grass community sometime in the past. Radiocarbon dating and correlation to lake and pollen records in nearby valley floors indicate that the higher $\delta^{13}C$ values in the subsoil were emplaced sometime between 3000 and 6000 yr B.P. The forest/savanna boundary must have shifted up-slope in the middle Holocene and then migrated back down after 3000 yr B.P. These data explained evidence for diminished agricultural activity during middle Holocene warming, when the forest/savanna boundary shifted above the study area, and for intensification of Neolithic agriculture after 3000 yr B.P., when conditions were cooler and the forest/savanna boundary had migrated below the study area.

Several isotope studies were carried out in the Old World in an attempt to better understand hominoid (including hominid) paleoenvironments. One of the first applications of stable isotope analyses of soils in an archaeological context was at Olduvai Gorge in East Africa (Cerling and Hay, 1986). The work focused on both stable C and stable O isotopes in soil carbonates preserved in the thick sequence of deposits (100 m) at this well-known hominid setting (Hay, 1976).

Most of the soils formed 2.2–1.17 Ma contained no carbonate, suggesting to the investigators that MAP was greater than 750–850 cm (Cerling and Hay, 1986, p. 74). This conclusion fits well with the above-noted data showing that the presence of a pedogenic carbonate horizon in a soil is indicative of an MAP <760 mm (Royer, 1999). A layer dated 1.62–1.67 Ma (the Lemuta beds) contains numerous soil calcretes. The O isotopes for this time are indicative of isotopically heavier meteoric waters. Increases in both $\delta^{13}C$ and $\delta^{18}O$ are used to infer warming and drying 0.62–1.17 Ma. Values of both $\delta^{13}C$ and $\delta^{18}O$ show continued enrichment from 0.62 Ma to the present, though some fluctuations are apparent. These data are interpreted to indicate continued warming and drying, culminating in the present arid climate, though with several intervals that were not as warm or as dry as today.

Some of the isotopic work in East Africa focused on the make-up and evolution of floral assemblages (forest vs. grasslands) because of divergent views on how they related to the evolution of early hominids and their hominoid ancestors. For example, Retallack et al. (1990) used soil morphology to propose that open grasslands had existed since the Miocene. Both Cerling (1992) and Kingston et al. (1994) studied the isotopic composition of pedogenic carbonates from buried soils formed in Miocene, Pliocene, and Quaternary deposits to better understand the development of plant communities. Both studies concluded that hominids probably evolved in a mixed grassland/dry woodland environment. Cerling (1992) further suggested that hominids may have coevolved with the spread of grasslands, though Ambrose and Sikes (1991) did not see the same evidence for grassland evolution. In any case, the data from both studies show that mixed C_3 and C_4 plant communities existed in the area since at least the Pliocene. This is particularly important given the lack of other paleobotanical and paleontological data and the inherent difficulties in using soils as climatic indicators.

The evolution of floral communities and hominoid paleoecology also is an issue in northern India, northern Pakistan, and Nepal. Carbon and oxygen isotopic analyses of soils from the Siwalik Group deposits record significant shifts in floral communities in the Miocene and Pliocene (fig. 8.12; Quade et al., 1995), including the time when the region was occupied by *Sivapithecus*. The carbon isotopic composition of both carbonate and organic matter from floodplain soils shifts dramatically starting ca 7.0 Ma (fig. 8.12), marking the displacement of largely C_3 vegetation, probably semideciduous forest, by C_4 grassland. This development is roughly coincident with the disappearance of hominoids in the region and also represents an important step in regional, long-term aridification.

On the Loess Plateau of China, Wang et al. (1997) analyzed the $\delta^{13}C$ and $\delta^{18}O$ signal of carbonates from buried soils to resolve issues of hominid paleoenvironments near the Lantian localities. As discussed above, at the Gongwangling site, *Homo erectus* remains from 1.15 Ma are associated with a vertebrate fauna indicative of a warm, forested environment similar to South China, but loess deposition is suggestive of a cold, arid environment (An et al., 1987, p. 201). The specific issue, moreover, is the dating of hominid adaptation to nontropical environments. To resolve the question of the environment at Gongwangling 1.15 Ma Wang et al. (1997) first examined the relationship between stable C and O isotopes in loess and buried soils from the last glacial and interglacial intervals as

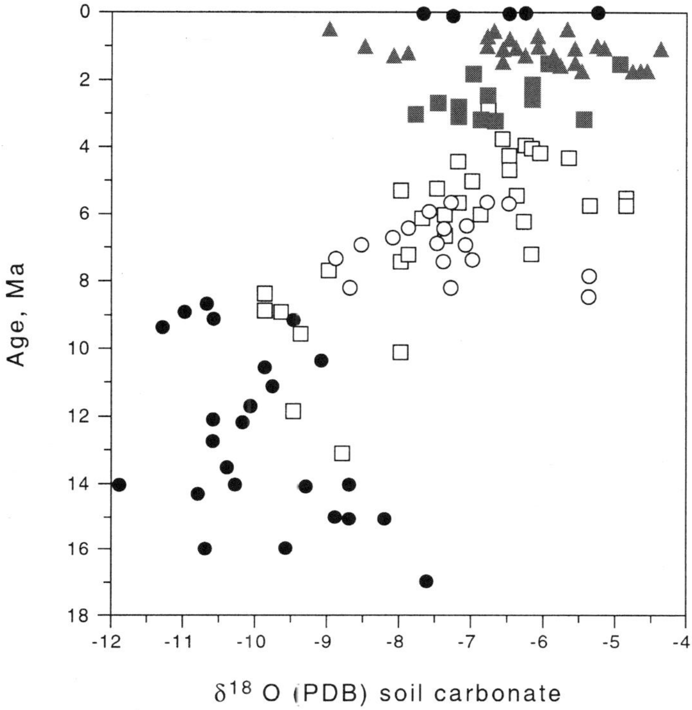

Figure 8.12 Changing levels of stable oxygen isotopes from soil carbonates over the last 18 Ma in Pakistan (modified from Quade et al., 1995, fig. 12B; reproduced with permission of the Geological Society of America).

well as the Holocene. Those data acted as analogues because the environments of those periods are well known. The data from the hominid level showed low $\delta^{18}O$ values in association with low soil organic-matter $\delta^{13}C$ values. On the basis of the analogues, this isotopic combination indicates that the region was influenced by Siberian-Mongolian winter and Indian summer monsoons with cold/cool dry winters and warm/mild, semihumid summer and fall. Thus, *Homo erectus* occupied a temperate climate regime 1.15 Ma in China.

9

Soils and Landscape Evolution

Soils and archaeological sites are intimately related to the landscape. Investigating soils across past and present landscapes provides a means of reconstructing and understanding the regional environmental and geomorphic context of archaeological site settings and specific site locations, regional site formation processes, and aspects of the resources available to people in a region. Archaeological sites tend to occupy small segments of the landscape, but human activity may affect a much larger area, and in any case, people wander far and wide from sites, interacting with the environment—including the landscape. Thus, no matter whether a site is just a lithic scatter or bone bed or if it is a tell, understanding the regional landscape is an important part of understanding a site and human behavior, and soils are an important means of understanding a landscape.

Soils are also important in reconstructing the evolution of landscapes and, consequently, the evolution of archaeological sites. That is, landscape evolution is an important external component of site-formation processes. Landscapes form the physical framework or underpinning for people and their activities and their resulting sites. As landscapes evolve, so do human activities and so do sites. Soils are key to recognizing and interpreting the evolutionary processes that shape the landscape and associated archaeological sites. Furthermore, the concept of landscape evolution also 1) is a logical continuation of the discussion of soil stratigraphy (chapters 5, 6) because it places soil stratigraphy in three or even four dimensions; 2) is a complement to the discussion of soils as environmental indicators (chapter 8), because landscape evolution can be linked to environmental change and because the evolution of the landscape itself, regardless of changes in other factors, represents a change in the environment from a human perspective; and 3) provides yet another means for predicting site locations. The discussion in this section, therefore, represents an integration of some of the principals outlined previously. Some of the studies presented in other chapters, such as the work on the Loess Plateau of China (chapters 6 and 8), and at Harappa and along the Ravi River (chapter 4), are good examples of landscape reconstructions for very large regions and are not repeated here. The chapter includes brief discussion of the concept of "landscape" in archaeology, followed by a section discussing catenas (and paleocatenas). Soil variability across a landscape is a unique aspect of soils, and therefore, the catena concept is an important means of identifying and interpreting soils both to understand landscape evolution and as indicators of past landscapes. The rest of the chapter presents case histories of soils used to reconstruct landscapes in archaeological contexts in a variety of settings and regions. The topic of human effects on soils and landscapes is discussed in chapter 11.

Landscape and Archaeology

Archaeological fieldwork traditionally focused on sites, and much geoarchaeological work still does. With the advent of interest in regional settlement patterns, environmental reconstruction, and site formation processes, however, the nature and evolution of regional landscapes became prominent topics in both archaeological and geoarchaeological research (e.g., Butzer, 1971, 1982; Renfrew, 1976). Because soils are important components of landscapes and because their genesis is intimately linked to and reflects the evolution of landscapes, soils are keys to reconstructing landscapes and their evolution. The record of landscape stability is archaeologically significant because archaeological debris should be associated with stable landscapes because of the likely human preference for stable surfaces and the likelihood for the concentration and preservation of artifacts and features on stable surfaces relative to aggrading or eroding ones (chapter 7). Moreover, an understanding of pedogenesis on and pedogenic relationships across a landscape is an important complement to the traditional geoarchaeological emphases on the more dynamic geomorphic processes of sedimentation and erosion. Wells (2001, p. 108) correctly observes that, "Without an appreciation of landscape evolution and geomorphic change, the potential for misinterpretation of archaeological survey data is immense." Indeed, her statement can be broadened to include most, if not all, types of archaeological field data.

The notion of "landscape" has been important in archaeology for decades, and "landscape archaeology" has been an explicit theme since at least the 1960s (Roberts, 1987), generating a sizeable literature (e.g., Wagstaff, 1987; Cherry et al., 1991; Rossignol and Wandsnider, 1992; Barker, 1995; Ashmore and Knapp, 1999). In most of that literature, the term "landscape" is used in a very broad sense, including cultural landscapes. Knapp and Ashmore (1999, p. 1) group what they call "archaeological thinking about the nature of landscape" into three general categories: a "minimalist" view of landscape as the backdrop against which archaeological remains are plotted; an economic and political perspective of landscapes as provider of resource, refuge, and risks "that both impel and impact on human actions and situations" (1999, p.1); and "socio-symbolic notions of landscapes as an entity that exists by virtue of its being perceived, experienced, and contextualized by people" (1999, p. 1). In this chapter and indeed throughout this volume, the term is used in a literal, geomorphic sense, probably falling into the both the "minimalist" and "economic political" categories of Knapp and Ashmore (1999). They, however, seem to underestimate or overlook the concept of the landscape as a dynamic component of the physical, natural environment, as a record of that environment and of environmental changes, and as an important influence on site formation processes. The landscape is indeed a "backdrop" for archaeological debris; a very important one whose processes and evolution must be understood to understand how a site evolved both during and after occupation.

Catenas and Paleocatenas

The variability of soils across a landscape caused by topographic and hydrologic variability makes soils uniquely suited as indicators of past landscapes. Understanding the processes that produce soil variability and result in catenas provides an approach to making interpretations regarding past landscapes.

The recognition of catenary relationships and formulation of the catena concept by Milne (1935a,b) was one of the most significant steps in the evolution of method and theory in soil geomorphology. The concept is an explicit statement of the relationship between pedogenic and geomorphic processes, describing the interaction of soils and landforms (fig. 2.4). The catena concept was originally described for soils formed on weathered bedrock ("residual soils") in the tropics of what is now Uganda, but the term has become widely applied in a variety of environmental settings and for soils formed in sediment (e.g., Gerrard, 1992; Birkeland, 1999). Catenary relationships can also be preserved in buried soils as paleocatenas and, as described in chapter 5, can be keys to recognizing buried soils. In the context of geoarchaeology and Quaternary stratigraphy, therefore, understanding catenary relationships is important in interpreting stratigraphic relationships (chapter 5)—particularly lateral variability in soil morphology—and in reconstructing landscapes and the record of landscape evolution.

Soil variability along a catena is related to the slope position and drainage characteristics of the soil. Ultimately these factors are related to water movement over and through the soil. Most simply put, the rate and amount of water moving across the upper and middle segments of a slope are important in determining the presence or absence and amount and type of erosion and sedimentation (in addition to determining vegetation characteristics) and the amount of water that moves into the soil for weathering and solute transport. On lower slopes or at the foot of slopes, these factors, combined with groundwater characteristics, strongly influence sediment accumulation and solute input, in addition to vegetation and drainage, and can contribute to soil cumulization (figs. 5.3 and 6.1; chapter 5). The relative importance of surface movement of sediment and subsurface movement of solutes is dependent on slope angle and distance from slope crest, in addition to climate. "Each catena is, therefore, the result of the complex interrelationships between soil and slope processes and will be governed by the differing ratio of erosion to deposition occurring on different parts of the slope. From the pedogenetic point of view, all country which has relief consists of zones of removal, transference, and accumulation, the limits of which can be peculiar to each transferable constituent or to each group of constituents of comparative mobility" (Gerrard, 1992, p. 32).

A common catenary association of soils can be found on floodplains, which are also a common setting for archaeological sites and the focus of much geoarchaeological research (chapters 5 and 6). Floodplain catenas are well described by Kraus and Aslan (1999, p. 309; fig. 6.1), providing the basis for the following description. The effects of surface and subsurface movement of water as well as groundwater are particularly apparent among floodplain soils, most obviously expressed by lateral changes in the quantity and distribution of parent material, soil organic matter, matrix and mottle colors, and soil solutes (e.g., carbonate, gypsum; fig. 6.1). The hydrological influences on pedogenesis and development of "hydromorphic soils" generally correlate with soil texture and floodplain topography. Hydromorphy is most pronounced in clayey, poorly drained floodplain depressions such as the flood basin, and least pronounced in sandy, moderately to well-drained soils on alluvial ridges. Sandy and silty soils formed on natural levees and channel bars, for example, typically have dark-brown to brown A and Bw horizons and low organic-matter contents, which reflect soil oxidation and moderate drainage. Gray soil colors are more abundant lower in the soil in Bg and Cg horizons and reflect reduced conditions and proximity to the groundwater. In contrast, poorly drained flood basin soils are gray, clayey, and higher in organic matter and contain abundant mottles and iron nodules (fig. 6.1). Prolonged saturation and poor drainage throughout the profile are reflected in a black, organic-rich, and mottled Ag horizon and a thick, gray Bg horizon with abundant mottles and nodules. The low topographic position and clayey texture result in poor soil drainage, anaerobic conditions, and the production and preservation of organic matter. Occasional arrival of floodwaters introduces more fines such as clay, which are mixed into the soil surface and obscured by organic-matter production and bioturbation (chapter 10), thereby overthickening the A horizon.

Two general categories of catenas, stable and unstable, can be recognized and described based on their slope processes (fig. 9.1; Birkeland, 1999, p. 238). Stable catenas have not experienced significant erosion, but slow, incremental particle movement down-slope may be important over time. Soil properties along the slope are mainly explained in terms of pedological process governed by topographic position and, thus, are predictable. Soils along a slope can be differentiated as follows: up-slope soils are drier, generally thinner (because of erosion and less water moving downward through the profile), better drained, and more oxidized (and hence redder), whereas down-slope soils are more moist, generally thicker (because of cumulization and, if not saturated, because of deeper translocation), and more poorly drained and sometimes (if saturated) reduced or gleyed, and thus grayer (fig. 9.1). The degree of erosion and cumulization will depend on a number of factors such as vegetation cover and steepness of slope. The drainage characteristics will be determined by parent material texture and groundwater characteristics (i.e., whether the water table intersects the down-slope soils and the degree of seasonal water table fluctuation).

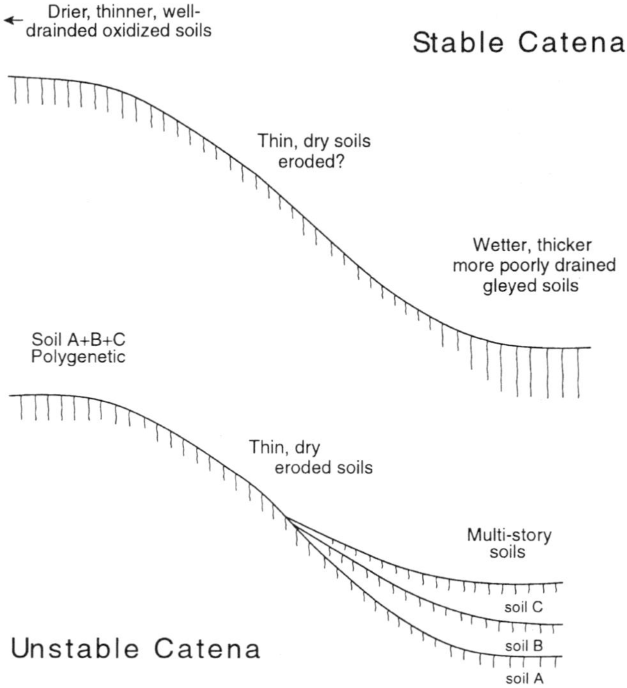

Figure 9.1 Schematic illustrations of stable versus unstable catenas (from *Soils and Geomorphology* by Peter W. Birkeland © 1984, fig. 9.7, Oxford University Press, Inc. Used by permission of Oxford University Press, Inc.). If the slope is stable, a predictable sequence of soils will be found. If the slope is unstable, a sequence of buried soils will be found in the lower slope position.

Catenary relationships produce marked soil variability in terrains with rolling topography (summarized by Birkeland, 1999, pp. 236–237). In Arctic climates, well-drained A-Bw profiles can grade down-slope to cryoturbated, poorly drained soils. In dry climates, saline and alkaline soils occupy the depressions, better-leached soils the slopes, and less-leached soils the summits. Different soil taxonomic units can be found at different positions along a catena. In aridic settings, a catenary association could include, for example, Orthids (A-Bw profile) on the summits, Argids (A-Bt profile) on the backslopes, and Natrargids (A-Bt profile with sodium) in the footslopes and toeslopes. In more humid environments, the summit Argiustoll (Ah-Bt profile) can give way (in the down-slope direction) to Haplustoll (Ah-Bw profile; shoulder), Argiustoll (Ah-Bt profile; backslope), and finally Aquoll (A-Bg profile; toeslope). If the landscape is well stabilized, the differences in soil properties with position could be the result of pedogenesis in place, resulting from differences in moisture, leaching, and vegetation over the rolling landscape. In this case, the various parts of the landscape are assumed to be approximately the same age. On landscapes with some incremental erosion, the soil variability can also be attributed to erosion up-slope and cumulization down-slope.

Catenas on unstable slopes will look like soils described in the K-cycle model of Butler (1959, 1982; chapter 3). Erosion on the steeper, upper part of a slope can result in deposition down-slope (fig. 9.1). Stability follows and is accompanied by soil formation along the entire slope (the erosion-deposition interval, and the duration of soil formation is a complete K-cycle). A subsequent period of

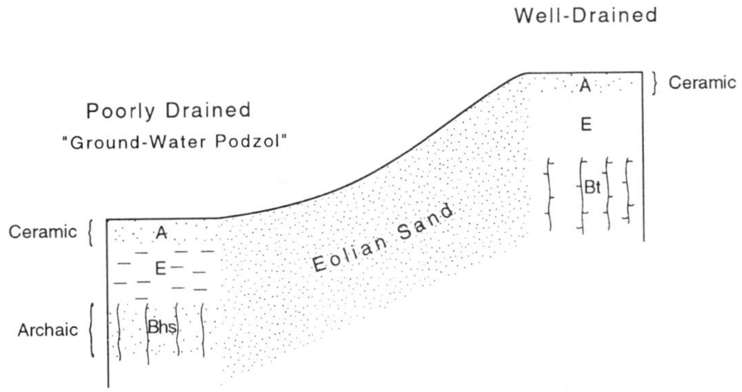

Figure 9.2 Catenary relationship between well-drained and poorly drained soils formed in eolian sand in west-central Florida and soil stratigraphic relationship of associated archaeology. Soils are approximately 1 m thick. (Based on Hunt and Hunt, 1957, figs. 3 and 4.)

instability could result in a second K-cycle. The position and extent of both the erosion and deposition zones can change with time. The end result could be a stable upland with soils representing the total duration of soil formation (a polygenetic soil?; chapter 5); an eroding slope with relatively poorly developed soils; and a lower slope receiving episodic deposition producing a sequence of weakly developed buried soils (fig. 9.1). An unstable setting such as this would have a catenary association of more weakly expressed soils, in contrast to better-expressed soils along a more stable catena.

Several examples illustrate catenas in geoarchaeological or potential geoarchaeological contexts, including a stable catena, an unstable catena, and a paleocatena. One of the earliest examples of explicit applications of pedology and soil geomorphology in archaeology was built around a catena. Hunt and Hunt (1957) recognized a repeated pattern of soil associations and archaeology during geologic reconnaissance in west-central Florida (fig. 9.2). In poorly drained lowland settings, "prepottery" (probably Archaic) sites were found within or below soils formed in sand with A-E-Bhs horizonation, termed "groundwater podzols" (probably Aquods of soil taxonomy). On uplands, well-drained soils in the same sand exhibited A-E-Bt horizonation. There, ceramic sites were found at the top of the soil. The eolian sediment, therefore, predated the ceramic occupation and probably was contemporaneous with some phase of Archaic habitation. Recognition of the catenary relationship of the soils and the stratigraphic relationship of the archaeology provided a simple model for prediction of archaeological sites using soils and their landscape setting.

Both a stable and an unstable catena are present in middle and late Holocene strata at the Lubbock Lake site, Texas. A soil formed in middle-Holocene eolian sediments illustrates a stable catena in which soil variability is the result of topographic and hydrologic variable (and some parent material variation; fig. 9.3). The soil, the "Lubbock Lake soil" (fig. 2.1; Holliday, 1985a,d; mentioned in chapters 6 and 8, and below) exhibits both valley axis and valley-margin facies. The

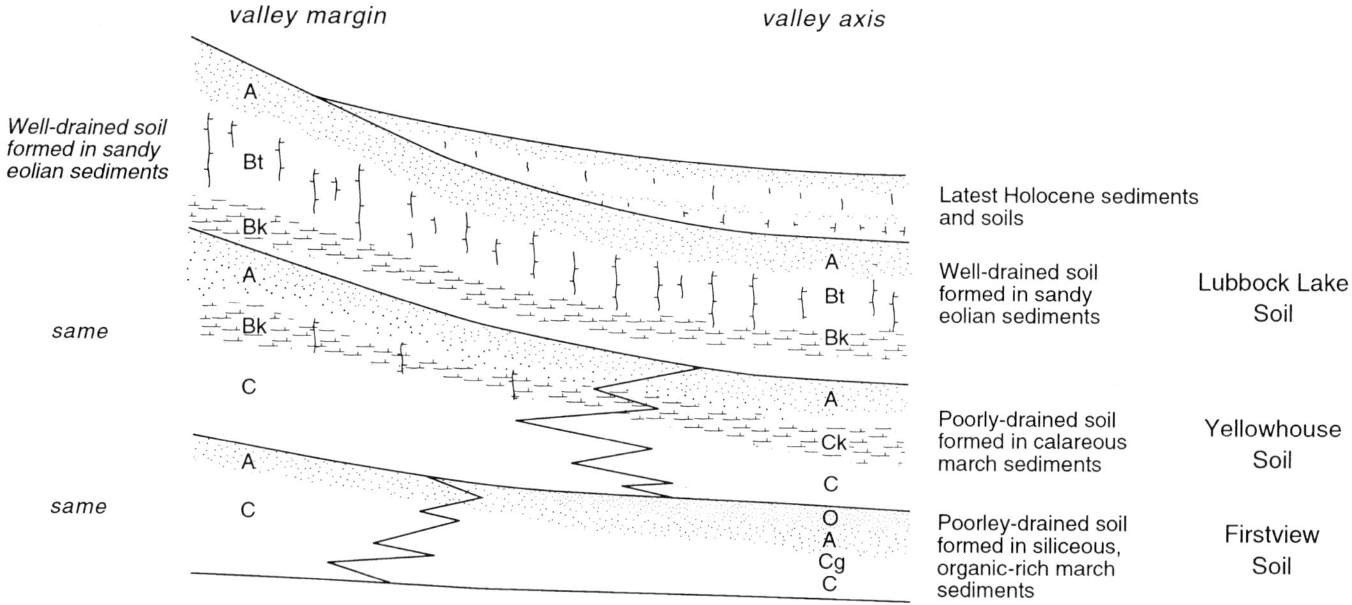

Figure 9.3 Generalized catenary relationships of soil stratigraphic units at the Lubbock Lake site, Texas (from Holliday, 1990b, fig. 4; reproduced with permission of the Geological Society of America). See also figs 2.1 and 7.15.

valley-axis facies averages 1 m in thickness. It has or had a mollic epipedon and Bt-Bk (Stage II) horizonation. The mollic epipedon is locally cumulic, due to the more-moist setting resulting from proximity to local marshes and high organic matter production, as well as local, slow aggradation. The valley margin facies, in contrast, is up to 2 m thick and has a more weakly expressed ochric epipedon over Bw-Bkw (Stage I) horizonation. The thickness of the soil is the result of rapid throughflow of water caused by the coarse texture of the parent material, but the lack of illuvial clay and weaker expression of the A horizon is because less water moved into the soil as a result of rapid runoff and, as noted below, because of some erosion.

The Lubbock Lake soil was buried along its middle and lower slopes in the late Holocene by slopewash and eolian deposits. The sedimentation was episodic, resulting in the formation of multistory buried soils with A-C, A-Bw, and weak A-Bt-Bk horizonation and representing several K-cycles (fig. 9.3). This situation is an example of an unstable catena. The best-expressed and most extensive of the buried soils (the "Apache soil" of Holliday, 1985a,d) has minimally expressed argillic and calcic horizons and, in proximity to valley axis marshes, a mollic epipedon. The surface soil in this sequence of late Holocene soils and sediments (the "Singer soil" of Holliday, 1985a,d) is expressed by an A-C profile. The transition in soil morphology from the soil in the up-slope position (the unburied Lubbock Lake soil) to the soils such as the Apache and Singer on the middle and lower slopes will be quite abrupt (as predicted by Birkeland, 1999, p. 238), because of the effects of erosion and deposition.

A soil geomorphic (and paleocatenary) relationship somewhat similar to the Lubbock Lake soil is apparent at the Lindenmeier site in Colorado. A dense Folsom occupation zone is associated with buried soils that formed on the floor of a small valley (Haynes and Agogino, 1960; Wilmsen and Roberts, 1978; Holliday, unpublished notes). Along the valley axis the archaeology is associated with a cumulic, mollic, welded pair of soils, each characterized by silty, very dark gray brown (10YR 3/2 m) ABtkb horizons, totaling over 1 m in thickness (fig. 9.4). Up-slope the soil complex thins significantly and the archaeology becomes compressed into a single, thin (5 cm), loamy Akb horizon, but it is still very dark gray brown (10YR 3/2 m). The valley axis setting was a slowly aggrading one and was probably heavily vegetated during the Folsom occupation, as indicated by the thickness of the soils, the compound ABt morphology (an aggrading A horizon that evolved into a Bt horizon), and the dark colors. The "k" horizon is the result of the welding of the b4, b5, and b6 soils (fig. 9.4). The dark colors and relatively high organic carbon content of the up-slope facies of the "Folsom soil" suggests relatively high biomass production, but the thinness of the up-slope facies indicates that the soil was subjected to erosion, which would be expected given its position high up on the paleolandscape on a sloping paleosurface.

Soils and Archaeological Landscapes

There are relatively few examples of the use of soils for reconstructing landscapes in an archaeological context. More commonly, soils have been used to aid in

Figure 9.4 Cross section of the valley fill at the Lindenmeier site, Colorado (looking east), approximately 120 m west (up-valley) of Smithsonian trench A (see Wilmsen and Roberts, 1978, map 1, or Haynes, 2003, fig. 3). The buried soil containing the Folsom occupation debris varies significantly in morphology as a function of slope position. Down-slope (left) along the valley axis, the soil is ~50 cm thick and cumulic (ABtkb horizonation). Up-slope, the soil is only 5 cm thick, with simple Ab–Cb morphology.

landscape reconstruction as a means of predicting site location, age, and preservation (see also chapters 4 and 7), typically in the context of cultural resources management studies (e.g., Hunt and Hunt, 1957; Bettis, 1992; Mandel, 1992). Furthermore, much landscape archaeology focuses on surface survey and archaeology on modern landscapes (e.g., Wagstaff, 1987a; Cherry et al., 1991; Wells, 2001). The discussion here deals more with buried landscapes and how soils can be used to reconstruct them, but it also addresses the evolution of contemporary landscapes as inferred from soils and the implications for interpreting surface archaeology.

Valleys of the Great Plains

Buried and surface Holocene soils in the dry valleys or "draws" of the Southern High Plains provide insights into the evolution of these drainage ways, which were also important settings for prehistoric and historic human activities (Haynes, 1975; Stafford, 1981; Holliday, 1995, 1997; Johnson and Holliday, 1995). The most intensive work on these soils and their paleogeomorphological implications was at the Lubbock Lake site in lower Yellowhouse Draw (Holliday, 1985a,b,d; Holliday and Allen, 1985). The paleoenvironmental implications of the soils were described in chapter 8. Three key soil stratigraphic units, each divisible into valley axis and valley margin facies, are indicative of much of the Holocene landscape history (fig. 9.3). The oldest buried soil is the Firstview soil, formed during the late Paleoindian and early Archaic occupation of the area

(~8500–6300 yr B.P.). The valley axis facies is a marsh soil (fig. 8.5A), described in chapter 8, formed in paludal deposits (stratum 2m). The valley margin facies, however, formed in sandy, eolian sediment (stratum 2s) and contains a thin A horizon low in organic matter and an oxidized Bw horizon with some calcium carbonate, indicating a relatively dry, well-drained setting. The late Paleoindian and early Archaic occupants of the area must have witnessed and occupied a marshy valley axis with dry, well-drained valley margins.

During the middle Holocene, highly calcareous marsh sediments (a marl; stratum 3m) buried the valley axis facies of the Firstview soil, whereas sandy eolian deposits (stratum 3s) buried the valley margin facies (fig. 9.3). The Yellowhouse soil then formed in strata 3m and 3s ~5500–5000 yr B.P. Along the valley axis in the marl, this soil has a thick, organic-rich A horizon and a calcareous C horizon. The morphology of the A horizon and lack of leaching of carbonates indicate that a high water table persisted along the valley axis. The valley margin facies of the soil is similar to the Firstview soil, with a weakly expressed A horizon and oxidized and calcareous C or Bw horizon, indicating dry, well-drained conditions. Thus, the middle Archaic inhabitants of the area lived on a landscape generally similar to that of the late Paleoindian and early Archaic people, though the valley axis wetland was a hard-water setting.

Stratum 3 and the Yellowhouse soil were buried by eolian sediments of stratum 4 (~5000–4500 yr B.P.). The Lubbock Lake soil, described above and in other chapters (fig. 2.1), formed in stratum 4 beginning ~4500 yr B.P., continuing to <1000 yr B.P. where buried by stratum 5, and exposed at the surface today where never buried. The Lubbock Lake soil contrasts with the older buried soils, being more strongly developed and better drained (fig. 9.3), providing several clues to substantial changes in the nature of the landscape during late Archaic, late Prehistoric, Protohistoric, and Historic occupations of the area. The soils in the sediments at Lubbock Lake show that through most of the Holocene most areas in the draw were dry and well drained. Episodic eolian and slopewash sedimentation characterized the valley margins during parts of the early and mid Holocene, but otherwise the landscape was stable. The most significant changes in the landscape through time were along the valley axis. Through the late Paleoindian and early Archaic occupation of the site, this area was marshy, but for the last 5000 radiocarbon yr, from middle Archaic to Historic time, most of the area was dry and well drained except for a narrow strip along the lowest part of the valley.

The Holocene soil geomorphic record for lower Yellowhouse Draw in the Lubbock Lake area is generally representative of the record for most reaches of the draws in the region (Holliday, 1995). The early Holocene paludal sediments and soils are rare, however, their presence determined by the presence of springs (Holliday, 1995).

The morphology and the geomorphic and stratigraphic relationships of soils buried in alluvium of rivers on the Great Plains provide a striking contrast in landscape evolution with the draws of the High Plains. Geoarchaeological research in central and southwestern Kansas resulted in a comprehensive picture of the evolution of the Pawnee and Smoky Hill rivers, based in large measure on soils (Mandel, 1992, 1994, 1995). The buried soils typically exhibit A-C, A-Bw, or

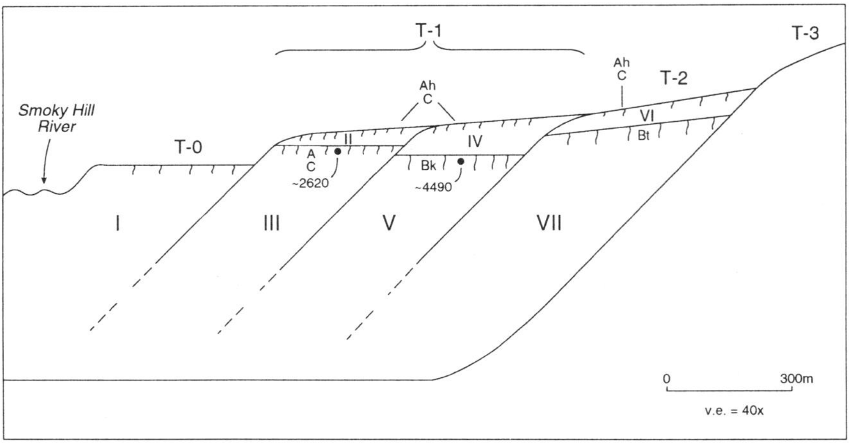

Figure 9.5 Valley fill along the Smoky River, Kansas, illustrating the complex history of alluvial landscape evolution (from Mandel, 1992, fig. 2–3; reproduced with permission of R. D. Mandel).

A-Bk morphologies (in which A horizons were not removed by erosion before burial), so differentiating them relies largely on radiocarbon dating and the stratigraphic relationships of alluvial fills (figs. 6.5 and 9.5). The late Quaternary alluvium is found below two terraces (T-2 and T-1) and the floodplain (T-0; fig. 9.5). Most of the alluvium (Unit I) below T-2 probably predates the human occupation of the region. Before ~10,000 yr B.P., the floodplains were stable and the associated alluvium was modified by pedogenesis. The soil associated with this surface exhibits A-Bt morphologies, indicative of stability under well-drained conditions for some amount of time. This soil was truncated and buried by a thin veneer of alluvium (Unit II) sometime during the Paleoindian occupation of the region. Most of the alluvium of both rivers is beneath the youngest terrace (T-1) or in alluvial fans. The T-2 and its associated alluvium were entrenched in the early Holocene. Throughout most of the rest of the human occupation of these landscapes, the rivers largely were aggrading, alternating between periods of sedimentation and periods of stability with soil formation. The one exception to this description is along the lower reaches of fourth-order tributaries of the Pawnee River. There, Holocene alluvium is found below two terraces (T-1 and T-2). Most of the archaeology in the region, therefore, is buried in alluvium.

During the period ~10,000–7000 yr B.P., the large valleys of the Pawnee River Basin were characterized by alternating phases of alluviation and floodplain stability with soil formation. The buried soils exhibit A-Bk profiles indicative of well-drained settings for relatively brief intervals of time. The A horizons are locally cumulic (and mollic), indicating that the vegetated floodplain landscape was at times slowly aggrading. A dateable record of early Holocene landscape evolution was not recovered from the Smoky Hill River.

Buried soils dating to the period ~7000–5000 yr B.P. were not found in the valley fill of the Pawnee and Lower Smoky Hill rivers. The sedimentology indi-

Figure 9.6 Stratigraphic section exposed at the Eagle's Roost alluvial fan, site 14EW174 (from Mandel, 1992, fig. 2–4; photo provided by and reproduced with permission of R. D. Mandel). The lower radiocarbon age is from a cumulic A horizon (Unit V) that was buried by rapid fan sedimentation. The upper radiocarbon age was determined on organic matter in the resistant Btk horizon (note the well-expressed structure), which was truncated by erosion across the fan surface.

cates that the valleys were undergoing cutting, filling, and in general, net transport of alluvium, with little or no long-term storage (i.e., with little stability). Along tributaries of the lower Smoky Hill River, however, sediment was carried out onto the floors of larger valleys where alluvial fans formed (fig. 9.6). Dating shows that these fans were forming by ~5700–5100 yr B.P. and probably began aggrading earlier in the Holocene. Buried soils in the fans exhibit well-drained A-C and A-Bt-Btk morphologies with thick, dark (mollic) surface horizons. The soils show that the fans built up episodically, but likely with some sedimentation even during relatively stable phases.

Beginning ~5000 yr B.P. the Pawnee and Smoky Hill river systems returned to the pattern of cyclic stability (with soil formation) and instability (with erosion and deposition). The bulk of the Holocene fill in the region (Units III–VII) as well as most buried soils date to the late Holocene. The soils have A-C and A-Bk profiles, indicating relatively brief phases of stability under well-drained conditions. The timing of late-Holocene floodplain stability and soil formation shows some variation according to the size of the streams. Radiocarbon dating indicates several discrete periods of alluviation, erosion, and stability with soil formation in large valleys: Deposition of Unit III >4400 yr B.P., followed by a brief phase of stability and soil formation, then burial (by Unit IV) and entrenchment. Another phase of sedimentation followed (Unit V) and along with several discrete periods stability and soil formation, one at 2750–2600 yr B.P. and another at 2000–1600 yr B.P., and then more alluviation (Unit VI). The older of these two episodes is

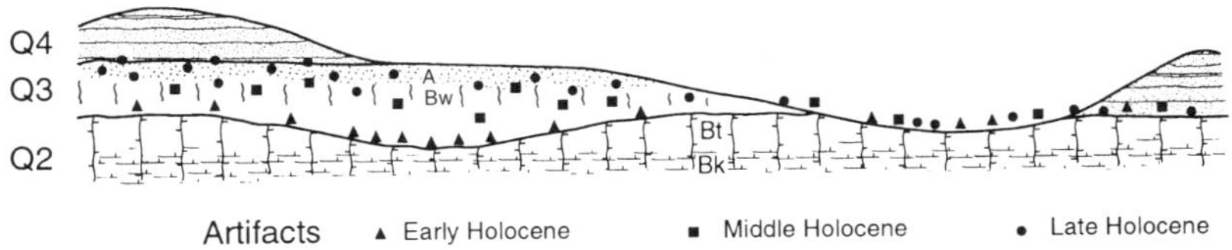

Figure 9.7 Schematic illustration of the relationship between lithostratigraphy (Q2, Q3, Q4), soil stratigraphy, and archaeology at White Sands, New Mexico (based on Blair et al., 1990). Q4 represents 20th century dunes with unweathered sand.

correlative with floodplain stability dated to ~2800–2000 yr B.P. in small valleys. However, the younger phase of stability with pedogenesis in large valleys overlaps only the beginning of the landscape stability dated ~1600–1000 yr B.P. in small valleys. Hence, late-Holocene deposition appears to have been time-transgressive throughout the drainage basins. Very late in the Holocene (within the last 1000 yr?) the streams entrenched, cutting through the older valley fill and creating the modern floodplains.

Dunes and Sand Sheets in the Southwestern United States

In the southwestern United States, soils have proven important in reconstructing the evolution of eolian landscapes that have been occupied since late Pleistocene time. The White Sands of south-central New Mexico accumulated episodically for the past 50,000+ radiocarbon yr (Fig. 9.7) (Blair et al., 1990). The oldest buried landscape likely to be contemporaneous with human activity was an extensive eolian sand sheet (Unit Q2) that aggraded ~15,000–9400 yr B.P. Associated with this landscape were arroyos and arroyo-mouth ponds. A soil with a Bt-Btk profile formed on Q2, probably under grassland vegetation and for at least 2000 yr. Little catenary variability was exhibited by this soil, indicating that the landscape was flat to gently rolling. The absence of an A horizon and a sharp upper boundary between the soil and overlying deposits shows that the landscape was scoured, probably by wind erosion, prior to burial. The surface was buried by either of two younger deposits (Q3, Q4), indicating that the Q2 surface was deflated several times. This scenario is also suggested by concentrations of artifacts on the eroded Q2 surface that vary widely in age. Unit Q3 buries Q2 and is sedimentologically and geomorphically identical to Q2. Unit Q3 accumulated and was modified by pedogenesis ~7300–100 yr B.P. A soil with an A-Bw-Bk horizon sequence formed in Q3. As with the older soil, the Q3 soil is horizontal to subhorizontal, indicative of a flat to gently rolling topography. Archaeological materials were found throughout Q3 as well as on it, indicating human use of the region during both active and stable phases of landscape development as well as during stability.

The landscape of the region today is characterized by mesquite-generated coppice dunes superimposed over a low-relief, hummocky topography consisting of deflation depressions and elongate eolian ridges, swales, and mounds. Much of the Holocene stratigraphy was destroyed in the deflation basins but preserved under the eolian dunes (fig. 9.7). This modern landscape is in strong contrast to the flat to gently rolling grassy landscape that characterized most of the Holocene. The buried soils proved to be the best clues to reconstructing the landscapes and to understanding where archaeological occupation zones are likely to be preserved and where they were destroyed. The model of landscape evolution and archaeological site evaluation proposed by Blair et al. (1990) provided a successful approach to mapping and site prediction by Monger (1995) working in a similar setting south of White Sands.

In northwestern New Mexico, soil stratigraphic, soil geomorphic, and geoarchaeologic investigations in the Chaco Dune Field revealed a record of landscape evolution broadly similar to that of the White Sands region (Hall, 1990; Smith and McFaul, 1997; Wells et al., 1990), though apparently with more cycles of deposition stability (fig. 6.12; see also chapter 6). Paleobotanical data (summarized by Smith and McFaul, 1997, pp. 129–133) provide additional insights into the character of the landscapes as they evolved through the late Quaternary. The oldest deposits (Qe1 of Wells et al., 1990) accumulated in the final millennia of the Pleistocene. Data on the timing of initial landscape stability are not available, but burial of the associated soil geomorphic surface (Tohatchi I of Smith and McFaul, 1997) may have been time-transgressive, with burial by eolian sediments beginning ~9300 yr B.P. on drier uplands and ~7800 yr B.P. in most lowlands. The duration of stability and pedogenesis was significant enough to produce a soil with well-expressed Bt and Bk (Stage II–II+) horizons, however. Several thousand years is probably a reasonable minimum estimate for the length of stability. The length of time likely available for occupation and the interpretation of a grassland-conifer cover, which should have been relatively attractive for habitation, indicates that the soil surface would be likely to produce Paleoindian archaeological debris. Little was found, however, probably because of erosion, deep burial "and/or . . . failure of researchers to identify or test sediments of Paleoindian age" (Smith and McFaul, 1997, p. 130).

Eolian and alluvial fan deposits and associated soils represent the interval 7800–4500 yr B.P. (Tohatchi II). Soils formed in the deposits are A-Bk and A-Bky with Stage I–I+ calcic horizons. Although as much as 3000 yr may have been available for pedogenesis, the presence of secondary gypsum is a clear indicator of drying conditions, relative to Tohatchi I, which would minimize pedogenic processes such as clay translocation and formation of Bt horizons.

The late Holocene is characterized by increased frequency of cycles of sedimentation/stability. They are characterized by eolian, alluvial, and playa deposition followed by formation of A-Bw soils or A-Bk soils with Stage I calcic horizons. The relatively weak soil development is the result of the relatively short duration of each interval of stability. The cycles are dated to 4500–3100 yr B.P. (Tohatchi III), 3100–2100 yr B.P. (Tohatchi IV), 2200–660 yr B.P. (Tohatchi V), and 660–100 yr B.P. (Tohatchi VI).

In the far western panhandle of Oklahoma, abundant materials from multiple Paleoindian occupations were recovered from stratified sand sheets at the Nall site (Baker et al., 1957; LaBelle et al., 2003). The site is along the sloping margin of a playa basin. Eolian sand sheets mantle the uplands around the playa and also drape over the playa margins. In the excavation area, the sand sheet is composed of multiple layers of eolian sand, the lower two layers each containing prominent buried soils and occupation debris. The lower "Nall soil" has a thick, dark, organic-rich (probably mollic) A horizon. Below the soil is coarse sand with redoximorphic features and manganese oxide staining. The soil probably formed under conditions of high organic-matter production and saturation. As the Nall soil is traced laterally and up-slope from the main excavations, it is better drained and exhibits an A-Btw profile. The variability in morphology of the Nall soil indicate that in the excavation area, the Paleoindian occupation was related to locally wet conditions, possibly a spring of seep. This would also account for the apparent intensity of the occupation.

Pacific Coast of Canada

Buried soils are proving to be significant in understanding the routes and timing of early human migrations from Asia to North America. Though an interior "ice-free corridor" route was long championed as the initial route for the peopling of the New World, a number of workers and a large array of data now cast doubt on that hypothesis (see review by Mandryk et al., 2001). Rather, attention has shifted to the West Coast of North America, where narrow shelf, now below sea level, may have provided access from Beringia to the interior of North America. Buried soils revealed in ocean cores are perhaps the best evidence for a landscape both unglaciated and exposed above sea level during early postglacial times (Luternauer et al., 1989; Fedje and Josenhans, 2000; Mandryk et al., 2001). The soil described by Luternauer et al. (1989) exhibits an oxidized zone (Bw horizon?) overlain by a "humified" zone with abundant plant detritus (Ah or O horizon?). Paleobotanical remains from this and other cores indicate a forested landscape (Josenhans et al., 1995; Mandryk et al., 2001). The data are limited so far, but buried soils provide unequivocal evidence for an exposed and habitable landscape.

Chunchucmil, Mexico

A recurring question in Maya archaeology is the degree of agricultural productivity necessary to support apparently high ancient Maya populations in an agriculturally limited environment. One component of the answer to this question is the nature of the soils in and around Maya sites (agricultural productivity and soils are dealt with in chapter 11). In the area of Chunchucmil, northwest Yucatán, Beach (1998b) approached the issue by reconstructing the soil geomorphic relationships during the Maya occupation. The region is a low-gradient plain. Soil variation results mostly from relatively minor topographic differences that divide the landscape into the coastal zone, the swamp estuary zone, the savanna-tzekel zone, and the karst plain. The regional soil diversity is described largely on the basis of microrelief, which determines local drainage conditions. The following description of soil–landscape relationships is from Beach (1998b, pp. 773–780). He essentially describes a series of catenas within each subdivision of the landscape (table 9.1). Descriptions of catenas, in or out of archaeological contexts, are rare for the tropical lowlands of Central America.

Table 9.1. Soil-geomorphic relationships in the Chunchucmil region, Yucatán, Mexico

Setting	Soil type	Horizon	Description[2]
A. Coastal and Swamp Catena[1]			
Swale	Tropaquept	A	20–14 cm; fSl; 10YR 2/2
		AB	14–20 cm; LS; 10YR 5/3
		C	20+ cm; LS; 10YR 6/3
Beach ridge	Psammaquent	A	0–20 cm; LS; 10YR 5/2
		C	20–26 cm; S; 10YR 7/3
		Ab	26–29 cm; LS; 10YR 5/2
		2Cb	29+ cm; LS; 10YR 7/2; large shells
B. Savanna and Tzekel Catena[3]			
Tzekel swale	Tropoaquept	A[4]	0–11 cm; CL; 10YR 2/1-3/2
		Cg	11–18 cm; CL; 10YR 5/2; calcite crystals
Tzekel knoll	Ustochrept	A	0–8 cm; CL; 10YR 2/1
		Bk	8–22 cm; CL; 10YR 7/3
		Cr	22+ cm; 10YR 8/2
Savanna swale	Paleustalf	A	0–7 cm; CL; 2.5YR 3/4
		Bt	7–21 cm; C; 2.5YR 3/6; clay films
		C	21+ cm; 10YR 8/2; calcrete
Knoll	Paleustoll	A	0–35 cm; CL; 10YR 2/1-3/2
		Cr	35+ cm; 10YR 8/2
C. Karst Plain Catena[5]			
Karst knoll	Haplustalf "kancab"	A	0–13 cm; SiCL; 5YR 3/2
		Bt	13–34 cm; CL; 5YR 4/4; clay films
Flat swale	Paleustalf "kancab"	A	0–21 cm; SiC; 10YR 2/1, 5YR 3/2
		Bt	21–30 cm; 2.5YR 3/6; clay films
		BCr	30–33 cm; CL; 2.5YR 4/6
	Paleustoll "kancab"	A	0–15 cm; SiCL; 5YR 3/2
		C	15+ cm; 10YR 8/2; calcrete
Knoll	Haplustoll "boxluum"	A	0–28 cm; CL; 10YR 2/2; 5YR 3/2
		AC	28–40 cm; CL; 10YR 4/3
		Cr	40+ cm; 10YR 8/2

[1] From Beach (1998, table 1).
[2] Abbreviations for soil textures: fSL = fine sandy loam; LS = loamy sand; S = sand; CL = clay loam; C = clay; SiCL = silty clay loam.
[3] From Beach (1998, table 2).
[4] Soil horizon sequence and description generalized from Beach (1998, Tables I, II, and III).
[5] From Beach (1998, table 3).

In the coastal zone, the topography consists of beach ridges and swales. The soils are weakly expressed, dominated by Entisols and Inceptisols. In part, the nature of the soils was determined by the youthfulness of the landscape (<5000 yr B.P.), which apparently is "rejuvenated" by severe storms that cause erosion and uprooting of vegetation. Otherwise, the topography determines the soil distribution, with Psamments, Psammaquents, Aquepts on the ridges and swales (table 9.1A), and some Histosols in low wetland areas. Inland from the coastal zone is an extensive wetland estuary with mangrove swamp. Histosols are found throughout the area.

Inland of the swamp estuary is a gradually rising zone of weathered limestone divided into a topographically lower and upper savanna characterized by swales and hillocks ("tzekeles"). In the lower savanna, groundwater seepage causes widespread precipitation of calcium carbonate and formation of a dense calcrete. Wet Inceptisols (Tropaquepts; table 9.1B) and some Histosols occur in this poorly drained setting. On the somewhat better-drained tzekel knolls, the soils are weathered directly on the limestone and consist of Ustochrepts (table 9.1B) with a few Mollisols. Thin but redder and more clayey "kancab" soils (Paleustalfs; table 9.1B) are common throughout the upper savanna, in both depressions and hillocks. Mollisols (*boxluum* or *chichluum* soils) also are found on some weathered bedrock hillocks (table 9.1B). Soil development is significantly stronger in the upper savanna, probably because the landscape is better drained and drier and because active precipitation of calcium carbonate, which maintains a relatively youthful landscape, is much less common.

On the upland karst plain, soils are generally similar to those of the upper savanna in being relatively well developed. Soil variability is the result of parent material lithology and topographic setting. On dense calcrete there is often no soil. On weathered knolls and hillocks of limestone are the relatively thin "boxluum" soils, characterized by Mollisols (Calciustolls and Paleustolls; table 9.1C). In swales and depressions, where weathering was more intense because of the more moist, heavily vegetated setting, soils are thicker and more strongly expressed (Paleustalfs and Paleustolls; table 9.1C).

The soil studies in the Chunchucmil region illustrate the severe limitations on agriculture (Beach, 1998b, pp. 786–787). About 50% of the landscape has no soil. Only about 20% of the soils are deep and fertile. Erosion probably was not and is not a factor in the soil patterns owing to the very low relief. If intensive agriculture was practiced, it may have been within walled mounds and fields, where water and fertilizer were easily accessible.

Greece

Soils have proven to be valuable for reconstructing the late Quaternary landscapes of the Argolid of southern Greece in regional archaeological contexts. The researchers were attempting to better understand the relationship between settlement and landscape dating from the earliest occupation of the region in the middle Paleolithic. A particular aim of the study was to test a well-known model by Vita-Finzi (1969), widely applied throughout the Mediterranean, that there were only two cycles of late Quaternary alluviation in Greece: one corresponding to the last glacial maximum and the other to the Little Ice Age (see summary discussion by Pope and Van Andel, 1984, p. 385, and Van Andel, 1998). The work on upper Pleistocene and lower Holocene stratigraphy (Pope and Van Andel, 1984; Van Andel and Zangger, 1990; Zangger, 1993; Van Andel, 1998) dealt with natural cycles of sedimentation and stability, whereas the focus of much of the research on upper Holocene landscapes dealt with assessment of human impacts (Van Andel et al., 1986, 1990; Van Andel and Zangger, 1990; Zangger, 1993). Most of the work focused on alluvium deposited on floodplains and in fans in valleys.

The geography of much of Greece during the middle Paleolithic occupation appears to have varied significantly, depending on local rates of sedimentation versus stability/pedogenesis. In the southern Argolid, soils, archaeology, and numerical dating methods show that the region was subjected to several cycles of alluviation and stability in the late Pleistocene and again in the late Holocene (Pope and Van Andel, 1984; Pope et al., 1984; Van Andel, 1998). Of interest here are the older alluvial units (members of the "Loutro alluvium"; discussed in chapters 6 and 7) dated to >250,000 yr B.P., >60,000 yr B.P., and 45,000–32,000 yr B.P. (table 6.1; Pope and Van Andel, 1984, p. 293). Each of these alluvial units is characterized by well-expressed soils (reddish-brown to red Bt horizons and nodular calcic horizons; table 6.1) denoting prolonged landscape stability. Indeed, the landscape on the Upper Loutro sediments remained stable from the latest Pleistocene through the early phases of agricultural activity; that is, the region was fairly stable in the Upper Paleolithic during full glacial time and on through the Neolithic and into the Early Bronze Age.

On the Argive Plain, soils were an integral component of geoarchaeological studies, especially the reconstruction of late Quaternary landscape evolution (Zangger, 1993). The plain is a constructional coastal landform built during the Pleistocene and Holocene. Sediments include both marine and alluvial deposits. Most depositional sequences are separated by buried soils. The earliest occupation of the area was in the early Holocene, indicated by Neolithic artifacts recovered from a prominent soil developed on the youngest Pleistocene sediments (Zangger, 1993, p. 50). The soil is characterized as a thick, dark brown A horizon that is also the most prominent stratigraphic marker in the late Quaternary sequence (Zangger, 1993, p. 50). The soil formed on the Argive Plain during the maximum lowering of sea level associated with the last glacial maximum. The soil is also described as a depositional unit and probably represents a slowly aggrading A horizon forming under a thick vegetation cover. The high clay content of the soil (which would inhibit rapid drainage), low relief and low elevation, and proximity of the sea probably resulted in a relatively wet soil with a lush vegetative cover for the Neolithic occupants who arrived in the early Holocene.

Coastal Plain of Israel

The red hamra and husmas soils found within the eolian sand and carbonate-cemented eolianites or "kurkar" of coastal Israel (see chapters 6 and 8) also provide an opportunity to reconstruct the middle and late Pleistocene landscapes of the region in an archaeological context. Paleolithic sites and artifacts are commonly associated with these sediments and soils (tables 6.3, 8.3, and 9.2; fig. 6.14; e.g., Farrand and Ronen, 1974; Ronen, 1975, 1977; Tchernov et al., 1994; Gvirtzman et al., 1999; Porat et al., 1999; Tsatskin and Ronen, 1999). Generally speaking, the eolian sediments that comprise the kurkar probably accumulated during high stands of sea level, whereas the soils formed on stable landscapes that existed during low stands and during parts of the regressional and transgressional phases (Ronen, 1975; Horowitz, 1979, p. 114; Gvirtzman et al., 1984, 1998, 1999; but see Goldberg, 1994, for a critique of this correlation). Gvirtzman et al. (1998, pp. 41–42) focus on the role of wet versus dry climate in forming the characteristics of hamra and other soils buried in the sands but still link the cycles of sedimentation and soil formation to sea level. Frechen et al. (2002) show a

Table 9.2. Morphology of a "Mousterian Hamra" paleocatena at Habonim, Israel

Unit	Description
Dune Crest[1]	
1	Upper Sandstone/Kurkar
2	Hamra: Sandy loam; reddish-brown (2.5YR 4/4) AB horizon; blocky structure with scattered $CaCO_3$ concretions; Mousterian artifacts; sandy loam; brown (5YR 4/6) Bk horizon with large $CaCO_3$ concretions
3	Lower Kurkar/Sandstone
Interdune Depression[2]	
Ia	Loose sandy clay; (10YR 7/8) with 1-m-thick lenticular calcrete layer.
Ib	Gleyed Soil: Sandy loam; yellow mottles (5Y 5/1) with blue-green root traces; prismatic structure; vertical cracks with sand (Ia) at top.
II	Vertisol: Sandy clay loam; (5Y 3/1 to 2.5Y 3/2); blocky and prismatic structure with slickensides; Fe/MN and $CaCO_3$ concretions; Mousterian artifacts
III	Truncated Soil: Sandy loam; (10YR 4/6) AB horizon; blocky structure with few slickensides; faint Mn coatings on ped faces; few $CaCO_3$ concretions
IV	Hamra: Loamy sand; (7.5YR 4/6); (A)B horizon; massive with $CaCO_3$ concretions, over (10YR 6/6-6/8) BCk horizon with $CaCO_3$ druses and pans

From Ronen et al. (1999) and Tsatskin and Ronen (1999).

[1] From Ronen et al. (1999, pp. 138, 139, 142).
[2] From Tsatskin and Ronen (1999, table 1 and pp. 369–370).

good correlation between phases of eolian sedimentation in the kurkar/hamra sequence and high sea level.

Farrand and Ronen (1974, p. 46) present the following model for the evolution of the kurkar and hamra landscapes (see also Gvirtzman et al., 1984). During a high stand of sea level, an abundant supply of littoral and offshore sand accumulated. As sea level began to fall, regressional dunes accumulated in the former supralittoral zone. Dunes continued to build until sea level fell below the zone of abundant offshore sand. Some of the shell fragments in the dune sands dissolved as the dunes built up and the $CaCO_3$ reprecipitated as cement, forming the eolianite. During the sea-level regression, both the supply of sand and shell fragments decreased, whereas rainfall and dust influx combined to produce a noncalcareous, clay-rich red (hamra) soil in the upper eolianite. Pedogenesis continued in all subaerially exposed sand throughout the low stand and through the following transgression and subsequent high sea-level stand.

The proposed relationship between sea level and soil formation has important implications for human occupation and the resulting archaeological record. First, the stabilized coastal plain would have been much wider than it is today, roughly double the present width during maximum sea-level lowering (following Horowitz, 1979, figs 2.1 and 2.2), providing a wide "Levantine corridor" for human migration along the eastern Mediterranean. Second, sorting out discrete occupation levels on the hamra will be problematic because individual sites will likely represent palimpsests of occupations on the long-stable hamra paleolandscapes.

Most of the Paleolithic sites of the Israeli coastal plain are associated with hamra. Ronen (1975, p. 230) unequivocally states that, "all cultural remains . . .

were in or on soils and never in sand or sandstone." Porat et al. (1999, fig. 2), however, illustrate the late Acheulian assemblage from the Holon site as coming from between two hamra. Nevertheless, most of the coastal plain sites do seem to be associated with the soils. Gvirtzman et al. (1998, p. 43) see this relationship as directly related to climate: The hamra formed under climatic conditions that were conducive to human occupation (i.e., more moist), but the eolian sand was deposited under a climate that did not attract humans (i.e., a drier climate). This interpretation does not take into account the sedimentological "dilution" of site occurrences in aggrading settings versus the likely concentration of sites on stable surfaces. To fully evaluate either scenario, however, requires numerical age control on the duration of sand deposition and the duration of pedogenesis.

To date, the hamra and other soils buried in the coastal eolian deposits have been used primarily as stratigraphic markers (Farrand and Ronen, 1974; Ronen, 1975; Horowitz, 1979, p. 114; Gvirtzman et al., 1984, 1998, 1999; discussed in Chapter 6). Few investigators have used the soils to better understand archaeological landscapes, but the opportunities are promising. In an early studies of hamra geoarchaeology, Ronen (1977) presented one of the few attempts at a broad reconstruction of Pleistocene landscapes using hamra exposed in newly opened road cuts. In the Mt. Carmel area, long famous for its Paleolithic sites, there are three kurkar ridges, each parallel to one another and the coast. "The Mousterian topography of the major [westernmost] sandstone ridge is clearly discernable and differs surprisingly little from today's configuration . . . the Mousterian soil on the north, south and east slopes parallel . . . the present slope of the hill and is covered by c. 4 m. of later sandstone, while at the crest of the hill it is buried by only 1 m. of sandstone. Hence, in Mousterian times, . . . [the site] was as prominent today above its surroundings" (Ronen, 1977, p. 186).

Variability in the morphology of buried hamra and related soils (chapter 6) is usually ascribed, at least in part, to climate variability through time (e.g., Dan and Yaalon, 1971; Ronen, 1975; Gvirtzman et al., 1998, 1999; Wieder and Gvirtzman, 1999), but other factors, which have important implications for interpreting archaeological landscapes, should be considered. The scenario for kurkar/hamra formation as a function of sea-level change, outlined above, should result in facies variations where the inland soils are more strongly expressed because they had more time to form. None are reported, however, probably because of the lack of data on regional variation of individual hamra. Time must be considered an important variable in the formation of the buried soils because differences in soil thickness, horizon sequences, clay content and rubification of Bt horizons, and degree of carbonate leaching can be time dependent. For example, the weakly expressed buried soils commonly observed between the Mousterian and Epi-paleolithic hamra are interpreted as the result of extremely low precipitation (Gvirtzman et al., 1998, pp. 41–42). In the interpretation of these same soils, Ronen (1975, p. 233) alludes more broadly to "conditions" that inhibited soil development. Alternatively, the presence of multiple, weakly expressed soils may be the result of episodic accretion of eolian sand, allowing only brief intervals for pedogenesis. In contrast, the characteristics of the strongly expressed Evron hamra are linked to both "intensive leaching" under a "wetter climate" and to "extended pedogenesis" (Sivan et al., 1999, p. 291). Better age control on periods of sedimentation and soil formation is necessary to adequately address the issue of time as a factor in the formation of the buried soils.

Catenary variation in the hamra and other buried soils as they formed across undulating dune topography also accounts for some morphological variability in the soils. Dan et al. (1968) and Dan and Yaalon (1971) describe both surface and buried catenas associated with hamra soils. A typical relationship from ridge-top to trough is hamra at the crest, sandy hamra on the slope, nazaz (pseudogley or Albaqualf) at the toeslope, and grumusol (Vertisol) in the lowland (table 9.2). This association of soils accounts for variation in color, clay content, and thickness of the soil. In regional stratigraphic studies of buried soils on the coastal plain of Israel, Gvitrzman et al. (1998, p. 37) provide one of the few references to buried catenas, describing the buried soils as thicker in the troughs of paleo-ridges and thinner on the crests (borne out in detailed investigations at the Habonim site, discussed below).

Two of the few detailed pedologic investigations of buried hamra and related soils are from archaeological sites. At both localities, slope position helps explain the setting and evolution of artifacts and soils. The soil stratigraphy and micromorphology of the lower Paleolithic site of Revadim were investigated by Gvirtzman et al. (1999) and Wieder and Gvirtzman (1999; fig. 9.8). The site consists of a large assemblage of stone tools (e.g., hand axes, choppers, cores, and flake tools) and bone (from extinct elephant, equids, cervids, suids, bovids, and felids) exposed in a quarry in a sequence of eolian deposits. The soil stratigraphy was traced far beyond the site in the walls of the quarry, in trenches, and in boreholes. These additional stratigraphic data allowed the site and associated soils to be placed in a broad geomorphic context that was not apparent in the quarry exposures. Three moderately to well-expressed soils are associated with these wind-derived sediments. The oldest soil is just below the archaeological assemblage. It is a reddish Btkb, a husmas variant of a red hamra soil. Above the Btkb is a "mixed zone," which produced the artifacts. It represents the eroded top of the husmas welded to the base of the next higher buried soil (B horizon of a "Quartzic Gray Brown" soil; QGB) by superimposition of a calcic horizon over both (fig. 9.8). The artifacts probably were left on the eroded surface of a hamra. Following burial they were mixed up into the base of the QGB soil. Calcium carbonate also was precipitated at the base of the QGB and in the upper hamra, welding the two and converting the hamra to husmas. The borehole data show that the occupation took place on or near the crest of a ridge. This probably explains the erosion of the occupation surface. The adjacent trough then filled, but on the flanks only a thin layer of sand covered the artifact zone.

The soil stratigraphy, soil geomorphology, and micromorphology of a "Mousterian paleohamra" at the Habonim site are discussed by Ronen et al. (1999) and Tsatskin and Ronen (1999) and provide a good paleotopographic contrast to the Revadim site. Habonim is on the northern (Carmel) coastal plain. The site exposed a buried soil formed low on the flanks of an ancient dune and across the adjacent interdune depression. The soil is a good example of a buried catena, with a truncated hamra 1.2 m thick on the slope of the paleodune, traceable to a welded pedocomplex 4.5 m thick in the interdune depression. These

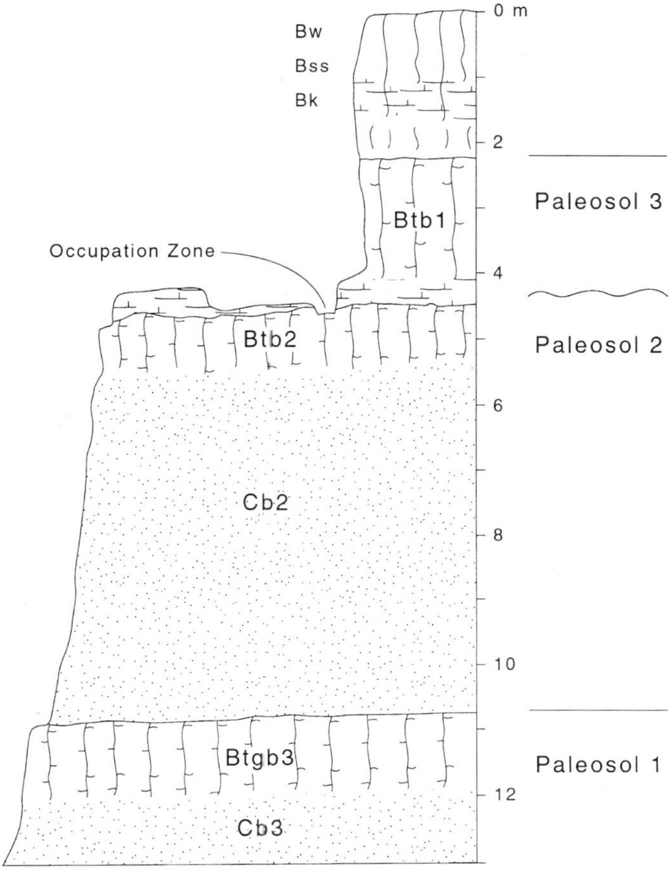

Figure 9.8 Soil stratigraphy of the Revadim Quarry (based on Gvirtzman et al., 1999, fig. 4). The base of Paleosol 3 (with the artifact zone) is a Btk horizon, locally welded to the Btb2 of Paleosol 2.

welded soils illustrate changing drainage conditions during the occupation of the area (table 9.2). The lowest soil (IV) is described as typical of the modern surface hamra (Tsatskin and Ronen, 1999, p. 369), though it contains a Bk horizon. The soil is indicative of well-drained conditions. Moving up the section, soils III and II become increasingly more clayey because of increased addition of eolian fines, show evidence of progressively poorer drainage conditions (darker colors caused by more organic matter production; mottling of colors, formation of Fe and Mn concretions) because of the increase in clay, and show progressive development of vertisolic (i.e., shrink–swell) features. Most Mousterian (Levallois) artifacts were recovered from Unit II; a few came from Unit III. Thus, although the kurkar ridges probably were always well drained, archaeological sites there were subjected to erosion. In contrast to the ridges, the groundwater conditions changed in the interdune depressions, which would have been attractive to people living in the area. The depressions also were more conducive to site preservation.

Yemen

Buried soils have proven useful in interpreting landscape evolution and correlating Holocene stratigraphic sequences in the southern Arabian Peninsula. Soils were used to place archaeological sites and artifact finds in the context of late Quaternary landscape evolution in the northern highlands of Yemen, both in the semiarid intermontane plateaus and plains (Wilkinson, 1997) as well as the drier desert basins (de Maigret et al., 1989; Fedele, 1990) and in the transition zone between the highlands and the desert dune fields of Saudi Arabia to the east (Grolier, 1988; Overstreet and Grolier, 1988, 1996). On the intermontane plateaus in the Dhamar area, 100 km south of Sana'a, research focused on desert basins, mountain valleys, and mountain slopes between the basins and valleys (Wilkinson, 1997). Much of the geoarchaeological work focused on the "Jahran palaeosol," a distinctive stratigraphic marker observed throughout the study area in a variety of landscape settings. The soil formed in a wide variety of late Pleistocene and early Holocene sediments, including eolian sand and silt, alluvial sand and gravel, volcanic ash, rocky slopewash, and paludal deposits. Though no details are provided, "the Jahran soil reflects the nature of the substratum within which it developed" (Wilkinson, 1997, p. 849). The soil is characterized by a dark-brown to very dark gray (7.5TR 3/2 to 10YR 3/1 dry) A horizon and a Bk horizon. Archaeological materials are rare below or in the Jahran soil, but the cultural debris that was recovered from the soil is Neolithic. The soil is buried by sediment with considerable archaeological debris, representing relatively intense sedentary settlement.

The presence of a distinctive soil formed in sediments representing very active geomorphic processes in a variety of landscape positions is indicative of marked stability in an otherwise unstable setting. The relatively high content of organic matter (>2%) and presence of abundant fine rootlet voids in the A horizon, and the presence of a Bk horizon, indicate a semiarid, perhaps grassland environment, also indicated by phytoliths. Burial of the soil and, hence, its presence as a stratigraphic marker are likely the result of anthropogenic disturbance of the landscape.

Stratigraphic equivalents to the Jahran soil are reported in more arid regions of Yemen and east into Saudi Arabia. In the Yemen Highlands just east of Dhamar, de Maigret et al. (1989) and Fedele (1990) describe a buried soil and a record of landscape evolution very similar to that presented by Wilkinson. The early Holocene is characterized by coarse- and fine-grained sediments of alluvial, colluvial, and eolian origin, interfingered with travertine, interpreted as indicative of alternating wet–dry environments (de Maigret, 1989, p. 240; Fedele, 1990, pp. 33–34). Superimposed on these sediments is a soil high in organic matter (up to 5%) and secondary $CaCO_3$ (up to 10%; Fedele, 1990, p. 33) dated to the middle Holocene and termed the "Thayyilah paleosol." This soil also contains archaeological debris from Neolithic settlements. The Thayyilah soil is thus a good stratigraphic correlate of the Jahran soil and is also indicative of a similar record of landscape evolution. The middle Holocene was characterized by stability caused by increased moisture and more dense vegetation cover. The soil is buried by eolian and colluvial deposits. Similar to the Jahran soil, burial of the Thayyilah soil seems to be associated with destabilization from human activity. In addition, some evidence indicates that the region was subjected to tectonic uplift that also destabilized the landscape and lowered the water table.

Farther to the northeast, in the al-Jadidah Basin in Wadi al-Jubah, where the Yemen Highlands drop down to the Arabian desert, buried soils were used to interpret a Holocene geoarchaeological record generally similar to that in the uplands (Grolier, 1988; Overstreet and Grolier, 1988, 1996; Brinkmann, 1996). Similar to the other areas, the early to middle Holocene was characterized by active erosion and deposition. Sedimentation was dominated by alluviation in the basin, with minor additions of slope-wash colluvium, eolian sand, and eolian silt. Unlike the Highland record, however, soils formed in these deposits as they aggraded episodically, producing sequences of buried, locally welded soils (Brinkmann, 1996). Two to three buried soils dated between 10,000 and ~5000 yr B.P. were identified in most sections, characterized by A-Bw-Bk profiles and classified as Mollisols because most A horizons are relatively dark gray and overthickened. The field data indicated that the early to middle Holocene was characterized geomorphically by periodic instability under arid conditions, which removed vegetative cover and allowed accelerated erosion, and by intervening periods of landscape stability under more moist conditions, which generated more plant growth and in turn stabilized the landscape and promoted soil development (Brinkmann, 1996, pp. 120–131, 174–175, 202–205; Overstreet and Grolier, 1996, pp. 363–364).

Around or shortly after ~5000 yr B.P., the moderately expressed Mollisols of the al-Jadidah Basin were buried by anthropogenic sediments. Deposition was related to the advent of farming and irrigation, which, along with a return to more arid conditions, destabilized the landscape through devegetation.

East African Rift Valley

The great tectonic rift that runs through east Africa contains some of the best known and most intensely studied hominid and Paleolithic sites known (e.g., Laetoli, Olduvai, Koobi Fora, and Hadar). The archaeological sites and paleontological localities are buried within volcanigenic, alluvial, and lacustrine sediments. Buried soils are common in these deposits; many are associated with the sites. Pedogenic and soil geomorphic details are rare, however; the buried soils are usually only mentioned in passing (e.g., Brown and Feibel, 1986, figs. 3, 4, 5; 1991, figs. 1.5, 1.6; Feibel et al., 1989, p. 601; 1991, p. 324; Kimbel et al., 1996, p. 551). Some information is available and provides clues to the evolution of the multiple late Tertiary and Pleistocene landscapes.

At Chesowanja, in the Northern Rift Valley of Ethiopia, human remains and artifacts were recovered from the Chemoigut Formation (early Pleistocene) and Chesowanja Formation (middle Pleistocene; Bishop et al., 1975, 1978). The Chemoigut beds consist of shallow-water lacustrine sediments with several "weathering zones of red clay with calcrete" and zones with roots casts. The horizons of weathering and plant growth appear to be buried soils, varying significantly in degree of development, but indicative of subaerial exposure resulting from fluctuating lake levels. If the calcretes are pedogenic, then they further indicate well-drained conditions for some time. The artifacts and fossils were found both in the lacustrine sediments and in association with the soils, indicative of land use at or near the body of water. The material from the soils may be *in situ*, but that from the lake deposits must be reworked. The Chemoigut Formation was folded and eroded before deposition of the Chesowanja Formation, which consists of two basalt layers. Acheulian artifacts were associated with a soil 3–4 m thick in the upper portion of the upper basalt. This occupation apparently was on a long-stable basaltic landscape, which was subsequently folded, eroded, and buried.

North of Chesowanja, in the Rift Valley of Ethiopia, Middle Paleolithic sites are associated with multiple buried soils of the Gademotta Formation (middle and late Pleistocene; Albritton, 1974; Wendorf and Schild, 1974a,b; Laury and Albritton, 1975; table 6.6; fig. 6.23). The Gademotta Formation is predated by a thick sequence of rhyolitic and volcaniclastic deposits (Kulkuletti Volcanics) resulting from repeated, nearby explosive eruptions. The Gademotta sediments were deposited under more quiescent conditions, but on a geomorphically unstable landscape—probably the result of dissection of the older, unstable volcanic deposits. The lithostratigraphy and soil morphology (table 6.6) are indicative of repeated cycles of slope instability, colluviation, stability, and soil formation, with occasional deposition of primary air-fall tephra. Most of the soils are fairly well expressed, based on the development of Bt horizons. How much of this development is the result of prolonged pedogenesis and how much is caused by rapid weathering of the volcanigenic sediment is unknown, however. The occupation of the area (or at least the preservation of the occupation debris) is largely associated with the stable cycles and pedogenesis. The sites associated with soils in Units 22, 30, and 34 are within the B horizon, indicating that the landscape was slowly aggrading, perhaps because of slow colluviation during the "stable" phases.

In Olduvai Gorge, Tanzania, a buried paleocatena was used to reconstruct the evolution of the local paleohydrology around an ancient lake occupied by hominids ~1.85–1.75 Ma (Ashley and Driese, 2000; Ashley, 2001; Ashley and Hay, 2002). The paleocatena was situated on a relatively stable landscape downslope of a more rapidly aggrading pyroclastic alluvial-fan complex, but inland from the paleolake. The paleocatena is within a "cumulative red paleosol" formed in volcaniclastic parent material and differentiated into an up-slope and a down-slope facies (over a distance of about 1 km). The cumulic soil was also divided stratigraphically into a lower paleosol and an upper paleosol. The up-slope facies of the lower soil is characterized by less evidence for plant rooting, greater clay translocation, and greater zeolitization (i.e., greater weathering of the volcaniclastic parent material) in comparison to the down-slope facies. The down-slope soil also exhibited strong redoximorphic mottling. Thus, the up-slope setting was probably a drier and better-drained site than the down-slope one, which was in proximity to wetlands along the paleolake margin. Furthermore, the down-slope soil exhibited evidence of two generations of redoximorphic features, separated by a phase of clay translocation, indicative of a shift from poorer drainage to better drainage and back to poor drainage. The changing drainage characteristics are attributed to fluctuations in lake level, among other factors.

Siwalik Group, India, Pakistan, and Nepal

The "Siwalik Group" is a broad, informal term used to describe the package of molassic sediment derived from the Himalaya as it was uplifted and deposited in a belt just south of the mountain system, extending from northwest Pakistan to northeast India. These deposits, exposed in the Himalaya foothills, are dominantly alluvial and include up to 7000 meters of floodplain sediments that span the upper Cenozoic (for overviews and summaries of the Siwalik Group, see the extensive sets of references in Johnson et al., 1981; Retallack, 1991a; Badgley and Behrensmeyer, 1995; and Quade et al., 1995). The evolution of the Siwalik Group has long been of interest because of its long record of continental sedimentation related to orogenesis, the associated mammalian fossil record, and in particular, the occurrence of hominoid fossils such as *Sivapithecus*, *Ramapithecus*, and *Gigantopithecus.* The Siwalik sediments also are notable because they contain scores of buried soils, which have been used to reconstruct the evolution of Miocene and Pliocene alluvial landscapes.

In Nepal, Quade et al. (1995) investigated buried soils in Siwalik sediments representing ~11 Ma of alluvial aggradation. Among other things, the soil morphologies, particularly the changing frequency of soil colors through time (fig. 9.9), are indicative of long-term changes in drainage characteristics. Soils in the Lower Siwalik (~12–8 Ma) are characterized by thoroughly bioturbated, pale-yellow (10YR 6/4 dry), reddish, or gray/green (10YR 5/2 to 5G 5/1) dry Bt or B horizons up to 2 m thick; Fe/Mn-oxide nodules are locally common in a few Bt horizons (Quade et al., 1995, p. 1385). These soils probably formed with a high but fluctuating water table. Soils in the Middle Siwalik (8–2.5 Ma) typically are gleyed and contain no carbonate nor Fe/Mn-oxide nodules; peat soils (lignite) also are locally common (Quade et al., 1995, pp. 1385, 1391). These characteristics are indicative of a poorly drained, swampy floodplain caused by a permanently high water table. The soils of the Upper Siwalik (<2.5 Ma) are more oxidized, with bright-yellow Bt horizons, dispersed carbonate nodules in a Bk horizon, and dark, organic-rich A horizons (Quade et al., 1995, pp. 1385, 1391). These soils reflect much better drained conditions than in the Middle Siwaliks, with a permanently lowered water table. The A horizons are indicative of grassland, an interpretation strongly supported by stable-C isotopes (see discussion in chapter 8).

Johnson (1977) conducted an investigation specifically focused on the soils of the Siwalik Group on far northwestern India. In particular, the work dealt with the Nagri Formation (~11–8.5 Ma), locally referred to as "middle Siwalik" but clearly older than the Middle Siwalik soils reported by Quade et al. (1995). Depending on conditions of outcrop, Johnson (1977) was able to reconstruct a Miocene catena across components of the Siwalik floodplain and to identify locations on the landscape relative to paleochannels. A recurring catenary pattern with two basic soil facies were identified: soils with primary stratification and abundant sand that were leached, relatively high in free Fe-oxides, and reddish colors (10YR and 7.5YR hues); and soils higher in clay with little stratification (indicative of bioturbation) or free Fe-oxides that were weakly leached, and yellowish gray to neutral colors (2.5Y–N). Soils with the former characteristics were

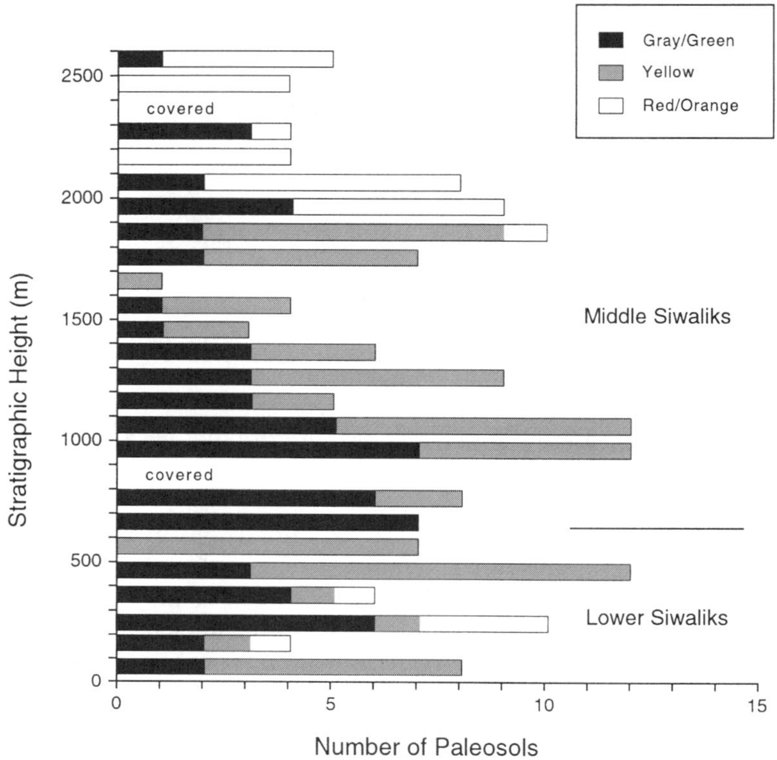

Figure 9.9 Frequency of colors in leached (usually Bt) horizons of buried soils in the Siwalik Group, Nepal (modified from Quade et al., 1995, fig. 6). Red, orange, and pale yellow colors are associated with well-drained floodplain settings. The gray/green soils formed in poorly drained floodplain positions.

proximal to the channel, associated with well-drained levees receiving frequent additions of sediment, and the latter characteristics were associated with soils in more distal, poorly drained settings in the flood basin that received finer-grained sediment more incrementally. These physical and chemical characteristics of the Siwalik soils support the interpretations of an aggrading floodplain setting.

North China Plain

The North China Plain of northeast China has been populated since the Paleolithic, and archaeological sites are commonly preserved in its deposits. Buried soils are also preserved and provide a means of reconstructing landscape evolution during the long human occupation. A good example of this application of soil geomorphology is provided by Jing et al. (1995) for the Laonanguan site in the Shangqui area (see table 7.1 and also the discussion in chapter 7). The plain is essentially a delta built up as the lower Yellow River dumped its high concentrations of silt (derived from the Loess Plateau to the west). Jing et al. (1995)

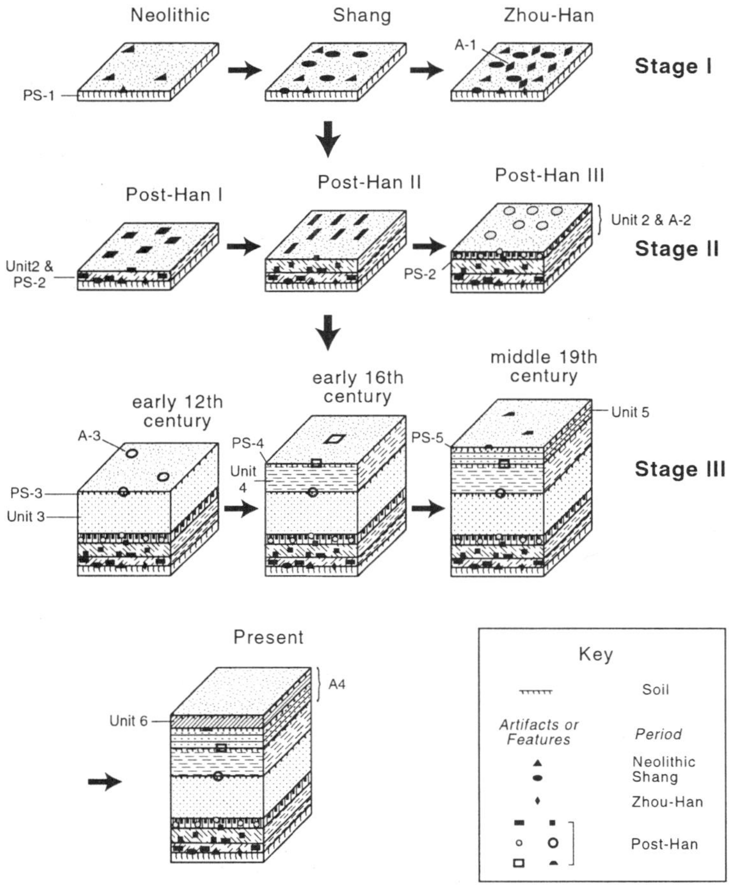

Figure 9.10 Holocene stratigraphic development and landscape evolution model for the Laonanguan site on the North China Plain. "PS" refers to soil stratigraphic units (see table 7.1). Modified from *Geoarchaeology* v. 10, pp. 481–513, fig. 9, by Z. Jing, G. Rapp, and T. Gao, © 1995, John Wiley & Sons, used by permission of John Wiley & Sons, Inc.

recognized three stages of landscape evolution through the Holocene, based on the morphology of the buried soils and the sedimentology (table 7.1; fig. 9.10). Stage I was a time of the earliest phase of human occupation at the site, from at least ~4500 to 2000 yr BP. The archaeology from this period, anthropogenic unit A-1, is mixed into the upper part of the relatively strongly expressed soil PS-1, indicative of prolonged floodplain stability. Because of the minimal flooding risk during this time, the investigators speculate that the Shanqui floodplain was perhaps the most favorable location for human occupation (Jing et al., 1995, p. 507). Stage II was a period of cumulization of PS-2, beginning ~2000 yr BP. Archaeological debris of A-2 is scattered throughout this cumulic soil, though discrete occupation zones are difficult to discern. These characteristics are indicative of the onset of slow sedimentation that buried A-1 and incased A-2.

Pedogenic processes were able to keep pace with sedimentation, however, and mixed most of the archaeological assemblages.

Stage III is characterized by formation of multiple, weakly expressed buried soils (PS-3–PS-5) within thick (up to 8 m) floodplain deposits (Units 3–6). Archaeological zones representing occupation from the early 12th century to the 20th century are found scattered throughout these sediments and soils. The geoarchaeological situation is indicative of a significant increase in sedimentation rate, probably because of increased flood frequency. This probably happened when the Yellow River shifted course in the 12th century and flowed more closely to the Shangqui area.

10

Soil Genesis and Site-Formation Processes

Pedogenic processes that produce or alter the soils associated with a landscape (buried or unburied) also modify the archaeological sites and other traces of human activity associated with that landscape and buried landscapes. The wide range of processes that form soils can profoundly affect the archaeological record. Pedogenesis, therefore, is an important component of the processes of archaeological site formation. Archaeological "site-formation processes" are those processes that modify artifacts and archaeological sites from the moment they were formed until they are uncovered by archaeologists (Stein, 2001b, pp. 37–38). Understanding formation processes is crucial in archaeology because archaeologists use the patterns of artifacts in the ground to infer behaviors. Formation processes identify patterns that are created by ancient behaviors and separate those patterns from the ones created by later cultural and natural processes (Stein, 2001b, p. 37).

In his influential volume *Formation Processes of the Archaeological Record*, Schiffer (1987, p. 7) notes that archaeologists try to infer past behavior based on the archaeological record, but the record "must be handled with great care by the investigator seeking to infer past behaviors, for the evidence that survives has been changed in many ways by a variety of processes." These processes introduce variability and ambiguity into the archaeological record. Schiffer (1987, p. 7) further distinguishes between cultural processes, in which the agency of transformation is human behavior, and noncultural processes, which stem from processes of the natural environment. Natural formation processes are many and varied and include plants, animals, wind, water, ice, and gravity, among others. Soil formation is also identified as an important process of site formation.

Schiffer (1987) provides a comprehensive discussion of natural site-formation processes, which are summarized by Stein (2001b). Nash and Petraglia (1987) and Goldberg et al. (1993) also provide a number of case histories of natural formation processes identified at archaeological sites.

Because soil formation represents the alteration of rock and sediment (chapter 1), pedogenic processes are important natural processes in the formation of archaeological sites. Other weathering processes that are significant in site formation can be grouped as "diagenetic alterations." "Diagenesis" refers to all the chemical, physical, and biologic changes undergone by a sediment after its initial deposition, exclusive of surficial alterations (weathering) and metamorphism (Bates and Jackson, 1980, p. 171). Retallack (1990, p. 129) includes pedogenesis as a kind of diagenesis, and Caple (2001, p. 588) refers to the interaction of the burial environment and archaeological evidence as diagenesis, thus lumping a variety of weathering processes that can include pedogenesis. However, most workers clearly differentiate diagenetic from pedogenic processes (e.g., Schaetzl and Sorenson, 1987; Catt, 1990, p. 65; 1998). Many diagenetic processes are similar to pedogenic processes, however, and distinguishing between the two may be difficult (Catt, 1998). For that reason, some diagenetic processes are also discussed below.

A succinct summary of soil-forming processes that is particularly apropos in the context of archaeological site formation, is provided by Buol et al. (1997, p. 132): "Pedogenic processes include gains and losses of materials from a soil body in accordance with the degradational, aggradational, or intermediate geomorphic character of the site, as well as translocations within a soil body." In an expressly archaeological context, Schiffer (1987, pp. 141–234) categorized many of these pedogenic processes at the scale of the artifact and the site. At the level of individual artifacts are chemical and biological processes that can modify or destroy the material record of the past. Schiffer (1987, pp. 141–198) did not specifically relate these processes to pedogenesis, but clearly they can be affected, and even driven, by soil formation. There are also a wide variety of mixing processes related to soil formation, grouped under the heading "pedoturbation" (Schiffer, 1987, pp. 206–217, following Wood and Johnson, 1978) that disturb artifact and site contexts. Between the chemical and biological processes that affect individual artifacts and mixing processes that disrupt sites, there are also the vertical translocation processes such as eluviation–illuviation that destroy or obscure stratigraphic relationships and introduce chemical weathering deeper in the soil profile.

In a more traditional soil science approach to soil formation, the various processes of site formation that result from pedogenic processes can be grouped into two broad and overlapping categories: horizonation and haploidization (Buol et al., 1997, pp. 50–51). The processes of horizonation probably are known to some extent to some archaeologists and to many geoarchaeologists. These are the processes that produce soil horizons (i.e., individual zones within a soil with similar physical and chemical characteristics). They make up most of the four broad categories of soil forming processes discussed by Simonson (1959, 1978) in his "multiple process model" (see also chapter 3): additions, losses or removals, transfers, and transformations (table 3.1, fig. 3.1).

In discussions of soil formation, the processes of horizonation and haploidization are sometimes set in opposition to one another; that is, formation of discrete horizons versus mixing of the entire soil (Buol et al., 1973, p. 91), and in other discussions soil mixing is emphasized over horizonation as a destructive force in archaeological sites (e.g., Wood and Johnson, 1978; Schiffer, 1987). The formation of discrete horizons, however, includes homogenization or mixing of the soil parent material and any contained artifacts or cultural features, which results in obscuring or destroying archaeological contexts. The destruction of bedding and stratification by the development of horizons is a relatively well-known phenomenon in archaeological contexts, as is the mistaken identification of soil horizons for geological stratification (chapters 1 and 5). Otherwise, however, the literature on the geochemical and biochemical aspects of horizonation as an archaeological site-formation processes is very sparse in contrast to discussions of the effects of specific soil chemical conditions (Schiffer, 1983, 1987) and mixing or haploidization (Wood and Johnson, 1978; Johnson and Watson-Stegner, 1990) on archaeological artifacts, features, and sites.

The following discussion is a look at the effects of pedogenesis on artifacts, deposits, and sites in terms of horizonation and haploidization, recognizing that significant overlap exists among these categories. The discussion is built around three broad categories. The first sections deal with the processes that form the individual soil horizons, organized on the basis of the master horizons and how those processes may affect archaeological materials and archaeological interpretations. The following subsection discusses the mixing or destruction of the whole soil via haploidization, organized around the soil-mixing categories of Wood and Johnson (1978; after Hole, 1961). The specific processes discussed under the topics of horizonation and haploidization include discussion of both the artifact- and site-scale processes of Schiffer (1987, pp. 143–234). The final subsection looks at the much broader issue of processes that may obscure soils after they are buried, which could influence their geoarchaeological interpretation.

Soil Horizonation Processes

A wide variety of processes produce soil horizons (chapter 3; table 3.1). All of the processes alter the soil parent material and result in homogenization or haploidization within relatively narrow zones within the soil. The processes of horizon development, therefore, can adversely affect archaeological sites. These processes can be grouped into the following categories (modified from Duchaufour, 1998, p. 2): accumulation and humification of organic matter at the surface (O horizon); incorporation and humification of organic matter within the mineral parent material, removal or eluviation of clay, surface and near-surface processes affecting at least the upper part of the soil (A horizon); leaching or eluviation of organic matter, clay, or sesquioxides—a near-surface process (E horizon); and weathering of primary minerals to secondary minerals, development of soil structure, or accumulation or illuviation of secondary constituents such as humus, clay, carbonates, and sesquioxides—all subsurface processes (B horizon). These processes produce significant differences in physical and chemical characteristics in the soils, and more important to this discussion, they have widely different effects on artifacts, archaeological features, and archaeological sites. The following discussion of soil horizonation and site formation is organized around the master soil horizons. There has been very little systematic research into the effect of soil horizonation or the broader burial environment on archaeological evidence (Caple, 2001), but Raiswell (2001) presents a useful summary of the postburial geochemical environment in an archaeological context.

The "O horizon" is dominated by SOM (Soil Survey Division Staff, 1993, pp. 118–119; Buol et al., 1997, p. 54). O horizons come in three basic morphologies: as leaf litter on a forest floor; as "peat bogs" (or "wet peats") in low, wetland settings; and as "blanket bogs" or "raised bogs" (or "dry peats") in upland settings. In bog settings the O horizons comprise Histosols, which are of more significance archaeologically than the simple forest litter. The terminology for peats and wetlands is not agreed on and can be confusing (see Zvelebil, 1987, pp. 96–100; Armentano, 1990; Moore, 1990; and Orme, 1990, for some differing approaches). In part this is because some terms are based on the nature of the organic matter and other terms and definitions are based on environmental or depositional setting. Shotyk (1992, fig. 9.1) provides a straightforward set of criteria based on SOM content: "organic soils" (or "carbonaceous sediment") contain at least 25% SOM by weight; "peats" contain at least 75% SOM by weight. Many, if not most, O horizons qualify as "histic" or "folist" epipedons in soil taxonomy (Soil Survey Staff, 1999, pp. 22–23). Dincauze (2000, pp. 314–315) describes the array of terrestrial wet sites as a kind of catenary sequence. Blanket bogs form in uplands under high (and typically cool) precipitation. Most lowland wetlands form where the water table intersects the ground surface. Sphagnum peats grow in acidic groundwater pools or shallow ponds and produce the classic peat bog. Fens are boggy landscapes formed in alkaline or neutral groundwater. Carrs are similar but also include woody swamp vegetation in addition to peat. Swamps form where woody vegetation alternates with open water. Marshes consist of grassy herbaceous vegetation that may be partly submerged. Coastal marshes are also low-energy depositional environments.

In wetlands, relatively rapid production of SOM and its preservation under anaerobic conditions results in net accumulation through time (the process of "paludization"; Buol et al., 1997, p. 137; see also Everett, 1983; Stevenson and Cole, 1999, pp. 34–40; and Rabenhorst and Swanson, 2000). If the SOM remains saturated, these settings can yield remarkably well-preserved organic archaeological materials, most commonly found in fens, blanket bogs, and coastal marshes (Dincauze, 2000, p. 315). These finds include a wide variety of archaeological sites and features such as the famous "bog bodies" of northern Europe (J. M. Coles, 1988; Coles and Coles, 1989) and other human remains (Doran and Dickel, 1988), extinct fauna (Dillehay, 1988, 1989; Overstreet, 1998), and a wide array of wood and plant remains including art objects, tools, and structures such as dwellings and trackways (fig. 10.1) (Coles and Coles, 1986; Hayen, 1987; B. Coles, 1987, 1988, 1990; Zvelebil, 1987; Purdy, 1991, 2001).

Preservation in peat settings varies widely, however, because of the specific geochemical conditions during the development and persistence of the peat (fig. 10.2; Stevenson and Cole, 1999, pp. 34–40; Cronyn, 2001). The situation is well

Figure 10.1 Preserved wood under peat representing the remains of the Sweet Track, Somerset Levels, England (from B. Coles, 1990, plate 5.4, provided by and published with permission of J. and B. Coles). The track was constructed ~5800 yr B.P., apparently for crossing a reed swamp. It was subsequently buried by formation of a peat bog.

described by J. M. Coles (1987, p. 10): "Each of these environments will have controlled the degree of preservation of materials, in seemingly infinite variety, so that wood may be immaculately preserved in one matrix but rotted out in another; and bones may be entirely disintegrated in some acidic peats but intact in muds and mucks. The bog bodies of the Iron Age and later periods from Denmark, Germany, Britain, and Ireland are variably preserved by their disposal in former pools and swamps, now peat-filled basins; bones may be almost entirely gone . . . or some may have gone and others remain in good condition . . . and yet

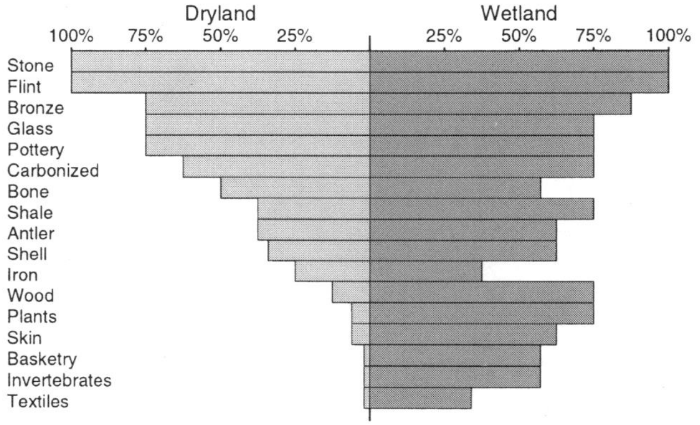

Figure 10.2 A generalized comparison of preservation potential for archaeological materials in dryland and wetland sites in Europe (modified from J. Coles, 1987, fig. 10).

skin, stomach contents, and eyeballs, for example, may have survived the percolating acidic waters."

Drainage of wetlands, either through natural environmental change or by human activity, results in decomposition or oxidation of the organic remains. These changes probably have a profoundly adverse effect on archaeological materials (e.g., Van Heeringen and Theunissen, 2001), but the literature on these processes and effects is sparse. Organic materials probably decay, whereas more resistant materials such as stone persist. Dewatering of the O horizon and subsequent compaction also help in destruction of archaeological materials via crushing. For example, skeletal remains of adult *Bison antiquus* in marsh sediments and soils at Lubbock Lake were crushed to ~2 cm thick (Johnson, 1987a).

The "A horizon" is that part of the mineral soil profile characterized by the addition of humified organic matter (Soil Survey Division Staff, 1993, p. 119; Buol et al., 1997, p. 54). The organic matter comes from both plants and animals living and then dying on or just below the soil surface, where the rate of SOM gains exceeds losses through decomposition (Birkeland, 1999, p. 107). Typically, the A horizon is the most dynamic part of the soil profile (e.g., Johnson and Watson-Stegner, 1990). As a result, evidence of human activity on or in an A horizon can be subjected to considerable modification and destruction.

Beyond humification and decomposition, several important processes work together to form an A horizon—processes that are also important in archaeological site formation. As an A horizon develops over time, it is subjected to eluviation (typically the removal of clay-sized particles) and decalcification and loss of sodium and other soluble salts. As a result of the addition of organic acids and the loss of basic ions, the A horizon can be acidic. In addition, A horizons are subjected to mixing by floral and faunal activity (Johnson and Watson-Stegner, 1990). For example, earthworms produce thick, organic-rich A horizons that are also thoroughly turbated (Johnson and Watson-Stegner, 1990), as are associated archaeological remains and features (e.g., Flegenheimer and Zárate, 1993; Davidson, 2002). Indeed, these A horizons are so distinctive that they are the basis for the "verm" great group of Mollisols, Alfisols, and Entisols in soil taxonomy (Soil Survey Staff, 1999). The growth of roots, a characteristic of A horizons, also contributes to mixing of this zone and disturbance of archaeological debris (Pyddoke, 1961, pp. 82–85; Johnson and Watson-Stegner, 1990; Flegenheimer and Zárate, 1993; Anderton, 1999) Lack of stratigraphic integrity and evidence of mixing archaeological contexts in A horizons is commonly reported (Janak and Martin, 1987; Holliday, 1992a; Flegenheimer and Zárate, 1993; Anderton, 1999). These processes, combined with the palimpsest nature of occupation zones on the stable surface of a soil (chapters 5, 6, and 8), further result in mixing of occupation debris.

The "E horizon" is characterized by the eluviation (or leaching) by water of clay, humus, iron, or aluminum. The result is a gray or white horizon characterized by a concentration of quartz and other resistant minerals in sand and silt sizes (Fanning and Fanning, 1989, pp. 42–47; Soil Survey Division Staff, 1993, pp. 119–120; Buol et al., 1997, p. 54). This intense eluviation results from the downward through-flow of water under the acidified conditions usually associated with trees, especially evergreens, and heather. The formation of an E horizon is most effective on weatherable parent materials such as tephra but will occur in a wide variety of sediments. Primary bedding is destroyed as these horizons develop. Only the most resistant, usually stone, archaeological materials survive in an environment of E horizon formation. Plant roots also penetrate E horizons and can disrupt archaeological features.

The formation of "B horizons" includes a wide range of pedogenic processes, but most germane to the discussion of site formation are the soil-forming processes of leaching, illuviation, weathering of primary minerals to release sesquioxides, and formation of soil structure (Soil Survey Division Staff, 1993, p. 120; Buol et al., 1997, pp. 54–55). One characteristic shared by most B horizons is the destruction of evidence for primary bedding and stratification, except perhaps where there is a strong contrast in the lithology of layers (e.g., a gravel lens interbedded with fines). This is a classic characteristic of a B horizon, and many processes can play a role, though rarely do all of them work together on a single soil. Mixing by plant and animal activity is probably one of the first processes that begin to obscure bedding in a C horizon as it evolves into a B horizon (fig. 5.5). Accumulation of translocated materials is also effective in obscuring or destroying stratification. In dry environments the formation of Bk and calcic horizons masks bedding. One of the first steps in B horizon development is reddening or "brunification," which results from incipient leaching, oxidation, and in cool, moist environments, the bonding of oxidized clay, iron, and humus. A particular process of B (and E) horizon development is "podzolization," which results from the mobilization and leaching of metal cations from the E horizon (usually qualifying as an "albic" horizon under such conditions) and precipitation as organic complexes in the underlying B horizon (a Bh horizon, usually qualifying as "spodic" in such settings). Illuviation of clay and formation of the Bt (and argillic) horizon and concomitant iron oxidation is probably the most widespread and best-known process of B horizon formation. These various processes of accumulation help to mask bedding, but the buildup in clay also contributes to minor shrink–swell and the formation of soil structure, which significantly disrupts bedding and stratification.

Processes of B horizon formation can have significant effects on archaeological sites. Podzolization is particularly damaging to artifacts and features. The development of the albic E and spodic Bh horizons obscure stratigraphy and feature boundaries (Callum, 1995; Anderton, 1999) and stain stone artifacts with iron oxides or coat them with iron encrustations (Wright and Roosa, 1966; Anderton, 1999). In some settings these coatings can have an adverse effect on materials such as metal (Gerwin and Baumhauer, 2000). Because of the intense leaching, podzolization also destroys floral and faunal remains and can even degrade some lithic and ceramic materials (Callum, 1995; Anderton, 1999; Gerwin and Baumhauer, 2000; Crowther, 2002). The precipitation of iron oxides in podzolizing and other well-drained pedogenic environments can also be difficult to differentiate from archaeologically relevant "ochre" (Thorson, 1990; Callum, 1995).

Iron pans, classified as placic horizons in soil taxonomy, are reported from Bronze Age burial mounds in Denmark (Breuning-Madsen et al., 2000, 2001). The pans in some sites encase the mound core and even help preserve encased materials such as wooden coffins and human remains. Immobile Fe^{3+} (ferric iron) and Mn^{4+} in the wet, anaerobic core of the mound apparently are reduced to mobile Fe^{2+} (ferrous iron) and Mn^{2+} and moved out to more aerobic, drier parts of the mound. Ferric iron then precipitates at the wet–dry boundary, forming the thin iron pan.

In dry environments, coatings of calcium carbonate encrust artifacts during the development of Bk horizons (Johnson and Holliday, 1986; Johnson, 1987a; Flegenheimer and Zárate, 1993; Hill et al., 1995; Gvirtzman et al., 1999). The carbonate adhering to bone will degrade the cortical surface because of the alkalinity. In areas with contrasting wet and dry seasons, shrinking–swelling in a clay-rich Bt horizon can disrupt artifact associations (Gvirtzman et al., 1999), discussed further in the text.

The weathering of artifacts by soil-forming processes varies significantly depending on artifact raw materials. The chemistry of bone, both organic and inorganic phases, can be altered dramatically by processes of leaching and exchange (Millard, 2001). Leaching of trace elements from bone is of particular concern in paleodiet studies (Nelson and Sauer, 1984).

Weathering of stone artifacts and formation of patinas is of wide interest in archaeology and can be influenced by soil-forming processes, especially in environments with acidic leaching or silica in solution (Purdy and Clark, 1987; Burroni et al., 2002). A variety of studies report on the effects of water chemistry on chert weathering and patina genesis (summarized in Burroni et al., 2002), but few specifically link water chemistry to the soil-forming environment. In India, the formation of ferricretes (roughly the equivalent of oxic horizons in soil taxonomy; table 2.1) is shown to have a deleterious effect on Paleolithic artifacts of quartzite and quartzitic sandstone (Pappu, 1999). The processes of iron encrustation of these artifacts results in weathering of the artifact surfaces. Weathering rinds also form on the quartzitic sandstone tools. Artifacts of both material types exhibit patinas, but the patination process is more intense on the quartiztic sandstones. In general, the type and rate of postburial processes affecting the artifacts in ferricrete are determined by artifact raw material, rate of burial, position of artifact in the soil, and local weathering conditions.

Ceramic artifacts also undergo a wide variety of weathering processes after they are lost or discarded and then buried (Rye, 1981, pp. 57, 120; Schiffer, 1987, pp. 158–162; Skibo, 1992, pp. 44–45; Freestone, 2001). Many or most of these processes are probably related to soil formation, particularly in the B horizon, but as with chert weathering, few of these postburial alterations have been explicitly linked to pedogenesis. Sherd exteriors are subjected to both physical and chemical weathering. In arid and semiarid environments, this can include coating by calcium carbonate or the precipitation of salts, which can discolor or obliterate the surface of sherds (Rye, 1981, pp. 57, 120; O'Brien, 1990). Because ceramic materials are porous, sherd interiors can also be weathered; ions can be leached from sherds or precipitated within them (Rye, 1981, p. 120; Bishop et al., 1982, pp. 295–296; Reid, 1984; Franklin and Vitali, 1985; Freestone, 2001). The growth of salt crystals can be especially deleterious (Rye, 1981, pp. 10, 35–36; O'Brien, 1990). Moisture moving through soils and sherds also weakens the pottery (Skibo and Schiffer, 1987). These changes in chemical composition hamper studies of raw material provenance because such analyses often rely on trace element concentrations (Franklin and Vitali, 1985). Furthermore, saturation of the B horizon, either permanent or periodic, causes sherds to expand and lose tensile strength.

Groundwater fluctuations in B and C horizons can effectively and rapidly degrade organic remains (Schiffer, 1987, pp. 148, 151; Cronyn, 2001) and obscure stratigraphic and probably archaeological feature boundaries. This important process is indicated (in the field and in thin section) by iron reduction or removal ("gleying"), sometimes co-occurring with iron oxidation and, as a group, referred to as "redoximorphic features" (Vepraskas, 1992). Redox depletions, caused by iron removal, include grey soil matrix and grey root mottles. Redox concentrations, formed by the reprecipitation of iron in the better-oxidized areas, include various iron oxide nodules and mottles. Reduced iron or mottles of oxidized and reduced iron can change the characteristic reddish-brown hues of a B horizon to dull blue-gray or olive-gray hues. Accompanying loss of structure by wetting can further compound the problem of identification. Kraus and Aslan (1999, pp. 308–309) provide a succinct and comprehensive summary of iron oxidation and reduction in floodplain settings. Alluvial soils that are saturated for several months of the year can undergo gleying, in which iron and manganese are reduced and mobilized. As the water table falls and the soil dries, the iron and manganese may be leached from the soil or concentrated in more-oxidized areas, either within peds or along ped faces and soil channels as mottles or nodules. In floodplain settings in which clay is abundant, seasonal rains and flooding produce perched water tables. Subsequent surface-water gleying produces grey horizons and mottles as a result of poor drainage, and these horizons overlie better-drained brown horizons. In contrast, groundwater gleying, caused by seasonal or periodic saturation of soil materials by groundwaters, is expressed by a downward increase in grey soil colors, reflecting proximity to the groundwater table.

Examples of the effects of the various horizon processes on archaeological sites and materials are not especially common, as indicated by the forgoing discussion. Even less common, however, are illustrations of the effects of overall horizonation on archaeological sites. Wilkinson (1990) provides a rare example

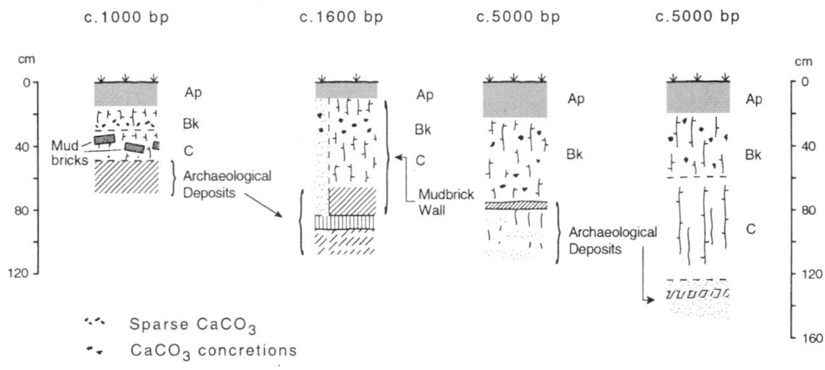

Figure 10.3 Soil development on archaeological mounds of increasing age in northern Iraq, illustrating the progressive alteration of the anthropogenic sediment (modified from Wilkinson, 1990, fig. 2).

of such effect using the long history of mound construction in Mesopotamia (northern Iraq). There, soil development was observed to destroy increasing amounts of the archaeological integrity of progressively older mounds (fig. 10.3). Mounds >7000 years old, if <1 m in height "may have been entirely homogenized by soil forming processes so that only the resistant artifacts remain. . . . [S]tratigraphy and mudbrick . . . architecture may have been lost" (Wilkinson, 1990, p. 92).

Haploidization and Other Mixing Processes

A broad array of physical and chemical processes, most related to those described above, can also destroy stratigraphy and partially or completely destroy or obscure horizonation. These processes can also destroy associations among artifacts and features in archaeological sites. The most common of these processes in archaeological and pedological contexts include mixing from bioturbation by animal activity (faunalturbation) and plant growth (floraturbation), freeze–thaw (cryoturbation), shrink–swell (argilliturbation), gravity (graviturbation), and air and wind (aeroturbation; Schiffer, 1987, pp. 206–217, following Wood and Johnson, 1978; Johnson and Watson-Stegner, 1990). The following discussion, built around the excellent reviews by Wood and Johnson (1978), Schiffer (1987), and Johnson and Watson-Stegner (1990), will focus on those processes that can be considered pedogenic or are of widespread and direct pedogenic significance: faunal- and floral-turbation, argilliturbation, cryoturbation, and graviturbation.

The mixing processes grouped under the broad heading of "bioturbation," including floralturbation and faunalturbation, are well known to most archaeologists, geoarchaeologists, and Quaternary geoscientists. The effects of bioturbation can lead to formation of a "biomantle," defined as one or more differentiated horizons in the upper part of soils produced largely by bioturbation (Johnson, 1990, pp. 84–85). The mixing processes result in movement of fines to the surface, their subsequent removal by rainwash and wind, and redistribution of coarser particles. In soils with large particles (gravel and coarser), mixing by animals produces a stone line at the base of the biomantle. Coarser clasts are moved up and form stone pavements in biomantles produced by plant mixing. The processes of mixing and the formation of the various kinds of stone zones in biomantle genesis all have wide-ranging implications for the interpretation of archaeological sites. In their comprehensive review, however, Mercader et al. (2002) point out that a variety of processes can produce "stone lines" (i.e., they are a good example of equifinality). Artifacts in stone lines are not necessarily devoid of archaeological contexts.

"Faunalturbation" includes the burrowing activities of a wide range of animals (particularly ants, termites, worms, crayfish, and small mammals) and can very effectively mix or churn a soil or archaeological site (Wood and Johnson, 1978; Johnson, 1990; Johnson and Watson-Stegner, 1990; references in Leigh, 2001, p. 283; fig. 10.4). The effects of pocket gophers (Pierce, 1992) and termites (McBreanty, 1990) on sites and soils can also apply to other agents of faunalturbation. They can significantly modify stratigraphic relationships, soil horizonation, and cultural features by movement of sediment; destroy fragile artifacts; disrupt sedimentary structures; and organically enrich the subsurface. Burrowing by soil fauna can homogenize surface and near-surface horizons (e.g., the A and E horizons; fig. 10.5) and lower large stones and artifacts from the surface into the soil to produce subsurface stone lines and "artifact horizons" (fig. 10.6; Erlandson, 1984; Johnson, 1989, 1990, 1994a; Johnson and Watson-Stegner, 1990; McBrearty, 1990). This process can effectively bury artifacts deposited on seemingly "stable" surfaces by mixing them down into the soil (Johnson, 1989, 1993; Balek, 2002). Experimental research by Mello Araujo and Marcelino (2003) indicates that larger burrowing animals (in this case armadillos) can move artifacts either up or down and thus they do not necessarily produce stone lines. McBrearty (1990) also notes that termites can hasten bone dissolution by increasing soil porosity. She concludes that termites have a profound impact on archaeological sites throughout the tropics.

Termite activity was proposed as a mechanism for convoluting the post-Acheulian artifact stratigraphy of Central Africa (Cahen and Moeyersons, 1977). Through the middle of the 20th century, the artifact chronology was based on limited stratigraphic information from a few sites. Only at Gombe Point, in Zaire, was a full post-Acheulian sequence reported. The artifacts were found in a "homogeneous sand-silt-clay mantle, 3–5 m thick" (Cahen and Moeyersons, 1977, p. 812); an eolian deposit derived from the Kalahari. Almost half of the artifacts at Gombe were found at the base of the eolian sediment (forming a stone line), which was a situation typical of other sites in the region. An analysis of artifact refits showed that contemporaneous stone tools were mixed throughout the section. The process of mixing was attributed to termites: The fauna move artifacts to the surface, and the weight of the artifacts collapses the termite burrows, which incorporates the lithic material back into the sediment. This process also accounted for the stone line. The investigators concluded that "[t]he post-Acheulian industries that have been defined in Central Africa probably represent nothing more than heterogeneous assemblages of stone artefacts, without

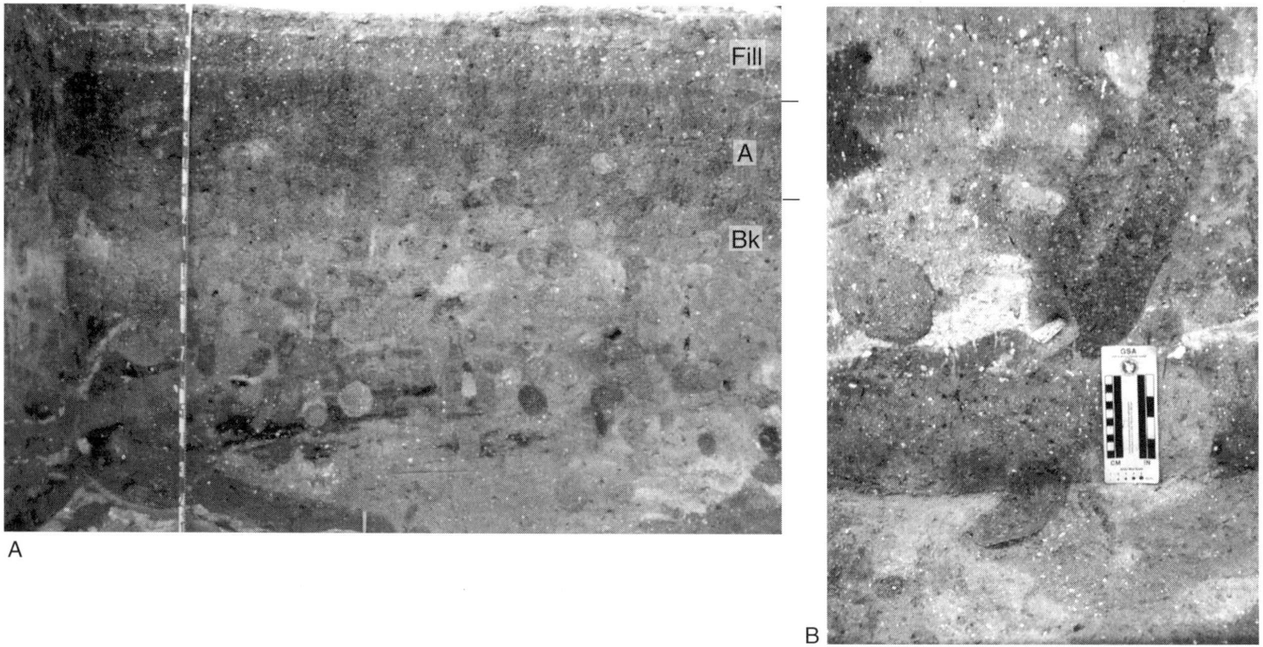

Figure 10.4 The effects of burrowing on soils. (A) "Chernozem" (Mollisol) formed in silt exposed in excavations at Kostenki 12, Russia (tape is scaled in decimeters and centimeters). Numerous krotovinas illustrate the effects of pervasive burrowing of the soil. (B) Detail of krotovinas at Kostenki 12. The fill from one of the krotovinas fell out of the wall, exposing a flint blade that rested in a burrow, brought down from above.

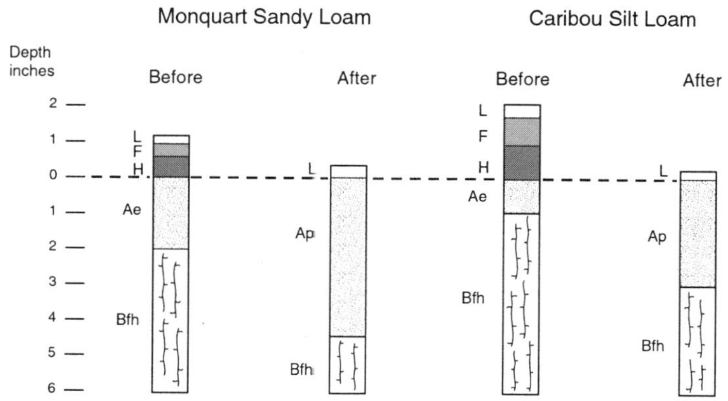

Figure 10.5 Two "podzols" (Spodosols) before invasion by earthworms and 3 years after invasion showing the effects of pervasive mixing (modified from Langmaid, 1964, fig. 1).

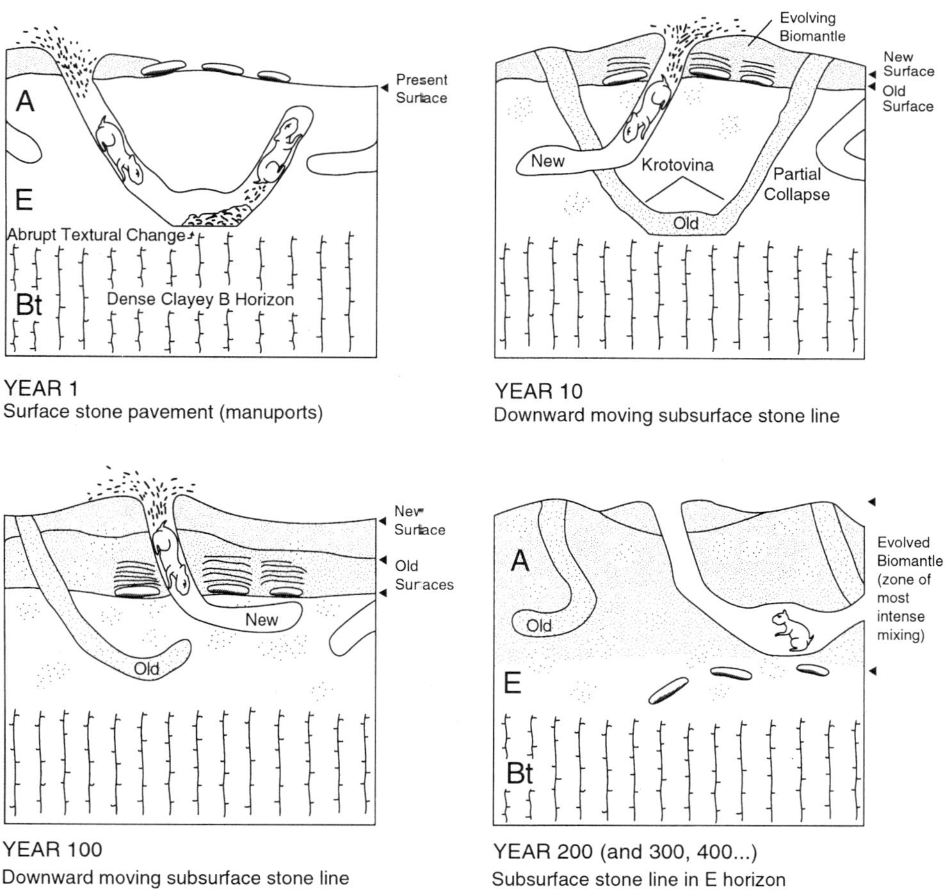

Figure 10.6 Sequential model (with a hypothetical time sequence) of how artifacts and other large surface stones are lowered through A to E horizons by rodent burrowing. The dense clay-rich Bt horizon is unburrowed. Stones smaller than burrow diameters are recycled to the surface by rodent action (modified from Johnson and Watson-Stegner, 1990, fig. 15; reproduced with permission of the Geological Society of America).

any stratigraphical, typological, chronological or ethnographical significance" (Cahen and Moeyersons, 1977, p. 815).

Earthworms are famous for the amount of work they can do in modifying the soil. They are probably the most important macroanimal of soil mixing (Wood and Johnson, 1978, p. 325; Stein, 1983, p. 278). Like rodents, worms can mix surface and near surface soil horizons (Limbrey, 1975, pp. 296–297, 315–316) and produce stone lines and artifact zones (Johnson, 1989, 1990; Johnson and Watson-Stegner, 1990). Stein (1983, pp. 280–281), lists the following effects of earthworms on soils and archaeological sites: obliteration of stratification, burial of surface objects by species that produce casts (waste), destruction of soil horizon and archaeological feature boundaries by burrowing, alteration of botanical assemblages by ingesting small plant remains (e.g., seeds), and alteration of soil chemistry by concentrating almost all soil elements. Davidson (2002) also notes that earthworms can effectively obliterate micromorphological evidence for cultivation within a few hundred years after field abandonment.

Earthworms are probably the most common and widespread agent for development of biomantles. Johnson (2002, p. 8) summarizes their role in biomantle genesis: "[E]arthworms ingest soil at depth then deposit it on the surface as fecal castings, a slow process of upward translocation of fine soil material. Large particles, such as gravels and artifacts that earthworms cannot ingest, will slowly sink, eventually to the lower zone of earthworm bioturbation. Given enough time [several decades], . . . earthworms will produce a biomantle composed of two texturally differentiated layers wherein fine material of worm-ingested and -translocated soil comes to overlie a coarser layer of non-worm-ingested materials. The coarser lower layer forms a stone-line or artifact-line, . . . which slowly settles downward as the overlying layer is biomechanically reorganized." In a classic study, Shaler (1891, p. 275) observed the effects of worms on archaeology, noting that "ancient implements, such as stone arrow-heads which the early peoples have dropped upon the earth, are soon covered over wherever the earthworms abound" (fig. 10.7). The effects of earthworms specifically on soil horizonation is vividly documented by Langmaid (1964) who studied changes in four podzols brought about in 3 yr by an invasion of earthworms (fig. 10.5). The worms completely obliterated the O, E, and part of the Bh horizons. In their place was a distinct, uniform horizon (a kind of Ap horizon) with an abrupt lower boundary.

Van Nest (2002) makes a case for "the good earthworm," however. Archaeological materials are common in soils formed on loess in the uplands of the Midwestern United States. The loess is largely late Pleistocene, so some mechanism must account for burial of the artifacts. Van Nest proposes that earthworm turbation is the mechanism for mixing artifacts from the surface down into the soil. This process can result in stratification of archaeological debris initially left on the surface if sufficient time passes between occupations. She concluded that enough time had elapsed to allow Archaic materials to be mixed down below the plow zone, but insufficient time had passed to allow Woodland materials to be moved to the level of the stone line of Archaic artifacts. This process enhanced the research potential of a site that otherwise would have been a mixed palimpsest of occupation debris at the surface. One significant drawback to this

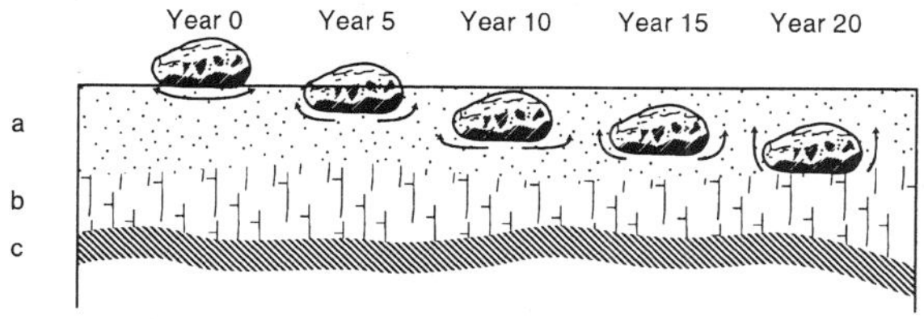

Figure 10.7 A model for stone (and artifact) burial by earthworm activity (topsoil [a] is 46 cm thick; [b] is subsoil; [c] is bedrock; modified from Shaler, 1891, fig. 10).

process of faunalturbation is that it reduces the visibility of archaeological sites by burying them.

"Floralturbation" refers to soil mixing by plants, largely through processes of or related to root growth (Wood and Johnson, 1978; Johnson, 1990; Johnson and Watson-Stegner, 1990). Deep penetration of soils and archaeological features by roots results in physical breakage or disruption of artifacts and features, introduction of organic matter, and formation of pathways or conduits for other materials (e.g., water; Johnson and Watson-Stegner, 1990; Flegenheimer and Zárate, 1993; Anderton, 1999).

One of the most destructive and dynamic types of floralturbation is tree uprooting, also known as "tree throw" (reviewed and summarized in archaeological contexts by Limbrey, 1975, pp. 288–291; Wood and Johnson, 1978, pp. 328–333; and Johnson and Watson-Stegner, 1990, pp. 544–549). This process does a tremendous amount of work on forested landscapes (fig. 10.8). For example, Thorson (1990, p. 416) speculates that tree throw affected archaeological sites throughout forested Beringia in Alaska. In southern Ontario, Canada, the effects of tree throw are so pervasive and destructive that special considerations had to be allowed in the description of soils for the purposes of soil survey (Fisher and Miller, 1980). Wood and Johnson (1978, p. 331) summarize data showing that most of the A and B soil horizons in the northern hardwood regions of North America will be floralturbated by tree throw within 500 yr. This process disrupts and mixes soil horizons, significantly disturbs archaeological features, and in some situations, can produce surface stone lines. Neumann (1978) describes an assemblage of artifact types spanning the past 2500+ yr mixed in a zone 25 cm thick because of tree throw as well as other forms of bioturbation. The growth of trees can also produce molds or "halos" that can be mistaken for archaeological features (Callum, 1995). Schaetzl (1986) documents tree throw as an agent of complete inversion of a soil horizon sequence. This process of "flipping over" a soil can also bury pockets of charcoal from a forest fire—pockets that could be misinterpreted as fire hearths.

"Argilliturbation" refers to mixing of soils because of shrinking and swelling as expandable clay minerals in the soil absorb and give up water. This is caused when soils high in fine clay are subject to seasonal wetting and drying, which forms significant cracks and inhibits or destroys soil horizonation. These processes cause significant engineering problems for roads, buildings, and other structures and lead to formation of an entire soil order in soil taxonomy: the Vertisol (Soil Survey Staff, 1999). Soils that have a tendency to shrink and swell, but that do not meet the qualifications of Vertisols, are described as "vertic."

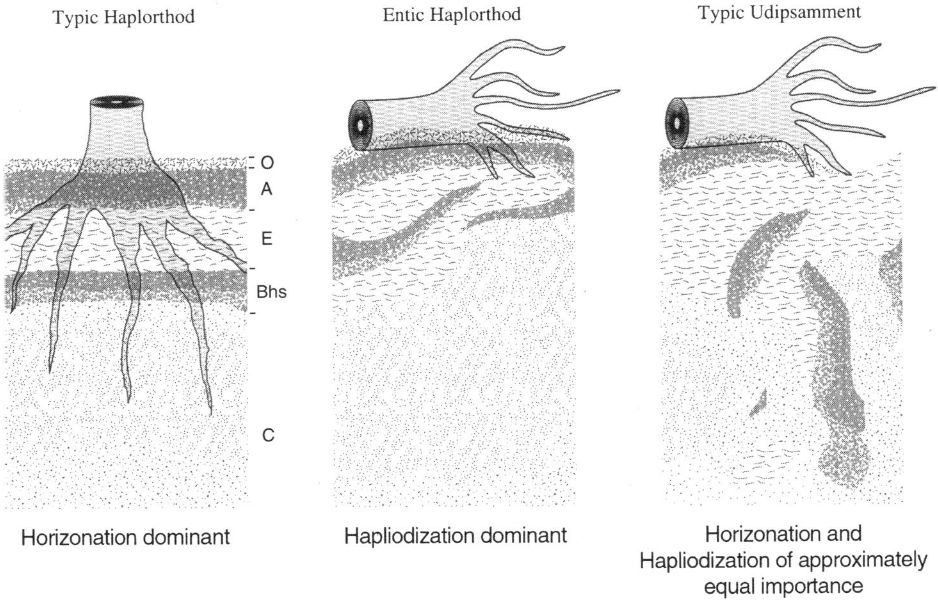

Figure 10.8 An example of the effects of treethrow on a Spodosol (modified from Johnson and Watson-Stegner, 1987, fig. 4; reproduced with permission of Lippincott, Williams, and Wilkins). The Typic Haplorthod illustrates "normal" Spodosol development. The Entic Haplorthod illustrates the effects of both podzolization as well as tree throw. The Typic Udipsamment is a soil almost completely homogenized by tree throw.

Vertisol dynamics are reviewed by Ahmad (1983) and Wilding and Tessier (1988). The opening of cracks allows materials at the surface or exposed along the walls of the cracks to fall down into the soils. The closing of cracks and the swelling process apparently can push material up through the soil and also induce lateral shifting (and formation of slickensides). The mixing process induced by Vertisols is vividly captured in their description as "self swallowing" soils (Oakes and Thorp, 1951). The vertical movement of shrinking–swelling soils also produces a distinctive hummocky or undulating surface called gilgai microrelief.

The mixing by Vertisols and vertic soils can have significant effects on archaeological artifacts, features, and sites (Duffield, 1970; Johnson and Hester, 1972; Wood and Johnson, 1978, pp. 352–358). Vertisols are capable of moving buried artifacts up to the surface or allowing surface materials to move down into the soil. Duffield (1970, p. 1055), an archaeologist who excavated in Vertisols, describes their potential effects: They can churn "archaeological features into a homogeneous mass and totally destroy the original context of a site" and have "characteristics that are capable of confusing any relative vertical stratigraphy that may have existed."

Although Vertisols and vertic soils are relatively widespread and can have a significant and adverse effect on archaeological sites, case studies of these effects are rare. Duffield (1970, pp. 1059–1060) describes several archaeological sites in east-central Texas (with the most extensive Vertisols in the United States) containing mixed artifacts assemblages, no clear stratigraphy, and artifacts associated with cracks. These contexts are attributed to shrinking–swelling processes. Wood and Johnson (1978, pp. 355–356; following Johnson and Hester, 1972) note how formerly buried boulders and cobbles in marine terraces of California Channel Islands are moved up through Vertisols, whereas artifacts that were left on the terrace surfaces are moved down into the soils.

Along the Duck River, Tennessee, clay-rich sediment comprises much of the Holocene alluvium (Brakenridge, 1984; Hofman, 1986). At the Cave Spring archaeological site, a buried soil within this clayey alluvium produced artifacts from top to bottom. This situation raised the question of whether the artifacts accumulated on a slowly aggrading surface or whether they were deposited on the surface of the soil and were then mixed down into it. Hofman (1986) conducted a careful excavation and artifact analysis and demonstrated, based on refitting of stone tools and manufacturing debris, that artifacts spanning 7000 yr of prehistory were moved vertically 20–40 cm. Several mixing processes likely were involved, but vertic characteristics were noted in the soils and the mixing was probably largely the result of the shrink–swell phenomenon.

Jacob (1995b) describes and illustrates the effects of vertic processes on buried soils in the Maya lowland. The buried A horizon is disrupted and locally is oriented almost vertically (fig. 10.9). In addition, subsoil material was locally observed extruded at the surface. Jacob also notes other similarly disrupted soils from sites in the region. The age of the landscape represented by the buried, disrupted soil is uncertain, but the effects of shrinking and swelling on archaeological features and sites could be dramatic. Jacob (1995b, p. 73) also notes that gilgai microtopography has been mistakenly interpreted as raised fields.

Artifact distributions in a Vertisol on Crete were investigated by Morris et al. (1997). The Vertisol was found in a sinkhole. Middle Minoan Period artifacts were found on the surface of the soil and at depths of 30–100 cm. The soil is high in clay and relatively homogeneous and exhibited vertical cracks, indicating that it is a Vertisol. To confirm that the artifact distribution was probably caused by argilliturbation, the soil was sampled and analyzed for particle-size distribution, total and organic carbon, and extractable metals to determine whether buried soils or erosional disconformities were present and could have contained buried living surfaces. The laboratory data yielded no evidence of buried surfaces, and the artifact distribution was ascribed to movement of artifacts down vertic cracks.

"Cryoturbation," similar to shrinking–swelling processes, can have a significant effect on soil horizonation and on artifact and archaeological-feature associations. The following summary of cryoturbation follows Wood and Johnson (1978, pp. 333–338). Cryoturbation is associated with permafrost and seasonally frozen ground and is said to affect as much as one-third of the soils on the planet (Johnson and Watson-Stegner, 1990, p. 553). "Permafrost" (perennially frozen ground) is widespread at high latitudes and covers over 25% of the Earth's land area. Permanently frozen ground imposes a broad array of problems for engineering, resulting in creation of the soil order Gelisol (Soil Survey Staff, 1999; Beecham and Tarnocai, 2000). "Seasonally frozen ground" (ground frozen by low seasonal temperatures that remains frozen only during the winter) affects nearly

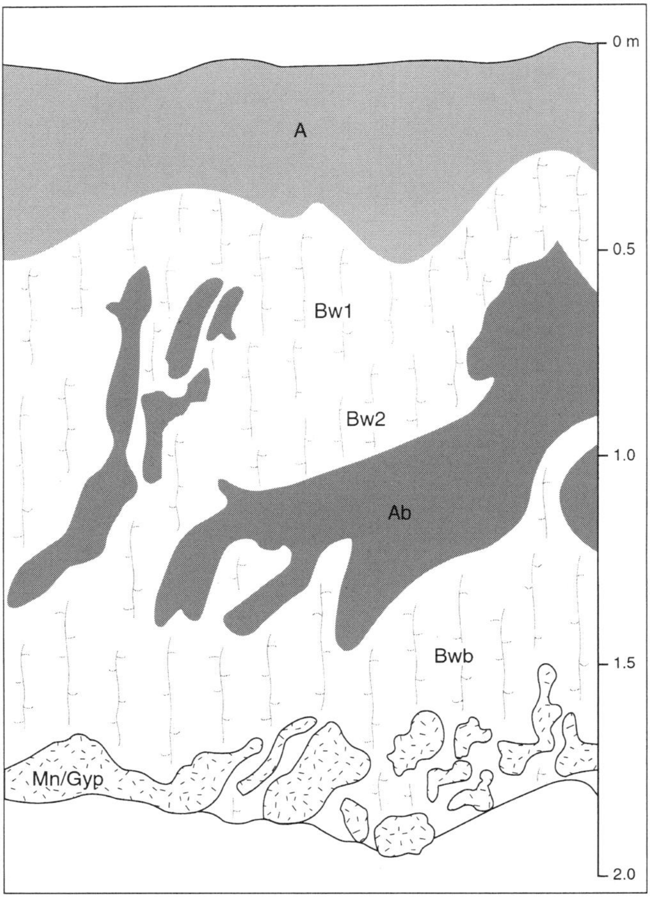

Figure 10.9 Soil profile from Nabke, Guatemala (modified from J. S. Jacob, 1995, "Archaeological pedology in the Maya lowlands," fig. 4–13. *In Pedological Perspectives in Archaeological Research*, M. E. Collins, B. J. Carter, B. G. Gladfelter, and R. J. Southard, Eds., pp. 51–80. Soil Science Society of America, Special Publication no. 44; reproduced with permission of the Soil Science Society of America). The buried A horizon was disrupted and distorted by shrink–swell processes in a Vertisol.

50% of the land area of the Earth (all high latitudes and most middle latitudes). The archaeological significance of cryoturbation is succinctly stated by Schweger (1985, p. 127): "the greatest handicap to northern archaeology is frost disturbance and its effects on archaeological matrix and the artifacts themselves." The principal effects of cryoturbation on soils in geoarchaeological contexts are frost-heave and thrust, frost cracking, involutions, and gelifluction (modified from Schiffer, 1987, pp. 213–215). Case histories of cryoturbation effects on soils in archaeological contexts are relatively rare, though the reviews by Bobrowsky et al. (1990), Thorson (1990), and Wilson (1990) do contain examples. A sizeable literature is also available on archaeologically based experimentation (see discussions in Johnson and Hansen, 1974; Johnson et al., 1977; Wood and Johnson, 1978; Bowers et al., 1983; Reid, 1984; Schweger, 1985; Skibo et al., 1989; Johnson and Watson-Stegner, 1990; Hilton, 2003).

"Frost heave" pushes artifacts and other large objects in the soil upward. "Because the movement is accumulative, the longer an object is buried, the greater will be its displacement. Frost-heave is a process which can mix stratigraphy [and horizonation] and cultural materials at archaeological sites" (Johnson and Hansen, 1974, p. 81). The process tends to most directly affect vertically oriented objects, and it also tends to force objects to a vertical orientation (Wood and Johnson, 1978, pp. 339–341). Other than verticality of artifacts and other large objects, diagnostic evidence of frost heave is rare (Thorson, 1990, p. 404). Because frost-heave so thoroughly mixes sites, and because many tundra sites are shallow, the result is the "paradox of the buried but archaeologically unstratified site" (Thorson, 1990, p. 416). This situation is described by Bowers (1982, p. 83): "although as many as five separate natural stratigraphic levels were recognized during the excavations [of the Lisbourne site], the extreme amount of cryogenic disturbance has resulted in the conclusion that the site should be treated essentially as a surface locality." Similarly, in archaeological investigations on the North Slope of Alaska, "soil layers served neither to define buried land surfaces or to provide definition of the internal structure of the site" (Gerlach, 1982, p. 29).

A good rule of thumb for the identification of artifacts moved by frost-heave is provided by Schweger (1985, p. 128): "A northern or alpine site with large-sized artifacts nearest the surface or with their long axes oriented vertically suggests frost heave." Frost-heave so heavily disturbed hearths at the Hungry Whistler site in the Rocky Mountains of Colorado that their original characteristics could not be identified (Benedict and Olson, 1978, p. 45). Frost-heave also thoroughly mixed features and living surfaces and displaced artifacts in some strata. Artifacts in a middle Holocene A horizon did not appear to be appreciably mixed, however. They were concentrated near the base of the soil zone, and this was used as evidence for postoccupation cumulization of the surface horizon (Benedict and Olson, 1978, pp. 38–39).

"Frost-cracks" and "involutions" produce obvious physical evidence in stratigraphic exposures. The effect of frost-cracking on soils probably is similar to that of cracking in Vertisols, which allows sediments and artifacts to be dropped down and mixed lower in the soil. The formation of involutions also creates problems of stratigraphic correlation and probably displaces artifacts.

"Gelifluction," the slow down-slope movement of water-saturated soil above permafrost (chapter 8), produces spectacular contortions of surface horizons of soils (figs. 8.1, 10.10, and 10.11) and is widely reported from cold-climate archaeological sites (e.g., Thorson and Hamilton, 1977; Schweger, 1985; Thorson, 1990). The process can significantly affect archaeological sites by removing material or complicating stratigraphic correlation. At the Iyatayet site on Cape Denbigh, on the west coast of Alaska, the type Denbigh Flint Complex was found buried in a heavily cryoturbated A-horizon of a podzolic soil (fig. 10.11; Hopkins and Giddings, 1953). The soil and contained archaeology were complexly folded by gelifluction lobes before burial. "Shear resulting from frictional drag between mobile, saturated, sandy silt above and immobile, frozen, rocky debris below

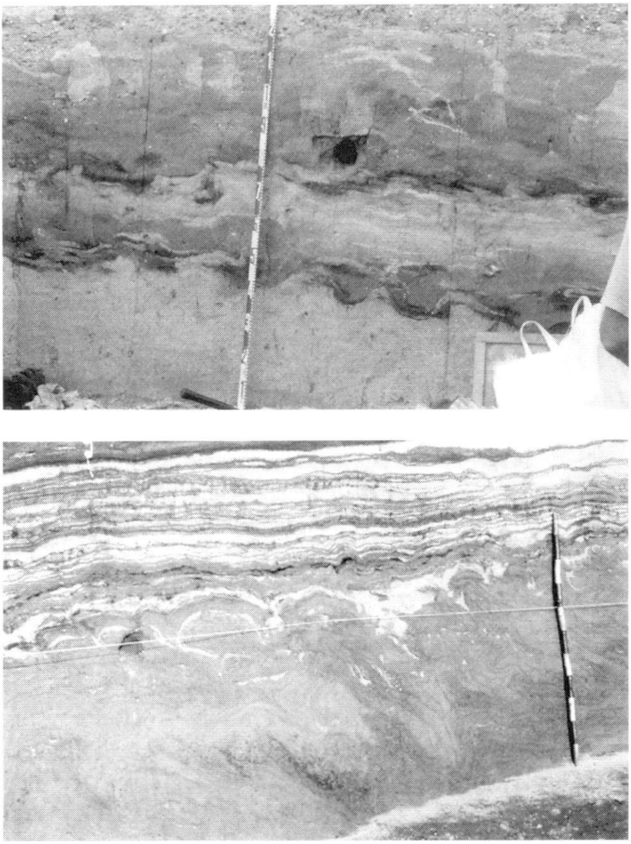

A

B

Figure 10.10 The effects of cryoturbation on soils. (A) Sediments and soils in the "Upper Humic Beds" (~30,000 yr B.P.) contorted by solifluction at Kostenki 14, Russia. These floodplain sediments, with Upper Paleolithic artifacts, initially were a Fluvent-type soil (the dark bands are floodplain A horizons) before cryoturbation and burial. (B) Stratigraphic section of the Onion Portage site, Alaska, showing the contorted sediments and soils of Bands 7 and 8 (see fig. 7.13) From Schweger (1985, fig. 3); photo provided by C. Schweger and reproduced with permission of C. Schweger, R. Bonnichsen, and the Center for the Study of the First Americans, Texas A&M University.

produced the intricate folds in the Denbigh culture layer. . . . During the folding, the total downslope length of the area occupied by the Denbigh flint layer was shortened," creating gaps in the soil and the occupation zone up-slope from the folds (Hopkins and Giddings, 1953, pp. 17, 19). The process and results can be likened to a carpet being folded, leaving the floor exposed where the carpet was pulled away.

The Engigstciak site in the Yukon Territory of far northwestern Canada was so heavily mixed by cryoturbation that stratigraphic relationships were reversed or destroyed altogether (Mackay et al., 1961). This is perhaps not surprising given the location of the site north of the Arctic Circle. Several archaeological zones

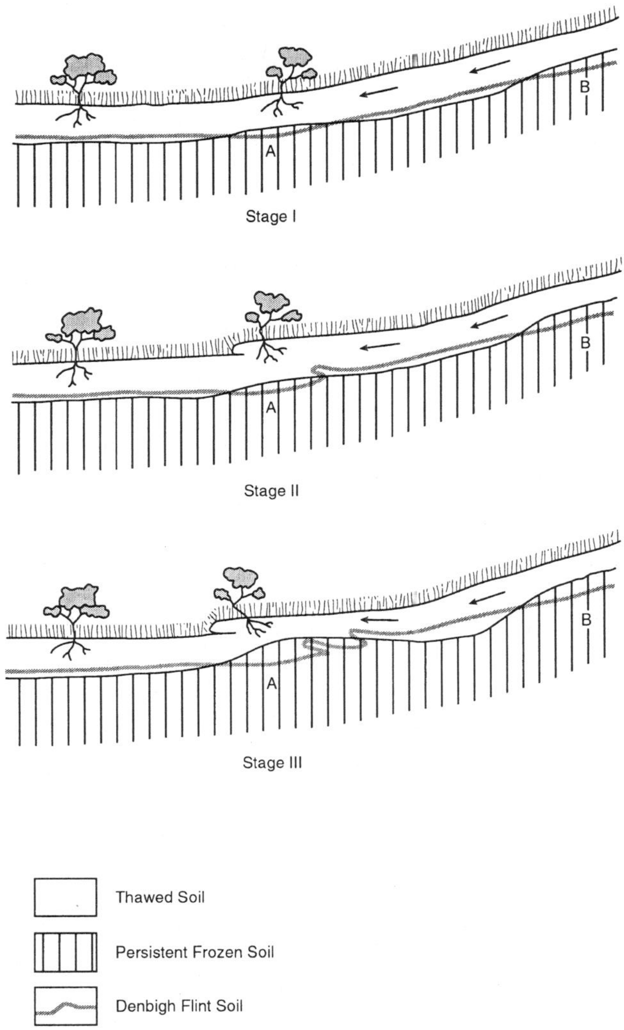

Figure 10.11 The effects of gelifluction on the buried occupation zone (Denbigh Flint Culture) at the Iyatayet site, Alaska (modified from Giddings and Hopkins, 1953, fig. 2).

dating to the middle and late Holocene (British Mountains Complex, Flint Creek, and Arctic Small Tool Tradition) were associated with buried soils, either weak A-C soils or O-C profiles. The processes affecting the site include growth and melting of ground ice, intense frost heaving, solifluction, and the formation of patterned ground (Mackay et al., 1961, p. 28).

Another but more subtle effect of ground freezing is formation of "pseudo-paleosols," described by Thorson (1990, p. 406) in an archaeological context: "Under moist, but not necessarily saturated conditions, dissolved ions and organic substances, humic colloids, and fine mineral particles are preferentially concentrated at the freezing point because they are selectively excluded during the phase change. This is a mechanism for creating organically stained subparallel bands of fine silt and clay that, once established, continue to trap downward-moving particles.... Oxidation bands may also form during thawing, when solutes from saturated sands migrate to still-frozen silt lenses where they precipitate upon refreezing." Such color banding was present and correctly interpreted at the Healy Lake site (McKennan and Cook, 1968) but could be mistaken for weakly expressed A and B horizons. This misinterpretation could lead to serious errors of dating and correlation (Thorson, 1990, p. 406).

Not all mixing processes destroy horizonation. Frolking and Lepper (2001) provide a somewhat unusual example of downward mixing of artifacts without haploidization at the Munson Springs site in central Ohio. The site yielded evidence of repeated occupation in the form of Paleoindian, Archaic, Early Woodland, and middle Woodland artifacts beneath, within, and near a low artificial mound in a footslope position. The soil buried by the mound exhibited an A-E-BE-Bt-C morphology typical of postglacial soils in the region. The artifacts were roughly in correct stratigraphic sequence within the buried soil, with the Paleoindian material the deepest and in the BE horizon. The soil stratigraphic data indicated, however, that there was no Holocene aggradation of the landscape. There is no evidence of accretionary upbuilding of the profile in terms of its thickness or the position of the maximum clay bulge in the Bt or evidence of erosion on the adjacent slope. The artifact distribution is explained by accumulation on the surface and mixing downward by bioturbation. The apparent artifact stratigraphy and absence of obvious evidence for mixing provides a cautionary note on the interpretation of artifact distribution within a soil profile.

Another issue regarding pedoturbation is determining whether an artifact or feature found below the surface was in fact buried by some mixing process or whether it was simply covered by a depositional process. This issue of equifinality of artifact burial was addressed by Leigh (1998, 2001), who regularly confronted the matter in sandy soils of the southeastern United States (see also Van Nest, 2002, working on loess-covered uplands of the midwestern United States, noted earlier). The loose nature of the soils combined with the high plant and animal activity regularly resulted in questions of artifact association and feature integrity. Leigh (2001) discussed several factors that must be taken into consideration in addressing the issue. The geomorphic setting must be assessed to determine the type and likelihood of sediment accumulation. In an example presented by Leigh (1998), the study area was on high stream divides, which precluded the possibility of alluvial or colluvial deposition. The physical characteristics displayed in exposures provided valuable clues. Primary sedimentary structures and stratigraphic discontinuities indicate burial by sedimentation. Evidence of burrowing (e.g., krotovinas) raises the likelihood of artifact mixing. Particle-size analysis is useful for documenting stratified sediments and for confirmation of specific types of depositional environments. Micromorphology can be an important aid in differentiating primary deposits from mixed ones. For example, thin sections can reveal microlaminae in unmixed sediments or disrupted microsedimentary structures in turbated soils. Surfaces compacted through human activity can also be identified in thin section (fig. 11.1). Numerical dating methods such as radiocarbon or luminescence techniques also can yield evidence of stratigraphic inversions or mixing.

Finally, the distribution and integrity of cultural materials and features themselves can be clues to the degree of pedoturbation. Assessing these characteristics is fairly straightforward, with extensive and complex features, but it is problematic with more-ephemeral sites and features such as lithic scatters. As listed by Leigh (2001, pp. 280–281, after Michie, 1990), mixing (largely through bioturbation) is indicated by a correlation between maximum artifact depth and the visibly apparent depth of bioturbation, by artifacts tilted at >0° to 90° to the horizontal, by the absence of intact features such as hearths, by the vertical and horizontal displacement of clustered artifacts that would otherwise be found together, and by evidence of a single behavioral activity (i.e., core reduction) or artifact (i.e., a broken pot) mixed throughout the soil.

Alteration of Buried Soils

When soils are buried, their physical and chemical characteristics undergo modification. The degree of modification will vary from minor to substantial, depending on a wide range of factors. The significance of the alterations is that the pedogenic characteristics that can be used for stratigraphic correlation or to infer age or paleoenvironmental conditions can be changed or destroyed. Postburial processes can also mask archaeological features. These alterations, therefore, can influence the use of buried soils in archaeological sites to answer archaeologically significant questions.

The processes that alter buried soils include additional pedogenesis and diagenesis (Catt, 1990, pp. 65–71; Retallack, 1991b; Olson and Nettleton, 1998). Buried soils commonly are affected by pedogenic processes that alter the sediment that buried the soil (figs. 2.1, 2.3, 8.6, and 10.12). Some buried soils may become part of the pedon of the overlying soil if they are shallowly buried or if soil formation in the overlying material is of sufficient duration or intensity to produce a thick profile. A profile in which a buried soil is "overlapped" by an upper soil is the "welded soil" of Ruhe and Olson (1980) or the "composite soil" of Morrison (1967, p. 25; 1978, pp. 83–84; chapter 5). More deeply buried soils not obviously welded to an overlying soil can be affected by some surficial processes, such as deep-wetting fronts or root penetration (Schaetzl and Sorenson, 1987). Welding can take many forms (figs. 5.3, 5.6, 10.12, and 10.13). A Bk horizon from the upper soil can form in the Ab or Bb of the buried soil, or the Ab can have a Bt horizon superimposed over it. The latter can happen through the "upbuilding" process (chapter 5), whereby slow accretion of the soil surface results in the upper part of the B growing upward into the lower former A. The welded or overlapped portion of the buried soil will exhibit characteristics of the original soil as well as the younger soil and "the complex integration of both older and younger soil properties may result in an obscure interpretation" (Olson and Nettleton, 1998, p. 186). For example, pedogenesis along the lower slopes of mounds in Baton Rouge, Louisiana, welded the mound soil to the submound soil by formation of a fragipan through the thin mound sediments and into the underlying soil (Homburg, 1988). This obscured the stratigraphy and hampered correlation of the base of the mound fill.

A significant factor in determining the appearance of a welded soil is whether the buried soil was eroded before burial. Erosion of the A horizon is common

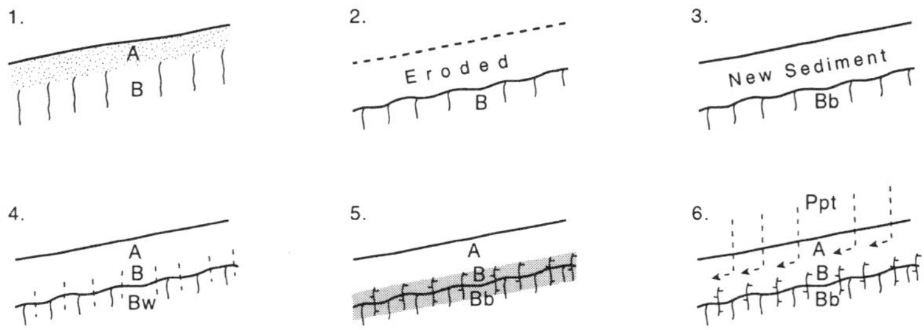

Figure 10.12 The effects of soil burial and soil welding (reprinted from *Quaternary International*, v. 51/52, C. G. Olson and W. D. Nettleton, "Paleosols and the effects of alteration," pp. 185–194, fig. 1, © 1998, with permission from Elsevier Science). The original ground soil (1) is truncated (2), and buried by fresh sediment (3). A new soil (4) forms in the younger deposit and, if the younger deposit is thin enough, develops down into the buried soil and becomes welded to it. The welding may result in induration (5), which controls downward water movement (6).

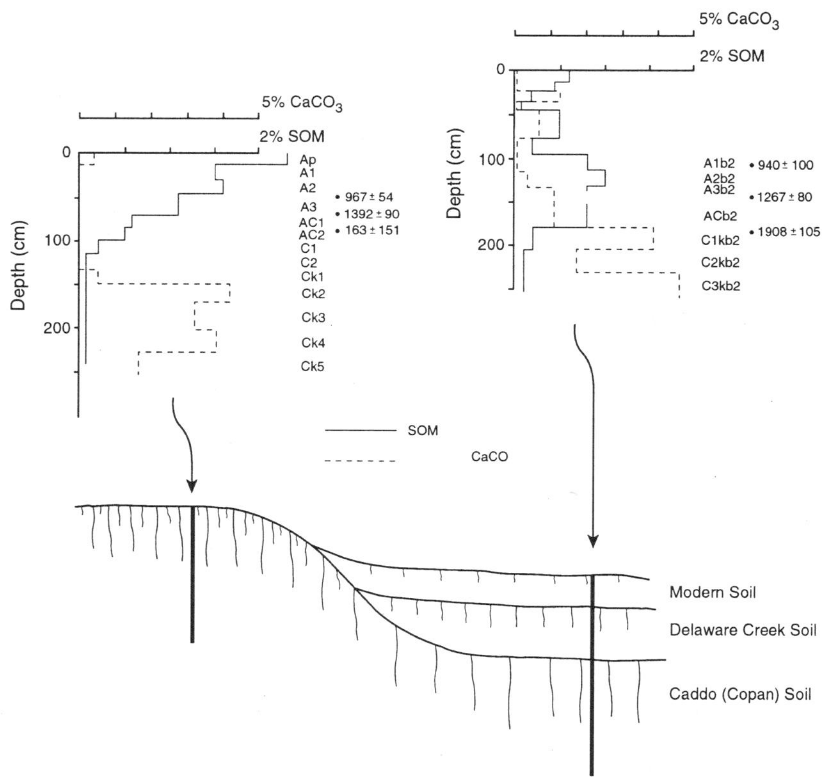

Figure 10.13 Late Holocene soils from Delaware Canyon, Oklahoma, illustrating stratigraphic relationships between multiple floodplain soils and unburied upland soil, the effects of burial on SOM content, and welding of buried soils by superimposition of calcic horizons (from Ferring, 1992, fig. 1–4; reproduced with permission of C. R. Ferring).

because of the loose, friable nature of A horizons exposed at the surface (chapter 5). Erosion may proceed to the top of any more-resistant horizon that may be present, such as a Bt, Bk, or Bx. Following burial, the presence of the more-resistant horizon within the new zone of pedogenesis will affect soil formation, particularly water movement (fig. 10.12). Precipitates will tend to "hang up" at or just above the boundary between the younger soil and the truncated one. This process will also operate where a buried soil was not truncated if there is a significant difference in permeability or porosity between the younger soil and the buried soil.

Identification of postburial pedogenic features is best accomplished by looking for "incompatible" horizon features (Catt, 1990, p. 70). For example, an A horizon is a zone of leaching and removal, whereas a Bk horizon is a zone of calcium carbonate accumulation, forming below an A horizon in a typical dry-land A-B-C soil. Presence of calcium carbonate films, threads or nodules in a buried A horizon, therefore, is indicative of translocation of carbonate into the A from an overlying soil.

Diagenesis can alter the morphology and chemistry of buried soils in two fundamental ways (modifying the terminology for pedogenic processes by Simonson, 1959, 1978): additions or the introduction of new morphological and chemical characteristics, and alterations or the destruction of characteristics of the original soil. Sediments can be affected by a wide range of diagenetic processes, but those most likely to alter late Cenozoic (i.e., archaeologically significant) buried soils include freeze–thaw (cryoturbation), compaction, cementation, mineral alteration, new mineral formation (authigenesis), dissolution, oxidation, and reduction (Nesbitt and Young, 1989; Catt, 1990, pp. 65–71; Retallack, 1990, pp. 129–145; Crowther et al., 1996; Olson and Nettleton, 1998). Olson and Nettleton (1998, p. 190), for example, note that in the middle latitudes, the frigid conditions that accompanied loess deposition during full glacial periods probably resulted in simultaneous frost-cracking of the underlying soil formed in the preceding interglacial period, allowing some silt to be mixed down into the soil as it was being buried. Diagenetic additions are particularly common in areas of fluctuating water tables or rising water tables, a common phenomenon in aggrading environments (e.g., floodplains). Processes of addition can include formation of groundwater and other carbonates, which can be mistaken for pedogenic carbonates (Machette, 1985; Tandon and Kumar, 1999), precipitation of silica in voids, and formation of iron oxides as nodules, plates, or filaments (Breuning-Madsen et al., 2000). Important alteration processes likely to affect buried soils include compaction and the reduction of iron and organic matter. The latter two processes can be particularly significant with respect to recognition of buried soils.

Postburial transformations can obscure buried A horizons, rendering them essentially invisible in the field (Limbrey, 1975, p. 313). This is caused by oxidation and microbial decomposition of organic matter, ultimately resulting in loss of the distinctive darker values and chromas of these surface horizons. A large body of literature shows that the organic matter content of A horizons drops rapidly once biota are no longer living in the zone following burial (figs. 10.13 and 10.14; e.g., Parsons et al., 1962; Stevenson, 1969; Bettis, 1988; Holliday, 1988;

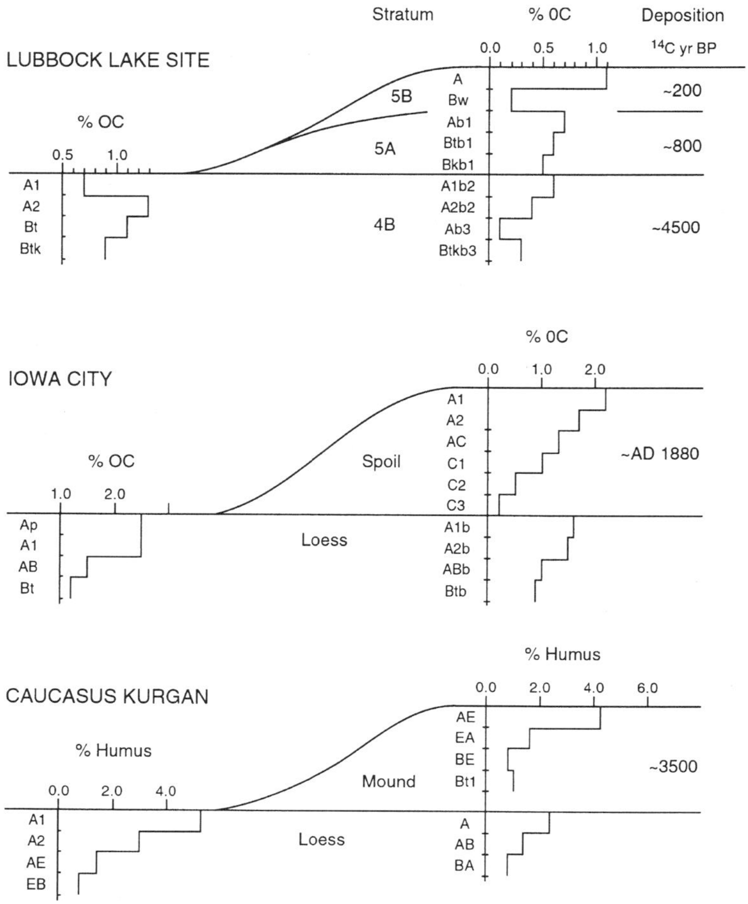

Figure 10.14 Examples of rates of SOM buildup in young soils and rates of SOM depletion following burial. Data for Lubbock Lake from Holliday (1985d, 1988, trench 104), Iowa City from Hallberg et al. (1978, pedon 2), and the Caucasus from Alexandrovskiy (2000, profile 166).

Kolb et al., 1990; Alexandrovskiy, 2000). At the Experimental Earthwork, Overtone Down, United Kingdom, Crowther et al. (1996) report a loss of 29% in organic C after just 33 yr of burial. Data on organic matter content, therefore, may not be effective in identifying buried A horizons. Cumulization or overthickening of soils also apparently obscures buried A horizons, particularly in aggrading eolian or alluvial settings (Johnson, 1977; McDonald and Busacca, 1992). The low values and chromas of A horizons can persist after burial, though few data are available on rates of persistence. Broadly speaking, the relative persistence of the darker colors in a buried A horizon may explain the relative abundance of identifiable buried A horizons in Holocene buried soils, but the ultimate diagenesis of the zone may be why they are relatively rare in Pleistocene sequences. One exception to the process of postburial alteration of A horizons is the black, organic-rich melanic epipedon of soils formed in tephra (Andisols). Frezzotti and Narcisi (1996) report soils buried in upper Pleistocene tephra and exhibiting black (2/2 Munsell value and chroma) A horizons with 4%–8% organic matter.

Soil burial may also contribute to subsurface iron reduction in a process Retallack (1991b) terms "burial gleization." As floodplains aggrade and soils are buried, the soils are submerged beneath the low, seasonal water table. In the presence of sufficient organic matter and reducing conditions, groundwaters can produce gley features in the buried soils. Limbrey (1975, p. 311) describes a similar process, but in the absence of a high water table. After burial, "the organic content falls [through oxidation] and the iron oxides are reduced and moved away [probably through chelation], the buried humic horizon takes on a grey colour."

Limbrey (1975, pp. 310–312) describes a series of postburial transformations in soils that had O horizons (largely podzolic soils in the British Isles). The organic-rich O horizon converts to a compact black mass with a waxy consistence when wet and a blocky structure on drying. One of the more striking postburial changes is the formation of an "iron pan" in these soils. This is apparently related to the translocation of oxidized iron out of the buried O horizon, noted above. The iron pan tends to form at the base of or below the O horizon. Mottles of manganese oxide sometimes form below the iron pan. The iron pans are common in podzolic soils buried below mounds and barrows and apparently were misinterpreted as premound anthropogenic features (Limbrey, 1975, p. 311).

Determining whether an A horizon was altered by postburial processes or by erosion before burial is important in archaeological contexts. If rates of human occupation and artifact dispersal are essentially the same through time, then the A horizon will have the highest concentration of archaeological materials (chapter 7). Erosion of the A horizon therefore removes a significant amount of the human record, whereas diagenic alteration of the A horizon simply renders the zone more difficult to identify but preserves the evidence of human occupation.

Valentine et al. (1980) provide an example of geoarchaeological research focused explicitly on deciphering alterations of buried soils. They also propose a system of soil horizon nomenclature that reflects postburial alteration and polygenesis of the buried soils (table 10.1). Seven buried soils were exposed at the HaRk1 site in a cliff-top dune on a terrace of the Peace River, Canada (chapter 7). Five of these soils contained evidence of human occupation (fig. 7.14). Micromorphological analysis combined with laboratory data showed that all buried A horizons were decalcified (i.e., there was no finely divided, primary calcite in thin section, and $CaCO_3$ content of several A horizons was low), but several exhibited secondary calcite along root channels and in voids. These data indicate that the surface horizon of each soil was leached before burial and that some soils were recalcified after burial. Though relatively minor, the cycles of leaching and recalcification may have implications for preservation of artifacts.

Soils formed in tephra can undergo significant physical and chemical changes following burial. In the cool, moist environment of northeastern Japan, soils formed in and buried by tephra go through predictable stages of alteration (Shoji

Table 10.1. Profile description of site HaRk1, British Columbia, Canda

Soil horizon[1]	Geomorphic	Event[2]	Depth (cm)	Pedogenic features[3] ascribed to pedogenic periods								
				1	2	3	4	5	6	7	8	9
Aph	I	Y.e	0–12	p,h,e								
Ahpk			12–16	h,p,e								
Bm			16–23	m,e								
Ahpb1	II 1530 and 1630 yr BP	Y.e	23–30	e	h,p,e							
Bmb1			30–36	e	m,e							
Ahpb2	III 2485 yr BP	Y.e	36–44	e	e	h,p,e						
Bmb2			44–57	e	e	m,e						
Ccab1			57–89	ca	ca	ca						
Ahpcab1	IV 4365 yr BP	Y.e	89–100			ca	h,p,e					
Ccab2			100–112				ca					
Ahpcab2	V	Y.e	112–123				ca	h,p,e				
Ccab3			123–136					ca				
Ahcab1	VI 5830 yr BP	Y.e	136–142					ca	h,e			
Ccab4			142–342						ca			
Ahcab2	VII	Y.e	342–348						ca	h,e		
Ccab5			348–353							ca		
Ahcab3	VIII	Y.e	353–393							ca	h,e	
Bmb3											m,e	
Ccab6			393–493								ca	
	IX	Z.a										
IlCcab	X	Y.f	493–525								ca	ca

From Valentine et al. (1980, table 3).

[1] After Canada Soil Survey Committee (1978).

[2] After Bos and Sevink (1975): Y.e: process of "eolian aggradation"; Y.f: process of "fluviatile aggradation"; Z.a: process of "absolute degradation."

[3] Horizon designations (Canada Soil Survey Committee, 1978) denote the following pedogenic processes: ca, the precipitation of secondary carbonates leached from horizons above; e, the leaching of primary and possibly secondary carbonates; h, the addition and incorporation of organic matter primarily from dead plants; m, the weathering of minerals by hydration, oxidation, or solution to give a color browner or redder than the parent material; p, human disturbance of a soil horizon.

et al., 1993, pp. 62–63). Unburied soils are heavily leached, resulting in formation iron-aluminum-humus complexes in the A horizon and formation of clay and iron-oxide minerals characteristic of weathered ash in the B horizon. With shallow burial, leaching continues and these same clay and iron-oxide complexes start to form in the A horizon, and because there is no more addition of organic matter, the pH rises. With additional tephra deposition, the deeper buried soils become zones of accumulation rather than leaching, and bases and silica illuviate. The boundary between the zones of leaching and accumulation is determined by a variety of factors including precipitation, evapotranspiration, water retention, vegetation, and drainage and would fluctuate with environmental changes.

In an archaeologically focused experiment in soil burial, a "Humic Rendzina" (in the British soil parlance; a Rendoll in the United States) was buried beneath a stack of turf and beneath chalk for 32 yr (Crowther et al., 1996). A number of changes in the buried soil were noted including compaction, changes in pH (more acidic under the turf, more alkaline under the chalk), earthworm activity (less under the turf, more under the chalk), and aeration (less under the turf, more under the chalk). These changes have implications for site preservation in terms of bioturbation (caused by earthworms) and artifact preservation (caused by compaction and changes in pH and aeration).

11

Human Impacts on Soils

Literally since they first set foot on the Earth's surface, humans and their hominid ancestors have affected soils. The degree of effect humans have on soils varies from the most subtle, which could include simply walking across the soil, to the most dramatic, such as wholesale removal, mixing, or burial associated with urbanization. Butzer (1982, pp. 123–156) and Davidson (1982) present very useful summaries of the history and nature of such influence in a geoarchaeological context. Geoarchaeologists can be confronted with soils subjected to a wide degree of anthropogenic alteration (tables 11.1, 11.2, and 11.3). The detection of these alterations and their differential distribution can, in theory, be used to determine site boundaries, define stratigraphic relationships, delimit intrasite activity areas and features, and aid in their functional interpretation (Woods, 1984, p. 67). The primary challenge is detecting the human-induced alteration and then, of course, interpreting it. In a very broad sense, the detectability of human impact is roughly proportional to the degree of impact; that is, very subtle alterations are difficult or impossible to detect, but more substantial changes are more obvious. As in most other aspects of archaeology, interpreting the meaning of anthropogenic effects on soils is much more problematic.

The study of human impacts on soils is one of the oldest applications of soil studies in archaeology, particularly in regions with a long history of significant human modification of the environment. The topic is also an important part of soil science and agriculture. As a result, there is a very large literature on the topic, especially for the Old World (e.g., table 11.4). There well may be more writing on this aspect of soils research in archaeology than on all others combined. The topic is of such interest because human impacts on soils can be so

Table 11.1. Examples of indirect human effects and catalysts on the weathering system

Weathering factor or catalyst and human impacts	Net effect
Chemical buffers	Neutralizes chemical weathering
Fertilizers	
Exposure	Exposes fresh parent material to weathering
Excavation	
Quarrying and mining	
Construction	
Agriculture	
Surface compaction and armoring	Protects subsurface from influx of weathering agents
Construction	
Paving	
Foot traffic	
Domestic animals	
Agricuture	
Painting and plastering	
Surface elevation or depression	Allows throughflow and runoff of weathering agents if elevated, or accumulation of weathering agents if depressed
Contruction	
Temperature	Chemical reaction catalyst
Fire	
Habitation	
Re-/devegetation[1]	
Climate change[1]	
Water balance	Increase/decrease in water-based weathering reactions
Water use	
Re-/devegetation[1]	
Climate change	
Biomass	Increase/decrease in biotic weathering reactions
Re-/devegetation[1]	
Climate change[1]	
Atmospheric gases	Increase/decrease in weathering reactions dependent on CO_2, O_2, SO_x, etc.
Air pollution	
Re-/devegetation[1]	
Climate change[1]	

From Pope and Rubenstein (1999, table 1). See original table for citations.

[1] Refers to human-induced vegetation change or climate change in these cases.

obvious and pervasive in archaeological contexts; recognizing human impacts is critical in sorting out artificial versus natural pedogenic and other geogenic processes; it provides another avenue of research into understanding the relationship between humans and their environment, especially the landscape; and it offers another means of getting at human behavior, either directly, as in studies of mound construction or agriculture, or more indirectly, as in studies of human-induced soil erosion. Regardless, the subject has long been of interest in archaeology and geoarchaeology.

The subject of human impacts on soils is not a traditional component of soil geomorphology, nor is it regularly dealt with as a component of geoarchaeology (at least not in North America). For example, there is little or no mention of the

Table 11.2. Examples of direct human impacts on weathering processes

Weathering process and human impacts
Abrasion
Foot traffic
Construction
Agriculture
Domestic animals
Mechanical breakage
Foot traffic
Construction
Quarrying/mining
Tool manufacture
Thermal and/or frost
Fire
Habitation
Construction
Biotic (mechanical)
Agriculture
Horticulture
Biochemical (organic acid dissolution, chelation, redox)
Habitation
Refuse
Agriculture
Hydration and hydrous solution
Irrigation, diversion, storage, groundwater use, and pollution of water resources
Salt (salt crystal growth, salt hydration, and salt dissolution)
Irrigation, diversion, storage, groundwater use, and pollution of water resources
Alkaline hydrolysis
Fire (ash)
Water and air pollution
Fertilizers
Acidic hydrolysis and carbonation[1]
Acid precipitation
Oxidation[1]
Water pollution
Fertilizers
Fire
Irrigation, diversion, storage, groundwater use, and pollution of water resources
Reduction
Water saturation

From Pope and Rubenstein (1999, table 2). See original citation for references.

[1] Excludes processes accounted for in biochemical.

topic in the standard works in soil geomorphology (Daniels and Hammer, 1992; Gerrard, 1992; Birkeland, 1999) or in geoarchaeology (Waters, 1992; Herz and Garrison, 1998; Rapp and Hill, 1998; Goldberg et al., 2001; a notable exception is the volume by Butzer, 1982, who is also a geographer and cultural ecologist). In part this is probably because the subject has drawn the interest and involvement of a wide variety of scientists not typically connected with soil geomorphology or geoarchaeology, such as a broader array of archaeologists, soil

Table 11.3. Some effects of human influence on the five factors of soil formation

Parent material

- Adding mineral fertilizers, lime, shell and bone, ash
- Removing excessive amounts of substances such as salts
- Removing through harvest more plant and animal nutrients than are replaced
- Adding substances in amounts toxic to plants or animals; compaction
- Altering soil constituents that depress plant growth

Topography

- Checking erosion through surface roughening, land forming, and structure building
- Raising land elevation by building up material; lowering elevation by leveling
- Causing subsidence by drainage of wetlands and by mining
- Accelerating erosion; excavating

Climate

- Adding water by irrigation; rain-making
- Release of CO_2 to atmosphere by industrialization (with possible climate warming)
- Subsurface warming of soil electrically, or by piped heat
- Change albedo by changing color of soil surface
- Removing water by drainage; heating air near the ground; creating smog
- Subject soil to excessive insolation, extended frost action, and exposure to wind
- Altering aspect by land forming; clearing and burning organic cover

Organisms

- Introducing and controlling populations of plants and animals
- Adding organic matter to soil directly as waste or fertilizer, or indirectly through introduced plants and animals
- Loosening soil by plowing to admit more oxygen; removing plants and animals
- Reducing organic matter content through burning, plowing, overgrazing, harvesting, accelerating oxidation, leaching
- Adding or fostering pathogenic organisms; adding radioactive substances

Time

- Rejuvenating soil through additions of fresh parent material or through exposure of local parent material by soil erosion
- Reclaiming land from under water; burying soil under solid fill or water
- Degrading the soil by accelerated removal of nutrients from soil and vegetative cover

Modified from Bidwell and Hole (1965, table 1).

chemists, soil geographers, agriculturalists, and cultural ecologists. This wide interest in human impacts on soils as a component of archaeology is largely because of the many archaeologists involved in the study of agricultural and complex societies (and the effect of such societies on soils and landscapes), the specialized nature of soil chemical analyses, and the broad interdisciplinary interest in the origins and evolution of agriculture. Because so many individuals not trained in pedology, soil geomorphology, or Quaternary stratigraphy are involved in the study of human impacts on soils, the terminology used in that literature often is

Table 11.4. Phosphorus studies in archaeology

Type of study and references
General articles
Phosphorus chemistry in soils
Chang and Jackson, 1957a, 1957b; Walker, 1964; Williams et al., 1967; Jackson, 1969; Syers and Walker, 1969a,b; Smeck, 1973, 1985; Walker and Syers, 1976; Dick and Tabatabai, 1977; Mehlich, 1978, 1984; Hedley et al., 1982; Harrison, 1987;[1] Sharpley et al., 1987; Barber, 1995 (chapter 9);[2] Stevenson and Cole, 1999 (chapter 9);[2] Pierzynski, 2000[2]
Early investigations
Lorch, 1930, 1939, 1940, 1954; Arrhenius, 1931, 1934, 1963; Solecki, 1951; Dauncy, 1952; Dietz, 1957; Mattingly and Williams, 1962; Eddy and Dregne, 1964; Cook and Heizer, 1965
Summaries and reviews
Cook and Heizer, 1965; Provan, 1971; Eidt and Woods, 1974; Woods, 1975,[1] 1977; Proudfoot, 1976;[2] Sjoberg, 1976;[2] Eidt, 1977, 1984,[2] 1985;[1] White, 1978; Carr, 1982 (chapter 4); Hamond, 1983;[2] Gurney, 1985;[1] Bethell and Máté, 1989;[1,2] Schuldenrein, 1995; Herz and Garrison, 1998 (chapter 9);[2] Leonardi et al., 1999
Methods for archaeology
Schwarz, 1967; Provan, 1971;[2] van der Merwe and Stein, 1972; Eidt, 1973, 1977, 1984,[2] 1985;[1] Eidt and Woods, 1974; Woods, 1975, 1977; White, 1978; Lewis, 1978;[3] Hassan, 1981; Keeley, 1981;[4] Hamond, 1983;[2] Linderholm and Lundberg, 1994; Bjelajac et al., 1996; Sanchez et al., 1996; Entwistle and Abrahams, 1997; Enwistle et al., 1998, 2000; Terry et al., 2000
Anthropic epipedons
Pettry and Bense, 1989; Kaufman and James, 1991; Sandor and Eash, 1995; Scudder, 1996
Case studies
North America
Eddy and Dregne, 1964; Cook and Heizer, 1965; Heidenreich et al., 1971; van der Merwe and Stein, 1972; Ahler, 1973; Heidenreich and Konrad, 1973; Heidenreich and Navratil, 1973; Berlin et al., 1977; Griffith, 1980, 1981; Carr, 1982 (chapter 8); Konrad et al., 1983; Woods, 1984; Craddock et al., 1985; Sandor et al., 1986; Skinner, 1986; Moore and Denton, 1988; McDowell, 1988; Kolb et al., 1990; Dormaar and Beaudoin, 1991; Lewis et al., 1994; Kerr, 1995; Schuldenrein, 1995; Bjelajac et al., 1996; Chaya, 1996; Scudder, 1996; Homburg and Sandor, 1997; Schlezinger and Howes, 2000; Sullivan, 2000
Central America
Cook and Heizer, 1965; Muhs et al., 1985; Dunning, 1994; Manzanilla, 1996a,b; Coultas, 1997; Terry et al., 2000;[5] Wells et al., 2000;[6] Parnell et al., 2001, 2002
South America
Eidt and Woods, 1974; Keeley, 1981; Eidt, 1984; Sandor, 1987; Lippi, 1988; Mora C. et al., 1991; Sandor and Eash, 1995; Lima et al., 2002
Europe
Provan, 1971; Sieveking et al., 1973; Davidson, 1973; Bakkevig, 1980; Keeley, 1981; Conway, 1983; Hamond, 1983; Ottaway, 1984; Craddock et al., 1985; Courty and Nørnberg, 1985; Gurney, 1985; Cavanagh et al., 1988; Prosch-Danielsen and Simonsen, 1988; Nunez, 1990; Nunez and Vinberg, 1990; Lillios, 1992; Dockrill and Simpson, 1994; Dockrill et al., 1994; Lewis et al., 1994; Linderholm and Lundberg, 1994; Quine, 1995; Engelmark and Linderholm, 1996; Sanchez et al., 1996; Entwistle and Abrahams, 1997; Simpson, 1997; Enwistle et al., 1998, 2000; Simpson et al., 1998; Leonardi et al., 1999; Vizcaíno and Cañabate, 1999; Macphail et al., 2000; Macphail and Cruise, 2001
Africa
Hassan, 1981; Lewis et al., 1994
Asia
Stimmell et al., 1984; Overstreet et al., 1988; Wilkinson, 1988, 1990; Brinkmann, 1996

[1] Extensive list of references.
[2] Good review discussion of soil P or archaeological P.
[3] Based largely on the work of Eidt (1973, 1977).
[4] Good historical review of methods.
[5] Good review of P studies in Mesoamerican archaeology.
[6] Good review of chemical analyses of anthrosols.

not standardized relative to the terminology used throughout the rest of this volume. In particular, the term "soil" often is applied to sediments. The traditional soil horizon nomenclature also is not always followed or used. Some examples cited in this chapter in fact deal with sediments, but they are mentioned because of the particular points made or issues raised. In such cases, terminological ambiguities are noted. More important, the topic of human impacts on soils is one of the oldest in geoarchaeology: It is broad and diverse and is clearly worthy of its own systematic book-length treatment. Because the subject is outside the mainstream of soil geomorphology, however, and because of its breadth, this chapter presents only a sketch of the topic.

For the field geoarchaeologist trying to interpret an archaeological site or region, and for the purposes of this chapter, a simple categorization of human impacts is whether they represent physical or chemical alterations of soils; that is, whether the alterations can be seen in the field (e.g., mound or ditch construction or compaction) or are only detectable in the lab (e.g., elevated levels of phosphorus). Of course some alterations are both physical and chemical (e.g., manuring and the formation of a "plaggen" is apparent in the field and is also detectable chemically). In the following section of this chapter, general discussion of physical and chemical impacts on soil by humans is presented. These discussions are followed by sections organized around the application of soil studies to understand human impacts as most commonly reported in the literature (topics that are not necessarily mutually exclusive): soil phosphorus and archaeology, the origin and nature of anthrosols, and agriculture. For the most part, discussion focuses on agricultural and preagricultural sites and on recognition, investigation, and interpretation of human impacts.

Anthropogenic Impacts: General Considerations

The means by which humans modify soils are many and varied. These modifications can come in an almost infinite array of scenarios at all scales of time, space, and intensity. In one of the first papers to deal expressly with the broad theme of artificial impacts on soils in a soil geomorphic context, Bidwell and Hole (1965) approached the topic from the standpoint of human impacts on the five factors of soil formation (table 11.3). This is a useful concept and categorization because much human impact on soils is arguably a matter of influencing the factors. Changing or removing vegetation, for example, affects the biotic factor; leaving behind shell, wood ash, or bone at a campsite changes the parent material. However, draining a wetland or creating a footpath or cart track across a soil are more direct impacts that are not easily categorized under the classical factors. For that reason, Amundson and Jenny (1991) reformulated the state factor equation to explicitly identify humans and human activity as separate factors. The factorial approach, though useful conceptually, is not entirely satisfactory for understanding human impacts because, as with the more general criticism of the state factors, it does not deal with processes.

A number of investigators among the many cited throughout this chapter inadvertently approached the issue of human impacts by applying a state factor approach, however, especially in studies of anthrosols and in applications of soil chemistry. In trying to detect human impacts, many researchers, usually pedologists and geographers, examined "natural" or "pristine" (or "virgin") soils; that is, those with no apparent human impact, along with analyses of the anthropogenically influenced soils. Hence, they examined a set of soils that varied only as a function of human inputs. In a sense, they are investigating what Yaalon and Yaron (1966) call "metapedogenesis," which refers to anthropogenic processes and the resulting soil profile changes. In this approach, the "natural" soil is the parent material for the human-induced modifications.

Pope and Rubenstein (1999), in a broader review of human impacts on weathering processes, provide a useful categorization of processes and effects that can be more specifically applied to soils (tables 11.1 and 11.2), somewhat akin to the grouping of (natural) soil forming processes according to the four-fold categorization of Simonson (1959, 1978; table 3.1), discussed in chapter 3. Pope and Rubenstein (1999) make an important distinction in the ways in which humans affect weathering processes by grouping them as "direct" and "indirect" (table 11.1 versus table 11.2). The specifics of their classifications could be debated, but essentially the same twofold classifications can apply to soil-forming processes per se. Direct impacts include excavation for a dwelling (e.g., a pit-house) or for borrow material (e.g., for mound construction) or the addition of fertilizer (e.g., manuring and formation of a "plaggen epipedon," discussed later). Most of the indirect impacts are only partially removed from the direct effects on soil genesis; they still influence weathering and pedogenesis. Compaction, for example, can impede water movement and thus indirectly influence soil genesis, but it still has a direct effect on the thickness of the horizon being compacted, and fertilizers still alter the soil chemistry even if they only indirectly affect soil-forming processes. In any case, the itemizations presented by Pope and Rubenstein (1999) are useful in highlighting the many ways in which people alter soils and soil-forming processes.

Physical anthropogenic alterations to soils include (but are not limited to) excavation of ditches; various kinds of pits, stake holes, and post holes; building of mounds, other earthworks, and roads; construction of dwellings; preparation and plowing of fields; and cultivation during crop growth. More complex societies with more dense populations and more construction activities have correspondingly more pervasive physical impacts on soils. For example, "Dark Earth" soils are a common component of the stratigraphic record of cities in Great Britain and Europe (e.g., Macphail, 1987, 1994; Courty et al., 1989). They are a direct result of and were buried by urban activity. Because of their ubiquity, they are described and discussed further later in this chapter.

One of the most obvious and widely reported physical human impacts on soils is burial caused by construction. This includes artificial burial of the original land-surface soil as well as formation of and burial by artificial deposits. Plowing is also a pervasive physical, anthropogenic alteration of soils and is discussed below along with the impacts of agriculture. The basic principals of interpretation of artificially buried soils are no different than those for soils buried in natural settings, discussed throughout this volume. The main differences are that the soils influenced by artificial burial usually are not laterally extensive and that soils within artificial fills typically are not particularly well expressed, because of the short duration of soil genesis between construction episodes. The effects of artificial burial are also similar, in most instances, to the postburial alterations described in chapter 10. Soils buried by mound (or barrow or kurgan) building are widely reported (Parsons et al., 1962; Limbrey, 1975; Hallberg et al., 1978; Bettis, 1988; Kolb et al., 1990; Ammons et al., 1992; Cremeens, 1995; Alexandrovskiy, 2000) and provide information on postburial alteration of soils (chapter 10). At the Crane site in Illinois, Carr (1982) concluded that midden deposits alter the underlying B horizon by adding organic matter such that the degree of aggregation and moisture retention is higher under the midden. In a study of soils along the coast of southwestern Florida, Scudder (1996) showed that middens had a minimal physical or chemical effect on the underlying "native" soil and also protected the buried soil from further surficial impacts.

The soils preserved below built landscapes sometimes preserve a record of preburial vegetation and land use (chapter 8 and below) and construction activities (e.g., Limbrey, 1975; Davidson, 1976; Butzer, 1981; Macphail, 1988, 1994; Kolb et al., 1990; Miller and Gleason, 1994a; Alexandrovskiy, 2000; Lang et al., 2003). Likewise, soils buried within mounds and other sorts of built archaeological features and landscapes can provide clues to construction and degradation (e.g., Cornwall, 1958; Limbrey, 1975, pp. 281–322; Davidson, 1976; Wilkinson, 1976; Evans, 1978, pp. 112–129; Butzer, 1981; Shackley, 1981, pp. 31–37; Kolb et al., 1990). Soils buried by backdirt and buried in fill also are useful for reconstructing and interpreting ditches and pits (e.g., Cornwall, 1958; Limbrey, 1975, pp. 281–322; Evans, 1978, pp. 112–129; Shackley, 1981, pp. 31–37; Weisler, 1999). The truncated solum of the soil into which these features were excavated help outline or identify the ditches and pits. Soils within the fill, as with soils in "natural" settings, are useful for understanding the sequence of filling, whether it was slow and continuous or episodic.

The recognition of pedogenic features has also proven useful for understanding the nature of mound fills. The use of mollic or mollic-like A horizons has been recognized in mounds in the midwestern United States and in kurgan in the North Caucasus of Russia. "Sod blocks" are identified as an important component of Hopewell mounds in Illinois (Van Nest et al., 2001). The block consist of the A horizon and a thin zone of unweathered silty parent material (loess). They appear to be derived from soils formed under grass due to their cohesiveness, fine structure, and relatively high organic carbon content (after 2000 yr of burial), which contrast with the surface horizons of woodland soils in the area today. Based on the unique characteristics of the sod blocks compared to the local soils, Van Nest et al. (2001) suggest that they were specifically sought after for mound construction. Few details are available for the kurgan fill other than that Alexandrovskiy (2000) describes much of the material in the mounds as a "Chernozemic pile," suggesting that it is composed of the dense, dark surface horizon of the regional grassland soil.

Chemical impacts on soils caused by human activity are generally much more subtle than physical impacts and usually require laboratory analyses for identification. Among agricultural and preagricultural groups, these anthropogenic additions to the soil come from human refuse and waste; burials; the products of animal husbandry in barns, pens, and on livestock paths; or intentional enrichment from soil fertilizer (modified from Eidt, 1984, pp. 29–30; see also Miller and Gleason, 1994b, for a review discussion of fertilizer in archaeological contexts; the topic of agricultural impacts is addressed in a separate section, following). With the advent of metallurgy and later industrialization, a much broader spectrum of chemicals and chemical compounds was added to the soil such as heavy metals and hydrocarbons. The most common chemical elements added to soils by human activity are carbon, nitrogen, phosphorus, and calcium, with lesser amounts of potassium, magnesium, sulphur, copper, and zinc (Cook and Heizer, 1965, pp. 1–3; Eidt, 1984, pp. 25–27; 1985). The quantities can be substantial. In their landmark study of soil chemistry and archaeology, Cook and Heizer (1965, p. 9) report the following annual anthropogenic additions of key elements (as a percentage of each element already present in the soil): 0.022%–0.439% of calcium, 0.672%–6.72% of nitrogen, and 0.495%–9.91% of phosphorus.

The most common chemical compound that is added to soils by humans in agricultural and preagricultural societies and that is also easily recognizable in the field is SOM. The archaeological interest and effort focused on anthropogenic SOM is second only to soil phosphate (Carr, 1982, p. 109). SOM is the organic fraction of the soil exclusive of undecayed plant and animal residue (Tabatabai, 1996). It is a complex mixture of chemical compounds that include carbon, nitrogen, phosphorus, and sulphur along with humus (Stein, 1992c; Duchaufour, 1998, p. 29; SOM typically is 1.7 to 2.0 times the content of organic carbon in the surface horizons of soils and up to 2.5 times organic carbon in subsurface horizons [Nelson and Sommers, 1982; Tabatabai, 1996]). The humus compounds impart the characteristic dark-brown to black coloration so characteristic of SOM. Human activity, largely through discard of organic waste (either in middens or as fertilizer), can add significant amounts of organic matter to the soil surface. Furthermore, additional SOM can be produced and added to the soil by stimulation of soil biota and aboveground biomass subsequent to human activity because of more favorable nutrient conditions often associated with anthropogenic changes. The result of these direct and indirect influences on SOM production is very low value and chroma in color and, often, overthickening. These are notable characteristics of the anthrosols described below. For example, the Terra Preta soils of the Brazilian Amazon have value and chroma <3 and at least 1.4% organic carbon (~2.5% SOM; Sombroek et al., 2002). Beyond SOM, chemical alterations of soils by human activity are rarely apparent in the field. Exceptions would include massive accumulations of materials such as shell (in a shell midden) or bone (in a bone bed) or some specialized activity areas (e.g., metal making) in urban sites.

Differentiating organic matter left by people from natural SOM is difficult, however, and generally must rely on recognition of associated anthropogenic contexts or materials such as artifacts, imported plant remains and residues, charcoal and ash (which can be identified as to plant type and heating intensity), and other chemical compounds such as phosphates and fatty acids. These constituents are identifiable on the basis of micromorphology (fig. 11.1; e.g., Courty et al., 1989, pp. 104–137; Macphail et al., 1990a,b) and chemical (e.g., phosphorus, discussed later) and biochemical (Bull et al., 2001) analyses. At a general level, SOM has

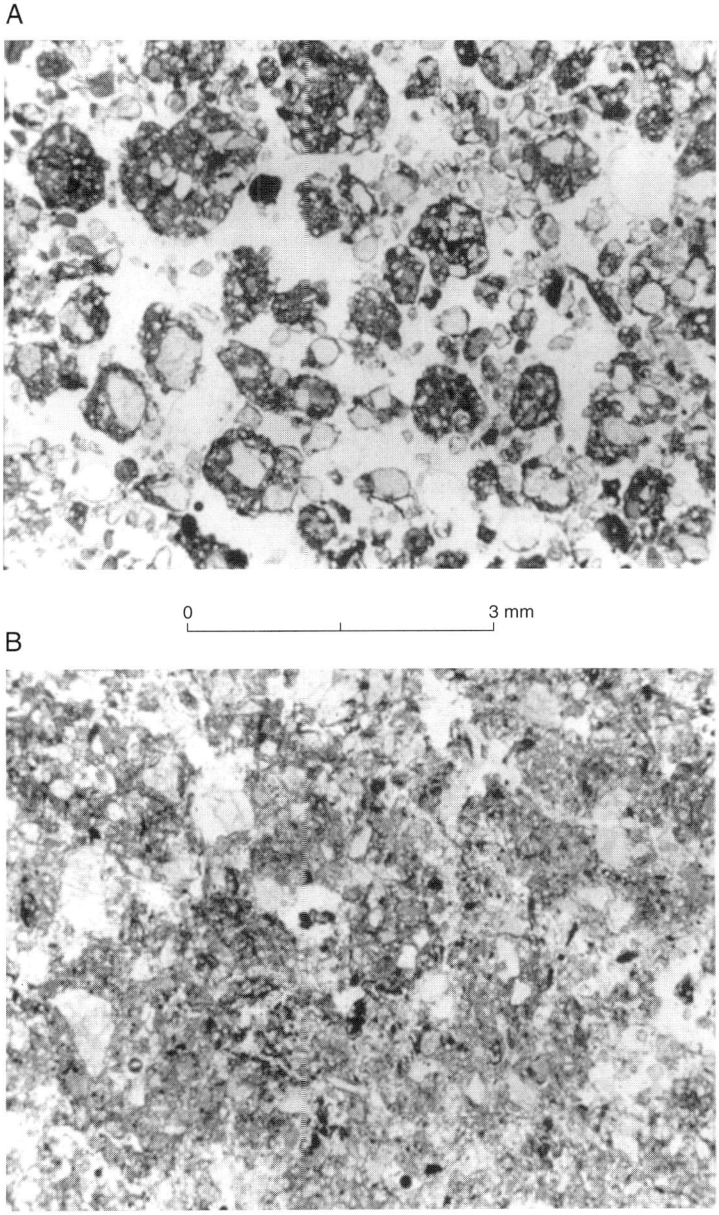

Figure 11.1 Photomicrographs of soil A horizons (0–10 cm depth) from the Zuni Reservation in northwestern New Mexico (modified from Homburg, 2000, fig. 3.6; photographs provided by and reproduced with permission of J. A. Homburg). (A) The granular structure of an uncultivated soil is well expressed in this thin section. (B) In contrast, the cultivated counterpart to (A) exhibits massive, compacted structure because of cultivation. Micrograph (A) also displays higher content of SOM coatings on grains compared with (B).

been used to aid in defining the limits of archaeological sites (Stein, 1992c). Carr (1982, pp. 68–69, 127–131), based on his own field work as well as a very useful (though now somewhat dated) review of the literature, notes that organic residues from human activity increase the humus content of organic matter and, therefore, the variability of humus content in soils may provide a means of identifying occupation zones or even activity areas within sites (e.g., Stein, 1984; Vizcaíno and Cañabate, 1999). Engelmark and Linderholm (1996) note that abandoned fields can retain elevated levels of SOM. An important caveat to add, however, is that levels of SOM in archaeological as well as natural soils can decline because of oxidation once organic detritus is no longer added (e.g., following abandonment or burial; chapter 10). In the examples discussed by Engelmark and Linderholm (1996), the high SOM levels in abandoned agricultural fields likely were maintained because the fields were manured, relatively young (late Holocene), and in the cool, moist climate of northwestern Europe. In a very different context, the massive infusion of organic matter on a soil surface under a bison kill apparently contributes to gleying in the soil under the resulting bone bed (Hill and Hofman, 1997).

Anthropogenic additions of carbon, nitrogen, phosphorus, calcium, potassium, magnesium, and sulfur must be identified on the basis of chemical analyses and, in theory, can be used as indicators of past human activity. Most of these elements are removed from soil more or less readily by leaching, oxidation, reduction, or plant uptake, however (Eidt, 1977; Carr, 1982, pp. 127–131). The nature and rates of the losses from the soil are determined by local biological and pedological processes. Phosphorus in its common form as phosphate, however, is stable and generally immobile in soils. Phosphate analysis in archaeological contexts is treated in a separate discussion later in this chapter.

Cook and Heizer (1965, pp. 2–3) argue that the analysis of carbon, nitrogen, and calcium will also contribute to an understanding of habitation history. Carr (1982, pp. 127–131) further suggests that because humus incorporates N, S, Mg, and Ca, in addition to P, anomalous increases in these elements in an archaeological site may be indicative of use-areas. Data on these elements in archaeological soils are slim, however, probably because they tend to be lost from or mobilized out of their sites of original anthropogenic deposition via the processes noted above. Calcium and metals such as aluminum, copper, manganese, mercury, iron, and zinc may be enriched in anthropogenic soils because of their binding with phosphorus as well as their tendency to bind with and be preserved in calcareous soils. In addition, calcium and $CaCO_3$, because of their association with bone and limestone processing, have been used as archaeological indicators of human activity (e.g., Hassan, 1981; Griffith, 1981; Woods, 1984; McDowell, 1988; Schuldenrein, 1995; Manzanilla, 1996a,b; Entwistle et al., 1998; Parnell et al., 2002). Potassium can be incorporated into soils from decomposing animal remains, burning, and wood ash, and appears to be a useful indicator of intense human activity (Davidson and Simpson, 1984; Schuldenrein, 1995; Middleton and Price, 1996; Entwistle et al., 1998, 2000). Nitrogen is abundant in human and animal waste, so it is probably initially abundant on sites with prolonged human use and in those where manuring was practiced. However, N, typically found as a component of SOM, is also subject to rapid decomposition and loss because of leaching and volatilization and does not appear to persist long in archaeological soils. Eddy and Dregne (1964) do, however, report elevated levels of nitrate (NO_3) in late prehistoric sites in southwestern Colorado and northwestern New Mexico.

Beginning in the late 1970s to early 1980s, a number of geoarchaeological studies included the analysis of a variety of soil chemicals. For example, at the Crane site, a Woodland occupation in Illinois, Carr (1982, chapter 8) concluded that anthropogenic enrichment of N, P, K, Ca, Mg, Na, and S was maintained in some portions of the site for hundreds of years. He compared the chemical signature of swept areas within the site to dumping areas outside the working localities. The soils of the swept areas were low in Ca and P, and the dump areas were high in these chemicals. This was almost certainly the result of the high content of bone and ash in the dump area. At the Benson site, a Huron village in Ontario, Canada, Griffith (1981) studied the utility of several chemicals for identifying archaeological features. Magnesium was very useful in distinguishing features created by wood ash and fish and bird bones. Calcium was high in middens and in cooking areas because of disposal of corn and animal tissue. Carbonates were also higher in the middens than elsewhere in the site. Phosphorus was high in most parts of the village, but particularly in middens, pits, and the insides of houses. Potassium levels were doubled in the middens and pits, but were low elsewhere. This site is not old, however, and the elevated levels of some of the chemicals may not persist through time because of weathering.

Schuldenrein (1995) investigated soil chemical trends at two very different kinds of sites in the United States. The Rucker's Bottom site, Georgia, is a Mississippian village site in stratified alluvium. The occupation zone was characterized by a "sheet midden" throughout the site as well as evidence of distinct pit fills. The sheet midden was stained with organic matter and ash, presumably all anthropogenic. Samples were collected from the sheet midden, occupation floors, post holes, trash middens, and palisade ditch fills as well as from nearby natural soils and were analyzed for Mg, Ca, K, Na, and P. The best indicators of anthropogenic chemical inputs in the late prehistoric are P and K. The P probably resulted from a wide variety of human activities (discussed later). The K is believed to result from decayed animal remains, burning of the soil, and wood ash. The Royel Goodenaw site, Nebraska, contains protohistoric (16th–18th century) hunting camps in stratified alluvium. The occupation zone consists of scattered artifacts and bone (i.e., limited human activity for brief periods) in a thin A horizon of a very weakly expressed soil. Samples were collected from areas with concentrations of bone as well as off-site natural soils and were analyzed for SOM, Ca, Mg, K, and P. Not surprisingly, Ca was present in the highest concentrations, indicative of bone weathering. Potassium registered the lowest levels. If K is indeed associated with burning, then low concentrations are expected because no evidence of burning was observed.

At the Munsungun Lake Thoroughfare in Maine, Konrad et al. (1983) used elevated levels of Mg, P, and Ca to locate occupation zones and to direct some excavations. The Ca and P were used to identify and delimit Paleoindian through Historic activity areas. The Ca was particularly useful for the later, more intense occupations; it was probably leached from older levels. High values of all three elements along with magnetic susceptibility anomalies were useful in locating hearths in later prehistoric levels, and Mg plus the magnetic data guided excavations to hearths in the oldest (Paleoindian) level. The Mg and magnetic anomalies also located a burned tree, "a tendency that reduces the predictive power of this technique in a wooded area" (Konrad et al., 1983, p. 24).

The proportional relationships of certain ions have also been investigated archaeologically. Soil pH, which is an expression of the proportion of H ions (or protons) to OH (hydroxyl) ions, has some sensitivity to anthropogenic inputs. The concentration of cations in the soil strongly influences pH (as cation content increases, pH increases; i.e., becomes more alkaline; see Birkeland, 1999, pp. 18–19 for a brief summary of soil pH). Prolonged or more intense occupations tend to release more cations to the soil; therefore, pH tends to be higher under longer or more-dense or more intensely occupied sites (Carr, 1982, p. 112). In some studies, variation in pH has been used to indicate use areas (Weide, 1966; Carr, 1982). Eddy and Dregne (1964), for example, used higher pH to show where bone (now weathered away) was deposited. At the Robitaille site, a Huron village on the acidic (podzolic) soils of Ontario, activity centers have higher pH values compared with site peripheries, further demonstrating that more intense activity raises pH (Heidenreich et al., 1971).

In studies of other elements, Cook and Heizer (1965) showed that total soil C, N, and Ca were also higher in site centers compared to peripheries. Heidenreich et al. (1971) noted that discrete midden deposits varied in Ca:Mg ratios. They proposed that these variations reflected different kinds of refuse, but they offered no interpretations of refuse types. They also noted discrete areas of high Mg in and near the center of the site, which they attributed to ash. In basic soils, Cook and Heizer (1965, p. 20) propose that Ca:P ratios work to distinguish areas with high bone (higher P) from areas with higher shell (higher Ca). Their data and discussion do not present a convincing case for this proposal (considerable data are presented, but the idea of Ca:P ratios to indicate shell vs. bone is asserted rather than demonstrated), but the basic assumption seems sound. Clearly, ethnoarchaeological or experimental data are needed.

Schlezinger and Howes (2000) looked at ratios of carbon, nitrogen, and various forms of phosphorus, rather than the absolute concentrations of these elements, at the Carns site in Massachusetts. Their approach has promise. Carbon and nitrogen are both depleted from soils relatively rapidly, but C:N ratios are higher (by a factor of 2) in an anthrosol than in an unoccupied control section. This may represent anthropogenic inputs of C (e.g., bone, shell) in addition to natural accumulation. The ratio of C to organic P is also elevated in occupation areas, probably because organic P rapidly mineralizes (discussed below and in appendix 2).

The advent of elemental analysis using inductively coupled plasma–mass spectrometry (ICP-MS) and ICP–atomic emission spectrometry provide a means of relatively rapid, simultaneous, and inexpensive determination of a wide range of soil chemicals including trace elements on a large number of samples. Linderholm and Lundberg (1994) show that elevated levels of Cu, P, Mn, Zn, and Ca are found in some archaeological sites using the ICP technique. This corresponds to the results of other studies that used more conventional laboratory analyses (summarized by Linderholm and Lundberg, 1994, p. 303). Other elements that may be useful indicators of human habitation using simultaneous elemental analyses include Ba, La, Ce, Pr, K, Ca, Th, and Rb (in historic farm sites in Scotland) (Entwistle et al., 1998) and K, Rb, and Th (in nonagricultural historic sites on the Isle of Skye, also in Scotland; Entwistle et al., 2000). Entwistle and Abrahams (1997), however, argue that major soil elements are difficult to measure by ICP-MS because of their high concentrations in anthropogenic soils.

ICP analysis has proven useful in examining chemical variations as indicators of activity areas within dwellings. Middleton and Price (1996) studied chemical variations within modern and archaeological household sites. The modern house (in the Oaxaca region of Mexico) showed elevated levels of K, Mg, and P in the hearth area, characterized by cooking ash. At the Ejutla site (200–800 A.D.; also in the Oaxaca area), analysis of a living floor in a residence revealed high levels of K and P in a cooking area, but otherwise no distinctive pattern, probably because of postdeposition alteration of the soil chemistry. At the Keatly Creek site (British Columbia, Canada; age of household unspecified), a hearth area likewise produced increased K and P along with Mn and Zn. Ca and P were also high in a food preparation area. This study illustrates the importance of comparative studies with modern settings. At the Mayan site of Piedras Negras, Guatemala, ICP was used to measure heavy metal concentrations, including sampling across patio groups, between house-mounds, and across middens (Wells et al., 2000) and within a dwelling (Parnell et al., 2002). The heavy metal and P analyses of samples from across the site combined with excavations indicate that individual trash deposits may be related to food refuse or to specific workshop, craft, or ceremonial activities and that some domestic structures may have been painted (Wells et al., 2000). Analysis of close-interval samples from within a dwelling suggests that elevated Ba and Mn, as well as P, are indicative of organic refuse disposal and that Hg and Pb concentrations are associated with craft production (Parnell et al., 2002).

Two case studies provide data and interpretations that further illustrate the complexities and ambiguities of soil chemical analyses on archaeological sites. McDowell (1988) reports the results of soil chemical studies at four archaeological sites (occupied 5000 yr B.P. to European contact) in the Elk Creek area of western Oregon (Cascade Mountains) and in two sites (occupied intermittently 9000 yr B.P. to European contact) along the western margin of the Willamette Valley, also in western Oregon. The sites in the Elk Creek area were long-term residential sites, possibly villages, in contrast to the study sites in the Willamette Valley, which were seasonal hunter-gatherer localities. The more-or-less permanent sites in the Elk Creek area showed enrichment in total P, Ca, and Mn. Areas with bone are enriched in bases and Fe. Areas without bone are not enriched in Mg, K, Na, and Fe, but may be enriched in Ca and Mn. The interpretation is that areas with high P but not significantly high levels of bases are probably human waste areas. In contrast, the hunter-gatherer sites in the Willamette Valley showed no enrichment in anthropogenic soil chemicals. This was attributed to the brief occupations that created the sites. If true, this indicates that many sites are not detectable by chemical prospection. Another possibility is that most of the anthropogenic chemicals were leached from the soils, particularly those chemicals associated with earlier occupations.

In an unusual but important study, Dormaar and Beaudoin (1991) looked at chemical signatures in the Calderwood bison jump site in Alberta, Canada. The study was unusual because it was carried out at a kill site rather than a habitation or agricultural site, where most chemical studies have been concentrated. In comparison to a natural Chernozem (Mollisol), the bone levels were enriched in SOM, soil P (total and available), N, and fatty acids. Moreover, these chemical constituents were also elevated in the zones between bone levels. This latter characteristic seemed anomalous and was attributed to downward leaching associated with translocated SOM and possibly mechanical mixing of the soil and offers a cautionary note for the application of soil chemical studies to vertically delimit discrete features.

There are several important aspects of the foregoing discussion of soil chemistry and archaeology. First, most of the sites or occupation zones that displayed some relationship between human activity and soil chemical characteristics are relatively young—typically late Holocene. Two explanations probably account for this relationship. Most of the anthropogenic elements, compounds, and other chemical characteristics (except soil P) probably do not persist in soils beyond a few thousand years. As noted above, this has been the conventional wisdom, and Middleton and Price (1996) note reduced levels of soluble elements such as Na and attribute this to leaching. Most of the sites investigated chemically also tend to be more complex and permanent (or at least of prolonged occupation), rather than hunter-gatherer sites from more transient occupations. A few studies (e.g., McDowell, 1988) indicate that the latter sites produce no chemical signatures, but data on this issue are scarce. More research is necessary to better understand the persistence of archaeologically significant chemicals in soils. Another significant issue is that on sites with long-term occupations or multiple, superimposed occupations, the chemical signatures of individual occupation periods are very difficult or impossible to sort out because of chemical translocations during and after occupation (i.e., a kind of anthropogenic soil welding; Entwistle et al., 1998).

The rest of this chapter focuses on three specific and interrelated aspects of human impacts on soils that are of considerable archaeological significance: soil P, anthrosols, and agriculture.

Soil Phosphorus and Archaeology

One of the most archaeologically significant anthropogenic alterations of soils is the addition of phosphorus (phosphorus exists in soils as the phosphate ion; some organic P compounds are not phosphates, so the term "phosphorus" should be used when referring to total soil P [Bethell and Máté, 1989, p. 5]). Phosphorus is unique among the elements in being a sensitive and persistent indicator of human activity. As noted above, a number of elements are left in the soil by humans, but only P leaves a prolonged signature of its human origins because natural and anthropogenic P tend to be strongly fixed in soils. The sources of anthropogenic phosphorus include human and animal waste; refuse derived from bone, meat, fish, and plants; burials; and manure used as fertilizer (Provan, 1971; Proudfoot, 1976; Eidt, 1984, pp. 29–30; Bethell and Máté, 1989). When people add P to the soil as organic products or inorganic compounds, the P quickly bonds with Fe, Al, or Ca ions (depending on local chemical conditions, particularly pH) to form relatively stable chemical compounds of inorganic phosphate minerals (Proudfoot, 1976; Bethell and Máté, 1989. In most soils, removal of these compounds cannot be stimulated by normal oxidation, reduction, or leaching processes, in contrast to other elements (Proudfoot, 1976; Eidt, 1977, 1984, 1985). When humans add P to the soil, therefore, it accumulates at the site of the deposition. With prolonged occupation, the accumulation of anthropogenic P can become quite large (by orders of magnitude) in comparison to the content of natural P in the soil. This is because P is the one element that is cycled mainly in geological time (Walker and Syers, 1976; Eidt, 1977, p. 1327). Other elements are cycled much more rapidly, assisted by microorganisms and plants in their cycling through the ecosystem, so the record of their association with people is lost.

The relationship between natural and anthropogenic P in the soil and the movement of P through natural and cultural environments is nicely encapsulated by Bethell and Máté (1989, p. 9): "Human activities can strongly redistribute P in soils. Plants take up P from the soil. They can be eaten by animals or harvested. The animals themselves can be moved or 'harvested'; they can be enfolded, concentrating P in a particular area. Dung residues can be collected and used as manures, respread over the fields; on the other hand they may be used as a fuel, as a walling material, or ignored. . . . As part of the produce of an economic system, P is very mobile; it's importance lies in the strong fixative powers of the soil. When P enters the soil system it is relatively immobile compared to other elements concentrated by the activities of humans."

Another factor that makes P suitable for geoarchaeological study is that anthropogenic P can exist in the pH range of most soils. Under acidic condition, P combines with iron and aluminum, whereas under basic conditions, P combines with calcium. As a consequence, soil P analysis can be used successfully in a wide variety of archaeological contexts. Indeed, where there is little or no surface evidence of human occupation, soil P analysis may be an appropriate tool for detecting traces of human activity and for determining the particular form and function associated with that presence.

The analysis of phosphorus has long been of interest to archaeologists as a means of detecting and interpreting evidence for human activity because of its relative stability in soils and its abundance in association with human activity. As a result, the literature on the topic of archaeological P is vast (e.g., table 11.4) and likely to be significantly larger than any other single aspect of soil science in archaeology. Moreover, phosphorus is important in plant growth and has been the topic of considerable research in soil science, resulting in another very large literature (e.g., table 11.4). The significance of phosphorus in archaeology and agriculture is well illustrated by accounts of Arab farmers in the Near East using soils excavated from archaeological sites to fertilize their fields (Herz and Garrison, 1998, p. 181). Understanding phosphorus chemistry is a key to understanding the principals, applications, and pitfalls of P analysis in archaeological contexts and is summarized in appendix 2. Discussion of some of the principal analytical procedures for determining soil P in archaeological and natural settings is provided below and also in appendix 2.

Soil P Methods in Archaeological Contexts

All soil phosphorus derives from the weathering of phosphate minerals, especially apatite, in soil parent material (Stevenson and Cole, 1999, pp. 282–283; appendix 2). The phosphorus then cycles from soil to plants to animals and back to soil, or it is lost by leaching. Human activity that disrupts these cycles leads to losses or gains in phosphorus relative to the P content of the local natural soils. Most geoarchaeological research on P deals with these losses and gains. In particular, because human activity imparts so much more P to soils than is usually found naturally, and because of the low solubility and relative stability and persistence of P, most archaeologically oriented P studies focus on gains in P as clues to site locations or on intrasite variability in P content as clues to site use.

Research by archaeologists, geoarchaeologists, and soil chemists since the middle of the 20th century has resulted in a bewildering array of terms for referring to soil P. In part this is because of the various forms of phosphorus in the soil, and in part it is because of the different chemical fractions that can be extracted both in field tests and in lab analyses. The resulting nomenclature refers to P in terms of its chemical make-up and its place in the geo- or biogeochemical environment (organic P, inorganic P, and total P) or in terms of chemical extractions or fractions of P (e.g., available P, occluded P). Understanding the difference between the forms of P and the extractions or fractions of P is a key to understanding soil P in archaeological contexts.

Phosphorus occurs in soils in various forms: as ions and compounds in the soil solution; adsorbed on the surfaces of inorganic soil constituents; in phosphate minerals, both crystalline and amorphous (these three forms of P are all "inorganic P"); and as a component of soil organic matter ("organic P") (Barber, 1995, p. 203; Stevenson and Cole, 1999, p. 292; appendix 2). "Total P" is inorganic P plus organic P. For the most part, the chemical extracts represent some portion of inorganic P; "most commonly used extractants do not yield clean separations nor discrete groupings of P forms in soils" (Smeck, 1985, p. 186). This is not always understood, and misstatements regarding the nature of soil P in archaeological contexts are all too common. As discussed below and in appendix 2, some chemical fractions measured in the laboratory appear to approximate specific forms of P (e.g., Chang and Jackson, 1957a; Eidt, 1984), but others clearly do not. A very important component of geoarchaeological P research is to spell out the procedures used, to understand what fraction is extracted and for comparison with the work of others.

The study of archaeological phosphorus evolved throughout the 20th century (well summarized by Woods, 1975, 2004; Eidt, 1984; and Bethell and Máté, 1989). Most of the early research was by O. Arrehnius and W. Lorch (table 11.4), working in northwestern Europe. After the Second World War, British and American investigators picked up the technique and applied some of the methods in their sites (e.g., R. Solecki, K. D. M. Dauncy, E. F. Dietz, and G. E. G. Mattingly and R. J. B. Williams; table 11.4). Following this pioneering work, there were several landmark studies that influenced most subsequent phosphorus work. Cook and Heizer (1965) published what still remains one of the most comprehensive and extensive studies of soil P and other elements in archaeological contexts. They published data on sites in the western United States and Mexico. Unfortunately, they do not discuss their results in terms of different forms or fractions of P, nor do they describe their methods. However, they provided the first systematic discussion of soil chemistry, including P chemistry, with an archaeological focus, and showed that P in archaeological sites must be considered relative to other elements and to the environments of deposition.

Archaeological P studies gained further attention in the 1970s with the rapid expansion of methods and applications, particularly the work of Eidt (1973, 1977, 1984, 1985), Eidt and Woods (1974), and Woods (1975, 1977), and the review paper by Proudfoot (1976). Of particular significance in this work was the recognition and incorporation of the extensive research into soil P and P fractionation by soil scientists, especially the work of S. C. Chang, M. L. Jackson, F. J. Stevenson, J. K. Syers, N. Smeck, and T. W. Walker (tables 11.4 and A2.1).

The various archaeological approaches to soil P analyses can be grouped into a number of basic categories, depending on how the methods are segregated (following Gurney, 1985, pp. 2–3; Bethell and Máté, 1989, pp. 10–13; and Terry et al., 2000, pp. 152–153; see also discussion in appendix 2): Available P (Pav); the spot test or ring test; total P (Ptot); inorganic P (Pin) extraction, sequential P fractionation, which separates and measures different compounds of soil P; and organic P (Porg). Much of the archaeological interest in soil P has focused on available P. This probably is because of the wide variety of relatively easy techniques for identifying or measuring Pav and because essentially all of the early work on P in archaeology focused on Pav (e.g., Arrehnius, Lorch, Dauncy, Solecki, Eddy and Dregne, Cook and Heizer in tables 11.4 and A2.1) Available P is a measure of plant-nutrient availability (i.e., it is an estimate of the small amount of P available for plant growth) developed in the agricultural sciences (Eidt, 1984, p. 35; Bethell and Máté, 1989, p. 6). Furthermore, it is not a measure of a single simple chemical, compound, or mineral (Bethell and Máté, 1989, p. 6). Soil scientists asked to analyze the P content of archaeological soils but who are not otherwise familiar with geoarchaeological techniques or questions typically provide data on the easily available P. It may provide a rough indication of P content, but it "does not correlate with any particular fraction which actually exists in nature" (Bethell and Máté, 1989, p. 6). The variety of methods for extracting Pav (appendix 2) also yield different amounts of P. For those reasons and several others (summarized by Hamond, 1983, pp. 61–62) Pav may not be the best archaeological indicator. It may work in drier environments, such as the southwestern United States (e.g., Homburg and Sandor, 1997), but not in wetter, leaching environments (e.g., Hamond, 1983).

The early geoarchaeological work on soil P dealt with citrate-soluble P, which is one of the more easily extractable forms of P (appendix 2). Some investigators have questioned whether easily extractable Pav data have any utility in geoarchaeological research (summarized by Bethell and Máté, 1989, p. 11). In spite of the reservations about the utility or meaning of the easily extractable P, it does seem to be broadly indicative of human activity. In one of the earlier applications, Eddy and Dregne (1964) measured the Pav extracted by a simple water wash (water-extractable P; see appendix 2). Their data correlate well with occupation zones in late prehistoric sites in southwestern Colorado and northwestern

New Mexico. Furthermore, the requirement for elevated levels of citrate-soluble P_2O_5 in the Anthropic epipedon of soil taxonomy is a measure of easily extractable Pav (Bethell and Máté, 1989, p. 11; Macphail et al., 2000, p. 72), and these can be quite high in archaeological contexts (P_2O_5 does not exist in nature; the formula is a convention used for fertilizer recommendations). Macphail et al. (2000, p. 72) illustrate this for several sites in Britain. At the Wilson-Leonard site in Central Texas (chapter 7), levels of citrate-soluble Pav from occupation zones were significantly higher than in nonoccupation zones (table 11.5).

An important step in the analysis of P was development of the spot test or ring test (or "Gundlach method," table 11.4) for quick field evaluation of P levels on archaeological sites (Bethell and Máté, 1989, p. 12). The method tests for easily extractable Pav. As a result, the meaning of the spot test can be ambiguous (summarized by Hamond, 1983, pp. 55–61), given the vagaries of Pav interpretation noted above and in appendix 2. Furthermore, the results are qualitative and not always reproducible (Eidt, 1977, 1984, pp. 36–38; Gurney, 1985, p. 2). These drawbacks have lead to outright rejection of the method by some archaeologists (e.g., Sjöberg, 1976, p. 451). Others have taken a more realistic approach and recognized the utility of the method given its simplicity and portability, but also its limitations, and view the method as an important component of field investigations (e.g., Hamond, 1983, p. 61; Gurney, 1985, p. 2; Bethell and Máté, 1989, p. 12). Lippi (1988), for example, applied the test at the remote site of Nambillo in Ecuador. A systematic coring strategy was used to establish the site stratigraphy and identify buried landforms. Soil samples recovered during coring were subjected to in-field phosphate analysis, and the results were used to identify areas of human activity. These data were then used to design an excavation strategy for the site. Bjelajac et al. (1996) also showed how the spot test could be calibrated at known sites to determine a minimum "site value" in a given region and then be used locally to aid in identifying or delimiting other sites.

Another significant improvement in P analysis was development of a quantitative procedure based on the extraction of P and measurement using colorimetry (Bethell and Máté, 1989, p. 12). Similar to the spot test, the method is relatively portable, quick, and easy, but it is also quantitative, and so is a popular technique for on-site analyses (e.g., Provan, 1971; Hassan, 1981; Craddock et al., 1985; Gurney, 1985; Terry et al., 2000; Wells et al., 2000; Parnell et al., 2001). A wide variety of procedures for extractions and measurement are now available (appendix 2). As a result, the forms of P extracted can vary considerably. Many of the methods measure Pav (Bethell and Máté, 1989, p. 12), well described by Terry et al. (2000, p. 153) as "soluble and readily labile P." Terry et al. (2000, p. 153) further note that their extraction, "is not always proportional to the total P of the soil; however, for archaeological prospection and activity area research, the spatial patterns of phosphate levels are important, rather than the absolute concentration." That is probably a fair comment, given the many variables that affect P levels in soil (e.g., Proudfoot, 1976; White, 1978; Bethell and Máté, 1989; Barber, 1995; Stevenson and Cole, 1999).

Once geoarchaeologists began following the P research from soil science, work on soil P in archaeology began to focus on total P (Bethell and Máté, 1989). Measurement of Ptot produces quantitative, comparable results, in contrast to many

Table 11.5. Citrate-extractable soil P and organic carbon from the Wilson-Leonard site, Texas

Stratum,[1] unit[2] and depth, cm	Soil horizon	Archaeology	P_2O_5[3] ppm	%O.C.[4]
Profile 2				
6				
IIIc				
0–9	Ap	Late Prehistoric	1145	0.90
9–28	A1	Late Archaic, with burned-	1145	0.78
28–50	A2	rock midden accumulation	870	0.67
IIIb				
50–78	BA	Middle Archaic, with burned-	1053	0.34
78–104	Bw1	rock midden accumulation,	756	0.23
104–129	Bw2	multiple phases	1351	0.18
129–158	Bw3		1443	0.20
5				
IIIa				
158–178	Ab1	Early Archaic, multiple	1328	0.20
IIb		phases[5]		
178–192	Bwb1		824	0.12
4				
IIa				
192–213	2Bw1b1	Late Paleoindian	870	0.14
213–230	2Bw2b1		756	0.16
3				
Ic,d				
230–260	3Bwb2		1145	0.14
2				
Ib				
260–276	4Ab3	Late Paleoindian (9500 yr	916	0.12
276+	4Cgb3	B.P.) burial	527	0.09
Profile 9 "Area B"				
6				
IIIc				
0–35	A1	Late Prehistoric	1145	1.86
35–64	A2	Late Archaic	1443	1.33
IIIb				
64–80	Ab1	Middle Archaic with multiple	1580	0.55
80–96	BAb1	occupations and burned rock	1489	0.49
96–115	Bw1b1	midden	1008	0.28
115–170	Bw2b1		710	0.14
5				
IIIa				
170–200	Bw3b1		458	0.12
IIb				
200–224	Bkb1		458	0.12
224–238	C1b1		504	0.05
4				
IIa				
238–260	2C2b1		344	0.07
260–298	3C3b1		424	0.20
2				
Ib				
298–313	4Ab2		504	0.17

From Goldberg and Holliday (1998).

[1] From Holliday (1992b).

[2] From Goldberg and Holliday (1998).

[3] P_2O_5 by citric acid extraction; expressed as P by Goldberg and Holliday (1998, table 6-1), which provide lower values than P_2O_5.

[4] O.C.(= organic carbon) by titration.

[5] Includes Gower, dated to 7400 yr B.P., associated with Unit IIIa, and Angostura, dated to 8800 yr B.P., associated with Unit IIb.

measures of Pav or the spot test, and may be the best indicator of human inputs of P when comparisons are made with natural soils (Bethell and Máté, 1989, p. 20). For example, following a comparison of methods, Skinner (1986) concluded that total P produced the highest correlation with anthrosols, but was positive only 60% of the time (appendix 2). Alternatively, Terry et al. (2000) present data that indicate that inorganic P extractions may be more sensitive to human inputs than Ptot. In a related study, Parnell et al. (2002) argue that extractable P is indeed more sensitive than Ptot. This may be because a much higher proportion of Pav will come from human input (see appendix 2), whereas total P includes all mineral P, which can be significantly higher than anthropogenic P. Until the late 1980s and 1990s, however, Ptot analysis in archaeology was limited because the procedures involve strong and dangerous reagents and are also time consuming (appendix 2; Conway, 1983; Gurney, 1985, p. 3; Bethell and Máté, 1989, pp. 12–13; Forster, 1995, p. 88). Furthermore, some methods described as extracting total P probably do not do so (appendix 2).

With increased availability of ICP spectrometry, however, measurement of Ptot became more common (e.g., McDowell, 1988; Linderholm and Lundberg, 1994; Entwistle and Abrahams, 1997; Entwistle et al., 1998, 2000). The ICP work is almost always done in the context of multielement analyses, and unfortunately, most of the methods discussions do not specify or otherwise deal with the form of P being analyzed, though some inferences are possible based on the extraction procedure (appendix 2). Methods based on digestion using hydroflouric acid (HF) or perchloric acid ($HClO_3$) (e.g., Linderholm and Lundberg, 1994; Entwistle and Abrahams, 1997; Entwistle et al., 1998, 2000) probably yield Ptot because the chemicals totally digest the sample and because perchloric extraction is a common method for Ptot (appendix 2). Thus, these measurements of P include total mineral P. Extracts made from nitric acid (HNO_3) or hydrochloric acid (HCl; e.g., Linderholm and Lundberg, 1994; Middleton and Price, 1996) probably yield some form of extractable P (appendix 2). ICP analysis is relatively straightforward and efficient and, with the growing availability of ICP spectrometers, an important component of soil chemical analysis. Those interested in using the ICP method for anthropogenic P analyses should be aware of potential problems, however, having to do with high levels of anthropogenic P and interference of P with other elements (appendix 2).

Few studies focus on the relationship of organic P (Porg) to human occupation. Porg represents a large part of the total P pool, and human activity can produce organic P. Of the archaeological studies that have determined both Porg and Ptot, Porg (as a percentage of Ptot) was low in the soils with archaeological contexts (summarized by Bethell and Máté, 1989, p. 18). This is attributed to the high content of P derived from bone (hence high inorganic P) in most archaeological sites. Porg also mineralizes relatively rapidly. Higher Porg to Pin ratios do seem to be associated with crop residues. Courty and Nørnberg (1985) and Engelmark and Linderholm (1996) provide two of the few studies of Porg. In comparisons with uncultivated soil they found elevated levels of Porg in abandoned agricultural fields.

The fractionation method developed by Eidt (1977, 1984) on the basis of the Chang and Jackson (1957a) procedure (as modified by Williams et al., 1971) probably generated more interest and controversy than any other single P procedure in archaeology (e.g., Hamond, 1983; Knapp, 1985; Bethell and Máté, 1989; Leonardi et al., 1999). The fractionation method also has had relatively minimal application, probably because it is labor-intensive, time consuming, and expensive (e.g., Gurney, 1985, p. 3; Homburg and Sandor, 1997, p. 142; Leonardi et al., 1999, p. 352) and because of questions concerning the meaning of the results. Eidt's (1977, p. 1328; 1984, pp. 40–42) approach (discussed further in appendix 2) is based on the idea that total inorganic P (Pti) is the best indicator of anthropogenic activity. The basic theory is probably sound: Human inputs of P-bearing materials are probably quickly converted to inorganic P. High levels of Pti are reported from archaeological sites in comparison to local natural soils (Lillios, 1992; Kerr, 1995; Schuldenrein, 1995). Eidt (1984, pp. 41, 43) further proposed that sequential extraction of various forms of inorganic P provide a measure of total inorganic P and that this measure may reveal clues to human activity based on a purported close correlation between land use and inorganic P levels. This issue is more problematic. The basic fractionation scheme involves extraction of: Fraction I, solution P and loosely bound Al and Fe phosphates; Fraction II, tightly bound or occluded forms of Al and Fe oxides and hydrous oxides; and Fraction III, occluded Ca phosphates (Eidt, 1984, p. 42). The sum of the three fractions should be total inorganic P (but probably is not; appendix 2).

Eidt (1984, p. 43), building on the work of Lorch (1930, 1939, 1940, 1954) proposed that low levels of Pti (10–220 ppm) corresponded to ranching and farming; moderate levels of Pti (200–2000 ppm) to more intense activities as would be found around dwellings, gardens, and manufacturing areas; and very high levels (>2000 ppm) to burials, garbage pits, slaughter areas, and urbanized zones. Eidt (1984, p. 43) also noted that the P data could be used to identify crop and forest types All of these correlations were asserted, not demonstrated. In field studies (Eidt, 1984, pp. 55–72, 87–106) data on Pti from archaeological zones for less than a dozen samples from contemporary gardens and residences were used to infer crop or plant types (e.g., manioc, yucca, or rice). Clearly, the sample sizes are too small to justify such conclusions. Much more information is needed on the range of variation of Pti for different types of activities and for natural soils in any given study area and on the nature of the soils associated with the contemporary and archaeological activity areas; for example, mineralogy and pH. Lillios (1992), for example, gathered a sizeable data set on Pti for contemporary vegetation before trying to interpret her archaeological Pti data.

One important issue raised by Eidt's work is that of changes in P forms through time. This characteristic of Pti is well documented in the soil chemistry literature (see discussion in appendix 2) and clearly shows that the assumption that P is stable in soils is not completely valid. At human and generational timescales P is stable, but at millennial timescale acid-extractable P (a form of inorganic P, including Pav) is depleted with time, thereby increasing total P. Edit (1977, 1984) proposed using a variation of this characteristic as a relative dating tool, specifically by comparing the ratio of Fraction II to Fraction I. Through time, the weakly bound (and therefore most easily weatherable) P is depleted, which increases the II/I ratio (it should also increase total P levels). Few archaeologists have used this approach, but the available examples (Lillios, 1992; Schuldenrein,

1995) show that it has promise. However, using changes in P forms for dating sites or for comparing the age of several occupations requires that anthropogenic additions to the soil body ceased throughout the site or occupation area at the same time, the additions were relatively consistent throughout the site area in their composition, and occupations are not superimposed or welded (W. Woods, personal communication, 2002).

Soil P Applications in Archaeological Contexts

The decades of soil P analyses in archaeology have resulted in a very large set of data from a variety of environments that illustrate or suggest a number of spatial relationships of P in archaeological contexts. Some of these relationships were noted in the above discussion, but further elaboration is necessary. Most studies of soil P in archaeology fall into three broad categories (summarized by Terry et al., 2000, pp. 152–153, and Parnell et al., 2001, p. 857, among others): 1) prospecting to locate or delimit sites, 2) the identification or delineation of features and activity areas, and 3) investigation of past agricultural practices (discussed in the section on "Agriculture" below). Soil P as a prospection tool is the oldest application of this method. Bethell and Máté (1989, p. 14) suggest that the pioneering work of Arrhenius (1934) is the only example of using soil P data to locate archaeological sites that were not identified by other means. They overstate the case (e.g., Schwarz, 1967; Thurston, 2001), but most research that includes analysis for soil P both off-site and on-site is for gauging the degree of human impacts or for defining the limits of sites or individual occupation zones (e.g., Mattingly and Williams, 1962; Provan, 1971; Davidson, 1973; Griffith, 1980; Hamond, 1983; Konrad et al., 1983; Woods, 1984; Gurney, 1985; Cavanagh et al., 1988; Dormarr and Beaudoin, 1991; Lillios, 1992). The most valuable studies in this regard have been the use of soil P to find "sites unseen." For example, in East Anglia, Craddock et al. (1985) used soil P analysis along with other more traditional methods of survey (e.g., field walking and soil magnetism) to locate shallowly buried sites. The P signal of the unseen site was preserved or mixed upwardly into the plow zone of the topsoil. A reversal of this approach was applied at the Iva Wright site in Ohio, a low-density, plow-disturbed site on an alluvial terrace (Skinner, 1986). Samples from below the plow zone that contained the site showed significantly elevated levels of soil P, allowing delimitation of a site that was otherwise very difficult to define spatially.

Vertical delimiting of occupations in stratified sediments using P has been successful in some situations (e.g., Konrad et al., 1983; McDowell, 1988), but not all. The Mabel Hall site, Ohio, is a multicomponent site in alluvium. Analysis of soil P could not differentiate occupation zones from sterile deposits (Skinner, 1986). This was in part attributed to frequent flooding of the site. Dormarr and Beaudoin (1991) noted a similar problem at the Calderwood bison jump site in Alberta, Canada, noted earlier. That site was not subjected to flooding. Davidson (1973) reported significantly elevated P levels from throughout a tell in Greece, even in zones with no obvious indication of occupation. The high P is attributed to human activity, some of which left no obvious signs, however. The data from Alberta and Ohio raise the possibility that the phosphorus signal from the tell may be caused by mixing or translocation. Schlezinger and Howes (2000), using data from the Carns site on Cape Cod, make a strong case for using Porg as an indicator of vertical site limits. This is because Porg is less mobile than Ptot and thus less likely to be translocated in a soil profile. These studies further indicate that investigators should not automatically assume that P is as stable and immobile in soils, as is sometimes touted.

Within-site variation of soil P has long been studied for identification and understanding of different kinds of activity areas and for directing excavation strategies (e.g., Dietz, 1957; Cook and Heizer, 1965; Heidenreich et al., 1971; Provan, 1971; Ahler, 1973; Griffith, 1981; Carr, 1982, pp. 112–115, 515–524; Conway, 1983; Hamond, 1983; Konrad et al., 1983; Cavanagh et al., 1988; Lippi, 1988; McDowell, 1988; Moore and Denton, 1988; Kerr, 1995; Schuldenrein, 1995; Manzanilla, 1996a,b; Middleton and Price, 1996; Sanchez et al., 1996; Parnell et al., 2001, 2002). The sensitivity of individual P fractions for detection of specific activity areas or types of features is unclear, however. Soil P seems well suited for detecting areas that were high in bone or for comparing swept versus dump areas (e.g., Dietz, 1957; Ahler, 1973; Carr, 1982, pp. 515–524; McDowell, 1988). At Piedras Negras, Guatemala, Parnell et al. (2002) found that total P and especially extractable P appear to be good indicators of food preparation and food consumption areas. In contrast, Kerr (1995) statistically analyzed the distribution of P fractions (based on Chang and Jackson, 1957a, and Eidt, 1977, 1984) at the Huntsville site, in Arkansas, and found no significant relationship with activity areas. As noted above, some evidence indicates that Porg is higher in old arable fields compared with uncultivated soils.

One promising approach to using P data for understanding activities is to look at ratios of various forms and fractions of P. The "P ratio" of Macphail et al. (2000) is the ratio of total P to extractable P (see also Engelmark and Linderholm, 1996). The ratio is $\leq$1.0 when the P is dominantly inorganic and $>$1.0 when the P is dominantly organic. Furthermore, ratios around 1.0 tend to be associated with dwelling sites, whereas ratios of 1.5 to 10 are found in stabling areas and manured fields. Engelmark and Linderholm (1996), Macphail (1998), Macphail et al. (2000), and Macphail and Cruise (2001) used citric acid for the extractable P and the ignition method for total P (table A2.1; though the ignition method probably does not measure true total P; appendix 2). There is no indication whether alternate methods of measuring P would yield different results. Further experimentation is warranted.

In summary, data on soil P has been widely used by archaeologists and geoarchaeologists for a variety of purposes. Bethell and Máté (1989, pp. 14, 16, 17) present a rather negative assessment of much of this work, noting that it has tended to support conclusions already drawn, has been used for a kind of "fishing expedition" (application of soil P analysis just to see what would turn up), and has rarely been used to find unseen and unknown sites. These are not unreasonable statements, but soil P analyses do seem to have some usefulness. For example, they can be used in helping to direct excavations by locating activity areas and for delimiting site areas. Levels of Ptot and Pin seem to be the best indicators of human activity. Quantitative field extractions are best for fieldwork, but rapid lab analyses (e.g., various ICP techniques) are now commonly available as well. Specific fractions of P may also be indicative of specific kinds of human activity, but a more promising approach is the use of P ratios. However, much more empirical data need to be gathered before specific relationships can be confidently offered. In particular, a variety of factors must be taken into consideration, including the chemistry of the original soils and sediments, the duration of pedogenesis, and the landscape position. The specific laboratory procedures used to extract and measure soil P must also be considered when assessing soil P data. Interpretations of soil P generally seem strongest when supported by other information such as elemental and SOM data and soil magnetism.

Anthrosols

Anthrosols of one sort or another can be found in most places on the Earth occupied by humans, but studies focused specifically on anthrosols tend to be in areas with landscapes more intensively modified than others. Probably the largest literature and most data come from Europe and Great Britain. Human impacts on soils in these regions are so pervasive that anthrosols are a significant component of local soil classification systems. In the soil survey of Britain, for example, "Man-Made Soils" are one of the major Soil Groups (equivalent to the Order level in the United States; Avery, 1990), and "Cultosols" are a significant subgroup. "Anthrosols" are also a primary subdivision of the soil classification system in China (e.g., Zhang et al., 2003). Given the wide geoarchaeological interest in anthrosols, there are surprisingly few systematic discussions of anthrosol research or characteristics.

Anthrosols vary widely in their physical and chemical characteristics. Few traits are universal. There are several characteristics that are common or that serve as clues to soils significantly modified by human activity. The most obvious is the presence of archaeological debris within the soil—in particular, organic detritus such as bone and charcoal in a surface horizon; that is, they tend to be associated with middens. Other physical features typical of surface horizons include abrupt, smooth boundaries between horizons or layers; abrupt, laterally discontinuous layers; and dark matrix colors (low value and chroma) extending to greater-than-expected depths for natural soils in the area (following Collins and Shapiro, 1987). The greater-than-expected depth is usually caused by artificial upbuilding and phosphorus. Chemical signatures include higher-than-expected values of organic matter relative to natural soils (discussed earlier and in appendix 2). Anthrosols also have been subjected to some form of pedogenic alteration, albeit relatively minor pedogenesis in many instances. The high content of organic matter, dark colors, and overthickened character of many anthrosols has resulted in their generic description as "Dark Earth" (e.g., Woods and McCann, 1999). As noted below, however, this term has been used in a wide variety of settings and means different things to different workers.

Although the category of anthrosols can include a wide array of soils, three types of anthrosols have been described at some length, if not more or less formally recognized: Plaggen, Dark Earths, and Terra Preta. Various other kinds of middens may also qualify as anthrosols. "Plaggen soils" are perhaps the most intensely studied of anthrosols, at least in a pedological context. Plaggen soils are those which "for centuries, have been fertilized with a mixture of manure and sods, litter or sand," such that, "the original soil has been buried under a humose sand cover of varying thickness" (Pape, 1970, p. 229). They are most common on the sandy landscapes of The Netherlands, Germany, and Belgium, but similar soils are reported from other parts of northern Europe and Great Britain (Conry, 1971, 1972; van de Westeringh, 1988; Simpson, 1997), Crete (Bull et al., 2001), Peru (Sandor, 1987; Sandor and Eash, 1995), and New Zealand (McFadgen, 1980a,b). Though not documented in North America, their ubiquity in Europe led to adoption of the "plaggen epipedon" in the U.S. soil taxonomy (Soil Survey Staff, 1975, 1999; Forbes, 1986, pp. 91–92). This diagnostic horizon is defined as a "human-made surface layer 50 cm or more thick that has been produced by long-continued manuring" (Soil Survey Staff, 1999, p. 26). This horizon is unusual in soil taxonomy in that the presence of artifacts is one of the defining characteristics. As they are defined in soil taxonomy, the plaggen epipedon is not as fertile as the anthropic epipedon. The anthropic horizon is viewed as originating from habitation debris high in organic matter and phosphates, whereas the plaggen is believed to result simply from manuring (Brasfield, 1984, pp. 13–14; Forbes, 1986, pp. 91–92, 101–102).

The plaggen and related soils share several characteristics that are used to identify them (table 11.6; Pape, 1970; van de Westeringh, 1988). They vary in color from dark gray or black to brown, probably because of different vegetation mixed with the manure from area to area (e.g., heather sod, grass sod, forest litter, or peat litter). Any one plaggen, however, will be homogeneous in color (the uniform color is believed related to a desire for uniform fertility). The plaggen horizon typically is sandy, but ultimately its texture is determined by the texture of the subsoil that produced the bedding. A buried soil typically is preserved below the plaggen, and the plaggen itself will be readily distinguishable by being significantly thicker (typically >50 cm) than the surface horizon (typically <30 cm) of nonanthropogenic soils in the same area. Chemically they will be relatively high in phosphate and have a low pH. Artifacts such as charcoal, brick and pottery fragments, and burned soil are common in plaggen. Recognition of plaggen soils and their geomorphic setting is a useful tool in local archaeological surveys (Dekker and De Weerd, 1973).

Several slightly different scenarios are proposed for the origin of plaggens. By all accounts, plaggen soils developed in the Middle Ages (probably around the 10th century; Pape, 1970; Heidinga, 1988; van de Westeringh, 1988). Manure was the preferred fertilizer, so to gather it, the floors of stables were strewn with forest litter, heather turves (slabs of heather cut from the ground), or grass sod to absorb the droppings from sheep and cattle. The mixture of manure, bedding, and mineral matter was then hauled out and strewn on fields. The mineral material brought in with the bedding sometimes provided additional nutrients. The mixture of manure, bedding, and mineral matter also increased the water-holding capacity of the otherwise excessively drained soils. The nutrients in manure alone were easily flushed through the sandy soils in the wet climate of northern Europe. The traditional interpretation of plaggen origins is that they resulted from a need

Table 11.6. Plaggen soils from Gelderland, The Netherlands

Horizon[1]	Depth, cm	Description
Black plaggen		
Aan1	0–45	Black (10YR 2/1), very humose fine sand; homogeneous with fine specks of charcoal; some bleached sand grains and small pieces of burned loam
Aan2	45–75	Very dark gray (10YR 2.5/1), very humose fine sand; homogeneous with some pieces of charcoal and more bleached sand grains than Aanl
Aan3	75–110	Very dark gray brown (10YR 3.5/1.5), very humose loamy fine sand; homogeneous with some charcoal, pieces of burned loam and sand bleached grains
Ab	110–135	Very dark gray brown (10YR 3/2), moderately humose, fine sand; abundant charcoal and bleached sand grains; buried A horizon of a weakly developed podzol
Bwb	135+	Light yellowish brown (10YR 6/4), fine sand; very poor in humus; buried, weakly developed B horizon of a humus-podzol
Brown plaggen		
Aan1	0–40	Very dark gray brown (10YR 3/2), moderately humose, loamy fine sand; homogeneous, somewhat clayey horizon with some charcoal and prominent grains of bleached sand
Aan2	40–75	Very dark gray brown (10YR 2.5/2), moderately humose, loamy fine sand; at the bottom some weak gray mottles, scattered charcoal and small iron concretions
Cg	75–100	Yellowish brown (10YR 5/6) sands; moderately poor in humus, with some rust spots (5YR 5/8)

Modified from Pape (1970, p. 236).

[1] Horizon nomenclature modified to conform with table 1.1 and appendix 1. Aan is anthropogenic A horizon (see appendix 1).

to restore or increase soil fertility as populations expanded in northern Europe (Pape, 1970; Heidinga, 1988; van de Westeringh, 1988). Heidinga (1988) notes, however, that the first appearance of plaggens is coincident with a time of significant drought across northern Europe. He argues that the development of the plaggen increased water-holding capacity and also deepened the plow zone, thus minimizing crop failure.

A variety of methods have been used to decipher plaggen genesis. Soil micromorphology combined with palynology has been used to investigate the sequential development of a plaggen in the Netherlands (Mücher et al., 1990). The work shows that the plaggen evolved for over 600 yr and was built of successive vegetation types (both wild and cultivated). The vegetation succession in the soil nicely reflects the vegetation succession recorded in a peat adjacent to the plaggen site, documenting prolonged local harvesting of the peat to be used as bedding. The thin-section analysis also showed that the only preserved plant remains are recent roots, indicating that most of the organic matter decomposition and humification of the plaggen material occurred before it was dumped on the field.

Stable carbon isotopes proved useful in unraveling the origins of two different kinds of plaggen-like soils on the Isle of Orkney, Scotland (Simpson, 1985). Two types of anthropogenic sedimentation were known from the island: "deep topsoils," which are essentially the same as the classic plaggen, and "farm mounds," composed largely of farm manure and thus genetically similar to plaggen. These two soils represent end members of two kinds of anthropogenic sedimentary systems, raising questions about early agricultural activity. Of particular interest was when or if the farmers added marine resources such as seaweed, which has been important in recent centuries. Simpson (1985) used stable $\delta^{13}C$ signatures from the two different soils to provide clues to their origins. Most of the samples from both the deep topsoil and the farm mounds yielded values in the range of –25‰ to –30‰, indicative of terrestrial sources (turf and manure), though some levels in the farm mound were around –19‰ and probably represent additions of seaweed (supported by recovery of weather shell from the same levels). These data indicate that reliance on marine resources is a relatively recent characteristic of local agricultural practices.

Another type of anthropogenic soil is the "Dark Earth," common in cities throughout much of Europe ("Urbic Anthrosols" of FAO-UNESCO, 1994). "Dark Earth" is a term applied to dark-colored, seemingly homogeneous urban deposits. In many ways they can be considered anthropogenic sediments rather than soil, but they have undergone surface weathering and are typically considered a soil (e.g., Macphail, 1987; Courty et al., 1989) and are so considered here. According to Courty et al. (1989, p. 261), "they are common to most cities with long histories, especially those in Europe of Medieval and earlier ancestry" (Courty, 1989, p. 261). In Britain these soils are linked to late- or post-Roman, Saxon, Viking, Medieval, and perhaps post-Medieval occupation (fig. 11.3). General characteristics of Dark Earths include "an exceedingly uniform color" of from dark grayish brown (10YR 4/2) dry to very dark gray (10YR 3/11) moist, mildly alkaline pH, some $CaCO_3$ (<10%), 1%–2% organic carbon, some phosphate, and abundant midden debris (table 11.7; figs. 11.2 and 11.3; Courty et al., 1989, p. 262).

Macphail (1983; 1987, p. 354) distinguishes two types of Dark Earth: organic-rich, sometimes waterlogged soils containing "cess material" probably representing refuse continuously dumped in a densely occupied urban environment; and a well-drained soil probably deliberately dumped for within-wall cultivation (fig. 11.3). Differentiating these types of Dark Earths and their genesis is important for understanding human activity in urban environments. Moreover, the late-Roman to Medieval periods are poorly represented archaeologically in many cities in Britain, and the Dark Earth soils provide the best clues to human activity during that interval (Macphail and Courty, 1985). The Dark Earths can contain exceptionally well-preserved archaeological materials, but environmental analyses have been "exceptionally unrewarding" (Courty et al., 1989, p. 262). In part this is because of poor preservation conditions (high alkalinity and repeated oxidation caused by water table fluctuations) and in part because of the homogeneous nature of the Dark Earth at a macromorphological scale.

Micromorphology has proven very useful in understanding the origins of Dark Earth (fig. 11.2) (Macphail, 1981, 1983, 1994; Macphail and Courty, 1985; Courty

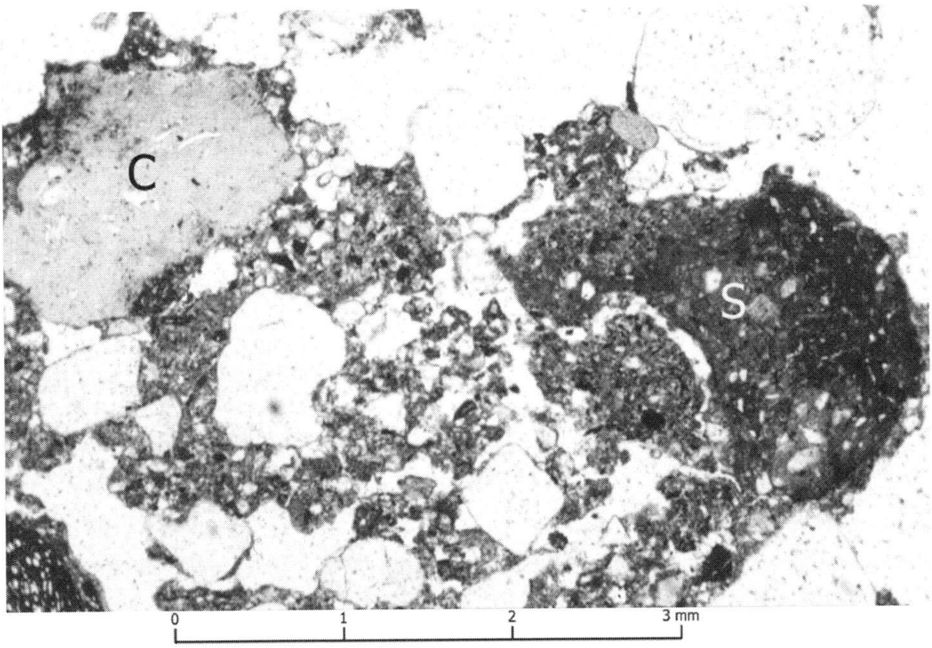

Figure 11.2 Thin-section photomicrograph (under plain polarized light) from a Late Roman (late 4th century AD) Dark Earth at Deansway, Worcester, containing midden debris (provided by and published with permission of R. I. Macphail). Portion of a human coprolite (C) is at left. A burned soil clast (S) is left. Phosphatized midden debris is scattered throughout (based on examination under blue-light autoflourescence).

Table 11.7. Dark Earth from the City of London

Unit	Depth, cm	Description
Dark earth	0–70	Very dark gray (10YR 3/1m 4/2d) sandy clay loam; weakly developed coarse subangular blocky structure; apparently humose; 20% gravel, oyster shell, bone, pottery, brickearth and mortar fragments; common charcoal; common root holes and earthworm channels; clear, generally smooth boundary
Pale dark earth	70–85	Very dark grayish-brown (10YRm 4/3d) massive sandy clay loam; abundant brickearth and mortar fragments; few biological channels; generally sharp, smooth boundary
Roman level	85–95	Strong brown (7.5YR 5/6) massive silt (brickearth) floor or walls; or pale brown (10YR 6/3) *opus signinum* or mortar foundation (for tesselated floor) or mixed collapse; generally sharp, smooth or wavy boundary
Alluvium	95+	Sand and gravel

Modified from Courty et al. (1989, p. 263). See also figure 11.3.

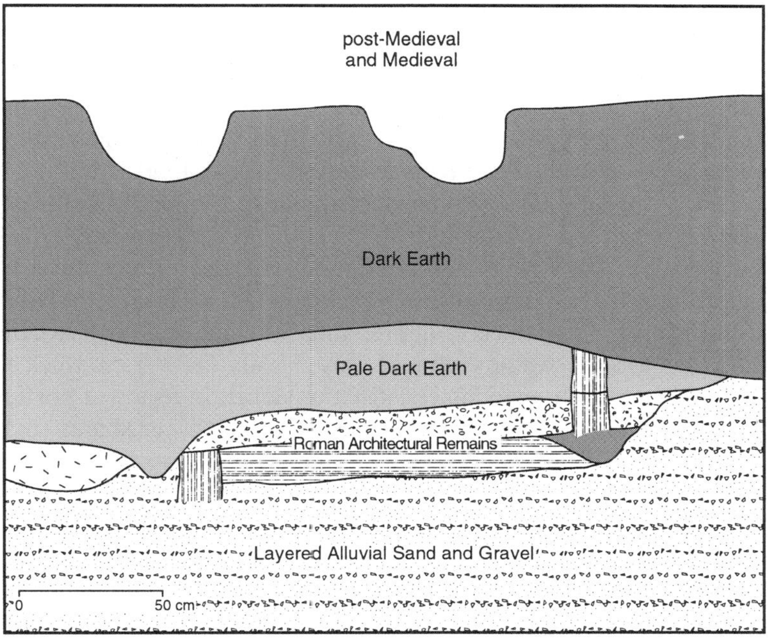

Figure 11.3 Hypothetical stratigraphic section of a Dark Earth in the City of London (modified from Courty et al., 1989, fig. 15.1).

et al., 1989; Davidson et al., 1992; Macphail and Goldberg, 1995; Macphail and Cruise, 2000, 2001). In a study of the Dark Earth in the City of London, Macphail and Courty (1985) differentiated a lower "pale Dark Earth" and an upper Dark Earth proper (table 11.7; fig. 11.3). The micromorphology showed that the lower, pale Dark Earth was a mixture of sand from the underlying alluvium and both fine and coarse fragments of human debris (e.g., charcoal, shell, bone, plaster, brick). There was relatively little evidence of mixing. The pale Dark Earth appears to be both debris from building decay and destruction and perhaps refuse from dumping. The Dark Earth proper has similar constituents, but also more organic matter and paleobotanical evidence of gardening and much more evidence for postdepositional mixing by worms and slugs. These investigations indicate that much of the character of the Dark Earth is the result of water logging and bioturbation, which account for its high degree of physical uniformity, regardless of age.

The micromorphological research on Dark Earth has several important archaeological implications. First, the Dark Earth is not particularly useful as a time-stratigraphic marker beyond the most general correlations. Though very similar in physical appearance from place to place, it formed at different times and probably at different rates. Second, the bioturbation of the Dark Earth may account for strata thought to be "missing" from underlying deposits. Yule (1990) noted that the Dark Earth was thought to be a distinct deposit of little significance that postdated the Roman occupation of London and other cities in Great

Britain. It was routinely removed before excavation of late Roman strata. The Dark Earth in some, if not many, settings represents homogenized late- and post-Roman occupations, however, and can therefore contribute to understanding the occupation history of the cities. In particular, "interpretations based [solely] on the surviving stratigraphy below the 'dark earth' may seriously underestimate the density of occupation in the later Roman period" (Yule, 1990, p. 627).

In Moscow, a "cultural layer" somewhat similar to the Dark Earths of northwest Europe is described by Alexandrovskaya and Alexandrovskiy (2000). Formation began about 1000 yr ago and continued until the 19th century. The unit displays more internal variability than the classic Dark Earth. It is typically 2–5 m thick but is up to 20 m thick in depressions. The cultural layer accumulated episodically and locally exhibits several buried soils (as 3–5-cm-thick "humus horizons"). It is stony with wooden construction debris in its lower portions preserved by an artificially raised water table. The cultural layer shares some similarities with the Dark Earth, however. Archaeological artifacts are abundant throughout. It is rich in organic matter (30 times the amount of natural surface soils in the region) and phosphate (20–200 times the surface soils). The cultural layer, like the Dark Earth, reflects the accumulated disposal of all manner of waste from human activities. In Moscow, perhaps because the unit is not fully waterlogged and was not bioturbated to the same degree as the Dark Earth, the cultural layer exhibits more physical variability through time and space.

A third widely investigated type of anthrosol is the *terra preta do Índio* ("black earth of the Indian") or simply *terra preta* soil of the Amazon Basin (Sombroek, 1966, pp. 174–176; Falesi, 1974; Smith, 1980; Zech et al., 1990; Woods, 1995; Denevan, 2001, pp. 104–110, 123–124; Glaser et al., 2001a,b; Glaser and Woods, 2004; Lehmann et al., 2004; Sombroek et al., 2002). A classical description of the Terra Preta is presented by Sombroek (1966, pp. 158–159): "Terra Preta soil is a well-drained soil characterized by the presence of a thick black, or dark grey, topsoil which contains pieces of artifacts. . . . [T]he Terra Preta soil is a kind of kitchen-midden, developed at the dwelling sites of pre-Columbian Indians" (table 11.8). Sombroek (1966, p. 174) further describes the Terra Preta occurring

Table 11.8. Terra Preta soil at Santarém, Brazil

Horizon[1]	Depth, cm	Description
Ap	0–35	Black (N 2/0) loamy sand; moderate, medium crumb structure; many roots; scattered pieces of old ceramics; gradual, smooth boundary.
AB	35–70	Very dark grayish brown (10YR 3/1) loamy sand; moderate to weak, medium crumb structure; many roots; diffuse, smooth boundary.
B1	70–100	Dark brown (10YR 3/3) heavy loamy sand; weak, coarse subangular blocky structure; slightly more compact than B2; diffuse, smooth boundary.
B2	100–160+	Dark brown to brown (10YR 4/3) light sandy loam; weak, coarse subangular blocky structure.

From Sombroek (1966, p. 159, Profile 53).

[1] Horizon nomenclature modified to conform with table 1.1 and appendix 1.

in small "patches," usually <1000 m^2. Though usually associated with the Brazilian Amazon, similar soils are reported from tributary basins of the Amazon in Columbia (Eden et al., 1984). Terra Preta soils are typically reported from upland areas adjacent to waterways on older terraces or upland bedrock (Sombroek, 1966, pp. 174–175; Eden et al., 1984; Denevan, 1996; Lima et al., 2002). More recent research shows that they are also common on interior uplands (Woods, 1995; Woods and McCann, 1999). In all settings, the dark Terra Preta contrasts strongly with underlying subsoils, which are red-to-yellow Ultisols, Oxisols, Spodosols, and eutrophic Oxisols (Sombroek, 1966; Smith, 1980; Lima et al., 2002). Indeed, the strong color contrasts led to their initial identification and to the realization that they likely accumulated on top of older soils. The Terra Preta date back to at least 2000 yr B.P. (Eden et al., 1984), and some are likely much older (W. Woods, personal communication, 2002).

Hypotheses regarding the genesis of the Terra Preta have evolved significantly (summarized by Smith, 1980; Woods and McCann, 1999). Initial thinking was that the soils developed in volcanic ash or were deposits of organic matter from lakes and ponds. The thinking was that associated archaeological debris was found in these areas because Native Americans were attracted to the soils because of their high fertility. Subsequent research clearly shows that the Terra Preta were created by human activity and thus are genuine anthrosols. The nature of the human activity involved in formation of Terra Preta remains unclear, however.

Continued research on the Terra Preta in the 1990s shows that they are considerably more variable in their distribution, morphology, and genesis. Much of this research was by Woods and McCann (1999; see also Woods, 1995, and Denevan, 1996), working in the lower Amazon, who conducted one of the few systematic regional studies of Terra Preta and whose data and interpretations form the basis for the following discussion. As noted above, Terra Preta are much more widespread than initially believed, and they are common away from waterways as well as along them. They also range in size significantly, from ~0.5 ha to >120 ha. Relatively few of the soils identified as Terra Preta show evidence for long-term habitation, in contrast to the long-believed "midden model" for their genesis. They also provide little evidence of any sort of direct habitation. "Ca and P levels are not significantly higher [than the background levels], cultural artifacts are rare, and the soil is typically not black but rather dark brown" (Woods and McCann, 1999, p. 9). Indeed, Woods and McCann (1999, p. 9) reintroduce the term *terra mulata* (after Sombroek, 1966, p. 175) to distinguish the brown "facies" from the classic black Terra Preta. The Terra Mulata also cover much larger areas than the Terra Preta (also noted in Sombroek, 1966, fig. 20). Woods and McCann (1999) propose that the classic black Terra Preta and associated midden debris represent household or near-household trash dumps, but the more ubiquitous dark brown Terra Mulata, largely devoid of artifacts or other obvious human debris, may represent agricultural soils modified by repeated mulching and frequent burning. This model of soil genesis has some important archaeological implications. It indicates long-standing habitation sustained by permanent gardens and fields. It also contradicts long-held models of settlement in the Amazon based on presumed agricultural limitations of upland and interior soils (see Denevan, 1996).

An alternative to the Woods-McCann model is presented by Lima et al. (2002). Their study showed that Terra Preta can be midden-like, with abundant archaeological debris and, in particular, elevated levels of Ca and especially P. They also argue that Terra Preta and related soils are much more restricted in distribution. Their study is based on only seven soil profiles and is, therefore, difficult to apply regionally, but Lima et al. (2002) clearly show that Terra Preta are much more variable in physical and chemical characteristics than indicated by the work in the lower Amazon.

Middens in various forms around the world may also qualify as anthrosols. These anthropogenic features include shell middens, more loosely termed "midden mounds," and burned-rock middens. Shell middens represent the artificial accumulation of a specific type of food waste—shell—usually mixed with other sorts of organic and inorganic detritus from human occupation. They are common in saltwater and freshwater littoral regions (and some riverine settings) on many continents in proximity to sources of shellfish. In the absence of shell, many probably would resemble Terra Preta soils (e.g., Scudder, 1996), but because of their abundant shell content, they have long been recognized as a particular and important type of archaeological feature or site (see the historical review of shell midden research in Stein, 1992a, and Claassen, 1998). Few shell middens have been subjected to pedogenic scrutiny or geoarchaeologic investigation, however (a notable exception is the work of Stein, 1982, 1992b; also Sawbridge and Bell, 1972; Scudder, 1996; and Morey and Crothers, 1998). They are rarely referred to as anthrosols, but they seem to share most of the characteristics of anthrosols outlined above. Kaufman and James (1991), for example, note that some shell middens contain anthropic epipedons. However, given the dearth of soil stratigraphic or soil geomorphic data on shell middens, they are not further discussed here.

"Midden mounds" are ubiquitous archaeological site types along stream valleys of the Ouachita Mountains in eastern Oklahoma and western Arkansas (Altschul, 1983, p. 9). They are low mounds (sometimes not even distinguishable topographically) that can cover hundreds of square meters. They are also characterized by a dark "greasy" midden soil, typically 1–2 m thick with abundant artifacts (fig. 11.4, upper panel; Altschul, 1983, p. 9). The dark midden soil is perhaps the most distinctive aspect of midden mounds, such that they are also known as "black mounds." From descriptions they share some similarities with Terra Preta. They contain late Archaic/Woodland ("Fourche Maline" focus) artifacts along with burials, hearths and fire pits, and remains of structures. The midden mounds have been studied by archaeologists since the 1930s, but little is known of their soil or other geoarchaeological characteristics. Colors are reported in the range of black (10YR 4/1 moist), very dark grayish brown (10YR 2/2 moist), very dark brown (10YR 3/2 moist), and dark brown (10YR 3/3 moist; e.g., Galm and Flynn, 1978; Altschul, 1983, pp. 39–41; Johnson, 1983). Correspondingly, organic matter levels are in the range of 2%–4% (Johnson, 1983). The midden mounds also seem to be devoid of natural stratification, presenting recurring problems of archaeological correlation (Altschul, 1983, pp. 326–327).

Burned-rock middens are ubiquitous aspects of the archaeological record in central and western Texas (Hester, 1991). These piles of fire-cracked limestone

Figure 11.4 Examples of midden soils in the south-central United States. (Upper panel) Excavations at site LaMf1 (the Mackey site, 34Lf-29) in Oklahoma, dug by the Works Progress Administration in 1940 (photo provided by and reproduced with permission of the Sam Noble Oklahoma Museum of Natural History, University of Oklahoma). Most of the crew is through the midden and beginning to dig in the sterile submound sediment. Note the thick, very dark character of the midden soil, typical of the midden mounds in the area. (Lower panel) Excavations in a burned rock midden (the Rogers Spring site) in Austin, Texas, showing the full thickness down to bedrock. The dark color of the matrix and the abundant fragments of burned limestone are well illustrated.

vary in size from a few decimeters thick by a few meters in diameter to several meters thick and tens of meters in length or diameter (fig. 11.4, lower panel). There are two dominant forms of burned-rock middens: mounds or domes of burned rock, found largely in central Texas and dating 5000–2500 yr B.P. (fig. 11.4, lower panel), and rings of burned rock, more common in west-central Texas and dating 2500–300 yr B.P. (Collins, 1991). Very little pedogenic data and even less soil chemical data are available for these features. A few personal observations can be made regarding the dome middens. The voids between the rock are filled with an ashy matrix that tends to be very dark gray or black (values and chromas 2/0, 2/1, 2/2, 3/1). This matrix typically includes small limestone fragments, charcoal, shell, bone, and lithic artifacts. The matrix is also relatively high in organic carbon (0–64 cm in Profile 9; table 11.5) and very high in P_2O_5 (0–78 cm in Profile 2, 0–64 cm in Profile 9; table 11.5), qualifying as an Anthropic epipedon in soil taxonomy. Although many of the burned-rock middens were excavated, their origins have long been debated (see papers in Hester, 1991). The ring middens fairly clearly represent food preparation areas, but the origins of the domed middens are more enigmatic, though they too are probably related to food preparation (see papers in Hester, 1991).

Pedologic research in Germany suggests that anthrosols not otherwise identified as Dark Earth, Plaggen, or middens, and not obviously anthropogenic, may be widespread in northern Europe (Kleber et al., 2003). The investigators were trying to understand the apparently random pattern of patches of Chernozemic soils (i.e., soils with thick, dark surface horizons high in organic matter). The study was limited to one known anthropogenic soil and a "natural" soil. They concluded that the chernozemic character was largely due to significant and widespread inputs of charred organic matter in the fine fraction due to burning. This charred particulate matter was mixed into the soil to a depth of almost 1 m, producing the thick, dark character of the soil, and also increasing the bulk density.

Agriculture

The development of agriculture probably has had more pervasive physical and chemical effects on soils than any other activity by preindustrial societies (Goudie, 2000, p. 29). Certainly no other preindustrial human activity has touched so much of the Earth's surface. At its most basic level, farming is the replacement of some natural vegetation community with some sort of artificial vegetation; that is, an agroecosystem. This relatively simple and straightforward activity has a host of far-reaching effects on the landscape and on soils (summarized and modified from Limbrey, 1978; Butzer, 1982, pp. 124–131; and Goudie, 2000, chapters 4 and 6; table 11.9, figs. 11.5 and 11.6). The original plant cover can be partially or completely removed, leaving the ground bare for at least some part of the year and subject to erosion by water or wind. Cultivation loosens the soil, and in the Old World the hooves of domesticated animals loosen or compact it. Devegetation alters soil moisture and can affect groundwater. Plowing, excavation of irrigation ditches, and construction of terraced fields all physically disturb soils as well. Devegetation, new kinds of plant residues (from burning and

Table 11.9. Common measures of soil quality in agroecological studies

Soil property	Characteristics and causes of degradation	Consequences of degradation
A horizon thickness	Decreased thickness caused by water or wind erosion	Reduces important organic matter-enriched surface layer that can be exploited by plants for water, nutrients, and oxygen. Shallower depth to possible root-limiting subsurface layers such as strongly developed argillic horizons.
	Increased thickness due to terracing	Formation of coarser structure, which reduces infiltration
Soil structure	Macromorphology: Lowered grade of granular or subangular blocky structure; trend toward massive especially in surface horizons. Due to compaction and organic matter decline.	Lowered infiltration capacity
	Micromorphology: Compare structure and pore characteristics of cultivated and uncultivated A horizons.	
Bulk density	Compaction (increase in bulk density above that of natural conditions) associated with soil structure degradation.	Compaction and structure degradation retard seed germination and root growth; reduce root access to water, oxygen, and nutrients; reduce diffusion of gases; and decrease water infiltration and available water capacity.
Organic carbon	Decrease in organic C due to accelerated microbial oxidation of organic matter in disrupted, exposed soil aggregates.	Limits soil physical, chemical, and biological properties important to plant growth.
Nitrogen	Decrease in total N accompanies declining organic matter in agricultural soils, though C:N ratio tends to decrease.	Nitrate and ammonium are plant-available forms of N, which is a key limiting factor for plant growth in all regions, including arid regions.
Phosphorus	P (both total and available) can decrease under plow-based agriculture.	P is a key ecological and soil indicator because of its low mobility, low availability to plants, and long-term stability of its forms in soils.
pH	Very high soil pH can indicate salt accumulation (measured by electrical conductivity), common in agricultural soils of arid and semiarid regions.	Detrimental impacts including crop species, occur both through direct chemical effects and through soil structural deterioration

Modified from Homburg (2000, table 3.1).

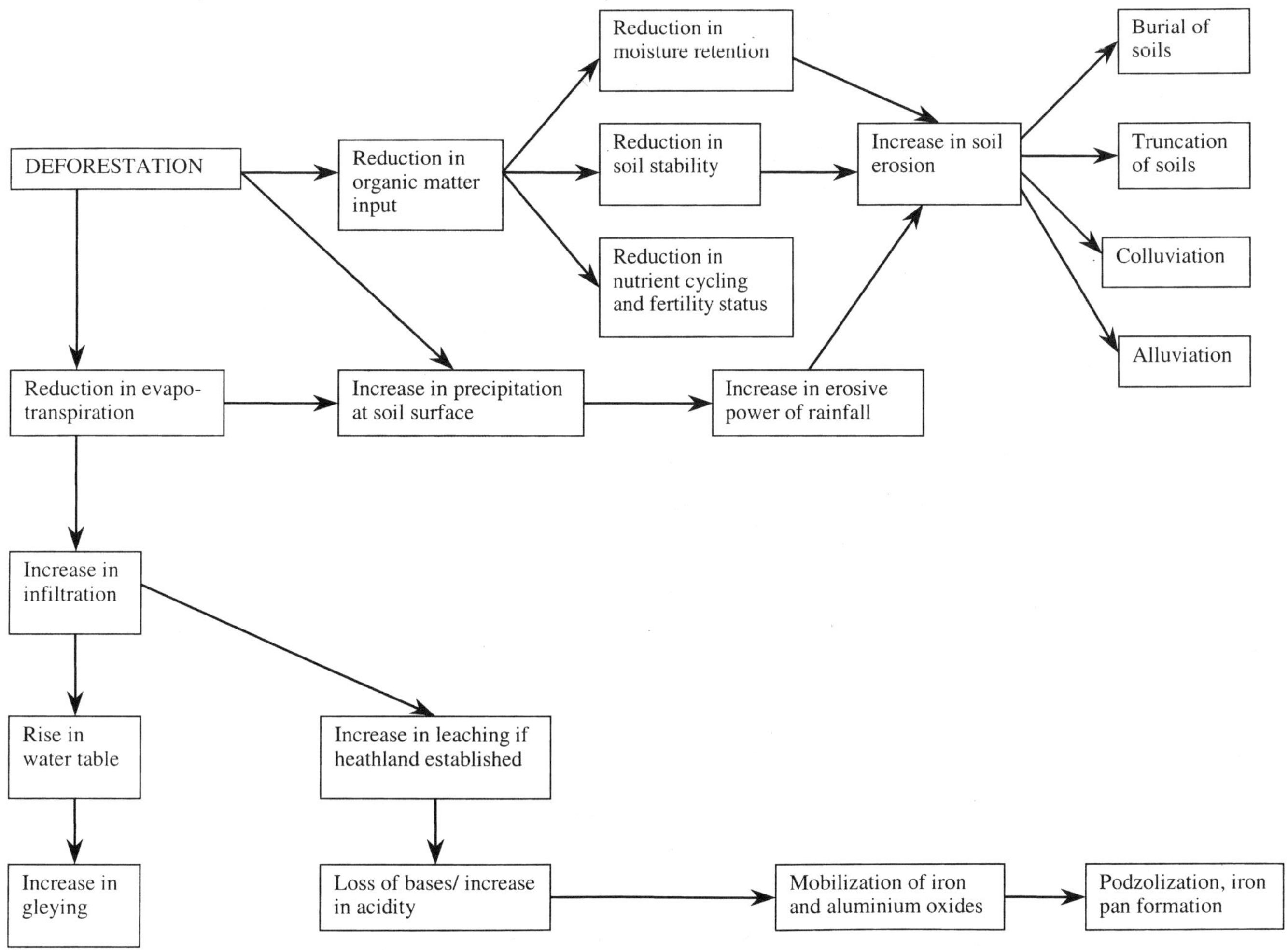

Figure 11.5 Soil changes that may result from deforestation (modified from Davidson, 1982, fig. 1.1).

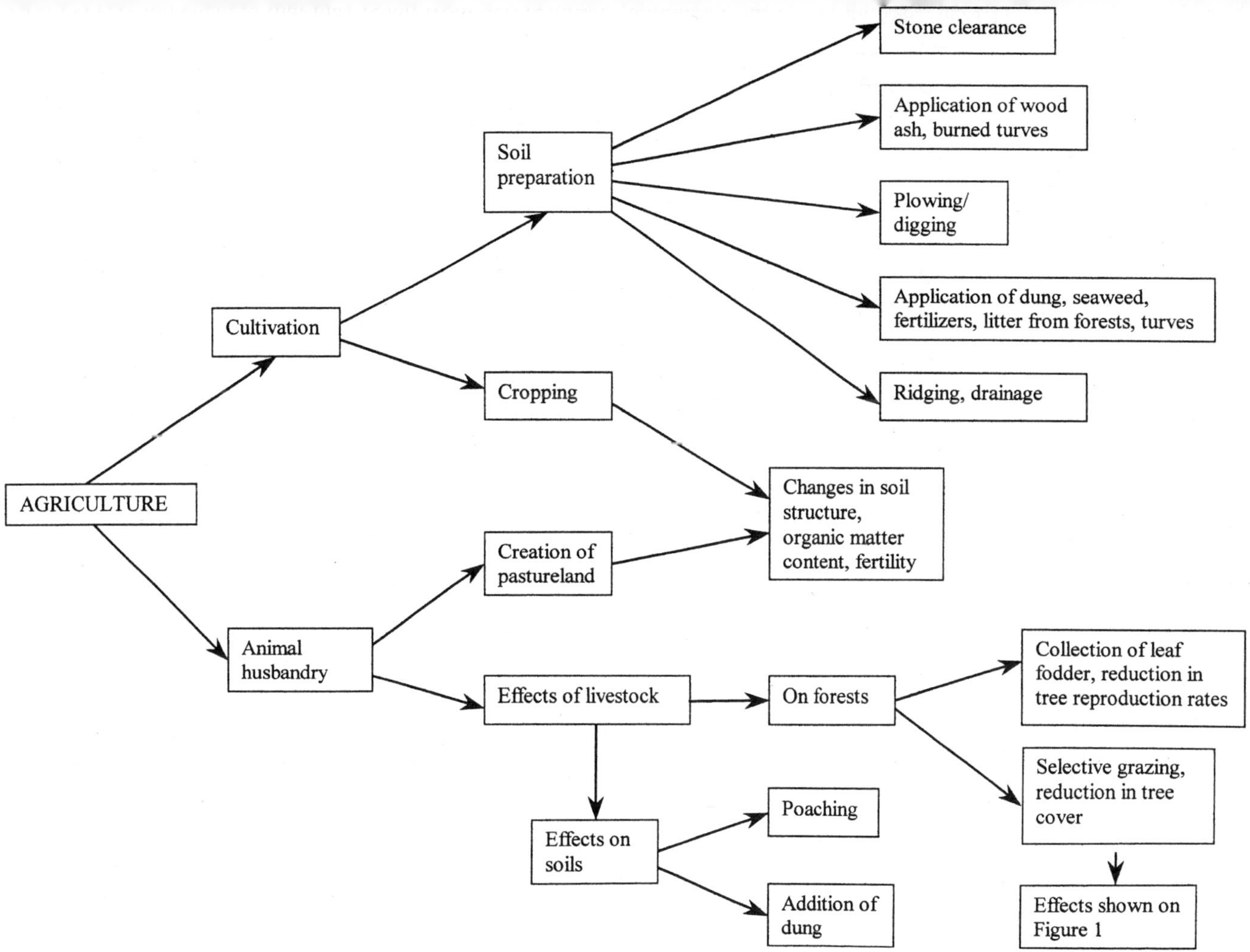

Figure 11.6 Possible effects on soils following the introduction of agriculture (modified from Davidson, 1982, fig. 1.2).

cropping), and additions of fertilizer can all alter soil chemistry. Changes in groundwater conditions can drastically affect the soil-forming environment. An elevated water table as well as irrigation also induces salinization if salts are present. As with soil P studies in archaeology, a very large literature is devoted to the archaeology and geoarchaeology of agriculture. Most of the following discussion is a brief sketch of various approaches to using soils for reconstructing agricultural activities.

Soil erosion is the most pervasive and most dramatic result of devegetation and is the subject of considerable geoarchaeological interest and research (e.g., Evans and Valentine, 1974; Pope and Van Andel, 1984; Van Andel et al., 1986, 1990; Dunning and Beach, 1994; Brinkmann, 1996; Overstreet and Grolier, 1996; Beach, 1998a; French and Whitelaw, 1999). Soil erosion deals largely with the absence of or loss of soils, however, with most research focusing on causes, processes, and results of erosion. Soils in and of themselves usually are not a component of this research. An exception to this generalization is the investigation of soils buried within the sediments that result from soil erosion (e.g., Pope and Van Andel, 1984; Van Andel et al., 1986, 1990; Brinkmann, 1996; French and Whitelaw, 1999). These buried soils provide clues to the timing of multiple phases of soil erosion. Otherwise, this topic and its vast literature will not be dealt with because it does not deal with soils per se, which is the aim of this volume and its focus on soil geomorphology in geoarchaeology. Interested readers and students are directed to the many papers, chapters, and books on the topic (e.g., Evans et al., 1975; Limbrey, 1975, pp. 112–126, 146–192; Limbrey and Evans, 1978; Butzer, 1982, pp. 123–156; Davidson, 1982; Thornes, 1987; Bell and Boardman, 1992; Brown, 1997, pp. 210–253; Dincauze, 2000, pp. 320–325; Goudie, 2000, pp. 185–199).

Beyond considering agriculture in terms of its effect on the landscape, the investigation of past agricultural activities is a key component of archaeological research focused on the evolution and maintenance of complex societies and on assessments of agricultural sustainability. In a very literal sense, the landscape is the archaeology. Despite the wide effect of agriculture on the landscape, the evidence for past agriculture is not always immediately obvious or easily recoverable (Sandor, 1995). Sandor (1995, p. 121) notes that two ways in which soils may indicate former farming sites are by their altered properties and by evaluation of their agricultural potential. Pedology and soil geomorphology are or should be important components of research into ancient agricultural societies, therefore, because of the intimate relationship between agriculture and soils. A wide array of approaches, both physical and chemical, has been used to study the geoarchaeology of agriculture (e.g., table 11.9). The work described earlier on Plaggen soils and Terra Preta illuminate a small component of this research. The discussion in the rest of this chapter focuses on detecting alterations to soils resulting from agricultural activity. Much of the research methodology for using soils to identify and understand agricultural practices and their effects comes from scientists working in Europe and Britain (e.g., Cornwall, 1958; Limbrey, 1975; Courty et al., 1989; Barham and Macphail, 1995), because of the long history, widespread use, and high-impact nature of farming in that part of the world. An emerging literature on the geoarchaeology of agriculture is also available for the New World (e.g., Turner and Harrison, 1983; Pohl, 1990; Sandor, 1992; Jacob, 1995a,b; Sandor and Eash, 1995; Homburg and Sandor, 1997; Denevan, 2001; Damp et al., 2002) because of the long interest in the archaeology of agriculture and complex societies there.

The unique morphological characteristics of soils provide an excellent backdrop against which agricultural activities may be identified. "Soil stratigraphy . . . is the most important and most complex evidence available for analyzing the structure of ancient gardens and fields; in many places it may be the only evidence" (Gleason, 1994, p. 14). The physical signatures of agriculture in soils are related to the disruption of the lateral continuity of and vertical gradations between soil horizons. These disruptions result largely from plowing and construction of ditches and furrows. Probably the most obvious initial effect of farming is mixing of the upper solum by plowing. This process is widely recognized today in the identification of the "Ap" plowzone horizon (table 1.1). A variety of possibilities exist as to the exact nature and degree of mixing, depending on the thickness of the A horizon, the character of the underlying horizon, and the depth of plowing. Shallow plowing of a thick A horizon will result only in mixing of part of the A horizon, whereas deep plowing of a shallow A over an E-B sequence might result in mixing of the A, the E, and the upper B horizon. In any case, the result is a homogeneous horizon of uniform thickness with a sharp lower boundary. Stones, artifacts, and other materials larger than sand size will be characterized by random orientation and roughly even distribution through the zone (Limbrey, 1975, p. 331). Cultivated A horizons, plowed or otherwise, are reported as being lighter in color than uncultivated A horizons, owing to the more limited input of SOM and increased oxidation of SOM because of increased surface area of the aggregates (fig. 11.1), and as having a blocky, almost massive structure, in contrast to the typical granular structure of uncultivated A horizons (fig. 11.1; Limbrey, 1975, p. 331; Courty and Nørnberg, 1985; Sandor et al., 1986; Sandor, 1992; Homburg and Sandor, 1997; Homburg, 2000).

Plowing can initiate or increase rates of accumulation of illuvial silt, clay, and humus just under the plow zone. Such zones are sufficiently prominent in association with modern agriculture that they are formally identified as "agric horizons" in soil taxonomy (Soil Survey Staff, 1999, pp. 28–29). The large pores in the Ap horizon and the absence of vegetation at the surface just after tilling promotes infiltration of muddy water to the base of the plowing, thereby translocating the fines. In thin section this illuviated fine-grained mineral and organic matter is apparent as "agricutans" (Courty et al., 1989, pp. 131–132). Illuvial clay and other materials are noted in association with plow zones in archaeological contexts in a wide variety of settings (e.g., Limbrey, 1975, pp. 332–334; Barnes, 1990), and some meet the qualifications of agric horizons (Sandor, 1987; Sandor and Eash, 1995). The various physical characteristics of Ap horizons can be preserved in buried soils but might be obliterated by subsequent plant and animal activity and other pedogenic processes in abandoned unburied agricultural soils.

Physical disruption of the lateral continuity of soil horizons can also provide clues to agricultural activity, applying a form of soil microstratigraphy. Plowing disrupts the A horizon and brings subsoil material up between the segments of topsoil. Such plow marks can be readily apparent and preserved in the upper portion of plowed soils under the right conditions (Fowler and Evans, 1967; Courty and Nørnberg, 1985; Macphail et al., 1990a). Limbrey (1975, pp. 163, 332–334) provides good illustrations and useful discussion of this phenomena. Plow marks will show up where the depth of plowing is greater than the depth of the A horizon, so that the plow digs down into the underlying (and lighter in color) E, B, or C horizon. Repeated plowing will eventually homogenize the plow zone and produce the typical Ap horizon. Plow marks will be preserved where a field was plowed only a few times, then abandoned or buried, or where a field is undergoing erosion so that successive plowing digs deeper into undisturbed soil until the field is abandoned or buried.

The excavation of ditches for drainage or irrigation and preparation of furrows in fields also disrupt soil horizonation and, where preserved, provide readily obvious indications of agriculture. Indeed, the soils are key references in the identification of ditches and furrows (fig. 11.7; e.g., Pohl et al., 1990, figs. 8.3–8.5; Erickson, 1995, fig. 3.12; Damp et al., 2002, figs. 2–5). Ditches are dealt with at

Figure 11.7 Cross section of one of the oldest irrigation canals in North America at the K'yana Chabina site (LA 48695), on the Zuni Reservation in northwestern New Mexico (from Damp et al., 2002, fig. 2; reproduced by permission of the Society for American Archaeology from *American Antiquity*, v. 67, no. 4, 2002; provided by J. E. Damp). The roman numerals indicate stratigraphic units described by Damp and coworkers. Unit III is a cumulic A horizon. The dark hues and fine structure show well and help to highlight the outline of the canal. The canal fill (part of Unit II) dates to ~2000 yr B.P.

some length by Cornwall (1958, pp. 49–60) and Limbrey (1975, pp. 290–304, 305–306), but their focus is more on ditch fills rather than the ditches themselves. There are relatively few studies that focus on the soil stratigraphic relationship of ditches and furrows to soils. In trying to understand Maya wetland agriculture, repeated, lateral truncations of a prominent, regional buried A horizon (the "Cobweb Clay paleosol") helped convince Jacob (1995a) that the Cobweb Swamp area of Belize was modified for water-management purposes. These modifications included enhancement of natural channels around low islands to improve drainage and digging of small ditches to better drain the surface of the islands. In the Rio Hondo/Pulltrouser Swamp area of Belize, Jacob (1995b) identified a buried soil likely correlative to the Cobweb Clay and modified by surface irregularities that probably represent tilling. Earlier work in the area also identified obvious evidence of drainage ditches cut into the soil (Bloom et al., 1983, Pohl et al., 1990, figs. 8.3–8.5).

Soils research in archaeological contexts in arid regions provides a good contrast with work in the more humid settings of Europe and Central and South America. Water availability is the overriding concern in dry regions (Dregne, 1963, p. 219) and results in a variety of agricultural practices focused on water conservation. In addition, the dry environment promotes preservation of soil properties reflecting cultivation. Arid lands also tend to be more sparsely populated today, providing opportunities for comparing cultivated and uncultivated soils not available in humid regions subjected to more pervasive human disturbance. Comparative studies have proven particularly useful in the Southwestern United States. Several such investigations in western New Mexico and central Arizona show that ancient cultivation did not necessarily "degrade" soils (Sandor et al., 1986; Sandor, 1992; Homburg and Sandor, 1997; Homburg, 2000). In these studies, for example, the upper sola of cultivated soils had somewhat reduced levels of P, N, and SOM, but they were still fertile soils, and in some cases, the lowering of nutrient levels was not statistically significant (e.g., Homburg and Sandor, 1997). The A horizons of these soils were also overthickened. The fields examined in these various investigations tended to be situated where they could receive runoff, which provided not only water but also sediment and nutrient-rich organic debris. Thickening of the A horizon also aids in meeting the soil volume requirements for crop roots. Homburg (2000) reports that Bt horizons tend to be deeper than average below the overthickened A horizons on the Zuni Reservation of northwestern New Mexico. This is probably related to surface aggradation (chapter 5). Cultivated soils also contain Bt horizons in other settings (Sandor, 1992; Homburg, 2000). Soils with Bt horizons apparently were crucial to successful farming in the southwestern United States because the clay held moisture in the rooting zone for a considerable time after rainfall or snowmelt.

In the arid Near East (southern Turkmenistan, southern Dakhistan) Lisitsina (1976) investigated cultivated and uncultivated soils to better understand the effect of ancient farming and irrigation associated with desert oases. The anthrosols were characterized by changes in soil structure similar to those mentioned above (i.e., a tendency toward coarser "cloddy" structure), but in contrast to those studies, the cultivated oasis soils exhibited higher humus content and evidence of increased biological activity. Humic enrichment of the A horizon in an irrigated arid soil is not surprising given the sensitivity of desert plant communities to increases in moisture.

Soil micromorphology has proven to be a particularly useful tool in the investigation of agricultural impacts on soils (tables 11.9 and 11.10). The micromorphology of pre-occupation natural soils as well as postoccupation pedogenesis must be considered in assessing anthropogenic impacts (fig. 11.1). Some disagreement exists over the degree to which micromorphology can fully confirm particular agricultural practices (Carter and Davidson, 1998, 2000; Macphail,

Table 11.10. Micromorphological indicators of land-use impacts

Land use activity	Micromorphological characteristics
Uprooting[1]	Mixed micro- or macrofabric composed of topsoil and subsoil fragments (e.g., A, Ah, E, Bt) to 20–100-cm depth; fissures infilled with dusty clay coatings; coarse wood charcoal, stone or other artifacts indicate anthropogenic uprooting; possibility of burned red soil Strongly and heterogeneously mixed fabric indicate rapid infilling. Homogeneous microfabric highly reworked biogenically indicates slow infilling
Clearance by burning	Microfabric with finely mixed and charred organic fragments and remnants of charcoal and burned wood; reddish brown fragments of rubified topsoil if clay was present; dusty clay and silt coatings in lower topsoil
Grazing	Evidence for grass-covered topsoil (Ah horizon showing intensive rooting and biological activity; finely dispersed, humified matter; randomly distributed phytoliths mixed with organic residues) combined with platy structure (elongated and platy pores within dense fabric); high concentration of fungal bodies
Stabling	Highly distinctive microfabric composed of compact crust of layered plant fragments (from hay, fodder, and bedding), dark-stained manure and amorphous organic material, trampled silt, dark-stained and phosphatized rock fragments; subsoil beneath stable heavily stained with phosphate
Plowing (including arding, hoeing, digging)	Mixing of topsoil and subsoil characterized by fragments of various horizons; light-textured soils will exhibit coatings of silt, clay, and impure clay; heavier-textured soils exhibit homogeneous coatings and infillings of clay and fine fragments of organic matter ("dusty clay"); coatings are concentrated at various levels in the topsoil; small aggregates of soil accumulate and pore space reduces at base of Ap; large aggregates and coarse voids concentrate near the surface; abundant evidence of high biological activity
Manuring and other additions	Higher amounts of organic fragments, especially phytoliths derived from dung; plaggen will exhibit fine, homogeneous mixture of organic matter, plant fragments, and phytoliths; turf exhibits common plant fragments, roots, and organo-mineral excrement from microfauna
Irrigation	Presence of Na indicated by dusty, poorly laminated, and weakly birefringent clays mixed with coarser particles; collapsed voids indicate unstable soil structure; high input of water favors translocation of clay and silt and induces formation of vesicles

Following Courty et al. (1989, pp. 126–137) and Macphail et al. (1990), with additional material from Cruise and Macphail (2000) and Macphail and Cruise (2001).

[1] Uprooting can be natural (chapter 10) or anthropogenic.

1998). Davidson (2002) also demonstrates that in cool, temperate upland settings, earthworms can effectively obliterate micromorphological evidence for cultivation within a few hundred years after field abandonment (also discussed in chapter 10). Most investigators, however, have found thin-section analyses to be invaluable in the study of ancient farming practices, particularly in the context of interdisciplinary research that might include physical and chemical soil analyses, paleobotany, and stable isotope analysis (e.g., Fisher and Macphail, 1985; Courty and Nørnberg, 1985; Macphail, 1986; Macphail et al., 1990a; Simpson, 1997; Simpson et al., 1998; Courty, 2001; Cruise and Macphail, 2000; Homburg, 2000; Macphail and Cruise, 2001). Few micromorphological characteristics of soils are definitive as to specific land-use activities, but sufficient data seem to be available to allow reasonable inferences (Courty et al., 1989, p. 126; Macphail et al., 1990a,b; Macphail, 1998). Experimental research on the anthropogenic impacts of agriculture has proven particularly useful (e.g., Macphail et al., 1990a,b; Gebhardt, 1995; Cruise and Macphail, 2000; Homburg, 2000). Micromorphology has been used to investigate and interpret land clearing, grazing, plowing, manuring and fertilization, and irrigation (table 11.10; Macphail, 1986; Macphail et al., 1987, 1990a,b; Courty et al., 1989, 126–137).

One of the most common microfeatures diagnostic of agricultural activity, particularly in finer-grained soils, is the "dusty clay and silt coating," which is essentially a type of agricutan, noted earlier (fig. 11.8). When a field is cleared by burning, the resulting ash (enriched with K, Ca, Mg, and charcoal) promotes translocation of clay plus fine particles of charred plants (Macphail, 1986; Courty et al., 1989, p. 129; Macphail et al., 1990a,b). The rapid infiltration of these fines is further enhanced by the loss of structural stability in the soil because of reduced SOM content and loss of vegetation (Macphail et al., 1990a; Macphail, 1992; Carter and Davidson, 1998). Fragments of burned soil can also be found, infiltrated into the subsoil (Macphail et al., 1990a,b; Macphail, 1992). Absence of fine charcoal and other micromorphological evidence of clearing in fill covering an ancient field has been used to suggest a hiatus (though perhaps temporary) in clearance and farming (Fisher and Macphail, 1985).

Mixing of components of individual soil horizons is also indicative of clearance and farming (fig. 11.8). Complex mixtures of microscopic fragments from A, E, and Bt horizons along with fine charcoal may be indicative of tree uprooting during clearance, whereas mixes of only surface or near-surface horizons are more likely caused by plowing (Macphail et al., 1987, 1990a,b; Courty et al., 1989, pp. 126–137; Gebhardt, 1995). Hollows created by clearing and tree uprooting can be differentiated from natural tree throw depressions (chapter 10) by the presence of fine charcoal in infill of the former, along with a more heterogeneous mix of horizons as a result of rapid burial by plowing and leveling, whereas the natural depressions will have a more homogenized infill because of slower filling and bioturbation (Macphail and Goldberg, 1990).

At the opposite end of the scale of impacts from the micromorphological are the dramatic regional-scale alterations of soils caused by ancient land clearance and farming. Many investigators, particularly in Europe and parts of Asia, have documented or summarized the drastic changes that occur when the natural vegetative cover is removed, altering the morphology and chemistry of the soils

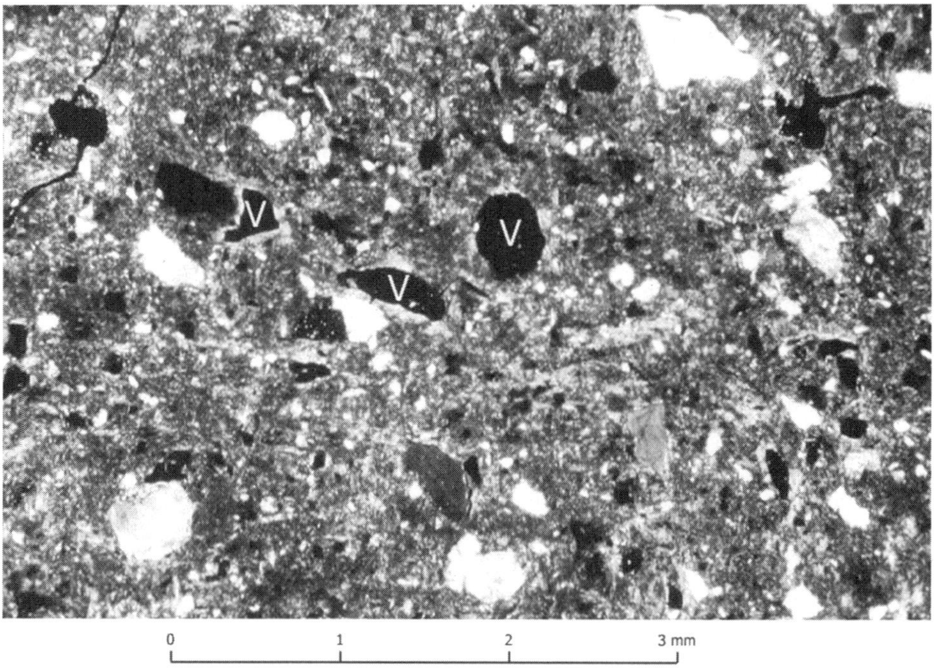

Figure 11.8 Thin-section photomicrograph (under cross-polarized light) from a soil buried beneath a Neolithic rampart at Maiden Castle, Dorset (provided by and published with permission of R. I. Macphail; Macphail, 1991). The random mix of particles of various sizes is indicative of mixing, probably by Neolithic clearance. Particles of clay and silt washed into voids (black areas; several large ones indicated by "V") to produce "dusty clay" coatings that show as bright areas under crossed polarizers.

(figs. 11.5 and 11.6; e.g., Limbrey, 1975, pp. 146–192; Butzer, 1982, pp. 123–156; Davidson, 1982; Macphail, 1986; Macphail et al., 1987). Macphail (1986, pp. 283–284) provides a good summary of these processes in northwest Europe:

> In Europe . . . man's effects have been shown to be more decisive locally on pedogenic trends than any climatic change during the Flandrian [Holocene]. By causing massive soil erosion, through woodland clearance and agriculture, he created large areas of shallow calcareous soils, thus reversing the natural trend of decalcification. By accelerating the leaching rates on acid substrates such clearance triggered podzolization. Deforestation leading to decreased evapo-transpiration rates also resulted in progressive upland hydromorphism and peat formation, at a much earlier time in some cases, than the onset of cooler and wetter conditions of the sub-Atlantic phase [late Holocene]. Thus limestones could have their decalcified topsoils removed by erosion, and the processes of podzolization and upland hydromorphism could be hastened by the destruction of natural vegetation.

Sorting out the chronology of these anthropogenic soil alterations often is difficult because they are usually manifested as a sequence of processes superimposed on the surface soil. The West Heslerton site in Yorkshire contains a

Table 11.11. Pedological evidence of environmental history at an archeological site in Yorkshire

Period	Event	Soil type[1]	Soil process
Modern		Calcareous brown sand	Recalcification
	—Burial by Eolian Sand—		(neutralization)
Saxon	Settlement and Cemetery		
(Late Anglian)	(Wolds Footslope)		
	Heath?	Podzol	Podzolization 3
	—Burial by Eolian Sand—		
Roman	Agriculture		
	—Eolian Erosion—		
	Heath?	Podzol	Podzolization 2
	—Burial by Eolian Sand—		
Early Iron Age	Agriculture		
Late Bronze Age			
	—Eolian Erosion—		
	Heath?	Humo-ferric Podzol	Podzolization 1
	Agriculture		
	Clearance?		
	Woodland regeneration?	Argillic brown earth	Acidification
			Decalcification
	—Burial by Eolian Sand—		
	—Eolian Erosion—		
Early Bronze Age	Barrow construction		
	—Burial by Eolian Sand—		
	—Eolian Erosion—		
Late Neolithic	Area already cleared?	Calcareous brown sand	

From Macphail (1986, table 2).

[1] Soil types in the U.K. system have roughly the following equivalents in the U.S. soil taxonomy: Calcareous Brown Sand = Psamment or Orthent; Podzol = Orthod or Aquod; Humo-ferric Podzol = Orthod. Argillic Brown Earth = Udalf.

sequence of sediments and soils spanning the Neolithic to Roman occupations, providing optimal conditions for establishing the soil history (table 11.11; Fisher and Macphail, 1985).

The importance of people in altering the landscape and associated soils can be overemphasized, however. For example, clay translocation and formation of soils with argillic horizons (*sols lessivés*) in south and east England are considered by some to be the result of forest clearance and cultivation (Limbrey, 1975). Fisher (1982) addressed this problem and convincingly shows that such an interpretation is not substantiated and that *sols lessivés* were probably the "modal" forest soils of the area before cultivation. Davidson (1982, p. 9) also questioned the idea that clay translocation must be related to devegetation.

Irrigation, cultivation, and the addition of manure and fertilizer can induce a variety of physical and chemical changes in soils, including overthickening, waterlogging, elevated (or reduced) levels of organic matter and associated darkening, and elevated (or reduced) levels of soil P and other elements such as Ca, K, N, and Na. Most of these processes and characteristics were described above, but a few additional comments are warranted. Irrigation and cropping can leave several kinds of chemical signals in the soil, depending on the specific nature of the farming activity. Irrigation increases the biotic activity of soils and thus raises their SOM content (summarized by Huckleberry, 1992, pp. 238–239). The combination of increased levels of SOM and waterlogging can result in gleying. As Huckleberry (1992) points out, however, in dry environments, once irrigation ceases and soil drainage improves, the SOM will oxidize and the gley colors may disappear. The evidence for irrigation in arid regions, therefore, may be transitory. Further, prolonged cultivation in drier environments can result in oxidation of SOM and increased compaction (e.g., Sandor, 1992, p. 228). The SOM helps maintain a porous, granular topsoil. Removal of SOM allows the soil to be compacted. In contrast, in more humid and cooler settings, and particularly where manure was used as fertilizer (e.g., northern Europe), elevated levels of SOM are found in cultivated fields long after abandonment (Courty and Nørnberg, 1985; Engelmark and Linderholm, 1996).

Another direct effect of growing crops is a loss of soil P. Plants and microorganisms mineralize the organic P and use it. Removal of the native plant cover and harvesting of crops removes the P that would otherwise return to the soil (Eidt, 1984, pp. 28–31; Sandor et al., 1986; Sandor, 1992; Homburg and Sandor, 1997; Leonardi et al., 1999). In some archaeological soils, therefore, levels of soil P are reduced. Manure can be added to soils to rectify this problem, as described above. This can result in high levels of Porg in cultivated fields, even those abandoned for centuries (Courty and Nørnberg, 1985; Engelmark and Linderholm, 1996). In the southwestern United States, some cultivated landscapes were in settings in which organic-rich debris accumulated, also noted earlier; this offset losses in P caused by crop uptake and harvest (e.g., Homburg and Sandor, 1997) Thus, both lowered and elevated levels of soil P can be chemical indicators of agriculture. Very generally, removal of crops (and soil P) without manuring or fertilizing is associated with slash and burn agriculture and nonpermanent settlement, whereas enrichment of soil P is associated with permanent farming settlements (Eidt, 1984, pp. 29–30). Leonardi et al. (1999) also show that both contemporary and buried Roman soils in northern Italy that were subjected to plowing displayed lower levels of organic P versus total P lower than soils not affected by humans. This ratio may be another indicator of plowing when used in conjunction with other indicators.

A long-standing issue in irrigation of soils is "salinization." Artzy and Hillel (1988, p. 235) provide a succinct summary of the process and problems of soil salinization:

> All waters used in irrigation . . . contain dissolved salts [typically varying combinations of chlorides and sulphates]. Since crop roots normally exclude most of the salts while extracting soil moisture, . . . the salts tend to accumulate in the soil and, unless leached, will in time poison the root zone. In arid regions, natural rainfall is generally insufficient for annual leaching, hence irrigation must be applied in excess of crop water requirements so as to remove harmful salts by downward percolation beyond the root zone. . . . [T]he practice of over-irrigation is only a temporary remedy. . . . The salt-laden water percolating below the root zone . . . tends to accumulate in the subsoil and to raise the water table. . . . As the water table rises

> progressively and eventually approaches the soil surface . . . [capillary action brings water up] from the water table through the root zone to the soil surface, where evaporation takes place, and this process tends to reinfuse the surface zone with salts.

The consequences of salinization can be severe (e.g., Goudie, 2000, chapter 4), and debates rage in the archaeological literature over the degree of its effect on ancient societies (e.g., Ackerly, 1988; Artzy and Hillel, 1988). Those issues aside, recognition of secondary salt in archaeological soils can be an important clue to ancient land use (e.g., Lisitsina, 1976; Courty et al., 1989, pp. 136, 170–172; Huckleberry, 1992). Because salts are so soluble, however, they are easily removed from soils. Direct evidence of ancient salinization, therefore, may not be preserved, but the former presence of sodium may be apparent in thin section (table 11.10; Courty et al., 1989, p. 136).

In east and southeast Asia, the process of "aquorization" is indicative of wet-rice production (Barnes, 1990; Zhang and Gong, 2003). The processes consist of the eluviation of iron and manganese from the plow zone when fields are flooded during the growing season and illuviation of the minerals in the subsoil during the dry season after the harvest. Absence of aquorized morphology is not indicative of the absence of rice paddy agriculture, however. The process apparently happens only in fields with particular drainage characteristics and also can be destroyed by prolonged field maintenance.

Appendix 1: Variations on U.S. Department of Agriculture Field Nomenclature

For this discussion I will write in the first person because these observations on the official nomenclature for designating soil horizons, subhorizons, and related attributes of soils as seen in the field are based largely on my own quarter-century of experience. The U.S. Department of Agriculture (USDA) developed a standard set of identifiers or shorthand notations for describing soils in the field, most of which are familiar to geoarchaeologists working in the United States (i.e., the O, A, E, B, and C horizons, the two dozen or so subhorizon symbols, and several other notations; table 1.1; Soil Survey Division Staff, 1993; Schoeneberger et al., 1998). The soil surveys and other governmental agencies of many other countries also developed their own, generally similar, sets of nomenclature. The United States's terms and other terms generally are for use in describing surface soils. For the most part they are also useful for describing buried soils and soil-stratigraphic units, but none are ideal. On the basis of my experience in the field and familiarity with the U.S. naming system and some other systems, several modifications of the USDA nomenclature have proven very useful, and I recommend them to my fellow geoarchaeologists. Others may find additional or alternative variations helpful, and some have described them (e.g., Birkeland, 1999, table 1.1, appendix 1). I encourage field workers to develop their own modifications and share them with the geoarchaeological community. My modifications are as follows.

Anthropogenic A horizon Soils with intensely occupied surfaces may develop artificially altered A horizons, characterized in the field by low value and chroma, sometimes overthickening, and presence of artifacts. In Europe

Table A1.1. Comparison of various nomenclatures for identifying buried soils and lithologic discontinuities

Lithology	Soil stratigraphy	U.S. Department of Agriculture	This volume[1]	Cremeens et al., 1998[1]	Kemp, 1999[1]
Silt	*Surface soil*	A	A	A	A
		Bt	Bt	Bt	Bt
		C	C	C	C
	First buried soil	Ab	Ab1	S1Ab	1Ab
		Bw1b	Bw1b1	S1Bwb1	1Bwb1
		Bw2b	Bw2b1	S1Bwb2	1Bwb2
Coarse sand		2C	2Cb1	S1C	1C
Gravel	*Second buried soil*	3Ab	3Ab2	S2Ab	2Ab
		3Btb	3Btb2	S2Btb	2Btb
		3C	3Cb2	S2C	2C

[1] This table is a comparison of approaches, not a comparison of specific profiles. Cremeens et al. (1998) and Kemp (1999) do not describe profiles with this horizon sequence.

these anthropogenic A horizons are referred to as "Aan" horizons (e.g., Pape, 1970; van de Westeringh, 1988).

Buried soils The "b" is used to identify buried soils in USDA terminology (table 1.1) and in other nomenclatures (e.g., McDonald et al., 1998, p. 108; Soil Classification Working Group, 1998, p. 12). Officially the symbol is added only to buried mineral horizons making up the solum (i.e., A, E, and B horizons). For consistency and clarity I also add it to O and C horizons (table A1.1). Moreover, when multiple buried soils are present, I find it useful to number the buried soils from the top down as a means of grouping and identifying individual buried soils (table A1.1). In discussion of the soils, they can then be referred to as b2, b3, and so forth. Some authors add numbers after the "b" to subdivide a buried horizon (e.g., Btb1, Btb2 as two subdivisions of a buried Bt horizon; e.g., Cremeens et al., 1998; Kemp, 1999). In my scheme, the subdivision nomenclature would precede the "b" (e.g., Bt1b1, Bt2b1, etc., to subdivide a Bt horizon within the first buried soil below the surface). Cremeens et al. (1998) further identified buried soils with an "S" prefix, numbered from the top down (table A1.1). Whatever the case, writers should clearly describe the nomenclature they are following (see also discussion of lithologic discontinuities in the following text).

Degree of pedogenic expression The USDA nomenclature is designed simply to describe the presence of specific pedogenic characteristics. For example, the "t" is used to indicate the presence of illuvial clay, as in a Bt horizon. However, a Bt horizon can have a few thin, patchy clay films or continuous, thick clay films. To convey this variability by glancing at the horizonation symbol, the "w" can be added to express a weakly developed feature; in the case of a weakly expressed Bt horizon, therefore, it can be indicated as "Btw." This nomenclature was developed from the USDA use of the "w" to denote a minimally expressed B horizon only (table 1.1), but I have found it very handy for noting any sort of weakly developed soil horizon (e.g., Bkw).

Humus-rich A horizon An A horizon with low values and chromas indicative of high content of organic matter (but not dominated by organic matter, as in an O horizon) is designated an "Ah" horizon in the Canadian and several European systems for describing soils (e.g., Hodgson, 1997, p. 86; Duchaufour, 1998, table 5; Soil Classification Working Group, 1998, p. 14). It is it roughly equivalent to a Mollic epipedon. This can be very useful for conveying the idea of a very dark surface or buried A horizon. The "h" in the Ah is not to be confused with the "h" indicating translocated humus in the Bh horizon (table 1.1).

Organic horizons The soil surveys of countries in climates generally cooler and wetter than the United States typically devote more attention to describing (and classifying) organic horizons and peats. The Soil Survey of England and Wales, for example, has three separate horizon notations for litter layers and seven kinds of O horizons (Hodgson, 1997, pp. 84–85). These distinctions are important. Litter layers are well drained and represent organic detritus that accumulated on forest floors. They are subdivided into L, F, and H horizons, roughly corresponding to Oi, Oe, and Oa horizons (table 1.1). The O horizon in this scheme is applied to only to peat layers, and subdivisions are based on the physical and chemical characteristics of the peat. Van Breemen and Buurman (2002, pp. 141, 365) described a slightly different scheme. The O horizon is used for well-drained forest floors and is subdivided into OL, OF, and OH. Peats are designated as H (histic) horizons. In Australia, peat soils are described as P horizons (McDonald et al., 1998, p. 104). These various notation schemes could be useful for geoarchaeological work under forests or in wetlands.

Vesicular A horizon Some A horizons in arid environments display rounded voids or vesicles. This type of A horizon is referred to as an Av horizon (McFadden, 1988).

Useful Variations Proposed by Others

Degree of pedogenic expression In the Canadian system for describing soils, a weakly expressed or juvenile horizon is indicated by "j" (Soil Classification Working Group, 1998, p. 14) Birkeland (1999, table 1.1) adapted this nomenclature for the USDA system, similar to the "w" described above, and found it quite useful (e.g., a weakly expressed or juvenile Bt horizon is designated "Btj").

Identification of lithologic discontinuities Numerical prefixes are used to denote lithologic discontinuities; that is, "a significant change in the particle-size distribution or mineralogy" of the parent material (Soil Survey Division Staff, 1993, p. 127; Schoeneberger et al., 1998, pp. 2–4). By convention, 1 is understood but is not shown (e.g., A-Bt-2C; table A1.1). Some geoarchaeologists use the numerical designator for lithologic discontinuities to indicate buried soils; that is, the numerical prefix changes for each buried soil regardless of the nature of the lithology (table A1.1; fig. 6.5; Kemp, 1999). As defined, however, the prefix is used and changes only in the case of a change in lithology, regardless of where in the profile such changes occur (e.g., "this volume" column in table A1.1). Profiles with no significant change in lithology should have no prefixes added to the horizonation, regardless of the presence of buried soils (e.g., in loess sections). Cremeens et al. (1998) modified this approach with an "S" prefix, noted earlier (table A1.1). The use of the "b" suffix (numbered, as described earlier) to identify buried soils obviates the need to use the numerical prefix for such identification. The *Soil Survey Manual* (Soil Survey Division Staff, 1993, p. 127) specifically indicates that the numerical prefix is not an indicator of a buried soil. In the Australian system (McDonald et al., 1998, pp. 110–113) the numerical prefix is used both for lithologic discontinuities and to number buried soils. There is no provision for numbering these components of soils when numerous buried soils and lithologic discontinuities are present, however.

Iron oxidation in the C horizon Iron oxidation is common in many unconsolidated parent materials that do not otherwise qualify as Bw horizons. In such situations, some investigators, particularly in the western United States, identify the zone as a "Cox" horizon (e.g., Birkeland, 1999, table 1.1). For clarity in stratigraphic work, where a Cox horizon is identified, any unweathered parent material is identified as a "Cu" horizon (adapted from European terminology; e.g., Hodgson, 1997, p. 93).

K as a master horizon The "K horizon" is a master horizon notation referring to a subsurface horizon so impregnated with pedogenic carbonate that its morphology is determined by the carbonate (Gile et al., 1965; Birkeland, 1999, table 1.1). The secondary carbonate coats or engulfs essentially all primary particles in a continuous medium. The zone is usually well-cemented. The K horizon was proposed as a result of USDA-sponsored research in south-central New Mexico (Gile et al., 1981) but was never adopted as USDA terminology. This was because it would set a precedent for elevating any horizon dominated by a single pedogenic processes to master horizon status (e.g., a zone very high in illuvial clay could be a "T" horizon; Holliday et al., 2002). Many investigators working in dry environments with calcareous soils have found the K = uc notation useful, however (e.g., Machette, 1985; Birkeland et al., 1991).

Welded and cumulic soils, and postburial alterations Few national soil survey programs have a system of field nomenclature that deals adequately with welded or cumulic soils. This is probably because most soil surveys do not deal with these aspects of soil genesis. One approach is to simply combine master and subhorizon symbols. For example, a Bk horizon welded to an underlying Ab horizon (or a cumulized A horizon converted to a B horizon, fig. 2.1B) could be identified as a BkAb or perhaps a BAkb. The notation "BA" in current USDA parlance, however, is used to indicate a transitional horizon between an A and a B, but more like a B. One innovative approach is used by Hajic et al. (1998). In their scheme, the secondary horizon (formed in the overlying sediment) is indicated in parentheses after the original buried horizon. Thus, an Ab(Bt) refers to a buried A horizon with a superimposed Bt horizon. A reverse of this approach also has some logic to it: Bt(Ab) for a Bt horizon superimposed over a buried A horizon. Valentine et al. (1980) propose a system of soil horizon nomenclature that reflects post-burial alteration and polygenesis of buried soils (table 10.1). This was developed in the context of geoarchaeological investigations of zones of human occupation and buried soils in a sand dune.

Appendix 2: Soil Phosphorus

Chemistry, Analytical Methods, and Chronosequences

Soil phosphorus chemistry is very complex, and many aspects of it are poorly understood. This has been a contributing factor to the large soil science literature on P. Soil P also occurs in several fractions, resulting in development of a wide array of analytical procedures to extract these fractions. Further, soil P will weather through time; a characteristic with significant archaeological implications, particularly in older sites. This appendix summarizes some of these basic characteristics of soil P. The first part of the discussion reviews the basics of P chemistry, focusing on its natural occurrence in soils and the sometimes confusing array of terms applied to the various forms of soil P. The next section summarizes the more common methods of extracting and measuring soil P in both soil chemistry and geoarchaeology and the relationships of the extractants to the known forms of P. Comparison of some of the geoarchaeological approaches to measuring soil P are further discussed in appendix 3. The final section of this appendix is a summary discussion of changes in soil P content through time.

Chemistry of Soil P

Most P in soils ultimately comes from rock weathering, particularly the weathering of apatite (Smeck, 1985, p. 187; Stevenson and Cole, 1999, pp. 282–283). This inorganic P is gradually taken up by plants, microorganisms, and animals as soils become established because P is an essential nutrient. These living organisms convert some P to organic forms within their cells. When the organisms die and decompose, both inorganic and organic forms of P are returned to the soil, initiating a biological cycle. The P that enters soils quickly becomes "fixed" or immobilized relative to other ions and compounds, because it has a strong tendency to be adsorbed onto surfaces of clay minerals or Fe- or Al-oxides; to form insoluble complexes such as Fe- and Al-phosphates (in acidic soils) and Ca-phosphates (in calcareous soils); or to be immobilized by microbes (Stevenson and Cole, 1999, pp. 290, 295–298).

Phosphorus occurs in two basic forms in soils: inorganic P and organic P. In both forms P occurs as one of two ions ($H_2PO_4{}_-$ or $HPO_4{}^{2-}$). Soil P is commonly subdivided and described in terms of its forms as extracted by various laboratory procedures. The nomenclature and even the categorization vary significantly, depending on the investigators, and with the passage of time, as more is learned about soil P. The scheme used by Stevenson and Cole (1999, p. 292, with minor modifications) is reasonably descriptive and comprehensive and partitions soil P into (1) insoluble organic forms of the soil biomass, in undecayed plant and animal residues, and as part of the SOM (humus); (2) soluble inorganic and organic compounds in the soil solution; (3) weakly adsorbed ("labile") inorganic P; (4) "sparingly soluble" or "nonoccluded" phosphates of Ca (in calcareous and alkaline soils of arid and semiarid regions), and Fe and Al (in acidic soils); (5) P strongly adsorbed or occluded by hydrous oxides of Fe and Al; and (6) fixed phosphates of silicate minerals.

Organic P is the P taken up by plants from the soil solution; this P is supplied by the soil from the labile P pool (Bethell and Máté, 1989, pp. 5, 6; Stevenson and Cole, 1999, p. 293). Phosphorus in organisms combines with organic compounds such as nucleic acids, phospholipids, phosphorylated polysaccharides, and phytin (Stevenson and Cole, 1999, p. 289). When a plant dies, the P it absorbed (the organic P) becomes mineralized or incorporated into soil microorganisms. The P in the microbes is also a component of the organic P. Some of this organic P is available to plants. The mineralized P will be found either in solution or will be fixed in the inorganic fraction, initially in the more labile pool (discussed later; Stevenson and Cole, 1999, fig. 9.1, pp. 292–293). As much as 50% of total P in A horizons is bound within SOM and, hence, is organic P (Barber, 1995, p. 210). Organic P is poorly understood and rarely dealt with in the archaeological literature, but given its abundance and considering that the source of some anthropogenic P is organic debris, it probably deserves more geoarchaeological scrutiny.

Within 24–48 hours after phosphorus is deposited in soil, some of it is changed into mineralized, inorganic P available for uptake by plants (Barber 1995, p. 206). This rapidly mineralized P also is termed "labile P" (Barber, 1995, p. 206; Stevenson and Cole, 1999, p. 293). The labile P "pool" also contains P that is weakly adsorbed to clay colloids (Stevenson and Cole, 1999, p. 293). Only a very small fraction (~1%) of the labile P is in solution and directly available for plant uptake. Most soil P (>90%) is insoluble or fixed as primary phosphate minerals, humus P, phosphates of Ca, Fe, and Al, and phosphate fixed by colloidal oxides and silicate minerals (Stevenson and Cole, 1999, p. 293). Phosphorus is essential for growth, but plants can only take it up in this weak solution. The soil P not quickly mineralized and moved into the labile pool moves into the inorganic nonlabile (occluded) pool.

In closing this section, a comment on terminology of soil P is in order. The terms "available P" and "labile P" are sometimes used interchangeably (e.g., Herz and Garrison, 1998, p. 182), along with terms such as "soluble P," "nonfixed P," and "nonoccluded P." These various forms of P are related but are not necessarily the same thing. Soluble P is the soil P in solution and is the direct source of P for plants. The soluble P comes from the much larger labile pool. The labile pool includes but is not necessarily the same thing as the nonfixed P or the nonoccluded P. Some of these forms of P are more labile or more easily converted to soluble P than other forms. The available P includes the soluble P and the labile P (Stevenson and Cole, 1999, pp. 292–293, 294–300).

Soil P Analyses

Macphail et al. (2000, p. 72) facetiously note that there "appears to be as many methods of extracting P from the soil as there have been workers in the field." They exaggerate, of course, but a grain of truth lies at the heart of their comment. Well over two dozen methods are published in the general soil chemistry and geoarchaeological literature (table A2.1) for determination of the several forms and fractions of P. Keeley (1981), Hamond (1983), Eidt (1984), Gurney (1985), Bethell and Máté (1989), and Macphail et al. (2000) provide discussions of some of the many methods used in archaeological contexts, but there is no comprehensive review. In soil science, several very useful compendia of laboratory methods are available, and all include discussion of P analyses. The American Society of Agronomy publishes a set of volumes on *Methods of Soil Analysis* that has gone through several editions (Black, 1965; Page et al., 1982; Sparks et al., 1996). The USDA also publishes the standard methods used by the National Resource Conservation Service (Soil Survey Laboratory Staff, 1996). In cooperation with the USDA and North Carolina State University, Pierzynski (2000) compiled a manual of standard methods of P analysis. The following discussion is a review commentary (building on that presented in chapter 11) on the methods used in soil P analysis, focusing on methods most commonly employed in archaeological contexts. Also included are published comparison studies of some of the methods and results.

There are two basic components to P analyses: the extraction of the P from the soil and measurement of the P in the extractant. Most research on soil P has focused on the extraction procedures because they get at the P added to the soil or available for plant uptake or both. The various approaches to extracting P from samples can be grouped into four to six basic categories (Gurney, 1985; Bethell and Máté, 1989, pp. 10–13; Terry et al., 2000, p. 153). The following groupings are made for this discussion: (1) extraction for available P (Pav); (2) chemical digestion of a soil sample for total P (Ptot); (3) measurements of organic P (Porg); (4) the spot test or ring test; (5) extractions of inorganic P (Pin) for fractionation studies and extractions to look at individual compounds of P; and (6) extractions for total elemental analysis. Approaches to measurement of P extractions fall into two basic categories: colorimetry and ICP. Most colorimetry is based on the technique of Murphy and Riley (1962; see discussion in Kuo, 1996, pp. 906–910) and

Table A2.1. Methods for analysis of soil phosphorus

Method[1]	References and comments[2,3]	Archaeological application
Total P (Colorimetry)		
Sodium carbonate (Na_2CO_3) fusion	Olsen and Sommers, 1982, pp. 404–406; Kuo, 1996; Bender and Wood, 2000	Davidson, 1973; Dormaar and Beaudoin, 1991; Simpson, 1997
Digest with perchloric acid ($HClO_4$)	Olsen and Sommers, 1982, pp. 406–407; Kuo, 1996; Bender and Wood, 2000	Mattingly and Williams, 1962; Ahler, 1973; Conway, 1983; Stimmel et al., 1984; Sandor et al., 1986; Skinner, 1986; Coultas, 1997; Sullivan, 2000; appendix 3, this volume
Digest with sulphuric acid (H_2SO_4)-hydrogen peroxide (H_2O_2)-hydroflouric acid (HF)	Bowman, 1988; Kuo, 1996	Leonardi et al., 1999; Lima et al., 2002
Oxidize with sodium hypobromite (NaOBr), dissolve in dilute sulphuric acid (H_2SO_4)	Dick and Tabatabai, 1977; Kuo, 1996; Bender and Wood, 2000	Keeley, 1981; Gurney, 1985; Sandor, 1987; Sandor and Eash, 1995
Total P? (Colorimetry)[4]		
Ignition at 240°C, extract with hydrochloric acid (HCl)	Olsen and Dean, 1965 Can be part of Porg procedure	Heidenreich et al., 1971,[5] Griffith, 1980
Ignition at 550°C, extract with sulphuric acid (H_2SO_4)	Olsen and Sommers, 1982, pp. 411–413 Part of Porg procedure	
Ignition at 550°C, extract with hydrochloric acid (HCl)	Hamond, 1983;[6] Meixner, 1986; Chaya, 1996[6] Bethel and Máté, 1989; use 550°C for noncalcarous soils, 400°C for any soil	Hamond, 1983; Chaya, 1996; Vizcaíno and Cañabate, 1999; appendix 3, this volume
Ignition at 550°C, extract with 2% citric acid ($C_6H_8O_7$) and HCl in calcareous soils (R. Macphail, personal communication, 2000)	Macphail et al., 2000	Engelmark and Linderholm, 1996; Macphail, 1998, table 1; Macphail et al., 2000
Extract by boiling in H_2SO_4	Cornwall, 1958, pp. 174–176; Shackley (1975, p. 69) describes this method as a measure of total P	
Extract with concentrated H_2SO_4 and concentrated HNO_3	Kolb and Homburg, 1991, adapted from Greenberg et al. (1992, pp. 4-108–4-117); see also Pote and Daniel, 2000	Homburg and Sandor, 1997

Total P (ICP)		
Dissolve with HF + aqua regia (HNO_3 + HCl)	SSLS 6S6a, 7C4a	
Digest with HCl + HNO_3		Lewis et al., 1994
Digest with HNO_3 + HF		McDowell, 1988
Total P? (ICP)		
Digest with hydroflouric acid (HF)	Linderholm and Lundberg, 1994	Linderholm and Lundberg, 1994
Extract with nitric acid (HNO_3)	Linderholm and Lundberg, 1994, developed in lieu of HF extraction for ICP-AES; results approach the total dissolution of HF	Linderholm and Lundberg, 1994
Digest with nitric acid-perchloric acid (HNO_3-$HClO_3$)	"Double-acid" extraction; Entwistle and Abrahams, 1997; Entwistle et al., 1998, 2000	Lewis et al., 1994; Entwistle and Abrahams, 1997; Entwistle et al., 1998, 2000; Parnell et al., 2002
Total Inorganic P by Fractionation (Colorimetry)		
"Chang & Jackson Fractionation"[7]	Chang and Jackson, 1957 (subsequently modified by Williams et al., 1967, and Syers et al., 1972; Bender and Wood, 2000; and others; see Kuo, 1996)[7]	Kerr, 1995; Leonardi et al., 1999
1. Extract with ammonium chloride	Water soluble/labile phosphate (easily soluble P)	
2. Extract with ammonium fluoride	Al-phosphate; exclude this step in calcareous soils (NH_4F-P)	
3. Extract with sodium hydroxide	Fe-phosphate (1st NaOH-P)	
(1 + 2 + 3)	(nonoccluded P)	
4. Extract with sodium citrate + sodium dithionite (extract with sodium citrate + sodium dithionite + sodium bicarbonate "CDB")	Reductant-soluble P (reductant-soluble P)	
5. Extract with sodium hydroxide	(2nd NaOH-P)	

continued

Table A2.1. (*cont.*)

Method[1]	References and comments[2,3]	Archaeological application
6. Extract with 0.5N sulphuric acid (H2SO4; extract with 0.5N HCl, then 1N HCl)	Ca-phosphate (acid-extractable Ca-P or Pca)	
	(residual organic P)	
7. Ignition at 550°C then extract with HCl	(residual inorganic P after 1–7; total P of original sample)	
8. Digest with Na_2CO_3 fusion	(occluded P)	
(4 + 5 + 8)	(total inorganic P)	
(1 + 2 + 3 + 4 + 5 + 6 + 8)	(organic P)	
(total P—total inorganic P)		
"Eidt Fractionation"	Eidt, 1984	Eidt, 1984; Sandor et al., 1986;[8] Moore and Denton, 1988; Overstreet et al., 1988; Lillios, 1992; Dunning, 1994; Schuldenrein, 1995; Brinkmann, 1996; Homburg and Sandor, 1997
Eidt I or NaOH + CB fraction: Extract with NaOH and sodium citrate-sodium bicarbonate	Easily extractable P; mainly loosely bound A1-phosphate and Fe-phosphate that is resorbed by $CaCO_3$, as well as the minute amount in solution; Pav	
Eidt II or CBD fraction: Extract with NaOH and sodium citrate-sodium bicarbonate-sodium dithionite	Tightly bound or occluded P; absorbed by diffusive penetration or by incorporation with Al and Fe oxides.	
Eidt III or HCl fraction: Extract with HCl	Fixed P within apatite or tightly bound to Ca-phosphate.	
"Hedley Fractionation"	Modified Chang and Jackson/Williams/Syers technique by Hedley et al., 1982	
Anion exchange resin	Extractable Pinorg	
Extract with $NaHCO_3$	Labile Porg and Pinorg, some microbial Pinorg	
Chloroform ($CHCl_3$) + $NaHCO_3$	Microbial P	
Extract with NaOH	Porg and Pinorg adsorbed to Fe and Al minerals	

Ultrasonification and extract with NaOH	Porg and Pinorg from internal surfaces of soil aggregates	
Extract with HCl	P from apatite; occluded P in weathered soils	
Digest with H_2SO_4 and oxidize with H_2O_2	Stable Porg and highly insoluble mineral P	
Organic P (Colorimetry)		
HCl + NaOH + NaOH (at 90°) for Pinorg; add $HClO_4$ for Ptot; Porg = Pt—Pinorg	Olsen and Sommers, 1982, pp. 408–411	
H_2SO_4 extract for Pinorg; ignite at 550°; add, H_2SO_4 for Pt; Porg = Pt—Pinorg	Olsen and Sommers, 1982, pp. 411–413	Courty and Nørnberg, 1985;[9] Schlezinger and Howes, 2000
Extract with dilute sodium hydroxide (NaOH), concentrated hydrochloric acid (HCl), and dilute sodium hydroxide (NaOH)	Kuo, 1996	
Extract with concentrated sulphuric acid (H_2SO_4) and dilute sodium hydroxide (NaOH) for Pinorg; digest extract in perchloric acid for Ptot; difference is Porg	Forster, 1995	Leonardi et al., 1999
Extract with acetylacetone	Kuo, 1996	
Available or Extractable P (Colorimetry)		
Extract with hydrochloric acid (HCl)	Chaya, 1996;[10] described as inorg P, used with ignition Ptot for Porg	Chaya, 1996
Extract by boiling in hydrochloric acid (HCl; total Pinorg?)	Provan, 1971; Sieveking et al., 1973; Craddock et al., 1985	Provan, 1971; Sieveking et al., 1973; Craddock et al., 1985; Gurney, 1985; Cavanagh et al., 1988; Wilkinson, 1988, 1990; Dockrill and Simpson, 1994; Dockrill et al., 1994
Extract with HCl in ultrasonic bath		Sanchez et al., 1996
Extract by boiling in H_2SO_4	Cornwall, 1958, pp. 174–176	Heidenreich et al., 1971;[11] Proudfoot, 1976
Extract with 0.002N H_2SO_4 buffered at pH 3 with $(NH_4)_2SO_4$	"Truog P"; Truog, 1930; Kamprath and Watson, 1980	Skinner, 1986
	Removes Ca-P; used in a widely available and popular field kit	

continued

Table A2.1. (*cont.*)

Method[1]	References and comments[2,3]	Archaeological application
Extract with sodium hydroxide (NaOH) and hydrochloric acid (HCl)	Unpublished except for Woods, 1984, p. 69, W. Woods, personal communication, 2002	Woods, 1984
Extract with acetic acid (CH_3COOH)	Conway, 1983	Conway, 1983
Extract with acetic acid (CH_3COOH) + sodium acetate ($NaC_2H_3O_2$); measure by comparison with color chips	Morgan "double acid" extract Morgan, 1941; Kamprath and Watson, 1980 Used in LaMotte STH series soil test kits; from www.lamotte.com; tends to extract less Pav than Olsen P or Bray l.	
Extract with sulphuric acid (H_2SO_4) + hydrochloric acid (HCl)	Mehlich-l or "double acid" or "North Carolina" soil test Mehlich, 1978; Kamprath and Watson, 1980; Olsen and Sommers, 1982, pp. 418–419; Kuo, 1996; Sims, 2000a Removes Ca-P and strongly fixed P in acid soils; extracts much more P than Bray-l; also used in LaMotte AST and DCL series soil test kits; www.lamotte.com	Hassan, 1981; Lima et al., 2002
Extract with acetic acid (CH_3COOH) + NH_4F + NH_4Cl + HCl Ammonium fluoride + ammonium chloride	Mehlich-2, dilute acid solution Mehlich, 1978 "Soluble and readily-labile P" of Terry et al. 2000, p. 155	Terry et al., 2000; Wells et al., 2000; Parnell et al., 2001, 2002; Fernández et al., 2002
Extract with acetic acid (CH_3COOH) + NH_4F + NH_4NO_3 + HNO_3 ammonium fluoride + ammonium nitrate + nitric acid	Mehlich-3 Mehlich, 1984; Sims, 2000b Results comparable to Mehlich-l, Bray-l and Olsen P	Coultas, 1997; Lima et al., 2002
Extract with water or dilute salt solution $CaCl_2$	P soluble in water (soil-solution P) Olsen and Sommers, 1982, pp. 419–420 Kuo, 1996; Self-Davis et al., 2000 Very small fraction of Pav	Eddy and Dregne, 1964[12]

Extract with sodium bicarbonate ($NaHCO_3$)	"Olsen P" Kamprath and Watson, 1980; Olsen and Sommers 1982, pp. 421–422; Kuo, 1996; Sims, 2000c Measures Al-P and Ca-P in calcareous, alkaline, or neutral soils; comparable to Bray 1	Dormaar and Beaudoin, 1991; Sandor and Eash, 1995; Leonardi et al., 1999
Extract with ammonium bicarbonate (NH_4HCO_3)	Olsen & Sommers, 1982, pp. 422–423 Kuo, 1996	
Extract with anion exchange resins	Olsen and Sommers, 1982, pp. 423–424 Kuo, 1996 Measures P uptake mechanisms by roots	
Extract with Fe oxide-impregnated filter paper	"P_i test" Kuo, 1996; Chardon, 2000 Gives results similar to anion exchange resin test but is less tedious	
Extract with isotopic dilution of ^{32}P	"Labile P" of Barber, 1995, p. 207 Olsen and Sommers, 1982, pp. 424–425 Kuo, 1996 Measures small amounts of water-soluble P	
Extract with 0.025N HCl + 0.03N ammonium fluoride (NH_4F) in 1:10 soil solution	"Bray & Kurtz P-l" or "Bray I" (absorbed P) SSLS 6S3,[13] McKeague, 1978, pp. 174 Olsen and Sommers, 1982, pp. 416–418 Sims, 2000d Easily acid-soluble P, largely Ca-phosphates and a portion of the Al- and Fe-phosphates; comparable to Olsen P	Ahler, 1973; Muhs et al., 1985; Sullivan, 2000
Extract with 0.1N HCl + 0.03N NH_4E in 1:17 soil/solution	"Bray II" or "medium strength" Bray McKeague, 1978, pp. 172–173 Easily acid-soluble P, largely Ca-phosphates and a portion of the Al- and Fe-phosphates	van der Merwe and Stein, 1972; Homburg, 1988

continued

Table A2.1. (*cont.*)

Method[1]	References and comments[2,3]	Archaeological application
Extract with 0.1N HCl + 0.05N NH_4F	"Strong Bray" McKeague, 1978, pp. 174–176 Easily acid-soluble P, largely Ca-phosphates and a portion of the Al- and Fe-phosphates	Moore and Denton, 1988
0.03N ammonium fluoride (NH_4F) + 0.03N sulphuric acid (H_2SO_4)	Modified "medium strength" Bray McKeague 1978, 4.44 Easily acid-soluble P, largely Ca-phosphates and a portion of the Al- and Fe-phosphates	
Extract with citric acid ($C_6H_8O_7$)	SSLS 6S5	Arrhenius, 1963; Engelmark and Linderholm, 1996; Macphail, 1998, Macphail et al., 2000; Lima et al., 2002; appendix 3, this volume
Extract with nitric acid (HNO_3) + ammonium molybdate; reduce with ascorbic acid ($C_6H_8O_6$)[14]	Spot test or Ring test or "Gundlach method"[14] Gundlach, 1961; Schwarz, 1967; Eidt, 1973; Woods, 1975	Schwarz, 1967; Eidt, 1973, 1977; Woods, 1975, 1984; Bakkevig, 1980; Keeley, 1981; Hamond, 1983; Gurney, 1985; Lippi, 1988; Bjelajac et al., 1996; Manzanilla, 1996b
Available or Extractable P (flow injection, automated ion analyzer)		
Extract with water, dilute salt extract ($CaCl_2$)	SSLS 6S7 "Soil-solution P" water-soluble fraction; principally consists of inorganic orthophosphate ions and possibly some small amount of organic P	

Extractable P (ICP)		
Extract with HCl	Middleton and Price, 1996 Probably yields some form of extractable P or Pav.	Middleton and Price, 1996; Thurston, 2001

[1] Categories are subdivided on the basis of measurement technique (mostly colorimetry or ICP). A variety of methods for colorimetry are available (e.g., Olsen and Sommers, 1982; Kuo, 1996; Soil Survey Laboratory Staff, 1996), most based on Murphy and Riley (1962). ICP extractions are for simultaneous measurement of a wide array of trace elements. Reagents: acetic acid (CH_3COOH), ammonium bicarbonate (NH_4HCO_3), ammonium chloride (NH_4Cl), ammonium fluoride (NH_4F), ammonium nitrate (NH_4NO_3), ascorbic acid ($C_6H_8O_6$), boric acid ($B(OH_3)$), chloroform ($CHCl_3$) ($C_6H_8O_7$), citric acid ($C_6H_8O_7$), hydrochloric acid (HCl), hydrogen fluoride (hydroflouric acid; HF), hydrogen peroxide (H_2O_2), nitric acid (HNO_3), perchloric acid ($HClO_4$), sodium acetate ($NaC_2H_3O_2$), sodium bicarbonate ($NaHCO_3$), sodium carbonate (Na_2CO_3), sodium citrate ($Na_3C_6H_5O_7$), sodium dithionite ($Na_2S_2O_4$), sodium hydroxide (NaOH), and sulphuric acid (H_2SO_4).

[2] Number following Olsen and Sommers (1982) is the identification system they used to refer to specific methods. Kuo (1996) does not follow this system.

[3] SSLS is the method described (and numbered) by Soil Survey Laboratory Staff (1996).

[4] These methods are described as measuring total P, but as discussed in appendix 2, they probably do not measure true total P.

[5] Heidenerich et al., 1971, do not indicate ignition temperature.

[6] Hamond (1983) and Chaya (1996) each used different colorimetry.

[7] Methods and terms in parentheses represent significant modifications of original Chang and Jackson (1957a) by Williams et al. (1967).

[8] Sandor et al. (1986) used Eidt I only for Pinorg or "moderately available P."

[9] Courty and Nørnberg (1985) do not indicate ignition temperature.

[10] Method not referenced, but possibly following Olsen and Sommers (1982, method 24-3.2.3).

[11] Heidenrich et al. (1971) note that their method is a modification of Cornwall's method but do not indicate the nature of the modification.

[12] Eddy and Dregne do not describe specific procedures for extraction or measurement; they may not be equivalent to the SSLS procedure.

[13] 6S3 Bray P-1 can be measured with a spectrophotometer or with a flow-injection automated ion analyzer.

[14] The original Gundlach (1961) procedure extracted with nitric acid. Eidt (1973) and Woods (1975) discuss substitution of nitric with hydrochloric acid for recovery of a greater variety and quantity of P compounds.

is not further discussed here. Several extraction techniques have been developed for total elemental analysis (including P) by ICP and are noted below.

Available P

Historically, considerable work on soil P in archaeology focused on available P (though some of the early work claimed to measure total P; chapter 11). This is probably because of the wide array of relatively simple methods for Pav analysis that have been available for decades (table A2.1). Kamprath and Watson (1980) present a useful though somewhat dated review discussion of testing soils for available P. In addition, the initial recognition of a relationship between human occupation and elevated levels of P was based on Pav (chapter 11). There are several significant problems in measuring and interpreting Pav. The varying methods for extracting Pav (table A2.1) will yield different amounts of P. This is largely because the different extractants get at different forms of Pav, depending on the strength of the extraction reagent and on the degree of solubility of the P (Kuo, 1996, p. 890). A simple water wash will get at the most easily extractable P, and progressively more vigorous techniques (e.g., citric acid extraction, boiling in HCl for 10 minutes, boiling in HCl for 2 hours) will yield progressively more P. The stronger reagents are probably extracting less-available inorganic P in addition to soluble and labile P.

Measurement of available P is traditionally done for measuring plant nutrient availability. Available P measures soil solution P and labile inorganic P, which is not a single simple chemical, compound, or mineral and represents only a minute portion (~1%–3%) of total P (Bethell and Máté, 1989, p. 6). Moreover, availability of P to plants depends on soil chemistry, water, texture, and structure and varies from soil to soil (Barber, 1995; Stevenson and Cole, 1999). Different plants also extract different amounts of P from the same soil. Determination of available P, therefore, may indicate roughly the P status of a soil but does not correlate with any particular P fraction that exists in nature (Eidt, 1984, p. 35; Bethell and Máté, 1989, p. 6). Estimates of easily extractable Pav are "extremely difficult to make because they attempt artificially to recreate chemical conditions around plant roots" (Eidt, 1984, p. 35).

Total P

Until the 1980s, analysis of Ptot in soils relied on several tedious techniques that included dangerous reagents such as perchloric acid and hydrofluoric acid (Dick and Tabatabai, 1977; Keeley 1981; Gurney, 1985; Chaya, 1996; Kuo 1996, p. 874; Macphail et al., 2000). Widely used methods included digestion in perchloric acid ($HClO_4$), fusion with sodium carbonate (Na_2CO_3), and sequential digestion in sulphuric acid (H_2SO_4), hydrogen peroxide (H_2O_2), and hydrofluoric acid (HF; table A2.1). The perchloric acid digestion does not extract all P, however, unless HF is added to the digestion solution to fully destroy minerals containing phosphate (Dick and Tabatabai, 1977). Dick and Tabatabai (1977) developed an alkali oxidation method that is simpler and safer than the others: boil with sodium hypobromite (NaOBr-NaOH), extract P with H_2SO_4, and determine P colorimetrically (table A2.1). They compared the results from this method with those from the other procedures, and their results compared favorably with $HClO_4$ digestion. The advent of ICP technology also provides a means of relatively simple and safe measurement of Ptot and is discussed in a separate section later.

Total P, as the name implies, measures all P in a sample, including anthropogenic, geogenic, and pedogenic P. Thus, soil parent materials high in P (e.g., high in apatite) will yield Ptot levels that overwhelm any signatures of human activity. This can be a significant drawback to soil geoarchaeological work in areas with high natural P.

Measurements of Ptot, along with Pin, are used in determination of organic P (Porg). The difference between Ptot and Pin is taken as Porg (discussed below). The total P is measured by colorimetry following ignition and then acid extraction. This procedure alone is rarely mentioned as a method of Ptot determination, probably because it is not a true measure of total P. Chaya (1996), however, used ignition and HCl extraction (table A2.1) to determine Ptot. A comparison with "total P" determined by electron microprobe showed generally similar results, but the microprobe method was not explained and is otherwise unknown.

A variety of analytical methods for soil P have been described as measuring total P, but clearly they are not. Most of these methods involve a strong acid such as hydrochloric (HCl) or sulphuric (H_2SO_4). Cornwall (1958, pp. 174–176) describes a method of phosphate analysis using 3 M H_2SO_4 and colorimetry; described by Shackley (1975, p. 69) as measuring "total P." Meixner (1986) also presents a method using HCl extraction described as measuring Ptot (similar to Chaya, 1996, noted earlier). The reagents employed in these methods are incapable of fully digesting all P-bearing minerals and compounds in the soil, however. For example, in a comparison (appendix 3), the HCl extraction produced significantly lower levels of P than did the perchloric digestion. Carr (1982, pp. 109–115) describes the P extractions from the early work of Arrehnius (1931, 1934), Lorch (1930, 1939, 1940, 1954), Dauncy (1952), Solecki (1951), Eddy and Dregne (1964), and Cook and Heizer (1965) (table 11.4) as total P, but most if not all of that early work dealt with easily extractable Pav.

In an archaeological context, an extraction procedure using concentrated sulphuric acid and concentrated nitric acid is described as yielding total P (Kolb and Homburg, 1991). In a comparison with perchloric extraction of samples from very low intensity occupations, the P levels from both methods are about the same, but in a similar comparison using samples from a shell midden, the P levels from perchloric extraction are fully an order of magnitude higher (appendix 3).

Organic P

Standard procedures for determination of Porg include ignition and acid-base extraction. Both are described by Olsen and Sommers (1982), but the acid-base extractions were later modified significantly based on the discussion by Kuo (1996). All of the methods vary in their efficiency and accuracy. A decision on choice of methods probably will depend on availability of lab facilities and desired accuracy.

Until the 1990s, Porg was measured indirectly (Kuo, 1996). In the ignition method, P is extracted with HCl following ignition (at 240°C, Olsen and Dean, 1965, or 550°C, Chaya, 1996; Meixner, 1986) or with H_2SO_4 following ignition at 550°C (Olsen and Sommers, 1982) and then measured by colorimetry. The extraction following ignition is assumed to represent total P. A nonignited sample is also extracted for inorganic P. Organic P is the difference between the Ptot and Pin. As noted above, the ignition and acid extraction probably is not a true measure of Ptot. Indeed, Walker (1964) uses the difference between H_2SO_4 extraction following ignition and without ignition as a means of estimating Porg, but uses the HF-HNO_3 digestion for Ptot. The ignition process probably does oxidize most Porg, and the difference between the ignited and nonignited sample probably provides an indirect means of Porg determination.

Spot Test

The spot test or ring test measures easily extractable P. In addition to the uncertainties over the meaning of Pav, the spot test results are qualitative and not always reproducible (Eidt, 1977, 1984, pp. 36–38; Gurney, 1985, p. 2). The original spot test described by Gundlach (1961) used HNO_3; but Eidt (1973) showed that HCl gives better results. The problems of using the spot test are well illustrated by Keeley (1981). In 20 applications of the spot test, there were 10 successful attempts at locating or defining occupation zones (in the United Kingdom and Sicily), four "partial successes" (i.e., slight enhancement of P allowed some inferences about occupation zones; in the United Kingdom and Peru), and six failures in which no relationship between soil P and occupations (in the United Kingdom) could be identified.

Inorganic P/P Fractions

There are many procedures for extracting various forms and fractions of inorganic P (table A2.1). Each method probably gets at different forms of P. In comparing results with that of other investigators, considerable care must be taken to reproduce the methods used. For example, Heidenreich et al. (1971) compared an extraction from a nonignited H_2SO_4 treatment (based on Cornwall, 1958, pp. 174–176; table A2.1) with an ignited HCl extraction. The ignited HCl extraction yielded higher P, in part because of the oxidation of Porg during ignition.

Inorganic P is sometimes divided into two fractions, based on where the P ion resides: "nonoccluded P" refers to P ions sorbed at the surfaces of Fe and Al oxides and hydrous oxides and CaCO3; "occluded P" refers to P ions within mineral matrixes. The terms "occluded" and "nonoccluded" refer to lab extractions, and researchers are unclear about the degree to which these fractions actually relate to their hypothesized position in or on minerals.

A well-known approach to measuring Pin fractions in both soil science and pedology is the sequential extraction procedure developed by Chang and Jackson (1957a), modified by a number of investigators (e.g., Williams et al., 1967, 1971; Hedley et al., 1982; table A2.1). This method has been used in soil chronosequence studies (discussed below). The best known of the fractionation methods in geoarchaeology is that of Eidt (1977, 1984), based on Williams et al. (1971; Overstreet et al., 1988, p. 123; chapter 11). The basic fractionation scheme involves extraction of Fraction I, solution P and loosely bound Al and Fe phosphates ("Pav" and non-occluded P); Fraction II, tightly bound or occluded forms of Al and Fe oxides and hydrous oxides; and Fraction III, occluded Ca phosphates (compare with other soil P factionations in table A2.1; Eidt, 1984, p. 42). Eidt (1984, pp. 41, 43) further proposed that sequential extraction of various forms of inorganic P provide a measure of total inorganic P. According to Sandor et al. (1986, p. 178), Eidt's inorganic P probably does not extract all inorganic phosphorus.

Inductively Coupled Plasma Techniques

The advent of ICP (inductively coupled plasma) spectrometry in the 1980s and 1990s provided a relatively rapid method of total elemental extraction and measurement, including P. Two approaches are used in ICP work: inductively coupled plasma–mass spectrometry (ICP–MS) and inductively coupled plasma–atomic emission spectrometry (ICP–AES). In ICP–AES the plasma heats the atom and a phototube measures the intensity of the color of the element. ICP–MS uses a plasma only to ionize the atoms and then sucks them into a magnet at high voltage, which bends them around a curve to an ion detector, sorting them by mass (the paths of light atoms bend more than the paths of heavier ones). The two methods do the same thing in different ways. In practice ICP–AES is best for elements that easily glow when hot; generally the top and left part of the periodic table. ICP–MS is more sensitive and can measure less-abundant elements, and it works better with heavy ions; generally the bottom and right side of the table (J. Burton, personal communication, 2002).

Some inferences about the form of P analyzed by ICP may be possible on the basis of the extraction procedure. Linderholm and Lundberg (1994) argue that trends in elemental concentration of P using total digestion of samples by hydroflouric acid (HF; i.e., probably yielding Ptot) can be reproduced more quickly by analyzing extracts made from nitric acid (HNO_3; i.e., some form of extractable P, rather than Ptot). Subsequent research by others (Entwistle and Abrahams, 1997; Entwistle et al., 1998, 2000) involved a nitric acid-perchloric acid (HNO_3—$HClO_3$) digestion for the extract, which may approximate Ptot given that perchloric extraction is a common method for Ptot. Entwistle et al. (2000) recovered low levels of P, however, possibly because the site they studied was not a farm; that is, no P-bearing manure was used on it. A simple HCl extract was used by Middleton and Price (1996), probably yielding some form of extractable P or Pav.

Several investigators have noted some problems or potential problems with P analysis by means of ICP spectrometry. One is that anthropogenic P may be present in too high a concentration to be determined along with trace elements (Entwistle and Abrahams, 1997). The P can be analyzed as a single element, but this may be difficult to justify economically. Entwistle and Abrahams (1997, p. 415), as part of their ICP research, determined P by one of the standard colorimetric methods (table A2.1). Another issue is chemical interaction among elements (McDowell, 1988). Phosphorus, for example, has a tendency to strongly bind with other elements such as Ca to form insoluble compounds. Depending on extraction procedures, therefore, elevated levels of P may or may not be detected.

P Comparisons

Considering the many methods available to extract and measure soil P, there have been surprisingly few attempts at comparing results to see which methods produce more or less similar values, which methods produce the highest levels of P, which methods yield results that best correlate with anthropogenicaly altered soils, and which methods yield sound results and also are the most efficient and suitable for the field. Appendix 3 presents the result of one of these studies. Only a handful of others are available.

The first method of P extraction used in archaeological contexts and one still employed is based on 2% citric acid. Though citric acid is traditionally viewed as an extractant for Pav, Macphail et al. (2000) argue that it gets at most of the inorganic P found in acid soils. Their summary discussion (Macphail et al., 2000, p. 72) certainly makes a good case that the method removes large amounts of P, but they do not clearly indicate that they are removing only easily extractable P and not including other forms of organic and inorganic P. Bakkevig (1980, p. 86) notes that 2N HCl extracts 10 times as much P as 2% citric acid.

Kuo (1996, pp. 870, 874) summarizes the results of the various methods of total P measurement. The perchloric acid and sodium hypobromite digestions yield comparable results, but both methods may underestimate total P in proportion to P embedded in the matrix of silicate minerals such as quartz. Sodium carbonate fusion and the H_2SO_4-H_2O_2-HF method both tend to extract more P than the perchloric acid and sodium hypobromite digestions. The difference between total P by fusion and that by perchloric acid digestion increases as the proportion of fine and coarse sands in the soil increases.

Ahler (1973) presents one of the earlier studies comparing P methods in an archaeological context. Samples were taken from fill at Rodgers Shelter, Missouri. The study compared the results of Ptot by perchloric acid digestion (table A2.1) to Pav by the Bray I acid extraction (table A2.1). He then compared both results to the density of occupation debris. Ptot varied as a function of artifact density through the stratigraphic sequence. Elevated Pav correlated with artifact density only in the lower levels, but not the upper deposits. The upper levels accumulated much more slowly than the lower ones, however, and the relatively low levels of Pav in the upper deposits were attributed to weathering.

Skinner (1986) compared a wide variety of P methods to determine which was most reliable for identification of anthropogenicaly modified soils in Ohio. The methods included the spot test, a Hellige-Truog kit (a commercially available kit for field analysis of soils; the P method is the Truog H_2SO_4 extraction for Pav; table A2.1), perchloric acid extraction for Ptot, and HCl extraction for Pin. The perchloric acid extraction produced the highest correlation with anthrosols, but was positive only 60% of the time.

Leonardi et al. (1999) compared different forms and extractions of P to determine which best supported the interpretation of ancient agricultural use of now buried soils. They looked at Porg versus Ptot as well as the fractionation of P. The results of both the Porg and Ptot analyses supported the interpretation based on archaeological indicators of agricultural use of the soils, but the latter method is quicker and less expensive than the fractionation scheme (Leonardi et al., 1999, p. 352).

Terry at al. (2000) evaluated a soil test kit for use in an archaeological field laboratory. The P method in the kit was based on the Mehlich-2 dilute acid procedure (table A2.1) and was compared to a bicarbonate extraction (for a Pin fraction; Olsen and Sommers, 1982, procedure 24-5.4) (table A2.1), perchloric acid digestion (for Ptot), and the ring test (rated on a 1–5 scale) of Eidt (1973). The trends in both the Mehlich and bicarbonate procedures are similar, but the Mehlich procedure produced more extractable P than did the bicarbonate method. The total P digestion produced roughly 100 times the P of the Mehlich procedure, which was expected because the method dissolves all P in the parent material. The values from the Mehlich method could not be used to estimate total P. Both of the inorganic fractions (Mehlich and bicarbonate) better reflected anthropogenic alterations of the soil than did the total P, however. The ring test results had only a moderate correlation with the Mehlich results. In the end, Terry et al. (2000) decided to use the Mehlich 2 kit because of its sensitivity to human inputs of P, its efficiency and cost, and its portability.

Several studies have compared methods of extracting and measuring available P. Kamprath and Watson (1980) show that the "Bray 1," "Olsen P," and "North Carolina" (or "Mehlich 1") tests (table A2.1) produce generally comparable results. In a geoarchaeological context, Proudfoot (1976, p. 95) compared available P levels based on a variety of extractants (using calcium lactate, hydrochloric acid, acetic acid, citric acid, and distilled water). The values generally ranged significantly, both because of P levels and because of chemical reactions during the extraction processes.

The Fate of Soil P Over Time

A key tenet of soil P research in archaeology is that at archaeological time scales (decadal to millennial) P remains "fixed" or "bound" (e.g., Eidt, 1984, pp. 26–28). Data from soil chronosequences show, however, that the P content of soils and, in particular, the relative proportion of Pin fractions can change over time. This process seems most apparent and dynamic in acidic soils or those in leaching environments (Walker, 1964; Smeck, 1985; Birkeland, 1984, p. 211, 1999, p. 185). Walker and Sayers (1976) and Birkeland (1984, p. 211; 1999, p. 185) summarize a number of studies indicating that P extractable in HCl ("Pca" of Walker and Sayers, 1976; roughly equivalent to Ca-phosphate, both mineralized labile and nonlabile; table A2.1) is depleted with time, converting to Ptot. In their pioneering study specifically focused on changes in soil P over time (fig. A2.1), Walker and Sayers (1976) looked at four nonanthropogenic chronosequences in New Zealand (table A2.2). Three of the soils formed under forest and one under grassland. Total P declined noticeably after 10,000–20,000 radiocarbon yr, but elevated levels were still apparent after as much as 130,000 yr.

Changes in soil P in the A horizons of the New Zealand chronosequences are particularly illustrative of alterations in forms of P over time and as a function

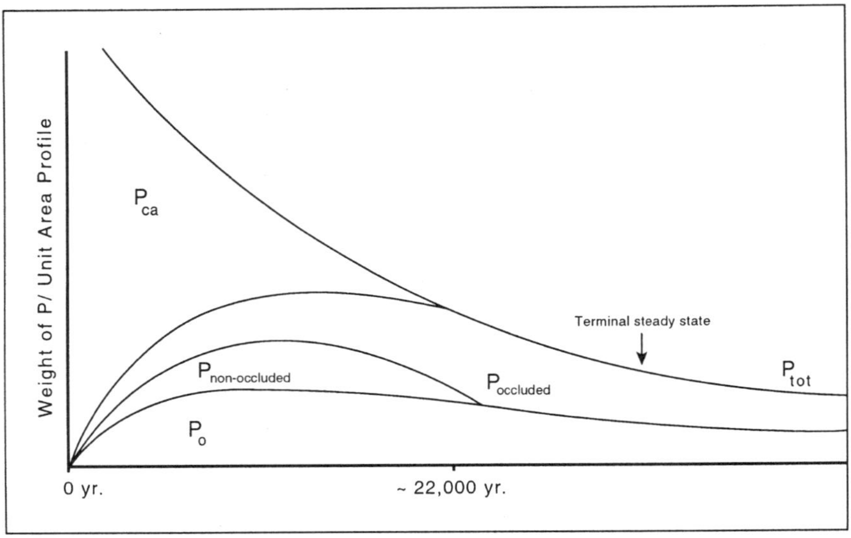

Figure A2.1 Changes in the forms and amounts of soil P fractions in a New Zealand chronosequence (reprinted from *Geoderma*, v. 15, T. W. Walker and J. K. Syers, "The fate of phosphorus during pedogenesis," pp. 1–19, fig. 1, © 1976, with permission from Elsevier Science). The forms of soil phosphorus are as follows: Ptot = total P (by Na_2CO_3 fusion); Pca = Ca-phosphate (extracted by HCl); Poccluded = reductant-soluble P (extracted with citrate-dithionate-bicarbonate) + 2nd NaOH extraction + residual inorganic P (by Na_2CO_3 fusion); Pnon-occluded = easily soluble P (extracted with NH_4Cl) + NH_4F-P + 1st NaOH-P; Porg = inorganic P-total P (see table A2.1, modifications to "Chang and Jackson Fractionation").

of environment (table A2.2). In the most acidic, intensely leached soils (Franz Joseph chronosequence), Ptot initially declines, then rises after 1000 radiocarbon yr because of the decomposition of plant SOM into Porg. The total P then steadily declines, though significant amounts remain after 22,000 radiocarbon yr. The next-drier soils (Reefton chronosequence) show a decrease over time but at a slower rate (and with a "plateau" of sorts at 14,000–16,000 yr B.P.). These soils have roughly the same amount of Ptot after 130,000 yr of weathering than the wetter Franz Josef soils had after 22,000 yr. The driest of the forest soils (Manawatu chronosequence) fluctuate in Ptot over the last 10,000 radiocarbon yr, with the oldest soil containing more total P than the youngest soil. In the dry grassland soils of the Canterbury chronosequence, Ptot remains high probably because there is no leaching, though only two profiles going back 6500 radiocarbon yr were available for study. In all of the chronosequences, levels of Pa drop rapidly, with very low levels remaining after just a few thousand years.

The persistence of total P in soils thousands of years old, even in leached ones, indicates that elevated levels of total anthropogenic P should also be preserved for thousands of years. Fractions of inorganic P (e.g., acid-extractable P) may not persist as long, especially in leaching environments, but the absolute levels of inorganic P fractions will undoubtedly depend on the amount of anthropogenic

Table A2.2. Phosphate in A horizons of New Zealand chronosequences

Sequence	Parent material	Mean annual precipitation (mm)	Mean annual temperature (°C)	Vegetation	Soil age[1]											
					Ptot	Pca	Ptot	Pca	Ptot	Pca	Ptot	Pca	Ptot	Pca	Ptot	Pca
Franz Josef	Glacial outwash	5090	11	Forest	0 yr		100 yr		1000 yr		5000 yr		12,000 yr		22,000 yr	
					766	719	652	368	867	25	373	3	328	0	163	0
Reefton	Alluvium	2030	11	Forest	Recent		14,000 yr		16,000 yr		23,000 yr		70,000 yr		130,000 yr	
					1013	439	719	33	719	11	694	9	278	0	179	0
Manawatu	Coastal dune	850	12	Forest	0 yr		50 yr		500 yr		3000 yr		10,000 yr			
					347	187	468	178	620	164	408	37	584	0		
Canterbury	Alluvium	650	10	Grassland	500 yr		6500 yr									
					654	221	797	52								

From Walker and Syers (1976, tables I and II).

[1] For surface horizons only; dates represent the age of the landscape and are in radiocarbon yr B.P. "Pca" is acid-extractable, Ca-bound P (Table A2.1). The pairs of numbers below each date is Ptot and Pca in ppm.

P added to the soil. Levels of inorganic P may remain higher, longer if initial amounts of anthropogenic P were high. In a study of the same chronosequence, Baker (1976) also showed that the ratios of total organic P and fractions of organic P changed through time and from the top of the soil down through the profile.

Woods (1975, p. 29) and Eidt (1977, p. 1328) report that both high- and low-pH soils retain P well through binding with either Ca or Fe/Al and further assert that soil P may be leached more rapidly in neutral soils. This trend is not reported by any other workers, however, and the data of Walker and Sayers (1976) and Birkeland (1984, 1999) clearly indicate that P is depleted more rapidly in soils that are more acid and more intensely leached. An important consideration here is that the factors influencing P retention and mobility are complex and varied. In contrast to sandy, podzolized soils mentioned above, in intensely weathered tropical soils high in clay, Al fixation of PO_4^{-3} is extremely strong, and almost no leaching of P occurs.

Appendix 3: Variability of Soil Laboratory Procedures and Results

VANCE T. HOLLIDAY
JULIE K. STEIN
WILLIAM G. GARTNER

This appendix (modified from Holliday and Stein, 1989) reviews several methods and results for some of the more commonly measured attributes of sediments and soils in geoarchaeological research. The purpose is not to recommend one method over another, but to illustrate the possible variability among techniques and results, to caution geoarchaeologists concerning comparisons with data from different investigators, and to point out some of the factors that must be taken into consideration when selecting a particular technique.

The samples analyzed are from two different kinds of archaeological sites in two different settings: the British Camp Site (45SJ24) on San Juan Island, Washington, and the Lubbock Lake Site (41LU1), Texas. The British Camp Site is a large, prehistoric, Northwest Coast shell midden underlying the historic site of British Camp. During the last 2000 yr, human occupation debris as well as marine and terrestrial fauna and flora and abundant mineral material were deposited along the shoreline (Stein, 1992b). Samples used in this study were collected from individual layers within the midden, as well as from exposures of non-cultural deposits located near the site. Lubbock Lake is in a dry valley on the semiarid Southern High Plains (Johnson, 1987b). The samples are from the profiles of several well-drained soils developed in sandy, eolian, and colluvial sediments making up much of the late Holocene valley fill (Holliday, 1985b,d).

The techniques compared are those for SOM, soil organic carbon (SOC), calcium carbonate ($CaCO_3$), and phosphorus (P). Particle-size analysis is also discussed, although not compared thoroughly. The soil geomorphic significance of organic matter and calcium carbonate content of soils and sediments is discussed in chapters 5, 7, 8, 10, and 11. Soil organic carbon is a component of soil organic matter, which includes plant, animal, and microbial residues (fresh and at all stages of decomposition), humus, and inert carbon forms such as charcoal, coal, and graphite (Nelson and Sommers, 1982; Stein, 1984, 1992c). SOM content of surface horizons should be 1.7 to 2.0 times that for SOC—maybe higher—and up to 2.5 times SOC in subsurface horizons (Nelson and Sommers, 1982; Tabatabai, 1996). Particle-size analysis (also referred to as grain-size analysis or granulometry) is a measure of the distribution of gravel, sand, silt, and clay content and is a basic part of pedological studies (e.g., clay content) as well as sedimentological analyses (including site-formation processes) (Timpson and Foss, 1994). Measurement of $CaCO_3$ provides an indication of leaching or secondary carbonate accumulation (Foss et al., 1993b). The analysis of archaeological sediments and soils for phosphorus content as an indicator of human presence and impacts is long established in archaeology and probably was the first widespread application of a standard soil analysis in archaeology, beginning in the 1930s (see discussion in chapter 11, table 11.4, and historical summaries by Woods, 1975; Eidt, 1984; and Bethell and Máté, 1989). This appendix reviews various methods applied to duplicate samples for OC, OM, $CaCO_3$, P, and particle-size analysis, followed by a discussion of results focusing on similarities in absolute values and trends.

Methods

In the following discussion a variety of analytical techniques are reviewed. Methods that are published are cited, and if the procedures are not published or are modified from published accounts, then they are described.

Organic Carbon and Organic Matter

Organic carbon and organic matter were determined on samples from British Camp following four different techniques; two of the techniques were employed on samples from Lubbock Lake. The SOC content for samples from both sites was measured using the Walkley-Black technique (Janitzky, 1986a). In this method a known quantity of a strong oxidizing agent is used to oxidize the SOC. The amount of SOC present is then determined by measuring the remaining oxidizing agent using reduction titration. Samples from British Camp were also analyzed for SOC by calculating the difference between total carbon and total inorganic carbon from carbonates. Total carbon was determined by dry combustion, a method based on the oxidation of SOC, thermal decomposition of carbonate minerals in a total-carbon analyzer (a medium-temperature resistance furnace), and measurement of the liberated CO_2 (Soil Survey Staff, 1972, method 6A2b; Nelson and Sommers, 1982). Total inorganic carbon from carbonates was measured using a Chittick apparatus (see following). A variation on this method of determining OC by difference is described by Foscolos and Barefoot (1970).

The SOM content of samples from Lubbock Lake and British Camp was determined using a loss-on-ignition technique (Stein, 1984), and the SOM in samples from British Camp was also measured, using a modified hydrogen peroxide weight-loss technique (Robinson, 1927; Soil Survey Staff, 1972, method

6A3a). In the loss-on-ignition technique the sample was crushed to pass a 2-mm screen, then dried at 100°C for 1 hour and ignited at 500°C to burn off organic matter. The weight loss calculated from before and after the 500°C burn represents the organic matter content of the sample. In the peroxide method the samples were dried and crushed to pass a 2-mm screen. Carbonates were removed using 10% HCl, and a carbonate-free weight was obtained after drying. The samples were placed in 500-mL beakers and warmed to medium temperature on a hot plate. Organic matter was oxidized by adding progressively more concentrated H_2O_2 (5%, 15%, 33%) over a period of 3 days. The samples were then dried and reweighed and the SOM content was calculated as the percentage weight difference before and after the peroxide treatment.

Calcium Carbonate

The $CaCO_3$ content of the samples was measured using three methods. Samples from British Camp and Lubbock Lake were analyzed using the acid-neutralization method (U.S. Salinity Laboratory Staff, 1954) and the Chittick apparatus (Machette, 1986). In both of these methods the carbonates were destroyed using HCl; in the acid-neutralization method the carbonate content was determined by measuring the amount of remaining HCl using titration, and in the Chittick-apparatus method the volume of CO_2 gas evolved from the reaction was measured. The carbonate content for the British Camp samples was also determined using loss-on-ignition (Stein, 1984). In this method samples are dried for 1 hour and burned at 1000°C. The weight loss calculated from before and after the 1000°C burn represents the carbonate content of the sample.

Particle-Size Analysis

Particle-size analysis is most often conducted by either sieving or settling or a combination of methods (e.g., Shackley, 1975; Folk, 1980; Soil Survey Laboratory Staff, 1996; Gee and Bauder, 1986; Janitzky, 1986b), including various pretreatments (see Shackley, 1975, and Timpson and Foss, 1994, for a succinct summary of pretreatment and measurement techniques in archaeological contexts). Sieving is usually used for fractionation of coarser material (sand-size and larger), and settling is used for particle-size analysis of the finer fraction. The settling method relies on the relationship that exists between settling velocity of particles in water and the particle diameter. This relationship is expressed by Stoke's Law, which essentially states that smaller particles settle at slower rates. There are two common methods of particle-size analysis by settling, hydrometer and pipet, and both were used in this study. In the hydrometer method a hydrometer is used to measure changes in the density of the water-sediment suspension at specific time intervals as the particles settle out. With the pipet, small subsamples of the sediment-water mixture are taken at specified depths at specified time intervals, dried, and weighed. Calculations of the particle-size distribution are made using Stoke's Law following both procedures. Samples for particle-size analysis are commonly treated for the removal of secondary constituents, especially organic matter and carbonates, before the analysis.

In this study, samples from Lubbock Lake were analyzed using a combination of sieve, hydrometer, and pipet techniques and pretreated with sodium acetate (NaOAc) and H_2O_2. Samples from British Camp were analyzed using sieve and pipet techniques and pretreated using sodium hypochlorite (NaOCl; e.g., commercial Chlorox bleach) and HCl. No samples were analyzed using all combinations of particle-size analysis and pretreatments. Thus, the comparisons suffer accordingly. The differences in results, however, based on methods of pretreatment, draw attention to the importance of carefully considering procedures before comparisons are made between sample results.

The particle classification for samples from Lubbock Lake is based on the U.S. Department of Agriculture system for soils (Soil Survey Staff, 1993), with particles divided into sand (2–.05 mm or 50 μm), silt (50–2 μm), and clay (≤2 μm). Although the sands can be further subdivided, they were not in this study. The particle classification system used for samples from British Camp follows the phi (ϕ) scale based on the Wentworth logarithmic scale (Folk, 1980, p. 23), both of which are commonly used in sedimentology (Blatt et al., 1980). The particles are classified at 1-ϕ intervals: –1 (2 mm), 0 (1 mm), 1 (.5 mm), 2 (.25 mm), 3 (.125 mm), 4 (.063 mm or 63 μm), 5 (32 μm), 6 (16 μm), 7 (8 μm), 8 (4 μm), 9 (2 μm), 10 (1 μm), and 11 (.5 μm).

Two approaches to particle-size analysis were employed on portions of the same samples from Lubbock Lake. In the first, organic matter and calcium carbonate were not removed from the samples. The samples were placed in a beaker with water and sodium hexametaphosphate for dispersing. The sand, silt, and clay content of the samples was then determined using the hydrometer method. In the second approach the samples were treated for removal of $CaCO_3$ and OM, using NaOAc and 30% H_2O_2, respectively (Gee and Bauder, 1986). The sand was then wet-screened, dried, and weighed. The silt and clay mixture was dispersed using buffered sodium pyrophosphate, and size fractions were determined using the pipet method.

Samples for particle-size analysis from the British Camp site were pretreated for organic matter using NaOCl, following Jackson (1969). The samples were then split into halves, one half pretreated for removal of carbonates (using HCl) and the other half left untreated. Samples were washed through a 4-ϕ screen. The sand fraction (greater than 63 μm) retained on the screen was dried, sieved, and weighed. The silt-clay fraction was dispersed in sodium hexametaphosphate and analyzed in 1-ϕ size intervals using the pipet method.

Phosphorus

Samples from Lubbock Lake and San Juan Island were subjected to four methods of soil P determination: perchloric acid digestion for total P (Olsen and Sommers, 1982; Kuo, 1996); sulphuric-nitric acid extraction for total P (Kolb and Homburg, 1991); hydrochloric acid extraction after ignition for extractable P (Meixner, 1986); and citric acid extraction for available P (Soil Survey Laboratory Staff, 1996, method 6S5; see also appendix 2 and table A2.1). All extractions were carried out at the University of Wisconsin–Milwaukee Soils Laboratory. The sulphuric acid-nitric acid extractant, the hydrochloric acid extractant, and the citric

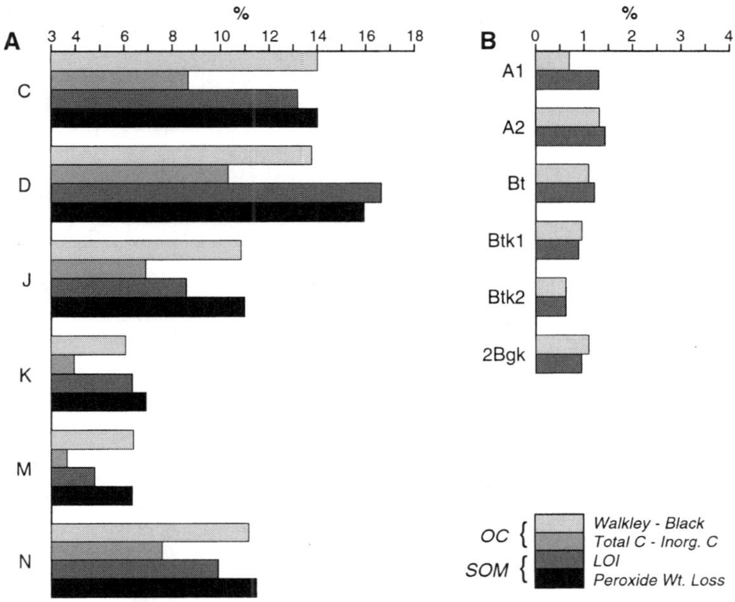

Figure A3.1 Comparisons of the results of analysis for samples from British Camp (A) and Lubbock Lake (B) for organic carbon (OC) and organic matter (OM) content (modified from *Geoarchaeology* v. 4, pp. 347–358, fig. 1, by V. T. Holliday and J. K. Stein, © 1989, John Wiley & Sons, used by permission of John Wiley & Sons, Inc.). Left-most column of each graph indicates field designations for samples (soil horizons at Lubbock Lake).

acid extractant were measured on a Spectronic 20 spectrophotometer. The perchloric acid extractant was measured on an ICP in the Soil and Plant Analysis Laboratory of the University of Wisconsin–Madison.

Results and Discussion

The four methods of SOC and SOM determination provide results with generally similar trends, but with some significant differences in absolute values (fig. A3.1). For the samples from British Camp, the results from both methods of SOM determination are generally similar. The SOC content of the British Camp samples based on total carbon is consistently and significantly lower than those for SOM. This relationship is to be expected because, as noted earlier and in chapter 11, SOC is one component of SOM. Considering this relationship, the levels of SOC by Walkley-Black are surprising because the results are generally similar to those for SOM.

In comparing the values of SOM determined by loss-on-ignition with the values of SOC determined by Walkley-Black, the SOM is slightly lower than SOC in four of the six samples from British Camp, and the SOM and SOC are about the same in five of the six samples from Lubbock Lake (fig. A3.1). These data are in contrast to other published results that suggest that the loss-on-ignition technique yields values that are consistently higher than Walkley-Black determined on calcareous and noncalcareous samples (Ball, 1964, ignition at 375°C and 850°C; Davies, 1974, ignition at 430°C; Tabatabai, 1996). Clay and carbonate content can affect loss-on-ignitions results (e.g., Davies, 1974; Tabatabai, 1996), and these factors are accounted for in the results for the British Camp samples, although not those from Lubbock Lake. Such corrections in the Lubbock Lake samples would result in even lower loss-on-ignition values, although the Lubbock Lake samples contain generally less than 20% clay (on a carbonate-free basis; fig. A2.2). The Lubbock Lake samples may simply have SOC levels below the accurate resolution of loss-on-ignition. Variations in sample size and in methods among laboratories also are shown to affect LOI results (Heiri et al., 2001). In general, loss-on-ignition is considered to be the "least reliable" method for OC determination (Tabatabai, 1996, pp. 1–2).

The mineralogy of the clay present in a sample can also affect loss-on-ignition results (Ball, 1964; Dean, 1974). The samples from both sites are lithologically homogeneous; the Lubbock Lake sediments are quartzose with clays dominated by illite, mixed-layer illite-smectite, and minor amounts of smectite and kaolinite (Holliday, 1985d). No analysis has been done on the clay mineralogy of the British Camp samples. The British Camp samples analyzed in this study contain relatively small amounts of clay (generally less than 20% on a carbonate-free basis; fig. A3.2).

The results of the $CaCO_3$ techniques are as variable as those for SOC and SOM. The acid-neutralization and Chittick methods yielded generally similar results for the samples from Lubbock Lake (fig. A3.3). For the British Camp samples the values from acid-neutralization are significantly higher than those from the Chittick method except for sample M (fig. A3.3). The results from loss-on-ignition are consistently 10%–15% higher than those from Chittick and, except for sample M, consistently just a few percent higher than acid-neutralization. The significance of the difference between the values from the Lubbock Lake and British Camp is not known. A significant difference does exist in the nature of the carbonates at the two sites, however. The British Camp carbonate is from shell and limestone (primarily $CaCO_3$), whereas the carbonate in the Lubbock Lake samples is pedogenically modified and derived from aerosolic dust and minor amounts of primary carbonate (also dominantly $CaCO_3$) in the sediment.

A number of studies of PSA indicate that the results from hydrometer and pipet analyses are usually in agreement (Liu et al., 1966; Walter et al., 1978), with the pipet method perhaps giving slightly better overall results (Sternberg and Creager, 1961). Because no one method of PSA was used on both the Lubbock Lake samples and the British Camp samples, the PSA results from the two sites cannot be compared. Within the Lubbock Lake samples different methods of PSA were also used before and after pretreatment. Even though this difference occurred, most of the variability in the Lubbock Lake results is probably caused by the effects of pretreatment, because both PSA methods have been tested and seen to give similar results. The PSA results from both Lubbock Lake and British Camp show marked differences between samples that were pretreated and those that were not (fig. A3.2). The comparisons from both sites show higher clay

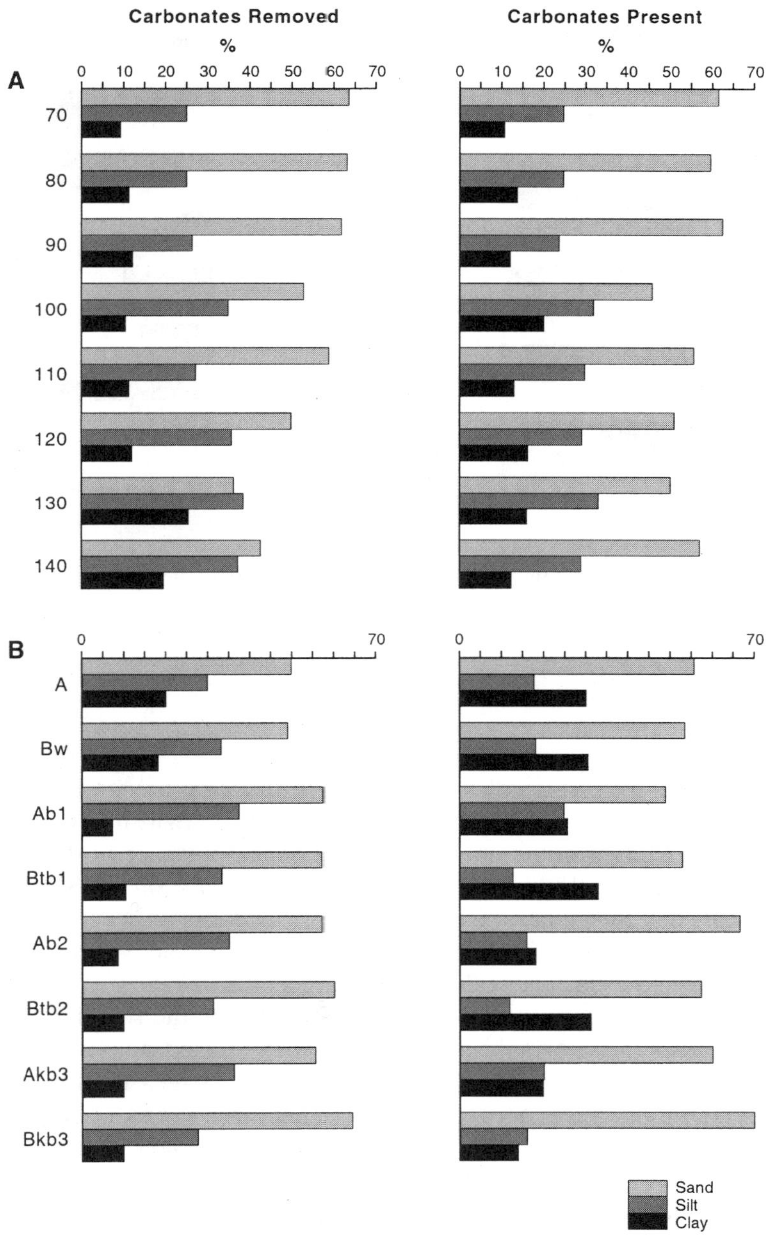

Figure A3.2 Comparisons of the results of particle-size analysis for samples pretreated for carbonate removal and samples not pretreated from British Camp (A) and Lubbock Lake (B). Modified from *Geoarchaeology* v. 4, pp. 347–358, fig. 2, by V. T. Holliday and J. K. Stein, © 1989, John Wiley & Sons, used by permission of John Wiley & Sons, Inc. Numbers and letters in left-most column refer to field designations of samples (depth below surface in cm for British Camp, and soil horizons for Lubbock Lake).

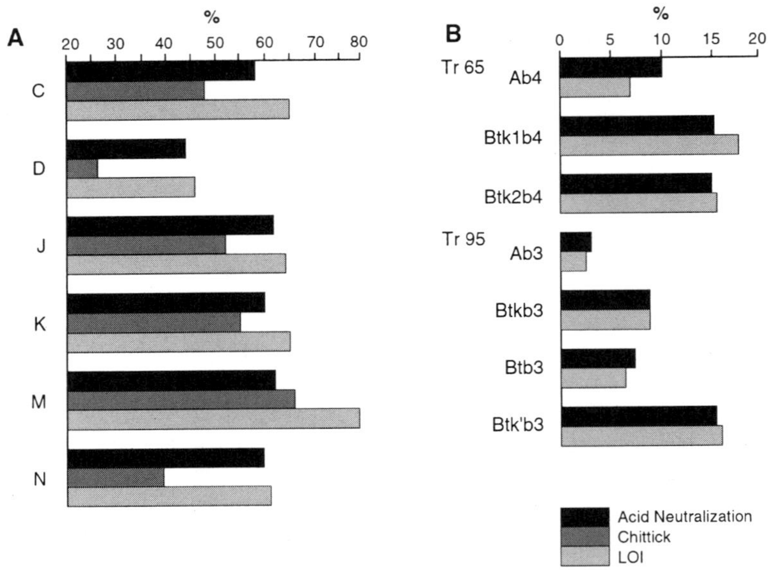

Figure A3.3 Comparisons of the results for three methods of analysis of carbonate content for samples from British Camp (A) and Lubbock Lake (B). Modified from *Geoarchaeology* v. 4, pp. 347–358, fig. 3, by V. T. Holliday and J. K. Stein, © 1989, John Wiley & Sons, used by permission of John Wiley & Sons, Inc. Left-most column of each graph indicates field designations for samples (soil horizons at Lubbock Lake).

contents in the untreated samples, although the differences are more pronounced in the Lubbock Lake values. The untreated samples from British Camp also show higher fine silt content in the untreated samples. The presence of the carbonate could have flocculated the clay and produced higher coarse silt and sand content, but it appears that the addition of a dispersant to the samples prevented this. The presence of increased fines in the untreated samples is probably a measure of the carbonate particles, which at both sites are largely illuvial.

Several trends are apparent in the analyses of P from the two sites. The most striking feature is the very high values of Ptot from British Camp (fig A3.4); fully an order of magnitude (or more) higher than the levels measured by the other extraction methods. This is perhaps not surprising given the nature of the site (a midden with abundant shell and bone), although the absolute levels are unusually high. Because the Ptot procedure extracts all mineral P, some may have come from primary minerals. No data are available on the mineralogy of the local sediment, however. After Ptot, the next-highest levels were measured using the citrate extraction and the sulphuric-nitric acid extraction, indicating that both methods may be getting at essentially the same forms of P. This is also somewhat surprising given that the double-acid method should be significantly more vigorous than the citric-acid extraction. Furthermore, the trends in values between the two methods are different (e.g., values by sulphuric-nitric increase from D to J, but decrease for M to N, yet D to J by citric acid decreases, and M to N increases). The trends among the various extractions vary for reasons that are unknown.

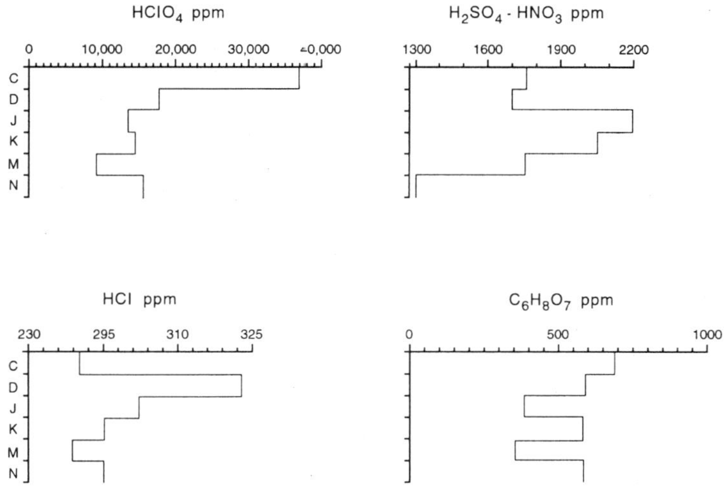

Figure A3.4 Comparisons of the results for four methods of P extraction for samples from British Camp ($HClO_4$ = perchloric acid; H_2SO_4—HNO_3 = sulphuric acid-nitric acid; HCl = hydrochloric acid; $C_6H_8O_7$ = citric acid). Left-most column indicates field designations for samples.

The measurements of P for the Lubbock Lake samples (fig. A3.5) provide a strong contrast with British Camp. The Lubbock Lake samples generally have much lower levels of soil P, which is not surprising given that occupations at Lubbock Lake were comparatively ephemeral. What is unusual, however, is the low level of P measured by the perchloric acid digestion. Indeed, some of the extractions by sulphuric-nitric acid yielded levels of P higher than the perchloric digestion. In theory, the perchloric acid digested all soil P, whereas the sulphuric and nitric acid extracts only some component of P. This incongruity would indicate some procedural problems, but the trends between the two sets of extractions are generally similar (e.g., highest levels of P are in the A horizons). These data may indicate that the samples had low levels of anthropogenic P and no geogenic P, resulting in similar levels of both fractions of P. Moreover, the precision of the two methods may not be sufficient to discriminate when levels of perchloric acid Ptot are similar to sulphuric-nitric acid Pav. This also points up another contrast with British Camp: The trends among all of the different extractions of the Lubbock Lake samples are generally similar, with the A horizons expressing higher levels of P than their respective subsoils. A horizons typically have the highest levels of P in soils because they are the zones of highest biological activity. This raises the question, however, of how much of the P signal at Lubbock Lake is natural and how much is anthropogenic. Archaeological remains were not apparent in Trench 95, but were in Trench 104, yet the P levels are roughly the same in both areas. This indicates that the occupations exposed in Trench 104 were not sufficiently intense to yield higher P levels. An alternative explanation is that there were occupations in the Trench 95 area, but they were just not obvious. The final point of comparison is that the lowest levels of

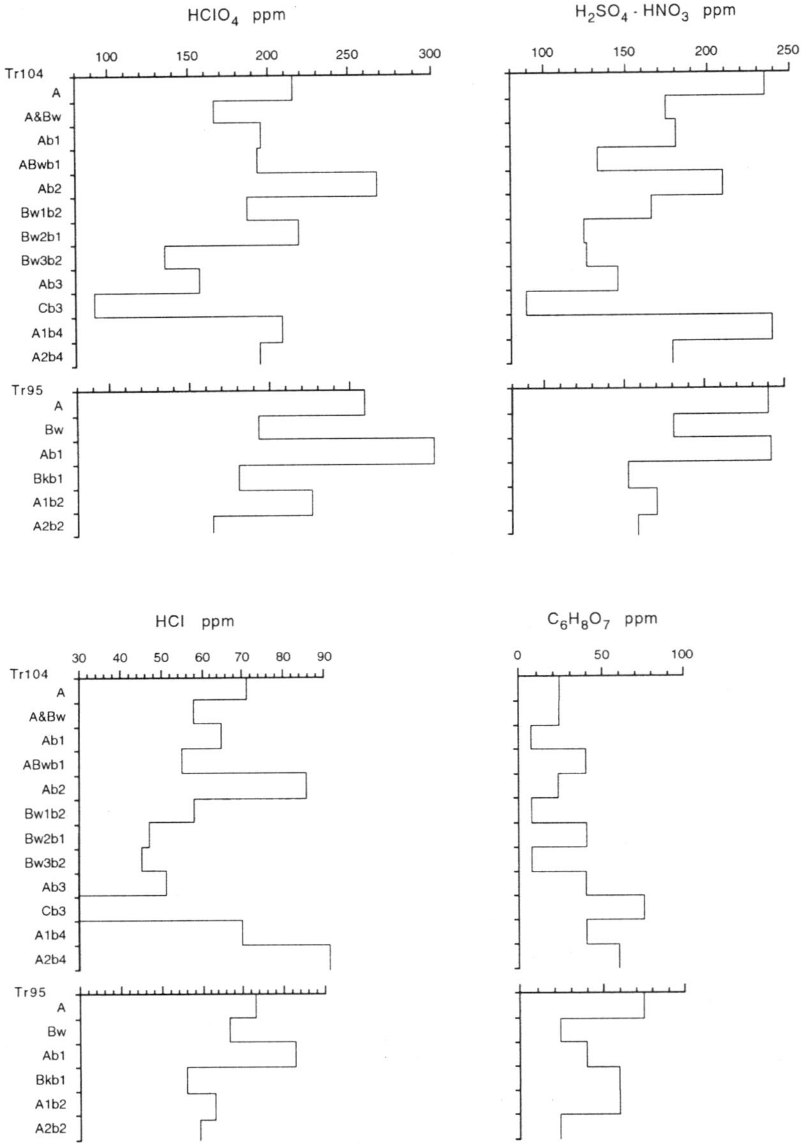

Figure A3.5 Comparisons of the results for four methods of P extraction for samples from Lubbock Lake ($HClO_4$ = perchloric acid; H_2SO_4—HNO_3 = sulphuric acid-nitric acid; HCl = hydrochloric acid; $C_6H_8O_7$ = citric acid). Left-most column indicates field designations (soil horizons) for samples.

P measurements were by HCl extraction at Lubbock Lake, just as at British Camp. The HCl at Lubbock Lake is not substantially lower, but the trend is consistent and seems not to yield as much P as the citric acid extraction.

Conclusions

The previous discussion shows that the results of analyses for organic carbon and organic matter content and calcium carbonate and P content can vary significantly as the analytical methods vary. The study also shows that methods of pretreatment are important considerations in PSA. Variability in results between laboratories is another potential problem, well documented for radiocarbon analyses (e.g., Martin and Johnson, 1995), but one not easily documented. The same methods may be followed, but minor differences in reagents and pretreatments, for example, may produce variability in results. Slight differences in procedures among laboratory technicians may also produce differences in results. Whenever possible, however, the same labs should be used and the same procedures should be followed for consistent results. Such variables among labs probably produce only minor variation in results, but still should be considered. The purpose of this study, however, is not to recommend any particular method or to address the determination of any specific characteristic of geoarchaeological sediments. Rather, it is to make several more general points concerning analytical methods in geoarchaeology.

There are a great variety of analytical techniques from which to choose in geoarchaeological studies, and there are often a number of methods available for analysis of a single attribute. As the methods (and labs) vary, however, so can the results. In choosing a particular technique, a number of factors must be considered, and these considerations are sometimes at odds. For example, in particle-size analysis, a decision to forego sample pretreatments makes for a quicker procedure, but one that yields results considerably less accurate if secondary (postdepositional) components are present. The pretreatments are simple but can be time consuming. This study, although not thoroughly rigorous, indicates that secondary accumulations of organic matter, carbonates, and salts should be removed from samples for PSA because they can induce erroneous readings (Gee and Bauder, 1986). If, however, constituents such as carbonates are primary components of the sediments (e.g., the sediments are derived from limestone or contain shell), then pretreatments should not be carried out. In either case the methods of pretreatment and PSA should be clearly stated to avoid inaccurate comparisons between data sets.

The decision of whether to pretreat samples also depends on interests concerning analysis of microartifacts (sand-sized artifacts). Some pretreatments can damage or destroy sand-sized artifacts (e.g., shell-tempered sherds, bone, and charcoal). If such analysis is desirable, then pretreatment is ignored. The factors involved in deciding which techniques to use should be included in published discussions of methods.

Accuracy, efficiency, and cost are usually the primary considerations in selecting among laboratory techniques, including those described in this appendix. All of the procedures discussed require accurate weighing to several decimal points, and therefore, an analytical balance is necessary. Otherwise the accuracy, speed, and cost, including equipment requirements, can vary considerably among the methods described. The methods of PSA are relatively inexpensive, requiring only moderate investment in glassware and reagents. The hydrometer method is significantly quicker than the pipet. As mentioned, however, there is evidence to indicate that PSA by pipet is somewhat more accurate.

Most of the OM and OC techniques yield similar trends, if not similar values. The Walkley-Black OC method can yield variable results (Nelson and Sommers, 1982) but is used widely because it is relatively simple and requires only a moderate investment in laboratory equipment. The analysis of OC by calculating the difference between total C and inorganic C is generally reliable but requires a total carbon analyzer, which is expensive. The loss-on-ignition and peroxide methods for OM give variable results, especially the latter (Nelson and Sommers, 1982), but both are simple, quick, and inexpensive techniques.

For carbonate analyses, loss-on-ignition is the simplest method, but the Chittick method probably yields more reliable results. Both methods appear to produce very similar trends, however, and the Chittick method is also very simple, although requiring a moderate investment in the apparatus and greater operator skill.

A wide variety of methods for P analyses are available to archaeologists and geoarchaeologists, fully outlined in chapter 11 and appendix 2. Only a few were examined here, including several widely used methods. The data presented here clearly show the variability in results among these methods. Determination of which method or methods are "best" will depend on the research questions being asked, the equipment, time, and funds available, and the work being compared with.

In deciding on laboratory procedures the research questions involved must be considered. In particular, if comparisons are to be made with the work of other investigators, then the same procedures must be used. This raises the final point: The methods used in any analytical study should be explicitly identified and referenced or described.

References

Ackerly, N. W., 1988. False causality in the Hohokam collapse. *The Kiva* 53, 305–319.

Aguado, E., and J. E. Burt, 1999. Understanding Weather and Climate. Prentice Hall, Upper Saddle River, N.J.

Ahler, S. A., 1973. A chemical analysis of deposits at Rogers Rock Shelter, Missouri. *Plains Anthropologist* 18, 116–131.

Ahmad, N., 1983. Vertisols. *In* Pedogenesis and Soil Taxonomy, Part 1, Concepts and Interactions (L. P. Wilding, N. E. Smeck, and G. F. Hall, Eds.), pp. 91–123. Developments in Soil Science 11A. Elsevier, Amsterdam.

Aigner, J. S., 1981. Archaeological Remains in Pleistocene China. C. H. Beck, Munich.

Albanese, J. P., 1982. Geologic investigations. *In* The Agate Basin Site (G. C. Frison and D. J. Stanford, Eds.), pp. 309–330. Academic Press, New York.

Al-Barrak, S., and D. T. Lewis, 1978. Soils of a grassland-forest ecotone in eastern Nebraska. *Soil Society of America Journal* 42, 334–338.

Albritton, C. C., Jr., 1974. Geological setting. *In* A Middle Stone Age Sequence from the Central Rift Valley, Ethiopia (F. Wendorf and R. Schild, Eds.), pp. 15–27. Polska Akademia Nauk Instytut Historii Kultury Materialnej, Warsaw.

Albritton, C. C., Jr., and K. Bryan, 1939. Quaternary stratigraphy in the Davis Mountains, trans-Pecos, Texas. *Geological Society of America Bulletin* 50, 1423–1474.

Alexandrovskaya, E., and A. L. Alexandrovskiy, 2000. History of the cultural layer in Moscow and accumulation of anthropogenic substances in it. *Catena* 41, 249–259.

Alexandrovskiy, A. L., 2000. Holocene development of soils in response to environmental changes: The Novosvobodnaya archaeological site, North Caucasus. *Catena* 41, 237–248.

Allaby, A., and M. Allaby, Eds., 1991. The Concise Oxford Dictionary of Earth Sciences. Oxford University Press, New York.

Allen, J. R. L., and V. P. Wright, 1989. Paleosols in Siliciclastic Sequences. Postgraduate Research Institute for Sedimentology, Short Course Notes 001. University of Reading, Reading.

Almy, M. M., 1978. The archeological potential of soil survey reports. *The Florida Anthropologist* 31, 75–91.

Altschul, J. A., 1983. Bug Hill: Excavation of a Multicomponent Midden Mound in the Jackfork Valley, Southeast Oklahoma. Report of Investigation no. 81-1. New World Research, Pollock, La.

Ambrose, S. H., and N. E. Sikes, 1991. Soil carbon isotope evidence for Holocene habitat change in the Kenya Rift Valley. *Science* 253, 1402–1405.

American Commission on Stratigraphic Nomenclature (ACSN), 1961. Code of stratigraphic nomenclature. *American Association of Petroleum Geologists Bulletin* 45, 645–665.

Ammons, J. T., D. L. Newton, J. E. Foss, and W. R. Lynn, 1992. Soil genesis of two Indian mounds in West Tennessee. *Soil Survey Horizons* 33 (summer), 38–45.

Amundson, R., Ed., 1994. Factors of Soil Formation: A Fiftieth Anniversary Retrospective. Special Publication no. 33. Soil Science Society of America, Madison, Wisc.

Amundson, R., and H. Jenny, 1991. The place of humans in the state factor theory of ecosystems and their soils. *Soil Science* 151, 99–109.

Amundson, R., and E. Pendall, 1991. Pedology and late Quaternary environments surrounding Harappa: A review and synthesis. *In* Harappa Excavations 1986–1990: A Multidisciplinary Approach to Third Millenium Urbansism (R. H. Meadow, Ed.), pp. 13–27. Monographs in World Archaeology no. 3. Prehistory Press, Madison, Wisc.

Amundson, R. G., Y. Wang, O. A. Chadwick, S. Trumbore, L. McFadden, E. McDonald, S. Wells, and M. DeNiro, 1994. Factors and processes governing the ^{14}C content of carbonate in desert soils. *Earth and Planetary Science Letters* 125, 385–405.

An, Z.-H., G. Kukla, S. C. Porter, and J. Xiao, 1991. Magnetic susceptibility evidence of monsoon variation on the Loess Plateau of Central China during the last 130,000 years. *Quaternary Research* 36, 29–36.

An, Z.-S., T.-S. Liu, X.-F. Kan, J.-Z. Sun, J. Wang, W.-Y. Kao, Y. Zhu, and M.-J. Wei, 1987. Loess-paleosol sequences and chronology at Lantian Man localities. *In* Aspects of Loess Research (T.-S. Liu, Ed.), pp. 192–203. China Ocean Press, Beijing.

An, Z.-H., T.-S. Liu, S. C. Porter, G. Kukla, X.-H. Wu, and Y.-M. Hua, 1990. The long-term paleomonsoon variation recorded by the loess-paleosol sequence in Central China. *Quaternary International* 7/8, 91–95.

An, Z.-S., and C.-K. Ho, 1989. New magnetostratigraphic dates of Lantian *Homo erectus*. *Quaternary Research* 32, 213–221.

Anderson, D. C., and H. A. Semken, Eds., 1980. The Cherokee Excavations. Academic Press, New York.

Anderson, D. D., 1968. A stone age campsite at the gateway to America. *Scientific American* 218 (6), 24–33.

Anderson, D. D., 1970. Akmak: An Early Archeological Assemblage from Onion Portage, Northwest Alaska. *Acta Arctica*, fascicle 16.

Anderson, D. D., 1988. Onion Portage: An Archaeological Site on the Kobuk River, Northwest Alaska. Anthropological Papers of the University of Alaska 22 (1–2).

Anderton, J. B., 1999. The soil-artifact context model: A geoarchaeological approach to paleoshoreline site dating in the Upper Peninsula of Michigan, USA. *Geoarchaeology* 14, 265–288.

Andrews, W. F., and R. O. Dideriksen, 1981. Soil Survey of Boone County, Iowa. Soil Conservation Service, U.S. Department of Agriculture, Washington, D.C.

Antevs, E., 1941. Age of the Cochise Culture stages. *In* The Cochise Culture (E. B. Sayles and E. Antevs, Eds.), pp. 31–55. Medallion Papers no. 29. Gila Pueblo, Globe, Ariz.

Antoine, P., D.-D. Rousseau, J.-P. Lautridou, and C. Hatté, 1999. Last interglacial-glacial climatic cycle in loess-palaeosol successions of north-western France. *Boreas* 28, 551–563.

Arkley, R. J., 1963. Calculation of carbonate and water movement in soil from climatic data. *Soil Science* 96, 239–248.

Armentano, T. V., 1990. Soils and ecology: Tropical wetlands. *In* Wetlands: A Threatened Landscape (M. Williams, Ed.), pp. 115–144. Basil Blackwell, Oxford.

Arrhenius, O., 1931. Die Bodenanalyse im Dienst der Archäologie. *Zeitschrift für Pflanzenernährung, Düngung, and Bodenkunde* Tiel B 10, 427–439.

Arrhenius, O., 1934. Fosfathalten i Skanska Jordar. *Sveriges Geologiska Undersokning* 28, 1–30.

Arrhenius, O., 1963. Investigation of soil from old Indian sites. *Ethnos* 28, 122–136.

Artz, J. A., 1985. A soil-geomorphic approach to locating buried late Archaic sites in northeast Oklahoma. *American Archeology* 5, 142–150.

Artz, J. A., 1995. Geological contexts of the early and middle Holocene archaeological record in North Dakota and adjoining areas of the northern Plains. *In* Archaeological Geology of the Archaic Period in North America (E. A. Bettis III, Ed.), pp. 67–86. Special Paper no. 297. Geological Society of America, Boulder, Colo.

Artzy, M., and D. Hillel, 1988. A defense of the theory of progressive soil salinization in ancient southern Mesopotamia. *Geoarchaeology* 3, 235–238.

Ashley, G. M., 2001. Archaeological sediments in springs and wetlands. *In* Sediments in Archaeological Contexts (J. K. Stein and W. R. Farrand, Eds.), pp. 183–210. University of Utah Press, Salt Lake City.

Ashley, G. M., and S. G. Driese, 2000. Paleopedology and paleohydrology of a volcaniclastic paleosol interval: Implications for early Pleistocene stratigraphy and paleoclimate record, Olduvai Gorge, Tanzania. *Journal of Sedimentary Research* 70, 1065–1080.

Ashley, G. M., and R. L. Hay, 2002. Sedimentation patterns in a Plio-Pleistocene volcaniclastic rift-platform basin, Olduvai Gorge, Tanzania. *In* Sedimentation in Continental Rifts (R. W. Renault and G. M. Ashley, Eds.), pp. 107–122. Society for Sedimentary Geology, Special Publication 73. Tulsa, Ok.

Ashmore, W., and A. B. Knapp, Eds., 1999. Archaeologies of Landscape. Blackwell, Malden, Mass.

Autin, W. J., 1992. Use of alloformations for definition of Holocene meander belts in the middle Amite River, southeastern Louisiana. *Geological Society of America Bulletin* 104, 233–241.

Avery, B. W., 1990. Soils of the British Isles. CAB International, Wallingford, U.K.

Badgley, C., and A. K. Behrensmeyer, 1995. Two long geological records of continental ecosystems. *Palaeogeography, Palaeoclimatology, Palaeoecology* 115, 1–11.

Baker, R. T., 1976. Changes in the chemical nature of soil organic phosphate during pedogenesis. *Journal of Soil Science* 27, 504–512.

Baker, W. E., G. L. Evans, and T. N. Campbell, 1957. The Nall site: evidence of early man in the Oklahoma Panhandle. *Bulletin of the Oklahoma Anthropological Society* 5, 1–20.

Bakkevig, S., 1980. Phosphate analysis in archaeology—Problems and recent progress. *Norwegian Archaeological Review* 13, 73–100.

Balek, C. L., 2002. Buried artifacts in stable upland sites and the role of bioturbation: A review. *Geoarchaeology* 17, 41–51.

Ball, D. F., 1964. Loss-on-ignition as an estimate of organic matter and organic carbon in non-calcareous soils. *Journal of Soil Science* 15, 84–92.

Balster, C. A., and R. B. Parsons, 1968. Geomorphology and Soils, Willamette Valley, Oregon. Agriculture Experiment Station, Special Report no. 265. Oregon State University, Corvallis, Ore.

Barber, S. A., 1995. Soil Nutrient Bioavailability: A Mechanistic Approach. 2nd ed. Wiley, New York.

Barham, A. J., and R. I. Macphail, Eds., 1995. Archaeological Sediments and Soils: Analysis, Interpretation and Management. Institute of Archaeology, University College London.

Barker, G., 1995. A Mediterranean Valley: Landscape Archaeology and *Annales* History in the Biferno Valley. Leicester University Press, London.

Barnes, G. L., 1990. Paddy soils now and then. *World Archaeology* 22, 1–17.

Barrett, L. R., and R. J. Schaetzl, 1992. An examination of podzolization near Lake Michigan using chronofunctions. *Canadian Journal of Soil Science* 72, 527–541.

Bartelli, L. J., 1984. Soil taxonomy: Its evolution, status, and future. *In* Soil Taxonomy: Achievements and Challenges (R. B. Grossman, R. H. Rust, and H. Eswaran, Eds.), pp. 7–13. Special Publication no. 14. Soil Science Society of America, Madison, Wisc.

Bates, M. R., A. J. Barham, C. A. Pine, and V. D. Williamson, 2000. Evaluation strategies for deeply stratified alluvial areas: The use of borehole stratigraphic logs. *In* Interpreting Stratigraphy: Site Evaluation, Recording Procedures and Stratigraphic Analysis (S. Roskams, Ed.), pp. 49–69. BAR International Series 910, Oxford.

Bates, R. L., and J. A. Jackson, Eds., 1980. Glossary of Geology. 2nd ed. American Geological Institute, Falls Church, Va.

Baümler, R., and R. Zech, 2000. Quaternary paleosols, tephra deposits and landscape history, South Kamchatka, Russia. *Catena* 41, 199–215.

Beach, T., 1998a. Soil catenas, tropical deforestation, and ancient and contemporary soil erosion in the Petén, Guatemala. *Physical Geography* 19, 378–405.

Beach, T., 1998b. Soil constraints on northwest Yucatán, Mexico: Pedoarchaeology and Maya subsistence at Chunchucmil. *Geoarchaeology* 13, 759–791.

Beatty, S. W., 2000. On the rocks; shaken not stirred. Book review forum on "Soils: A New Global View." *Annals of the Association of American Geographers* 90, 785–787.

Belcher, W. R., and W. R. Belcher, 2000. Geologic constraints on the Harappa archaeological site, Punjab Province, Pakistan. *Geoarchaeology* 15, 679–713.

Bell, M., and J. Boardman, Eds., 1992. Past and Present Soil Erosion: Archaeological and Geographical Perspectives. Oxbow Monographs no. 22. Oxbow Books, Oxford.

Bender, M. R., and C. W. Wood, 2000. Total phosphorus in soil. *In* Methods of Phosphorous Analysis for Soils, Sediments, Residuals, and Waters (G. M. Pierzynski, Ed.), pp. 45–49. Southern Cooperative Series Bulletin no. 396. Southern Extension/Research Activity-Information Exchange Group 17, U.S. Department of Agriculture and North Carolina State University, Raleigh.

Benedict, J. B., 1976. Frost creep and gelifluction features: A review. *Quaternary Research* 6, 55–76.

Benedict, J. B., and B. L. Olson, 1978. The Mount Albion Complex: A Study of Prehistoric Man and the Altithermal. Research Report no. 1. Center for Mountain Archeology, Ward, Colo.

Bergman, C. A., and J. F. Doershuk, 1995. OSHA regulations and the excavation of the deeply stratified Sandts Eddy site (36NM12). *Journal of Middle Atlantic Archaeology* 11, 17–29.

Berlin, C. L., J. R. Ambler, R. H. Hevly, and G. G. Schaber, 1977. Identification of a Sinagua agricultural field by aerial thermography, soil chemistry, pollen/plant analysis, and archaeology. *American Antiquity* 42, 588–600.

Bethell, P., and I. Máté, 1989. The use of soil phosphate analysis in archaeology: A critique. *In* Scientific Analysis in Archaeology and its Interpretation (J. Henderson, Ed.), pp. 1–29. Institute of Archaeology, Archaeological Research Tools no. 5. University of California–Los Angeles, Los Angeles.

Bettis, E. A., III, 1988. Pedogenesis in late Prehistoric Indian mounds, upper Mississippi valley. *Physical Geography* 9, 263–279.

Bettis, E. A., III, 1992. Soil morphologic properties and weathering zone characteristics as age indicators in Holocene alluvium in the Upper Midwest. *In* Soils and Landscape Evolution (V. T. Holliday, Ed.), pp. 119–144. Smithsonian Institution Press, Washington, D.C.

Bettis, E. A., III, 2000. A brief history of geoarchaeology in the eastern Plains and Prairies. *In* Geoarchaeology in the Great Plains (R. D. Mandel, Ed.), pp. 137–165. University of Oklahoma Press, Norman.

Bettis, E. A., III, and E. R. Hajic, 1995. Landscape development and the location of evidence of Archaic cultures in the Upper Midwest. *In* Archaeological Geology of the Archaic Period in North America (E. A. Bettis, III, Ed.), pp. 87–113. Special Paper no. 297. Geological Society of America, Boulder, Colo.

Bidwell, O. W., and F. D. Hole, 1965. Man as a factor of soil formation. *Soil Science* 99, 65–72.

Bilzi, A. F., and E. J. Ciolkosz, 1977a. Time as a factor in the genesis of four soils developed in recent alluvium in Pennsylvania. *Soil Science Society of America Journal* 41, 122–127.

Bilzi, A. F., and E. J. Ciolkosz, 1977b. A field morphology rating scale for calculating pedogenic development. *Soil Science* 124, 45–48.

Binford, L. R., 1968. Review of "A Guide to Field Methods in Archaeology: Approaches to the Anthropology of the Dead" by R. F. Heizer and J. A. Graham. *American Anthropologist* 70, 806–808.

Birkeland, P. W., 1974. Pedology, Weathering and Geomorphological Research. Oxford University Press, New York.

Birkeland, P. W., 1984. Soils and Geomorphology. 2nd ed. Oxford University Press, New York.

Birkeland, P. W., 1992. Quaternary soil chronosequences in various environments—Extremely arid to humid tropical. *In* Weathering, Soils, and Palaeosols (I. P. Martini and W. Chesworth, Eds.), pp. 261–281. Elsevier, New York.

Birkeland, P. W., 1994. Review of "Soils in Archaeology." *Geoderma* 61, 143–144.

Birkeland, P. W., 1999. Soils and Geomorphology. 3rd ed. Oxford University Press, New York.

Birkeland, P. W., and R. Gerson, 1991. Soil-catena development with time in a hot desert, southern Israel—Field data and salt distribution. *Journal of Arid Environments* 21, 267–281.

Birkeland, P. W., M. N. Machette, and K. M. Haller, 1991. Soils as a tool for applied Quaternary geology. Miscellaneous Publication no. 91-3. Utah Geological and Mineral Survey, Salt Lake City.

Birkeland, P. W., and R. R. Shroba, 1974. The status of the concept of Quaternary soil-forming intervals in the western United States. *In* Quaternary Environments: Proceedings of a Symposium (W. C. Mahaney, Ed.), pp. 241–276. York University Geographical Monographs no. 5. York University, Toronto.

Bischoff, J. L., R. J. Shlemon, T. L. Ku, R. D. Simpson, R. J. Rosenbauer, and F. E. Budinger, 1981. Uranium-series and soil-geomorphic dating of the Calico archaeological site, California. *Geology* 9, 576–582.

Bishop, R. L., R. L. Rands, and G. R. Holley, 1982. Ceramic compositional analysis in archaeological perspective. *In* Advances in Archaeological Method and Theory, v. 5 (M. B. Schiffer, Ed.), pp. 275–330. Academic Press, New York.

Bishop, W. W., Ed., 1978. Geological Background to Fossil Man: Recent Research in the Great Rift Valley, East Africa. University of Toronto Press, Toronto.

Bishop, W. W., A. Hill, and M. Pickford, 1978. Chesowanja: A revised geological interpretation. *In* Geological Background to Fossil Man: Recent Research in the Great Rift Valley, East Africa (W. W. Bishop, Ed.), pp. 309–327. University of Toronto Press, Toronto.

Bishop, W. W., M. Pickford, and A. Hill, 1975. New evidence regarding the Quaternary geology, archaeology, and hominids of Chesowanja, Kenya. *Nature* 258, 204–208.

Bjelajac, V., E. M. Luby, and R. Ray, 1996. A validation test of a field-based phosphate analysis technique. *Journal of Archaeological Science* 23, 243–248.

Black, C. A., Ed., 1965. Methods of Soil Analysis. Monograph no. 9. American Society of Agronomy, Madison, Wisc.

Blackstock, D. A., 1979. Soil Survey of Lubbock County, Texas. Soil Conservation Service, U.S. Department of Agriculture, Washington, D.C.

Blackwell, B., and H. P. Schwarcz, 1995. The Uranium series disequilibrium dating methods. *In* Dating Methods for Quaternary Deposits (N. W. Rutter and N. R. Catto, Eds.), pp. 167–208. GEOTEXT 2, Geological Association of Canada, St. John's, Newfoundland.

Blair, T. C., J. S. Clark, and S. G. Wells, 1990. Quaternary continental stratigraphy, landscape evolution, and application to archaeology: Jarilla piedmont and Tularosa graben floor, White Sands Missile Range, New Mexico. *Geological Society of America Bulletin* 102, 749–759.

Blatt, H., G. Middleton, and R. Murray, 1980. Origin of Sedimentary Rocks. 2nd ed. Prentice Hall, Englewood Cliffs, N.J.

Bloom, P. R., M. Pohl, C. Buttleman, F. Wiseman, A. Covich, C. Miksicek, J. Ball, and J. Strin, 1983. Prehistoric Maya wetland agriculture and the alluvial soils near San Antonio Rio Hondo, Belize. *Nature* 301, 417–419.

Blum, M. D., J. T. Abbot, and S. S. Valastro, Jr., 1992. Evolution of landscapes on the Double Mountain Fork of the Brazos River, West Texas: Implications for preservation and visibility of the archaeological record. *Geoarchaeology* 7, 339–370.

Blum, M. D., and S. Valastro, Jr., 1992. Quaternary stratigraphy and geoarchaeology of the Colorado and Concho Rivers, West Texas. *Geoarchaeology* 7, 419–448.

Blum, M. D., and S. Valastro, Jr., 1994. Late Quaternary sedimentation, lower Colorado River, Gulf Coastal Plain of Texas. *Geological Society of America Bulletin* 106, 1002–1016.

Boardman, J., 1985a. Comparison of soils in midwestern United States and western Europe with the interglacial record. *Quaternary Research* 23, 62–75.

Boardman, J., Ed., 1985b. Soils and Quaternary Landscape Evolution. Wiley, Chichester, 391 pp.

Bobrowsky, P. T., N. R. Catto, J. W. Brink, B. E. Spurling, T. H. Gibson, and N. W. Rutter, 1990. Archaeological geology of sites in western and northwestern Canada. *In* Archaeological Geology of North America (N. P. Lasca and J. Donahue, Eds.), pp. 87–122. Centennial Volume no. 4, Geological Society of America, Boulder, Colo.

Bockheim, J. G., 1980. Solution and use of chronofunctions in studying soil development. *Geoderma* 24, 71–85.

Bockheim, J. G., and C. Tarnocai, 2000. Gelisols. *In* Handbook of Soil Science (M. E. Sumner, Ed.), pp. E.256–E.269. CRC Press, Boca Raton, Fla.

Bonadonna, F. P., D. Brocchini, M. A. Laurenzi, C. Principe, and G. Ferrara, 1998. Stratigraphical and chronological correlations between Monte Vulture volcanics and sedimentary deposits of the Venosa basin. *Quaternary International* 47/48, 87–96.

Bookidis, N., J. Hansen, L. Snyder, and P. Goldberg, 1999. Dining in the Sanctuary of Demeter and Kore at Corinth. *Hesperia* 68, 1–54.

Bos, R. H. G., and J. Sevink, 1975. Introduction of gradational and pedomorphic features in descriptions of soils: A discussion of the soil horizon concept with special reference to paleosols. *Journal of Soil Science* 26, 223–233.

Bousman, C. B., 1998. Late Paleoindian archeology. *In* Wilson-Leonard: An 11,000-year Archeological Record of Hunter-Gatherers in Central Texas; Volume I: Introduction, Background, and Synthesis (M. B. Collins, Ed.), pp. 161–210. Environmental Affairs Division, Archeology Studies Program, Report no. 10. Texas Department of Transportation, Austin.

Bowen, D. Q., 1978. Quaternary Geology. Pergamon Press, Oxford.

Bowen, R., 1991. Isotopes and Climates. Elsevier Applied Science, New York.

Bowers, P. M., 1982. The Lisbourne site: Analysis and cultural history of multi-component lithic workshop in the Iteriak Velley, arctic foothills, Alaska. Fairbanks. *Anthropological Papers of the University of Alaska* 20, 79–112.

Bowers, P. M., R. Bonnichsen, and D. M. Hoch, 1983. Flake dispersal experiments: Noncultural transformation of the archaeological record. *American Antiquity* 48, 553–572.

Bowman, R. A., 1988. A rapid method to determine total phosphorous in soils. *Soil Science Society of America Journal* 52, 1301–1304.

Bown, T. M., and M. J. Kraus, 1987. Integration of channel and floodplain suites in aggrading fluvial systems: I. Developmental sequence and lateral relations of lower Eocene alluvial paleosols, Willwood Formation, Bighorn Basin, Wyoming. *Journal of Sedimentary Petrology* 57, 587–601.

Bradley, R. S., 1999. Paleoclimatology: Reconstructing Climates of the Quaternary. 2nd ed. Academic Press, San Diego.

Brakenridge, G. R., 1984. Alluvial stratigraphy and radiocarbon dating along the Duck River, Tennessee: Implications regarding floodplain origin. *Geological Society of America Bulletin* 95, 9–25.

Brasfield, J. F., Ed., 1984. Guy D. Smith Discusses Soil Taxonomy. Reprinted from "Soil Survey Horizons." Soil Science Society of America, Madison, Wisc.

Breuning-Madsen, H., M. K. Holst, and M. Rasmussen, 2001. The chemical environment in a burial mound shortly after construction—An archaeological-pedological experiment. *Journal of Archaeological Science* 28, 691–697.

Breuning-Madsen, H., J. Rønsbo, and M. K. Holst, 2000. Comparison of the composition of iron pans in Danish burial mounds with bog iron and spodic material. *Catena* 39, 1–9.

Brevik, E. C., and T. E. Fenton, 1999. Improved mapping of the Lake Agassiz Herman strandline by integrating geological and soil maps. *Journal of Paleolimnology* 22, 253–257.

Brewer, R., 1972. Use of macro- and micromorphological data in soil stratigraphy to elucidate surficial geology and soil genesis. *Journal of the Geological Society of Australia* 19, 331–344.

Brewer, R., 1976. Fabric and Mineral Analysis of Soils (reprint with supplement). Robert E. Krieger, Huntington, N.Y.

Brewer, R., Crook, K. A. W., and J. G. Speight, 1970. Proposal for soil stratigraphic units in the Australian stratigraphic code. Report by the subcommittee for soil-stratigraphic nomenclature. *Journal of the Geological Society of Australia* 17, 103–111.

Briggs, D., 1977. Sources and Methods in Geography: Sediments. Butterworths, Boston.

Brinkmann, R., 1996. Pedological characteristics of anthrosols in the al-Jadidah Basin of Wadi al-Jubah, and native sediments in Wadi al-Ajwirah, Yemen Arab Republic. *In* Environmental Research in Support of Archaeological Investigations in the Yemen Arab Republic, 1982–1987 (M. J. Grolier, R. Brinkmann, and J. A. Blakely, Eds.), pp. 45–211. American Foundation for the Study of Man, Washington, D.C.

Broderson, W. D., 1994. From the surface down: An introduction to soil surveys for agronomic use (revised). U.S. Department of Agriculture, Soil Conservation Service, Washington, D.C.

Bronger, A., and J. A. Catt, 1998a. The position of paleopedology in geosciences and agricultural sciences. *Quaternary International* 51/52, 87–93.

Bronger, A., and J. A. Catt, 1998b. Summary outline and recommendations on paleopedological issues. *Quaternary International* 51/52, 5–6.

Bronger, A., and T. Heinkele, 1989a. Micromorphology and genesis of paleosols in the Luochuan loess section, China: Pedostratigraphic and environmental implications. *Geoderma* 45, 123–143.

Bronger, A., and T. Heinkele, 1989b. Paleosol sequences as witnesses of Pleistocene climatic history. *In* Paleopedology: Nature and Applications of Paleosols (A. Bronger and J. Catt, Eds.), pp. 163–186. Catena Supplement 16. Catena Verlag, Cremlingen, Germany.

Bronger, A., R. Winter, O. Derevjanko, and S. Aldag, 1995. Loess-paleosol sequences in Tadjikistan as a palaeoclimatic record of the Quaternary in Central Asia. *Quaternary Proceedings* 4, 69–81.

Bronger, A., R. Winter, and S. Sedov, 1998. Weathering and clay mineral formation in two Holocene soils and in buried paleosols in Tadjikistan: Towards a Quaternary paleoclimatic record in Central Asia. *Catena* 34, 19–34.

Brown, A. G., 1997. Alluvial Geoarchaeology: Floodplain Archaeology and Environmental Change. Cambridge University Press, Cambridge.

Brown, C. N., 1956. The origin of caliche in the northeast Llano Estacado, Texas. *Journal of Geology* 46, 1–15.

Brown, D. E., 1982. Great Basin conifer woodland. *Desert Plants* 4, 52–57.

Brown, F. H., and C. S. Feibel, 1986. Revision of lithostratigraphic nomenclature in the Koobi Fora region, Kenya. *Journal of the Geological Society, London* 143, 297–310.

Brown, F. H., and C. S. Feibel, 1991. Stratigraphy and depositional environments and palaeogeography of the Koobi Fora Formation. *In* Koobi Fora Research Project, v. 3: The Fossil Ungulates: Geology, Fossil Artiodactyls, and Palaeoenvironments (J. M. Harris, Ed.), pp. 1–30. Clarendon Press, Oxford.

Bruins, H. J., and D. H. Yaalon, 1979. Stratigraphy of the Netivot section of the Negev (Israel). *Acta Geologica Academiae Scientiarium Hungaricae* 22, 161–169.

Bryan, K., 1941a. Correlation of the deposits of Sandia Cave, New Mexico, with the glacial chronology (appendix). *In* Evidences of Early Occupation in Sandia Cave, New Mexico, and Other Sites in the Sandia-Manzano Region (F. C. Hibben, Ed.), pp. 45–64. Smithsonian Miscellaneous Collections no. 99. Smithsonian Institution, Washington, D.C.

Bryan, K., 1941b. Pre-Columbian agriculture in the Southwest as conditioned by periods of alluviation. *Association of American Geographers Annals* 31, 219–242.

Bryan, K., 1948. Los suelos complejos y fosiles de la altiplanicie de Mexico en relacion a los cambios climaticos. *Sociedad Geológia, Mexicana, Boletin*, 13, 1–20.

Bryan, K., and C. C. Albritton, Jr., 1943. Soil phenomena as evidence of climatic changes. *American Journal of Science* 241, 469–490.

Buck, B. J., J. Kipp, Jr., and H. C. Monger, 2002. Inverted clast stratigraphy in an eolian archaeological environment. *Geoarchaeology* 17, 665–687.

Bull, I. D., P. P. Betancourt, and R. P. Evershed, 2001. An organic geochemical investigation of the practice of manuring at a Minoan site on Pseira, Crete. *Geoarchaeology* 16, 223–242.

Bull, W. B., 1991. Geomorphic Responses to Climatic Change. Oxford University Press, New York.

Bullock, P., N. Fedoroff, A. Jongerius, G. Stoops, and T. Tursina, 1985. Handbook for Soil Thin Section Description. Waine Research Publications, Wolverhampton.

Bunting, B. T., 1965. The Geography of Soil. Aldine Publishing Company, Chicago.

Buol, S. W., F. D. Hole, and R. J. McCracken, 1973. Soil Genesis and Classification. Iowa State University Press, Ames.

Buol, S. W., F. D. Hole, and R. J. McCracken, 1984. Soil Genesis and Classification. 3rd ed. Iowa State University Press, Ames.

Buol, S. W., F. D. Hole, R. J. McCracken, and R. J. Southard, 1997. Soil Genesis and Classification. 4th ed. Iowa State University Press, Ames.

Burbank, D. W., 1985. The age of the Karewas, Kashmir as determined from fission-track dating and magnetostratigraphies. *In* Climate and Geology of Kasmir and Central Asia (D. P. Agrawal, S. Kusumgar, and R. V. Krishnamurthy, Eds.), pp. 19–26. Current Trends in Geology, v. VI. Today and Tomorrow's Printers and Publishers, New Delhi.

Burroni, D., R. E. Donahue, A. M. Pollard, and M. Mussi, 2002. The surface alteration features of flint artefacts as a record of environmental processes. *Journal of Archaeological Science* 29, 1277–1287.

Busacca, A. J., 1987. Pedogenesis of a chronosequence in the Sacramento Valley, California, U.S.A., I. Application of a soil development index. *Geoderma* 41, 123–148.

Busacca, A. J., and M. Cremaschi, 1998. The role of time versus climate in the formation of deep soils of the Apennine fringe of the Po Valley, Italy. *Quaternary International* 51/52, 95–107.

Butler, B. E., 1959. Periodic Phenomena in Landscapes as a Basis for Soil Studies. Publication no. 14. CSIRO, Melbourne.

Butler, B. E., 1980. Soil Classification for Soil Survey. Clarendon Press, Oxford.

Butler, B. E., 1982. A new system for soil studies. *Journal of Soil Science* 33, 581–595.

Butzer, K. W., 1971. Environment and Archaeology. 2nd ed. Aldine, Chicago.

Butzer, K. W., 1973. Spring sediments from the Acheulian site at Amanzi (Uitenhage District, South Africa). *Quaternaria* 17, 299–319.

Butzer, K. W., 1977. Geomorphology of the Lower Illinois Valley as a Spatial-Temporal Context for the Koster Archaic Site. Reports of Investigations no. 34. Illinois State Museum, Springfield.

Butzer, K. W., 1981. Rise and fall of Axum, Ethiopia: A geo-archaeological interpretation. *American Antiquity* 46, 471–495.

Butzer, K. W., 1982. Archaeology as Human Ecology. Cambridge University Press, Cambridge.

Cahen, D., and J. Moeyersons, 1977. Subsurface movements of stone artefacts and their implications for the prehistory of Central Africa. *Nature* 266, 812–815.
Caldwell, R. E., O. C. Olson, J. B. Crombrite, and R. G. Leighty, 1958. Soil Survey of Manatee County, Florida. U.S. Department of Agriculture, Washington, D.C.
Callum, K. E., 1995. Archaeology in a region of Spodosols, part 2. *In* Pedological Perspectives in Archaeological Research (M. E. Collins, B. J. Carter, B. J. Gladfelter, and R. J. Southard, Eds.), pp. 81–94. Special Publication no. 44. Soil Science Society of America, Madison, Wisc.
Campbell, C. A., E. A. Paul, D. A. Rennie, and K. J. McCallum, 1967a. Applicability of the carbon-dating method of analysis to soil humus studies. *Soil Science* 104, 217–224.
Campbell, C. A., E. A. Paul, D. A. Rennie, and K. J. McCallum, 1967b. Factors affecting the accuracy of the carbon-dating method in soil humus studies. *Soil Science* 104, 81–85.
Campbell, I. B., 1986. Recognition of paleosols in Quaternary periglacial and volcanic environments in New Zealand. *In* Paleosols: Their Recognition and Interpretation (V. P. Wright, Ed.), pp. 208–241. Princeton University Press, Princeton, N.J.
Cannon, A., 2000. Settlement and sea levels on the central coast of British Columbia: Evidence from shell midden cores. *American Antiquity* 65, 67–77.
Canti, M. G., and F. M. Meddens, 1998. Mechanical coring as an aid to archaeological projects. *Journal of Field Archaeology* 25, 97–105.
Cantin, M., and C. R. Stafford, 1997. Phase I Surface and Subsurface Archaeological Investigations of the Proposed Mulzer Stone Quarry in Titus Bottoms, Harrison County, Indiana. Technical Report no. 33. Indiana State University Anthropology Laboratory, Terre Haute.
Caple, C., 2001. Overview—Degradation, investigation and preservation of archaeological evidence. *In* Handbook of Archaeological Sciences (D. R. Brothwell and A. M. Pollard, Eds.), pp. 587–593. Wiley, New York.
Carr, C., 1982. Handbook on Soil Resistivity Surveying: Interpretation of Data from Earthen Archaeological Sites. Center for American Archaeology Press, Evanston, Ill.
Carter, S. P., and D. A. Davidson, 1998. An evaluation of the contribution of soil micromorphology to the study of ancient arable agriculture. *Geoarchaeology* 13, 535–547.
Carter, S. P., and D. A. Davidson, 2000. A reply to Macphail's comments on "An evaluation of the contribution of soil micromorphology to the study of ancient arable agriculture." *Geoarchaeology* 15, 499–502.
Catt, J. A., 1979. Soils and Quaternary geology in Britain. *Journal of Soil Science* 30, 607–642.
Catt, J. A., 1986. Soils and Quaternary Geology: A Handbook for Field Scientists. Oxford Science Publications, Oxford.
Catt, J. A., 1988. Soils of the Plio-Pleistocene: Do they distinguish types of interglacial? *In* The Past Three Million Years: Evolution of Climatic Variability in the North Atlantic Region (N. J. Shackleton, R. G. West, and D. Q. Bowen, Eds.), pp. 539–557. Philosophical Transactions of the Royal Society of London, B 318, London.
Catt, J. A., 1990. Paleopedology manual. *Quaternary International* 6, 1–95.
Catt, J. A., 1991. Soils as indicators of Quaternary climatic change in mid-latitude regions. *Geoderma* 51, 167–187.
Catt, J. A., 1996. Review of "Soils: A New Global View." *Quaternary Science Reviews* 15, 92–93.
Catt, J. A., 1998. Report from Working Group on Definitions used in Paleopedology. *Quaternary International* 51/52, 84.
Cavanagh, W. G., S. Hirst, and C. D. Litton, 1988. Soil phosphate, site boundaries and charge point analysis. *Journal of Field Archaeology* 15, 67–83.
Cavarretta, G., M. Fornaseri, M. Follieri, O. Girotti, and C. Turner, Eds., 1998. Quaternary stratigraphy in volcanic areas. *Quaternary International* 47/48, 1–182.
Cerling, T. E., 1992. Development of grasslands and savannas in East Africa during the Neogene. *Palaeogeography, Palaeoclimatology, Palaeoecology* 97, 241–247.
Cerling, T. E., 1999. Stable carbon isotopes in palaeosol carbonates. *In* Palaeoweathering, Palaeosurfaces and Related Continental Deposits (M. Thiry and R. Simon-Coinçon,

Eds.), pp. 43–60. International Association of Sedimentologists, Special Publication no. 27. Blackwell Science, London.

Cerling, T. E. and R. L. Hay, 1986. An isotopic study of paleosol carbonates from Olduvai Gorge. *Quaternary Research* 25, 63–78.

Cerling, T. E., and J. Quade, 1993. Stable carbon and oxygen isotopes in soil carbonates. *In* Climate Change in Continental Isotopic Records (P. K. Swart, K. C. Lohmann, J. McKenzie, and S. Sarin, Eds.), pp. 217–231. Geophysical Monograph no. 78. American Geophysical Union, Washington, D.C.

Chang, S. W., and M. L. Jackson, 1957a. Fractionation of soil phosphorus. *Soil Science* 84 133–144.

Chang, S. C., and M. L. Jackson, 1957b. Soil phosphorus fractions in some representative soils. *Journal of Soil Science* 9, 109–119.

Chardon, W. J., 2000. Phosphorus extraction with iron oxide-impregnated filter paper (P_i test). *In* Methods of Phosphorous Analysis for Soils, Sediments, Residuals, and Waters (G. M. Pierzynski, Ed.), pp. 15–16. Southern Cooperative Series Bulletin no. 396. Southern Extension/Research Activity-Information Exchange Group 17, U.S. Department of Agriculture and North Carolina State University, Raleigh.

Chartres, C. J., 1980. A Quaternary soil sequence in the Kennet Valley, central southern England. *Geoderma* 23, 125–146.

Chatters, J. C., and K. A. Hoover, 1986. Changing late Holocene flooding frequencies on the Columbia River, Washington. *Quaternary Research* 26, 309–320.

Chaya, H. J., 1996. Studies of soils from an Aleutian Island site. *In* Archaeological Chemistry: Organic, Inorganic, and Biochemical Analysis (M. V. Orna, Ed.), pp. 131–138. Symposium Series no. 625, American Chemical Society, Washington, D.C.

Cherry, J. F., J. L. Davis, and E. Mantzourani, 1991. Landscape Archaeology as Long-Term History: Northern Keos in the Cycladic Islands from Earliest Settlement until Modern Times. Monumenta Archaeologica no. 16. University of Calfornia–Los Angeles Institute of Archaeology, Los Angeles.

Claassen, C., 1998. Shells. Cambridge University Press, New York.

Clayton, L., S. R. Moran, and W. B. Bickley, Jr., 1976. Stratigraphy, Origin and Climatic Implications of Late Quaternary Upland Silt in North Dakota. Miscellaneous Series no. 54. North Dakota Geological Survey, Grand Forks.

Clayton, L., S. R. Moran, and J. P. Bluemle, 1980. Explanatory Text to Accompany the Geologic Map of North Dakota. Report of Investigations no. 69. North Dakota Geological Survey, Grand Forks.

Cole, D. R., and H. C. Monger, 1994. Influence of atmospheric CO_2 on the decline of C_4 plants during the last deglaciation. *Nature* 368, 533–536.

Coles, B., 1987. Tracks across the wetlands: Multi-disciplinary studies in the Somerset levels of England. *In* European Wetlands in Prehistory (J. M. Coles and A. J. Lawson, Eds.), pp. 145–167. Clarendon Press, Oxford.

Coles, B., 1988. The Somerset levels: Multidisciplinary investigations and a wealth of results. *In* Wet Site Archaeology (B. A. Purdy, Ed.), pp. 89–102. The Telford Press, Caldwell, N.J.

Coles, B., 1990. Wetland archaeology: A wealth of evidence. *In* Wetlands: A Threatened Landscape (M. Williams, Ed.), pp. 145–180. Basil Blackwell, London.

Coles, B., and J. M. Coles, 1986. Sweet Track to Glastonbury: The Somerset Levels in Prehistory. Thames and Hudson Publishers, New York.

Coles, B., and J. M. Coles, 1989. People of the Wetlands: Bogs, Bodies and Lake-Dwellers. London. Thames and Hudson Publishers, New York.

Coles, J. M., 1987. Preservation of the past: The case for wet archaeology. *In* European Wetlands in Prehistory (J. M. Coles and A. J. Lawson, Eds.), pp. 1–21. Clarendon Press, Oxford.

Coles, J. M., 1988. An assembly of death: Bog bodies of northern and western Europe. *In* Wet Site Archaeology (B. A. Purdy, Ed.), pp. 219–235. The Telford Press, Caldwell, N.J.

Coles, J. M., and A. J. Lawson, Eds., 1987. European Wetlands in Prehistory. Clarendon Press, Oxford.

Collins, M. B., Ed., 1998a. Wilson-Leonard: An 11,000-Year Archeological Record of Hunter-Gatherers in Central Texas. Austin, Environmental Affairs Division, Archeology Studies Program, Report no. 10 (5 volumes). Texas Department of Transportation, Austin.

Collins, M. B., 1998b. The Late Prehistoric components. *In* Wilson-Leonard: An 11,000-year Archeological Record of Hunter-Gatherers in Central Texas; Volume I: Introduction, Background, and Synthesis (M. B. Collins, Ed.), pp. 271–276. Environmental Affairs Division, Archeology Studies Program, Report no. 10. Texas Department of Transportation, Austin.

Collins, M. B., J. Guy, and S. W. Dial, 1998. The Archaic period, 8800–1300 B.P. *In* Wilson-Leonard: An 11,000-year Archeological Record of Hunter-Gatherers in Central Texas; Volume I: Introduction, Background, and Synthesis (M. B. Collins, Ed.), pp. 211–270. Environmental Affairs Division, Archeology Studies Program, Report no. 10. Texas Department of Transportation, Austin.

Collins, M. E., B. J. Carter, B. G. Gladfelter, and R. J. Southard, Eds., 1995. Pedological Perspectives in Archaeological Research. Special Publication no. 44. Soil Science Society of America, Madison, Wisc.

Collins, M. E., and G. Shapiro, 1987. Comparisons of human-influenced and natural soils at the San Luis archaeological site, Florida. *Soil Science Society of America Journal* 51, 171–176.

Colman, S. M., and K. L. Pierce, 2000. Classifications of Quaternary geochronologic methods. *In* Quaternary Geochronology: Methods and Applications (J. S. Noller, J. M. Sowers, and W. R. Lettis, Eds.), pp. 2–5. AGU Reference Shelf no. 4, American Geophysical Union, Washington, D.C.

Colman, S. M., K. L. Pierce, and P. W. Birkeland, 1987. Suggested terminology for Quaternary dating methods, *Quaternary Research* 28, 314–319.

Conry, M. J., 1971. Irish Plaggen soils: Their distribution, origin, and properties. *Journal of Soil Science* 22, 401–416.

Conry, M. J., 1972. Pedological evidence of man's role in soil profile modification in Ireland. *Geoderma* 8, 139–146.

Conway, J. S., 1983. An investigation of soil phosphorous distribution within occupation deposits from a Romano-British hut group. *Journal of Archaeological Science* 10, 177–128.

Cook, S. F., and R. F. Heizer, 1965. Studies on the Chemical Analysis of Archaeological Sites. University of California Publications in Anthropology no. 2. University of California Press, Berkeley.

Cornwall, I. W., 1958. Soils for the Archaeologist. Phoenix House, London.

Cornwall, I. W., 1960. Soil investigations in the service of archaeology. *Viking Fund Publications in Anthropology* 28, 265–299.

Costin, A. B., and H. A. Polach, 1969. Dating soil organic matter. *Atomic Energy in Australia* 12, 13–17.

Coultas, C. L., 1997. Soils of Caracol, Belize and their significance to agriculture and land use. *In* Proceedings of the Second International Conference on Pedo-Archaeology (A. C. Goodyear, J. E. Foss, and K. E. Sassaman, Eds.), pp. 103–109. Occasional Papers of the South Carolina Institute of Archaeology and Anthropology, Anthropology Studies no. 10. University of South Carolina, Columbia.

Courchesne, F., 2000. Breaking the barrier of conceptual locks. Book review forum on "Soils: A New Global View." *Annals of the Association of American Geographers* 90, 782–785.

Courty, M.-A., 1992. Soil micromorphology in archaeology. *In* New Developments in Archaeological Science (M. Pollard, Ed.), pp. 39–62. Oxford University Press, New York.

Courty, M.-A., 2001. Microfacies analysis assisting archaeological stratigraphy. *In* Earth Sciences and Archaeology (P. Goldberg, V. T. Holliday, and C. R. Ferring, Eds.), pp. 205–239. Kluwer Academic/Plenum Publishers, New York.

Courty, M.-A., and N. Fedoroff, 1985. Micromorphology of recent and buried soils in a semiarid region of Northwestern India. *Geoderma* 35, 287–332.

Courty, M.-A., P. Goldberg, and R. I. Macphail, 1989. Soils and Micromorphology in Archaeology. Cambridge University Press, New York.

Courty, M.-A., and P. Nørnberg, 1985. Comparison between buried uncultivated and cultivated Iron Age soils of the west coast of Jutland, Denmark. *Iskos* 5, 57–69.

Courty, M.-A., and H. Weiss, 1997. The scenario of environmental degradation in the Tell Leilan region (N. E. Syria) during the late Third Millennium abrupt climate change. NATO ASI Series, I 49, 107–148. Springer, New York.

Craddock, P. T., D. Gurney, F. Pryor, and M. J. Hughes, 1985. The application of phosphate analysis to the location and interpretation of archaeological sites. *Archaeological Journal* 142, 361–376.

Cremaschi, M., 1987. Paleosols and Vetusols in the Central Po Plains (Northern Italy): A Study in Quaternary Geology and Soil Development. Studi e ricerche sul territorio no. 28. Pubblicazione fuori serie, Università di Milano. Sezione di geologia e paleontologia, Edizioni Unicopli, Milan.

Cremaschi, M., 1990. Loess, aeolian deposits, and related palaeosols in the Mediterranean region. *Quaternary International* 5, 1–137.

Cremeens, D. L., 1995. Pedogenesis of Cotiga Mound, a 2100-year-old Woodland mound in southwest Virginia. *Soil Science Society of America Journal* 59, 1377–1388.

Cremeens, D. L., 2003. Geoarchaeology of soils on stable geomorphic surfaces: Mature soil model for the glaciated Northeast. *In* Geoarchaeology of Landscapes in the Glaciated Northeast (D. L. Cremeens and J. P. Hart, Eds.), pp. 49–60. New York State Museum Bulletin 497, The University of the State of New York, Albany.

Cremeens, D. L., and J. P. Hart, 1995. On chronostratigraphy, pedostratigraphy, and archaeological context. *In* Pedological Perspectives in Archaeological Research (M. E. Collins, B. J. Carter, B. G. Gladfelter, and R. J. Southard, Eds.), pp. 15–33. Special Publication 44. Soil Science Society of America, Madison, Wisc.

Cremeens, D. L., J. P. Hart, and R. G. Darmody, 1998. Complex pedostratigraphy of a terrace fragipan at the Memorial Park site, central Pennsylvania. *Geoarchaeology* 13, 339–359.

Cronyn, J. M., 2001. The deterioration of organic materials. *In* Handbook of Archaeological Sciences (D. R. Brothwell and A. M. Pollard, Eds.), pp. 627–636. Wiley, New York.

Crook, K. A. W., and R. J. Coventry, 1967. Climatically controlled Quaternary sedimentation and soils in Ryans Creek Valley, N.S.W. *Australia New Zealand Association for Advanced Science, Section C Abstracts of Proceedings*, T9–11.

Crowther, J., 2002. The experimental earthwork at Wareham, Dorset after 33 Years: Retention and leaching of phosphate released in the decomposition of buried bone. *Journal of Archaeological Science* 29, 405–411.

Crowther, J., R. I. Macphail, and G. M. Cruise, 1996. Short-term, post-burial change in a humic rendzina soil, Overtone Down experimental earthwork, Wiltshire, England. *Geoarchaeology* 11, 95–117.

Cruise, G., and R. I. Macphail, 2000. Microstratigraphical signatures of experimental rural occupation deposits and archaeological sites. *In* Interpreting Stratigraphy: Site Evaluation, Recording Procedures and Stratigraphic Analysis (S. Roskams, Ed.), pp. 183–191. BAR International Series no. 910, Oxford.

Curry, B. B., and M. Pavich, 1996. Absence of glaciation in Illinois during marine oxygen isotope stages 3 through 5. *Quaternary Research* 46, 19–26.

Dahlgren, R., S. Shoji, and M. Nanzyo, 1993. Mineralogical characteristics of volcanic ash soils. *In* Volcanic Ash Soils: Genesis, Properties and Utilization (S. Shoji, M. Nanzyo, and R. Dahlgren, Eds.), pp. 101–143. Amsterdam, Elsevier.

Dahms, D., and V. T. Holliday, 1998. Paleosols as environmental indicators: A critical review. *Quaternary International* 51/52, 109–114.

Dalrymple, J. B., 1958. The application of soil micromorphology to fossil soils and other deposits from archaeological sites. *Journal of Soil Science* 9, 199–209.

Dalrymple, J. B., 1967. A study of paleosols in volcanic ash-fall deposits from northern North Island, New Zealand, and the evaluation of soil micromorphology for establishing their stratigraphic correlation. Quaternary Soils (R. B. Morrison and H. E.

Wright, Jr., Eds.), pp. 103–122. Desert Research Institute, University of Nevada, Reno.

Damp, J. E., S. A. Hall, and S. J. Smith, 2002. Early irrigation on the Colorado Plateau near Zuni Pueblo, New Mexico. *American Antiquity* 67, 665–676.

Dan, J., and D. H. Yaalon, 1971. On the origin and nature of the paleopedological formations in the coastal desert fringe areas of Israel. *In* Paleopedology: Origin, Nature and Dating of Paleosols (D. H. Yaalon, Ed.), pp. 245–260. University of Israel Press, Jerusalem.

Dan, J., D. H. Yaalon, and H. Koyumdjinsky, 1968. Catenary soil relationships in Israel. A. The Netanya catena on coastal dunes of the Sharon. *Geoderma* 2, 95–120.

Dane, J. H., and G. C. Topp, Eds., 2002. Methods of Soil Analysis: Part 4—Physical Methods. Book Series no. 5. Soil Science Society of America, Madison, Wisc.

Daniels, R. B., and R. D. Hammer, 1992. Soil Geomorphology. Wiley, New York.

Daniels, R. B., and L. A. Nelson, 1987. Soil variability and productivity: Future developments. *In* Future Developments in Soil Science Research (L. L. Boersma, Ed.), pp. 279–291. Soil Science Society of America, Madison, Wisc.

Dauncy, K. D. M., 1952. Phosphate content of soils on archaeological sites. *Advancement of Science* 9, 33–37.

Davidson, D. A., 1973. Particle size and phosphate analysis—Evidence for the evolution of a Tell. *Archaeometry* 15, 143–152.

Davidson, D. A., 1976. Processes of tell formation and erosion. *In* Geoarchaeology (D. A. Davidson and M. L. Shackley, Eds.), pp. 255–266. Westview Press, Boulder, Colo.

Davidson, D. A., 1982. Soils and man in the past. *In* Principles and Applications of Soil Geography (E. M. Bridges and D. A. Davidson, Eds.), pp. 1–28. Longman, London.

Davidson, D. A., 2002. Bioturbation in old arable soils: Quantitative evidence from soil micromorphology. *Journal of Archaeological Science* 29, 1247–1253.

Davidson, D. A., S. P. Carter, and T. A. Quine, 1992. An evaluation of micromorphology as an aid to archaeological interpretation. *Geoarchaeology* 7, 55–65.

Davidson, D. A., and I. A. Simpson, 1984. The formation of deep topsoils in Orkney. *Earth Surface Processes and Landforms* 9, 75–81.

Davies, B. E., 1974. Loss-on-ignition as an estimate of soil organic matter. *Soil Science Society of America Proceedings* 38, 150–151.

Davis, A. M., 1986. Identifying soil units in archaeological sites with complex stratigraphy. *The Canadian Geographer* 30, 85–87.

Davis, E. M., 1953. Recent data from two Paleo-Indian sites on Medicine Creek, Nebraska. *American Antiquity* 18, 380–386.

Davis, E. M., 1962. Archeology of the Lime Creek Site in Southwestern Nebraska. Special Publication of the University of Nebraska State Museum no. 3. University of Nebraska, Lincoln.

Davis, R. S., and V. A. Ranov, 1999. Recent work on the Paleolithic of Central Asia. *Evolutionary Anthropology* 9, 186–193.

Dean, W. E., Jr., 1974. Determination of carbonate and organic matter in calcareous sediments and sedimentary rocks by loss on ignition: Comparison with other methods. *Journal of Sedimentary Petrology* 44, 242–248.

Dekker, L. W., and M. D. De Weerd, 1973. The value of soil survey for archaeology. *Geoderma* 10, 169–178.

de Maigret, A., C. Azzi, B. Marcolongo, and A. M. Palmieri, 1989. Recent pedogenesis and neotectonics affecting archaeological sites in North Yemen. *Paleorient* 15, 239–243.

Denevan, W. M., 1996. A bluff model of riverine settlement in prehistoric Amazonia. *Annals of the Association of American Geographers* 86, 654–681.

Denevan, W. M., 2001. Cultivated Landscapes of Native Amazonia and the Andes. Oxford University Press, New York.

Derbyshire, E., 2001. Recent research on loess and palaeosols, pure and applied. *Earth-Science Reviews* 54, 1–260.

Derbyshire, E., D. Keen, R. A. Kemp, T. A. Rolph, J. Shaw, and X. M. Meng, 1995a. Loess-paleosol sequences as recorders of palaeoclimatic variations during the last glacial-interglacial cycle: Some problems of correlation in north-central China. *Quaternary Proceedings* 4, 7–18.

Derbyshire, E., R. Kemp, and X. M. Meng, 1995b. Variations in loess and paleosol properties as indicators of paleoclimatic gradients across the Loess Plateau of North China. *Quaternary Science Reviews* 14, 681–697.

Dethier, D. P., 1988. The soil chronosequence along the Cowlitz River, Washington. U.S. Geological Survey Bulletin 1590-F. Government Printing Office, Washington, D.C.

Dick, W. A., and M. A. Tabatabai, 1977. An alkaline oxidation method for determination of total phosphorus in soils. *Soil Science Society of America Journal* 41, 511–514.

Dietz, E. F., 1957. Phosphorus accumulation in soil of an Indian habitation site. *American Antiquity* 22, 405–409.

Dijkerman, J. C., M. C. Kline, and G. W. Olson, 1967. Properties and genesis of textural subsoil lamellae. *Soil Science* 104, 7–16.

Dillehay, T. D., 1988. Early rainforest archaeology in southwestern South America: Research context, design and data at Monte Verde. *In* Wet Site Archaeology (B. A. Purdy, Ed.), pp. 177–206. The Telford Press, Caldwell, N.J.

Dillehay, T. D., 1989. Monte Verde, A Late Pleistocene Settlement in Chile. Smithsonian Institution Press, Washington, D.C.

Dimbleby, G. W., 1985. The Palynology of Archaeological Sites. Academic Press, London.

Dincauze, D. F., 2000. Environmental Archaeology: Principles and Practice. Cambridge University Press, New York.

Dixon, E. J., and G. S. Smith, 1990. A regional application of tephrochronology in Alaska. *In* Archaeological Geology of North America (N. P. Lasca and J. Donahue, Eds.), pp. 383–398. Centennial Volume no. 4, Geological Society of America, Boulder, Colo.

Dockrill, S. J., and I. A. Simpson, 1994. The identification of prehistoric anthropogenic soils in the Northern Isles using an integrated sampling methodology. *Archaeological Prospection* 1, 75–92.

Dockrill, S. J., J. M. Bond, A. Milles, I. Simpson, and J. Ambers, 1994. Tofts Ness, Sanday, Orkney: An integrated study of a buried Orcadian landscape. *In* Whither Environmental Archaeology (R. Luff and P. Rowley-Conwy, Eds.), pp. 115–132. Oxbow Books, Oxford.

Dodonov, A. E., 1984. Stratigraphy and correlation of upper Pliocene-Quaternary deposits of Central Asia. Lithology and Stratigraphy of Loess and Paleosols (M. Pécsi, Ed.), pp. 201–211. Geographical Research Institute, Hungarian Academy of Sciences, Budapest.

Dodonov, A. E., 1991. Loess of Central Asia. *GeoJournal* 24, 185–194.

Dodonov, A. E., and L. L. Baiguzina, 1995. Loess stratigraphy of Central Asia: Palaeoclimatic and palaeoenvironmental aspects. *Quaternary Science Reviews* 14, 707–720.

Doran, G. H., and D. N. Dickel, 1988. Multidisciplinary investigations at the Windover site. *In* Wet Site Archaeology (B. A. Purdy, Ed.), pp. 263–289. The Telford Press, Caldwell, N.J.

Dormaar, J. F., 1987. Are palaeosols recognizable? *In* Man and the Mid-Holocene Climatic Optimum (N. A. McKinnon and G. S. L. Stuart, Eds.), pp. 53–57. The University of Calgary Archaeological Association, Calgary.

Dormaar, J. F., and A. B. Beaudoin, 1991. Application of soil chemistry to interpret cultural events at the Calderwood Buffalo Jump (DkPj-27), southern Alberta, Canada. *Geoarchaeology* 6, 85–98.

Dormaar, J. F., and L. E. Lutwick, 1969. Infrared spectra of humic acids and opal phytoliths as indicators of paleosols. *Canadian Journal of Soil Science* 49, 29–37.

Dorronsoro, C., and P. Alonso, 1994. Chronosequence in Almar River fluvial-terrace soil. *Soil Science Society of America Journal* 58, 910–925.

Dregne, H. E., 1963. Soils of the arid west. *In* Aridity and Man: The Challenge of the Arid Lands in the United States (C. Hodge and P. C. Duisberg, Eds.), pp. 215–238. Publication no. 78. American Association for the Advancement of Science, Washington, D.C.

Driessen, P. M., and R. Dudal, Eds., 1991. The Major Soils of the World. Department of Soil Science, Agricultural University, Wageningen, The Netherlands; Institute for Land and Water Management, Katholiecke Universteit, Leuven.

Duchaufour, P., 1982. Pedology, Pedogenesis, and Classification (T. R. Paton, Trans.). Allen and Unwin, Boston.

Duchaufour, P., 1998. Handbook of Pedology: Soils, Vegetation, Environment. A.A. Balkema, Rotterdam.

Duffield, L. F., 1970. Vertisols and their implications for archaeological research. *American Anthropologist* 72, 1055–1062.

Dumond, D. E., 1963. A practical field method for the preservation of soil profiles from archaeological cuts. *American Antiquity* 29, 116–118.

Dunning, N. P., 1994. Ancient Maya anthrosols: Soil phosphate testing and land use. *In* Proceedings of the First International Conference on Pedo-Archaeology (J. E. Foss, M. E. Timpson, and M. W. Morris, Eds.), pp. 201–210. The University of Tennessee Agricultural Experiment Station, Knoxville.

Dunning, N. P., and T. Beach, 1994. Soil erosion, slope management, and ancient terracing in the Maya lowlands. *Latin American Antiquity* 5, 51–69.

Eash, N. S., and J. A. Sandor, 1995. Soil chronosequence and geomorphology in a semi-arid valley in the Andes of southern Peru. *Geoderma* 65, 59–79.

Eddy, F. W., and H. E. Dregne, 1964. Soil tests on alluvial and archaeological deposits, Navajo Reservoir District. *El Palccio* 71, 5–21.

Eden, M. J., W. Bray, L. Herrera, and C. McEwan, 1984. *Terra preta* soils and their archaeological context in the Caquetí basin of southeast Colombia. *American Antiquity* 49, 125–140.

Eidt, R. C., 1973. A rapid chemical field test for archaeological site surveying. *American Antiquity* 38, 206–210.

Eidt, R. C., 1977. Detection and examination of anthrosols by phosphate analysis. *Science* 197, 1327–1333.

Eidt, R. C., 1984. Advances in Abandoned Settlement Analysis: Application to Prehistoric Anthrosols in Columbia, South America. University of Wisconsin-Milwaukee, Center for Latin America, Milwaukee.

Eidt, R. C., 1985. Theoretical and practical considerations in the analysis of anthrosols. *In* Archaeological Geology (G. Rapp and J. A. Gifford, Eds.), pp. 155–190. Yale University Press, New Haven, Conn.

Eidt, R. C., and W. I. Woods, 1974. Abandoned Settlement Analysis: Theory and Practice. Field Test Associates, Shorewood, Wisc.

Engel, S. A., T. W. Gardner, and E. J. Ciolkosz, 1996. Quaternary soil chronosequences on terraces of the Susquehanna River, Pennsylvania. *Geomorphology* 17, 273–294.

Engel-DiMauro, S., 1995. Constructing the palaeovegetational record for the buried soils in the Hungarian young loess sequence: A view from phytolith analysis. *Loess in Form* 3, 79–94.

Engelmann, A., A. Neber, M. Frechen, W. Boenigk, and A. Ronen, 2001. Luminescence chronology of Upper Pleistocene and Holocene aeolianites from Netanya South, Sharon coastal plain, Israel. *Quaternary Science Reviews* 20, 799–804.

Engelmark, R., and J. Linderholm, 1996. Prehistoric land management and cultivation: A soil chemical study. *In* Proceedings from the 6th Nordic Conference on the Application of Scientific Methods in Archaeology, *Arkaeologiske Rapporter* 1, 315–322.

Entwistle, J. A., and P. W. Abrahams, 1997. Multi-element analysis of soils and sediments from Scottish historical sites. The potential of inductively coupled plasma-mass spectrometry for rapid site investigation. *Journal of Archaeological Science* 24, 407–416.

Entwistle, J. A., P. W. Abrahams, and R. A. Dodgshon, 1998. Multi-element analysis of soils from Scottish historical sites. Interpreting land-use history through physical and geochemical analysis of soil. *Journal of Archaeological Science* 25, 53–68.

Entwistle, J. A., P. W. Abrahams, and R. A. Dodgshon, 2000. The geoarchaeological significance and spatial variability of a range of physical and chemical soil properties from a former habitation site, Isle of Skye. *Journal of Archaeological Science* 27, 287–303.

Erickson, C. L., 1995. Archaeological methods for the study of ancient landscapes of the Llano de Mojos in the Bolivian Amazon. *In* Archaeology in the Lowland American Tropics (P. W. Stahl, Ed.), pp. 66–95. Cambridge University Press, New York.

Erlandson, J. M., 1984. A case study in faunalturbation: Delineating the effects of the burrowing pocket gopher on the distribution of archaeological materials. *American Antiquity* 49, 785–790.

Evans, J. G., 1978. An Introduction to Environmental Archaeology. Cornell University Press, Ithaca, N.Y.

Evans, J. G., S. Limbrey, and H. Cleere, Eds., 1975. The Effect of Man on the Landscape: The Highland Zone. Report no. 11. The Council for British Archaeology Research, London.

Evans, J. G., and K. W. G. Valentine, 1974. Ecological changes induced by prehistoric man at Pitstone, Buckinghamshire. *Journal of Archaeological Science* 1, 343–351.

Evans, L. J., 1995. The use of paleosols in dating Quaternary deposits. *In* Dating Methods for Quaternary Deposits (N. W. Rutter and N. R. Catto, Eds.), pp. 269–281. GEOTEXT 2, Geological Association of Canada, St. John's, Newfoundland.

Evans, M. E., and F. Heller, 2001. Magnetism of loess/palaeosol sequences: Recent developments. *Earth Science Reviews* 54, 129–144.

Everett, K. R., 1983. Histosols. *In* Pedogenesis and Soil Taxonomy, part 2, The Soil Orders (L. P. Wilding, N. E. Smeck, and G. F. Hall, Eds.), pp. 1–53. Developments in Soil Science 11B, Elsevier, Amsterdam.

Eyre, S. R., 1968. Vegetation and Soils: A World Picture. 2nd ed. Aldine, Chicago.

Falesi, I. C., 1974. Soils of the Brazilian Amazon. *In* Man in the Amazon (C. Wagley, Ed.), pp. 201–229. University Presses of Florida, Gainesville.

Fanning, D. S., and M. C. B. Fanning, 1989. Soil Morphology, Genesis, and Classification. Wiley, New York.

Farrand, W. R., 1994. Confrontation of geological stratigraphy and radiometric dates from upper Pleistocene sites in the Levant. *In* Late Quaternary Chronology and Paleoclimates of the Eastern Mediterranean (O. Bar-Yosef and R. S. Kra, Eds.), pp. 33–53. American School of Prehistoric Research, Cambridge.

Farrand, W. R., and A. Ronen, 1974. Observations on the kurkar-hamra succession on the Carmel coastal plain. *Tel Aviv (Journal of the Institute of Archaeology of Tel Aviv University)* 1, 45–54.

Fastovsky, D. E., 1991. Review of "Soils of the Past: An Introduction to Paleopedology." *Sedimentology* 38, 181–184.

Fedele, F. G., 1976. Sediments as palaeo-land segments: The excavation side of study. *In* Geoarchaeology (D. A. Davidson and M. L. Shackley, Eds.), pp. 23–48. Westview Press, Boulder, Colo.

Fedele, F. G., 1990. Man, land and climate: Emerging interactions from the Holocene of the Yemen Highlands. *In* Man's Role in the Shaping of the Eastern Mediterranean Landscape (S. Bottema, G. Entjes-Nieborg, and W. Van Zeist, Eds.), pp. 31–42. A. A. Balkema Publishers, Rotterdam.

Fedje, D. W., and H. Josenhans, 2000. Drowned forests and archaeology on the continental shelf of British Columbia, Canada. *Geology* 28, 99–102.

Fedoroff, N., M. A. Courty, and M. L. Thompson, 1990. Micromorphological evidence of paleoenvironmental change in Pleistocene and Holocene paleosols. *In* Soil Micromorphology: A Basic and Applied Science (L. A. Douglas, Ed.), pp. 653–665. Elsevier, New York.

Fehrenbacher, J. B., K. R. Olson, and I. J. Jansen, 1986. Loess thickness in Illinois. *Soil Science* 141, 423–431.

Feibel, C. S., F. H. Brown, and I. McDougall, 1989. Stratigraphic context of fossil hominids from the Omo Group deposits: Northern Turkana Basin, Kenya and Ethiopia. *American Journal of Physical Anthropology* 78, 595–622.

Feibel, C. S., J. M. Harris, and F. H. Brown, 1991. Palaeoenvironmental context for the late Neogene of the Turkana Basin. *In* Koobi Fora Research Project, v. 3: The Fossil Ungulates: Geology, Fossil Artiodactyls, and Palaeoenvironments (J. M. Harris, Ed.), pp. 321–370. Clarendon Press, Oxford.

Fenton, T. E., 1983. Mollisols. *In* Pedogenesis and Soil Taxonomy, part 2, The Soil Orders (L. P. Wilding, N. E. Smeck, and G. F. Hall, Eds.), pp. 125–163. Developments in Soil Science 11B, Elsevier, Amsterdam.

Fenwick, I. M., 1985. Paleosols: problems of recognition and interpretation. *In* Soils and Quaternary Landscape Evolution (J. Boardman, Ed.), pp. 3–21. John Wiley, Chichester.

Fernández, F. G., R. E. Terry, T. Inomata, and M. Eberl, 2002. An ethnoarchaeological study of chemical residues in the floors and soils of Q'eqchi' Maya houses at Las Pozas, Guatemala. *Geoarchaeology* 17, 487–519.

Ferring, C. R., 1986. Rate of fluvial sedimentation: Implications for archaeological variability. *Geoarchaeology* 1, 259–274.

Ferring, C. R., 1990. Archaeological geology of the Southern Plains. *In* Archaeological Geology of North America (N. P. Lasca and J. Donahue, Eds.), pp. 253–266. Centennial Volume no. 4. Geological Society of America, Boulder, Colo.

Ferring, C. R., 1992. Alluvial pedology and geoarchaeological research. *In* Soils in Archaeology (V. T. Holliday, Ed.), pp. 1–40. Smithsonian Institution Press, Washington, D.C.

Ferring, C. R., 1995a. The late Quaternary geology and archaeology of the Aubrey Clovis site, Texas. *In* Ancient Peoples and Landscapes (E. Johnson, Ed.), pp. 273–281. Museum of Texas Tech University, Lubbock.

Ferring, C. R., 1995b. Middle Holocene environments, geology, and archaeology in the Southern Plains. *In* Archaeological Geology of the Archaic Period in North America (E. A. Bettis, III, Ed.), pp. 21–35. Special Paper no. 297. Geological Society of America, Boulder, Colo.

Ferring, C. R., 2001. Geoarchaeology in alluvial landscapes. *In* Earth Sciences and Archaeology (P. Goldberg, V. T. Holliday, and C. R. Ferring, Eds.), pp. 77–106. Kluwer Academic/Plenum Publishers, New York.

Finkl, C. W., 1980. Stratigraphic principles and practices as related to soil mantles. *Catena* 7, 169–194.

Fisher, P. F., 1982. A review of lessivage and Neolithic cultivation in southern England. *Journal of Archaeological Science* 9, 299–304.

Fisher, P. F., and R. I. Macphail, 1985. Studies of archaeological soils and deposits by micromorphological techniques. *In* Paleoenvironmental Investigations: Research Design, Methods and Data Analysis (N. R. J. Fieller, D. D. Gilbertson, and N. G. A. Ralph, Eds.), pp. 93–125. BAR International Series no. 258, Oxford.

Fisher, R. F., and R. J. Miller, 1980. Description of soils that have disturbed or discontinuous horizons. *Soil Science Society of America Journal* 44, 1279–1281.

Fitzpatrick, E. A., 1971. Pedology: A Systematic Approach to Soil Science. Oliver and Boyd, Edinburgh.

Fitzpatrick, E. A., 1979. Soils: Their Formation, Classification, and Distribution. Longman, New York.

Fitzpatrick, E. A., 1993. Soil Microscopy and Micromorphology. Wiley, New York.

Flegenheimer, N., and M. Zárate, 1993. The archaeological record in Pampean loess deposits. *Quaternary International* 17, 95–100.

Folk, R. L., 1980. Petrology of Sedimentary Rocks. Hemphill, Austin, Tex.

Folks, J. J., 1975. Soil Survey of Santa Fe Area, New Mexico. U.S. Department of Agriculture, Soil Conservation Service, Washington, D.C.

Follmer, L. R., 1978. The Sangamon Soil in its type area—A review. *In* Quaternary Soils (W. C. Mahaney, Ed.), pp. 125–165. Geo Abstracts, Norwich.

Follmer, L. R., 1982. The geomorphology of the Sangamon surfaces: Its spatial and temporal attributes. *In* Space and Time in Geomorphology (C. E. Thorn, Ed.), pp. 117–146. Allen and Unwin, London.

Follmer, L. R., 1983. Sangamonian and Wisconsinan pedogenesis in the midwestern United States. *In* Late Quaternary Environments of the United States, Vol. 1: The Late Pleistocene (S. C. Porter, Ed.), pp. 138–144. University of Minnesota Press, Minneapolis.

Follmer, L. R., 1998. Preface. *Quaternary International* 51/52, 1–3.

Follmer, L. R., D. L. Johnson, and J. A. Catt, Eds., 1998. Revisitation of Concepts of Paleopedology. *Quaternary International* 51/52, 1–221.

Food and Agriculture Organization of the United Nations (FAO-UNESCO), 1974. Soil Map of the World, 1:500,000. 10 volumes. UNESCO, Paris.

Food and Agriculture Organization of the United Nations (FAO/UNESCO), 1994. Soil Map of the World, Revised Legend with Corrections. ISRIC Technical Paper no. 20. UNESCO, Wageningen.

Forbes, T. R., Ed., 1986. The Guy Smith Interviews: Rationale for Concepts in Soil Taxonomy. Soil Management Support Services Technical Monograph no. 11. U.S. Department of Agriculture, Soil Conservation Service, Washington, D.C.

Forman, S. L., 1991. Late Pleistocene chronology of loess deposition near Luochuan, China. *Quaternary Research* 36, 19–28

Forster, J. C., 1995. Soil phosphorus. *In* Methods in Applied Soil Microbiology and Biochemistry (K. Alef and P. Nannipieri, Eds.), pp. 88–93. Academic Press, London.

Foscolos, A. E., and R. R. Barefoot, 1970. A rapid determination of total organic and inorganic carbon in shales and carbonates. *Geological Survey of Canada Paper* 70, 1–8.

Foss, J. E., and M. E. Collins, 1987. Future users of soil genesis and morphology in allied science. *In* Future Developments in Soil Science Research (L. L. Boersma, Ed.), pp. 293–299. Soil Science Society of America, Madison, Wisc.

Foss, J. E., R. J. Lewis, and M. E. Timpson, 1995. Soils in alluvial sequences: Some archaeological implications. *In* Pedological Perspectives in Archaeological Research (M. E. Collins, B. J. Carter, B. G. Gladfelter, and R. J. Southard, Eds.), pp. 1–14. Special Publication no. 44. Soil Science Society of America, Madison, Wisc.

Foss, J. E., R. J. Lewis, M. E. Timpson, M. W. Morris, and J. T. Ammons, 1993a. Pedologic approaches to archaeological sites of contrasting environments and age. *In* Proceedings of the First International Conference on Pedo-archaeology (J. E. Foss, M. W. Morris, and M. E. Timpson, Eds.), pp. 19–22. Special Publication, Agricultural Experiment Station. University of Tennessee, Knoxville.

Foss, J. E., M. W. Morris, and M. E. Timpson, Eds., 1993b. Proceedings of the First International Conference on Pedo-archaeology. Special Publication, Agricultural Experiment Station. University of Tennessee, Knoxville.

Foss, J. E., and A. V. Segovia, 1984. Rates of soil formation. *In* Groundwater as a Geomorphic Agent (R. G. LaFleur, Ed.), pp. 1–17. Binghamton Symposium in Geomorphology no. 13. Allen and Unwin, Boston.

Fowler, P. J., and J. G. Evans, 1967. Plough marks, lynchets and early fields. *Antiquity* 41, 289–301.

Franklin, U. M., and V. Vitali, 1985. The environmental stability of ancient ceramics. *Archaeometry* 27, 3–15.

Frechen, M., and A. E. Dodonov, 1998. Loess chronology of the middle and upper Pleistocene in Tadjikistan. *Geologische Rundschau* 87, 2–20.

Frechen, M., B. Dermann, W. Boenigk, and A. Ronen, 2001, Luminescence chronology of aeolianites from the section at Givat Olga, coastal plain of Israel. *Quaternary Science Reviews* 20, 805–809.

Frechen, M., A. Neber, B. Dermann, A. Tsatskin, W. Boenigk, and A. Ronen, 2002. Chronostratigraphy of aeolianites from the Sharon coastal plain of Israel. *Quaternary International* 89, 31–44.

Freestone, I. C., 2001. Post-depositional changes in archaeological ceramics and glasses. *In* Handbook of Archaeological Sciences (D. R. Brothwell and A. M. Pollard, Eds.), pp. 615–625. Wiley, New York.

French, C. A. I., 2003. Geoarchaeology in Action: Studies in Soil Micromorphology and Landscape Evolution. Routledge Press, New York. 291 pp.

French, C. A. I., and T. M. Whitelaw, 1999. Soil erosion, agricultural terracing and site formation processes at Markiani, Amorgos, Greece: The micromorphological perspective. *Geoarchaeology* 14, 151–189.

Frezzotti, M., and B. Narcisi, 1996. Late Quaternary tephra-derived paleosols in central Italy's carbonate Apennine Range: Stratigraphical and paleoclimatological implications. *Quaternary International* 34–36, 147–153.

Frolking, T. A., and B. T. Lepper, 2001. Geomorphic and pedogenic evidence for bioturbation of artifacts at a multicomponent site in Licking County, Ohio, U.S.A. *Geoarchaeology* 16, 243–262.

Gale, S. J., and P. G. Hoare, 1991. Quaternary Sediments: Petrographic Methods for the Study of Unlithified Rocks. Wiley. New York.

Galm, J. R., and P. Flynn, 1978. The Cultural Sequences at the Scott (34Lf-11) and Wann (34Lf-27) Sites and Prehistory of the Wister Valley. Research Series no. 3, The University of Oklahoma Archaeological Research and Management Center, Norman.

Garrison, E. G., 2003. Techniques in Archaeological Geology. Springer, New York. 304 pp.

Gebhardt, A., 1995. Soil micromorphological data from traditional and experimental agriculture. *In* Archaeological Sediments and Soils: Analysis, Interpretation and Management (A. J. Barham and R. I. Macphail, Eds.), pp. 25–39. Institute of Archaeology, London.

Gee, G. W., and J. W. Bauder, 1986. Particle-size analysis. *In* Methods of Soil Analysis, Part 1—Physical and Mineralogical Methods. 2nd ed. (A. Klute, Ed.), pp. 383–411. Monograph no. 9. American Society of Agronomy, Madison, Wisc.

Gerig, A. J., 1985. Soil Survey of the Clackamas County Area, Oregon. U.S. Department of Agriculture, Soil Conservation Service, Washington D.C.

Gerlach, S. C., 1982. Small site archaeology in northern Alaska: The shoreline bluff survey. *Anthropological Papers of the University of Alaska* 20, 15–49.

Gerrard, J., 1992. Soils and Geomorphology. Chapman and Hall, London.

Gerrard, J., 2000. Fundamentals of Soils. Routledge, London.

Gerwin, W., and R. Baumhauer, 2000. Effect of soil parameters on the corrosion of archaeological metal finds. *Geoderma* 96, 63–80.

Gibbard, P. L., 1994. Pleistocene History of the Lower Thames Valley. Cambridge University Press, Cambridge.

Giddings, J. L., 1962. Onion Portage and other flint sites of the Kobuk River. *Arctic Anthropology* 1, 6–27.

Gifford, J. A., 1969. A hydraulic corer for use in water saturated sediment. *Journal of Sedimentary Petrology* 39, 1619–1621.

Gile, L. H., 1979. Holocene soils in eolian sediments of Bailey County, Texas. *Soil Science Society of America Journal* 43, 994–1003.

Gile, L. H., 1985. The Sandhills Project Soil Monograph. Rio Grande Historical Collections, New Mexico State University, Las Cruces.

Gile, L. H., J. W. Hawley, and R. B. Grossman, 1981. Soils and Geomorphology in the Basin and Range Area of Southern New Mexico—Guidebook to the Desert Project. Memoir no. 39. New Mexico Bureau of Mines and Mineral Resources, Socorro.

Gile, L. H., F. F. Peterson, and R. B. Grossman, 1965. The K horizon—A master soil horizon of carbonate accumulation. *Soil Science* 99, 74–82.

Gile, L. H., F. F. Peterson, and R. B. Grossman, 1966. Morphological and genetic sequences of carbonate accumulation in desert soils. *Soil Science* 101, 347–360.

Gladfelter, B. G., 1977. Geoarchaeology: The geomorphologist and archaeology. *American Antiquity* 42, 519–538.

Gladfelter, B. G., 1985. On the interpretation of archaeological sites in alluvial settings. *In* Archaeological Sediments in Context (J. K. Stein and W. R. Farrand, Eds.), pp. 41–52. Center for the Study of Early Man, Peopling of the Americas Edited Volume Series no. 1. University of Maine, Orono.

Gladfelter, B. G., 1992. Soil properties of sediments in Wadi Feiran, Sinai: A geoarchaeological interpretation. *In* Soils and Landscape Evolution (V. T. Holliday, Ed.), pp. 169–192. Smithsonian Institution Press, Washington, D.C.

Gladfelter, B. G., 2001. Humid alluvial systems. *In* Sediments in Archaeological Contexts (J. K. Stein and W. R. Farrand, Eds.), pp. 93–125. University of Utah Press, Salt Lake City.

Glaser, B., G. Guggenberger, L. Haumaier, and W. Zech, 2001a. Burning residues as conditioner to sustainably improve fertility in highly weathered soils of the Brazilian Amazon Region. *In* Sustainable Management of Soil Organic Matter (R. M. Rees, B. Ball, C. Watson, and C. Campbell, Eds.), pp. 190–194. CAB International, Wallingford, U.K.

Glaser, B., L. Haumaier, G. Guggenberger, and W. Zech, 2001b. The "Terra Preta" phenomenon: A model for sustainable agriculture in the humid tropics. *Naturwissenschaften* 88, 37–41.

Glaser, B., and W. I. Woods, Eds., 2004. Explorations in Amazonian Dark Earths. Springer-Verlag, Heidelberg, Germany.

Gleason, K. L., 1994. To bound and to cultivate: An introduction to the archaeology of gardens and fields. *In* The Archaeology of Garden and Field (N. F. Miller and K. L. Gleason, Eds.), pp. 1–24. University of Pennsylvania Press, Philadelphia.

Goh, K. M., 1980. Dynamics and stability of soil organic matter. *In* Soils with Variable Charge (B. K. G. Theng, Ed.), pp. 373–393. New Zealand Society of Soil Science, Lower Hutt.

Goldberg, P., 1979. Micromorphology of Pech-de-l'Azé II sediments. *Journal of Archaeological Science* 6, 17–47.

Goldberg, P., 1983. Applications of micromorphology in archaeology. *In* Soil Micromorphology, vol. 1 (P. Bullock and C. P. Murphy, Eds.), pp. 139–150. A B Academic Publishers, Berkhamsted.

Goldberg, P., 1987. Soils, sediments and Acheulian artifacts at Berekhat Ram, Golan Heights. *In* Soil Micromorphology (N. Fedoroff, L. M. Bresson, and M. A. Courty, Eds.), pp. 583–589. Association Française pour l'Etude du Sol, Plaisir.

Goldberg, P., 1988. The archaeologist as viewed by the geologist. *Biblical Archaeologist* 51, 197–202.

Goldberg, P., 1992. Micromorphology, soils, and archaeological sites. *In* Soils and Landscape Evolution (V. T. Holliday, Ed.), pp. 145–167. Smithsonian Institution Press, Washington, D.C.

Goldberg, P., 1994. Interpreting late Quaternary continental sequences in Israel. *In* Late Quaternary Chronology and Paleoclimates of the Eastern Mediterranean (O. Bar-Yosef and R. S. Kra, Eds.), pp. 89–102. American School of Prehistoric Research, Cambridge, Mass.

Goldberg, P., 1998. Micromorphological analysis of Wilson-Leonard sediments. *In* Wilson-Leonard: An 11,000-year Archeological Record of Hunter-Gatherers in Central Texas; Volume IV: Archaeological Features and Technical Analyses (M. B. Collins, Ed.), pp. 1343–1363. Texas Archeological Research Laboratory Studies in Archaeology no. 31. The University of Texas at Austin, Austin.

Goldberg, P., 2000. Micromorphology and site formation at Die Kelders Cave 1, South Africa. *Journal of Human Evolution* 38, 43–90.

Goldberg, P., and T. Arpin, 2000. Micromorphological analysis of sediments from Meadowcroft Rockshelter, Pennsylvania: Implications for radiocarbon dating. *Journal of Field Archaeology* 26, 325–342.

Goldberg, P., and V. T. Holliday, 1998. Geology and stratigraphy. *In* Wilson-Leonard: An 11,000-year Archeological Record of Hunter-Gatherers in Central Texas; Volume I: Introduction, Background, and Synthesis (M. B. Collins, Ed.), pp. 77–121. Texas Archeological Research Laboratory Studies in Archaeology no. 31. The University of Texas at Austin, Austin.

Goldberg, P., V. T. Holliday, and C. R. Ferring, Eds., 2001. Earth Sciences and Archaeology. Kluwer Academic/Plenum Publishers, New York.

Goldberg, P., and R. I. Macphail, 2003. Strategies and techniques in collecting micromorphology samples. *Geoarchaeology* 18, 571–578.

Goldberg, P., D. T. Nash, and M. D. Petraglia, 1993. Formation Processes in Archaeological Context. Prehistory Press, Madison, Wisc.

Goldberg, P., and S. Sherwood, 1994. Micromorphology of Dust Cave sediments: Some preliminary results. *Journal of Alabama Archaeology* 40, 56–64.

Goldberg, P., and I. Whitbread, 1993. Micromorphological study of a Bedouin tent floor. *In* Formation Processes in Archaeological Contexts (P. Goldberg, D. T. Nash, and M. D. Petraglia, Eds.), pp. 165–188. Monographs in World Archaeology no. 17. Prehistory Press, Madison, Wisc.

Goodyear, A. C., J. E. Foss, and K. E. Sassaman, Eds., 1997. Proceedings of the Second International Conference on Pedo-Archaeology. Occasional Papers of the South Carolina Institute of Archaeology and Anthropology. University of South Carolina, Columbia.

Gordon, B. C., 1978. Chemical and pedological delimiting of deeply stratified archaeological sites in frozen ground. *Journal of Field Archaeology* 5, 331–338.

Goring-Morris, A. N., and P. Goldberg, 1991. Late Quaternary dune incursions in the southern Levant: Archaeology, chronology and paleoenvironments. *Quarternary International* 5, 115–137.

Goudie, A. S., 1983. Calcretes. *In* Chemical Sediments and Geomorphology (A. S. Goudie and K. Pye, Eds.), pp. 93–131. Academic Press, London.

Goudie, A. S., 2000. The Human Impact on the Natural Environment. 5th ed. The MIT Press, Cambridge, Mass.

Green, F. E., 1962. The Lubbock Reservoir site. *The Museum Journal (West Texas Museum Association)* 6, 83–123.

Green, W., and J. F. Doershuk, 1998. Cultural resource management and American archaeology. *Journal of Archaeological Research* 6, 121–167.

Greenberg, A. E, L. S. Clesceri, and A. D. Eaton, Eds., 1992. Standard Methods for the Examination of Water and Wastewater. 18th ed. American Public Health Association, American Water Works Association, and Water Environment Federation. Victor Graphics, Baltimore, Md.

Gremillion, K. J., Ed., 1997. People, Plants, and Landscapes: Studies in Paleoethnobotany. University of Alabama Press, Tuscaloosa.

Griffith, M. A., 1980. A pedological investigation of an archaeological site in Ontario, Canada, I: An examination of the soils in and adjacent to a former village. *Geoderma* 24, 327–336.

Griffith, M. A., 1981. A pedological investigation of an archaeological site in Ontario, Canada, II: Use of chemical data to discriminate features of the Benson site. *Geoderma* 25, 27–34.

Groenman-van Waateringe, W., and M. Robinson, Eds., 1988. Man-Made Soils. BAR International Series 410, London.

Grolier, M. J., 1988. Geomorphology of the al-Jubah quadrangle, Yemen Arab Republic: A guide to the interpretation of landforms in the basin of Wadi al-Jubah. *In* Geological and Archaeological Reconnaissance in the Yemen Arab Republic, 1985 (W. C. Overstreet, M. J. Grolier, and M. R. Toplyn, Eds.), pp. 289–357. American Foundation for the Study of Man, Washington, D.C.

Gubin, S. V., 1984. Diagenesis of dry steppe soils buried by artificial fill. *Pochvovedenie* 6, 5–13.

Gundlach, H., 1961. Tüpfelmethode auf Phosphat, angewandt in prähistorischer Forschung (als Feldmethode). *Mikrochimica Acta* 5, 735–737.

Gurney, D. A., 1985. Phosphate Analysis of Soils: A Guide for the Field Archaeologist. Technical Paper no. 3. The Institute of Field Archaeologists, Birmingham.

Gvirtzman, G., M. Netser, and E. Katsav, 1998. Last-glacial to Holocene kurkar ridges, hamra soils, and dune fields in the coastal belt of central Israel. *Israel Journal of Earth Sciences* 47, 29–46.

Gvirtzman, G., E. Shachnai, N. Bakler, and S. Ilani, 1984. Stratigraphy of the Kurkar Group (Quaternary) of the coastal plain of Israel. *Geological Survey of Israel Current Research* 1983–1984, 70–82.

Gvirtzman, G., and M. Wieder, 2001. Climate of the last 53,000 years in the eastern Mediterranean, based on soil-sequence stratigraphy in the coastal plain of Israel. *Quaternary Science Reviews* 20, 1827–1849.

Gvirtzman, G., M. Wieder, O. Marder, H. Khalaily, R. Rabinovich, and Hagai Ron, 1999. Geological and pedological aspects of an early-Paleolithic site: Revadim, central coastal plain, Israel. *Geoarchaeology* 14, 101–126.

Haas, H., V. T. Holliday, and R. Stuckenrath, 1986. Dating of Holocene stratigraphy with soluble and insoluble organic fractions at the Lubbock Lake site, Texas: An ideal case study. *Radiocarbon* 28, 473–485.

Hageman, B. P., 1972. Reports of the International Quaternary Association Subcommission on the Study of the Holocene, Bulletin 6.

Haidouti, C., and N. J. Yassoglou, 1982. The genesis of certain soil profiles on an archaeological site in southern Greece. *Soil Science Society of America Journal* 46, 1047–1051.

Hajic, E. R., 1990. Koster Site Archeology I: Stratigraphy and Landscape Evolution. Research Series no. 8. Center for American Archeology, Kampsville, Ill.

Hajic, E. R., R. D. Mandel, J. H. Ray, and N. H. Lopinot, 1998. Geomorphology and geoarchaeology. *In* The 1997 Excavations at the Big Eddy Site (23CE426) in Southwest Missouri (N. H. Lopinot, J. H. Ray, and M. D. Conner, Eds.), pp. 74–110. Center for Archaeological Research, Special Publication no. 2. Southwest Missouri State University, Springfield, Mo.

Hall, G. F., 1983. Pedology and geomorphology. *In* Pedogenesis and Soil Taxonomy, part 1, Concepts and Interactions (L. P. Wilding, N. E. Smeck, and G. F. Hall, Eds.), pp. 117–140. Elsevier, Developments in Soil Science 11A, Amsterdam.

Hall, G. F., and C. G. Olson, 1991. Predicting variability of soils from landscape models. *In* Spatial Variabilities of Soils and Landforms (M. J. Mausbach and L. P. Wilding, Eds.), pp. 9–24. Special Publication no. 28. Soil Science Society of America, Madison, Wisc.

Hall, R. D., 1999. A comparison of surface soils and buried soils: Factors of soil development. *Soil Science* 164, 264–287.

Hall, S. A., 1990. Holocene landscapes of the San Juan Basin, New Mexico: Geomorphic, climatic, and cultural dynamics. *In* Archaeological Geology of North America (N. P. Lasca and J. Donahue, Eds.), pp. 323–334. Centennial Volume no. 4. Geological Society of America, Boulder, Colo.

Hall, S. A., 2002. Field Guide to the Geoarchaeology of the Mescalero Sands, Southeastern New Mexico. State of New Mexico Historic Preservation Division and New Mexico Bureau of Land Management, Santa Fe.

Hall, S. A., and S. Valastro, 1995. Grassland vegetation in the Southern Great Plains during the last glacial maximum. *Quaternary Research* 44, 237–245.

Hallberg, G. R., 1984. The U. S. system of Soil Taxonomy: From the outside looking in. *In* Soil Taxonomy-Achievements and Challenges (R. B. Grossman, H. Eswaran, and R. H. Rust, Eds.), pp. 45–59. Soil Science Society of America, Madison, Wisc.

Hallberg, G. R., N. C. Wollenhaupt, and G. A. Miler, 1978. A century of soil development in spoil derived from loess in Iowa. *Soil Science Society of America Proceedings* 42, 339–343.

Hamilton, T. D., 1970. Appendix: Geologic relations of the Akmak assemblages, Onion Portage area. *Acta Arctica*, fascicle 16, 71–80.

Hamond, F. W., 1983. Phosphate analysis of archaeology sediments. *In* Landscape Archaeology in Ireland (T. Reeves-Smyth and F. Hamond, Eds.), pp. 47–80. BAR British Series no. 116, Oxford.

Han, J. T., W. S. Fyfe, and F. J. Longstaffe, 1998. Climatic implications of the S5 paleosol complex on the southernmost Chinese Loess Plateau. *Quaternary Research* 50, 21–33.

Harden, J. W., 1982. A quantitative index of soil development from field descriptions: Examples from a chronosequence in central California. *Geoderma* 28, 1–28.

Harden, J. W., and E. M. Taylor, 1983. A quantitative comparison of soil development in four climatic regimes. *Quaternary Research* 20, 342–359.

Harland, W. B., R. L. Armstrong, A. V. Cox, L. E. Craig, A. G. Smith, and D. G. Smith, 1990. A Geologic Time Scale 1989. Cambridge University Press, Cambridge.

Harrison, A. F., 1987. Soil Organic Phosphorus: A Review of World Literature. CAB International, Wallingford, U.K.

Hartung, S. L., S. A. Scheinhost, and R. J. Ahrens, 1991. Scientific methodology of the National Cooperative Soil Survey. *In* Spatial Variabilities of Soils and Landforms (M. J. Mausbach and L. P. Wilding, Eds.), pp. 39–48. Special Publication no. 28. Soil Science Society of America, Madison, Wisc.

Hassan, F. A., 1981. Rapid quantitative determination of phosphate in archaeological sediments. *Journal of Field Archaeology* 8, 384–387.

Hassan, F. A., 1985. Fluvial systems and geoarchaeology in arid lands: With examples from North Africa, the Near East and the American Southwest. *In* Archaeological Sediments in Context (J. K. Stein and W. R. Farrand, Eds.), pp. 53–68. Center for the Study of Early Man, Peopling of the Americas Edited Volume Series no. 1. University of Maine, Orono.

Hay, R. L., 1976. Geology of the Olduvai Gorge. University of California Press, Berkeley.

Hayen, H., 1987. Peatbog archaeology in Lower Saxony, West Germany. *In* European Wetlands in Prehistory (J. M. Coles and A. J. Lawson, Eds.), pp. 117–136. Clarendon Press, Oxford.

Haynes, C. V., 1968. Geochronology of late-Quaternary alluvium. *In* Means of Correlation of Quaternary Successions (R. B. Morrison and H. E. Wright, Jr., Eds.), pp. 591–631. University of Utah Press, Salt Lake City.

Haynes, C. V., 1975. Pleistocene and Recent stratigraphy. *In* Late Pleistocene Environments of the Southern High Plains (F. Wendorf and J. J. Hester, Eds.), pp. 59–96. Publication no. 9. Ft. Burgwin Research Center, Taos, N.M.

Haynes, C. V., 1982. Great Sand Sea and Selima sand sheet, eastern Sahara: Geochronology of desertification. *Science* 217, 629–633.

Haynes, C. V., 1987. Curry Draw, Cochise County, Arizona: A late Quaternary stratigraphic record of Pleistocene extinction and paleo-Indian activities. *In* Cordilleran Section (M. L. Hill, Ed.), pp. 23–28. Centennial Field Guide no. 1. Geological Society of America, Boulder, Colo.

Haynes, C. V., 1989. Bagnold's barchan: A 57-yr record of dune movement in the eastern Sahara and implications for dune origin and paleoclimate since Neolithic times. *Quaternary Research* 32, 153–167.

Haynes, C. V., 1990. The Antevs-Bryan years and the legacy for Paleoindian geochronology. *In* Establishment of a Geologic Framework for Paleoanthropology (L. F. Laporte, Ed.), pp. 55–68. Special Paper no. 242. Geological Society of America, Boulder, Colo.

Haynes, C. V., 1991. Geoarchaeological and paleohydrological evidence for a Clovis-age drought in North America and its bearing on extinction. *Quaternary Research* 35, 438–450.

Haynes, C. V., 1995. Geochronology of paleoenvironmental change, Clovis type site, Blackwater Draw, New Mexico. *Geoarchaeology* 10, 317–388.

Haynes, C. V., 2001. Geochronology and climate change of the Pleistocene-Holocene transition in the Darb el Arba'in Desert, eastern Sahara. *Geoarchaeology* 16, 119–141.

Haynes, C. V., 2003. Dating the Lindenmeier Folsom site, Colorado, U.S.A., before the radiocarbon revolution. *Geoarchaeology* 18, 161–174.

Haynes, C. V., and G. A. Agogino, 1966. Prehistoric springs and geochronology of the Clovis site, New Mexico. *American Antiquity* 31, 812–821.

Haynes, C. V., and D. L. Johnson, 1984. A provisional soil chronosequence for the Quaternary of the eastern Sahara. *Geological Society of America Abstracts with Programs* 19, 762–763.

Haynes, C. V., T. A. Maxwell, and D. L. Johnson, 1993. Stratigraphy, geochronology, and origin of the Selima Sand Sheet, eastern Sahara, Egypt and Sudan. *In* Geoscientific Research in Northeast Africa (U. Thorweihe and H. Schandelmeier, Eds.), pp. 621–626. A. A. Balkema Press, Rotterdam.

Head, M. J., 1999. Radiocarbon dating of arid zone deposits. *In* Paleoenvironmental Reconstruction in Arid Lands (A. K. Singhvi and E. Derbyshire, Eds.), pp. 293–326. A. A. Balkema, Rotterdam.

Hedley, M. J., J. W. B. Stewart, and B. S. Chauhan, 1982. Changes in inorganic and organic phosphorus fractions induced by cultivation practices and by laboratory incubations. *Soil Science Society of America Journal* 46, 970–976.

Heidenreich, C. E., A. R. Hill, D. M. Lapp, and S. Navratil, 1971. Soil and environmental analysis at the Robitaille site. *In* Palaeoecology and Ontario Prehistory—II (W. M. Hurley and C. E. Heidenreich, Eds.) pp. 179–237. Department of Anthropology Research Report no. 2. University of Toronto, Toronto.

Heidenreich, C. E., and V. A. Konrad, 1973. Soil analysis at the Robitaille site, part II: A method useful in determining the location of longhouse patterns. *Ontario Archaeology* 20, 33–62.

Heidenreich, C. E., and S. Navratil, 1973. Soil analysis at the Robitaille site, part I: Determining the perimeter of the village. *Ontario Archaeology* 20, 25–32.

Heidinga, H. A., 1988. Climate and plaggen soils. *In* Man-Made Soils (W. Groennmann-van Waateringe and M. Robinson, Eds.), pp. 21–33. BAR International Series no. 410, Oxford.

Heiri, O., A. F. Lotter, and G. Lemcke, 2001. Loss on ignition as a method for estimating organic and carbonate content in sediments: Reproducibility and comparability of results. *Journal of Paleolimnology* 25, 101–110.

Heizer, R. F., 1966. A Guide to Archaeological Field Methods. 3rd ed. The National Press, Palo Alto, Calif.

Heizer, R. F., and J. F. Graham, 1967. A Guide to Field Methods in Archaeology: Approaches to the Anthropology of the Dead. The National Press, Palo Alto, Calif.

Herz, N., and E. G. Garrison, 1998. Geological Methods for Archaeology. Oxford University Press, New York.

Hester, T. R., Ed., 1991. The Burned Rock Middens of Texas: An Archeological Symposium. Studies in Archeology no. 13. The University of Texas at Austin Texas Archeological Research Laboratory, Austin.

Hester, T. R., H. J. Shafer, and K. L. Feder, Eds., 1997. Field Methods in Archaeology. 7th ed. Mayfield, Mountain View, Calif.

Hill, M. E., Jr., and J. L. Hofman, 1997. The Waugh site: A Folsom-age bison bonebed in northwestern Oklahoma. *Plains Anthropologist* 42, 63–83.

Hill, M. G., V. T. Holliday, and D. J. Stanford, 1995. A further evaluation of the San Jon site, New Mexico. *Plains Anthropologist* 40, 369–390.

Hilton, M. R., 2003. Quantifying postdepositional redistribution of the archaeological record produced by freeze-thaw and other mechanisms: An experimental approach. *Journal of Archaeological Method and Theory* 10, 165–202.

Hodgson, J. M., Ed., 1997. Soil Survey Field Handbook. Technical Monograph no. 5. Soil Survey and Land Research Centre, Silsoe.

Hoffecker, J. F., 1987. Upper Pleistocene loess stratigraphy and Paleolithic site chronology on the Russian Plain. *Geoarchaeology* 2, 259–284.

Hoffecker, J. F., 1988. Early Upper Paleolithic sites of the European USSR. *In* The Early Upper Paleolithic: Evidence from Europe and the Near East (J. F. Hoffecker and C. A. Wolf, Eds.), pp. 237–272. BAR International Series no. 437, Oxford.

Hoffecker, J. F., 2002. Desolate Landscapes: Ice-Age Settlement in Eastern Europe. Rutgers University Press, New Brunswick.

Hofman, J. L., 1986. Vertical movements of artifacts in alluvial and stratified deposits. *Current Anthropology* 27, 163–171.

Hole, F. D., 1961. A classification of pedoturbations and some other processes and factors of soil formation in relation to isotropism and anisotropism. *Soil Science* 91, 375–377.

Hole, F. D., and J. B. Campbell, 1985. Soil Landscape Analysis. Rowman and Allanheld, Tontawa, N.J.

Holliday, V. T., 1985a. Archaeological geology of the Lubbock Lake site, Southern High Plains of Texas. *Geological Society of America Bulletin* 96, 1483–1492.

Holliday, V. T., 1985b. Early Holocene soils at the Lubbock Lake archaeological site, Texas. *Catena* 12, 61–78.

Holliday, V. T., 1985c. Holocene soil-geomorphological relations in a semi-arid environment: The Southern High Plains of Texas. *In* Soils and Quaternary Landscape Evolution (J. Boardman, Ed.), pp. 325–357. Wiley, New York.

Holliday, V. T., 1985d. Morphology of late Holocene soils at the Lubbock Lake archaeological site, Texas. *Soil Science Society of America Journal* 49, 938–946.

Holliday, V. T., 1985e. New data on the stratigraphy and pedology of the Clovis and Plainview sites, Southern High Plains. *Quaternary Research* 23, 388–402.

Holliday, V. T., 1987a. Geoarchaeology and late Quaternary geomorphology of the middle South Platte River, northeastern Colorado. *Geoarchaeology* 2, 317–329.

Holliday, V. T., 1987b. A reexamination of late-Pleistocene boreal forest reconstructions for the Southern High Plains. *Quaternary Research* 28, 238–244.

Holliday, V. T., 1988. Genesis of a late Holocene soil chronosequence at the Lubbock Lake archaeological site, Texas. *Annals of the Association of American Geographers* 78, 594–610.

Holliday, V. T., 1989a. The Blackwater Draw Formation (Quaternary): A 1.4-plus m.y. record of eolian sedimentation and soil formation on the Southern High Plains. *Geological Society of America Bulletin* 101, 1598–1607.

Holliday, V. T., 1989b. Middle Holocene drought on the Southern High Plains. *Quaternary Research* 31, 74–82.

Holliday, V. T., 1989c. Paleopedology in archeology. *In* Paleopedology: Nature and Applications of Paleosols (A. Bronger and J. Catt, Eds.), pp. 187–206. Catena Supplement 16. Catena Verlag, Cremlingen, Germany.

Holliday, V. T., 1990a. Pedology in Archaeology. *In* Archaeological Geology of North America (N. P. Lasca and J. Donahue, Eds.), pp. 525–540. Centennial Volume no. 4. Geological Society of America, Boulder, Colo.

Holliday, V. T., 1990b. Soils and landscape evolution of eolian plains: The Southern High Plains of Texas and New Mexico. *In* Soils and Landscape Evolution (P. L. K. Knuepfer and L. D. McFadden, Eds.), pp. 489–515. Elsevier, New York.

Holliday, V. T., 1992a. Soil formation, time, and archaeology. *In* Soils in Archaeology: Landscape Evolution and Human Occupation (V. T. Holliday, Ed.), pp. 101–117. Smithsonian Institution Press, Washington, D.C.

Holliday, V. T., Ed., 1992b. Soils in Archaeology: Landscape Evolution and Human Occupation. Smithsonian Institution Press, Washington, D.C.

Holliday, V. T., 1994. The "State Factor" approach in geoarchaeology. *In* Factors of Soil Formation: A Fiftieth Anniversary Retrospective (R. Amundson, Ed.), pp. 65–86. Special Publication no. 33. Soil Science Society of America, Madison, Wisc.

Holliday, V. T., 1995. Stratigraphy and Paleoenvironments of Late Quaternary Valley Fills on the Southern High Plains. Memoir no. 186. Geological Society of America, Boulder, Colo.

Holliday, V. T., 1997. Paleoindian Geoarchaeology of the Southern High Plains. University of Texas Press, Austin.

Holliday, V. T., 2000. Folsom drought and episodic drying on the Southern High Plains from 10,900-10,200 ^{14}C yr B.P. *Quaternary Research* 53, 1–12.

Holliday, V. T., 2001a. Quaternary geoscience in archaeology. *In* Earth Science in Archaeology (P. Goldberg, V. T. Holliday, and C. R. Ferring, Eds.), pp. 3–35. Kluwer Academic/Plenum Publishers, New York.

Holliday, V. T., 2001b. Stratigraphy and geochronology of upper Quaternary eolian sand on the Southern High Plains of Texas and New Mexico, United States. *Geological Society of America Bulletin* 113, 88–108.

Holliday, V. T., and B. L. Allen, 1987. Geology and soils. *In* Lubbock Lake: Late Quaternary Studies on the Southern High Plains (E. Johnson, Ed.), pp. 14–21. Texas A&M University Press, College Station.

Holliday, V. T., C. R. Ferring, and P. Goldberg, 1993. The scale of soil investigations in archaeology. *In* Effects of Scale on Archaeological and Geoscientific Perspectives (J. K. Stein and A. R. Linse, Eds.), pp. 29–37. Special Paper no. 283. Geological Society of America, Boulder, Colo.

Holliday, V. T., C. V. Haynes, J. L. Hofman, and D. J. Meltzer, 1994. Geoarchaeology and geochronology of the Miami (Clovis) site, Southern High Plains of Texas. *Quaternary Research* 41, 234–244.

Holliday, V. T., L. D. McFadden, E. A. Bettis, and P. W. Birkeland, 2002. The soil survey and soil geomorphology. *In* Profiles in the History of the U.S. Soil Survey (D. Helms, A. B. W. Effland, and P. J. Durana, Eds.), pp. 233–274. Iowa State Press and Blackwell Publishers, Ames, Ia.

Holliday, V. T., and J. K. Stein, 1989. Variability of laboratory procedures and results in geoarchaeology. *Geoarchaeology* 4, 347–358.

Homburg, J. A., 1988. Archaeological investigations at the LSU Campus Mounds site. *Louisiana Archaeology* 15, 31–204.

Homburg, J. A., 2000. Anthropogenic Influences on American Indian Agricultural Soils of the Southwestern United States. Ph.D. Dissertation, Department of Agronomy, Iowa State University, Ames.

Homburg, J. A., and J. A. Sandor, 1997. An agronomic study of two Classic Period agricultural fields in the Horseshoe Basin. *In* Vanishing River: Landscapes and Lives of the lower Verde Valley: The Lower Verde Archaeological Project, vol 2. Agricultural, Subsistence, and Environmental Studies (CD-ROM) (J. A. Homburg and R. Ciolek-Torrello, Eds.), pp. 127–147. SRI Press, Tucson, Ariz.

Hopkins, D. M., 1975, Time-stratigraphic nomenclature for the Holocene epoch. *Geology* 3, 10.

Hopkins, D. M., and J. L. Giddings, 1953. Geological Background of the Iyatayet Archeological Site, Cape Denbigh, Alaska. Smithsonian Miscellaneous Collections no. 121. Smithsonian Institution, Washington, D.C.

Horowitz, A., 1979. The Quaternary of Israel. Academic Press, New York.

Hoyer, B. E., 1980. The geology of the Cherokee Sewer site. The Cherokee Excavations (D. C. Anderson and H. A. Semken, Eds.), pp. 21–66. Academic Press, New York.

Huckleberry, G., 1992. Soil evidence of Hohokam irrigation in the Salt River Valley, Arizona. *The Kiva* 57, 237–249.

Huckleberry, G., 2001. Archaeological sediments in dryland alluvial environments. *In* Sediments in Archaeological Contexts (J. K. Stein and W. R. Farrand, Eds.), pp. 67–92. University of Utah Press, Salt Lake City.

Huggett, R. J., 1998. Soil chronosequences, soil development, and soil evolution: A critical review. *Catena* 32, 155–172.

Humphrey, J. D., and C. R. Ferring, 1994. Stable isotopic evidence for latest Pleistocene and Holocene climatic change in north-central Texas. *Quaternary Research* 41, 200–213.

Hunt, C. B., 1972. Geology of Soils. W. H. Freeman, San Francisco.

Hunt, C. B., and A. P. Hunt, 1957. Stratigraphy and archeology of some Florida soils. *Geological Society of America Bulletin* 68, 797–806.

Imbellone, P. A., and M. E. Teruggi, 1993. Paleosols in loess deposits of the Argentine Pampas. *Quaternary International* 17, 49–55.

International Atomic Energy Agency (IAEA), 1998. Isotope Techniques in the Study of Environmental Change. Proceedings Series, International Atomic Energy Agency, Vienna.

Imbrie, J., and K. P. Imbrie, 1979. Ice Ages: Solving the Mystery. Enslow Publishers, Hillside, N.J.

Iriondo, M., Ed., 1999. Loess in South America and Related Topics. *Quaternary International* 62, 1–117.

Ito, T., S. Shoji, Y. Shirato, and E. Ono, 1991. Differentiation of a Spodic horizon from a buried A horizon. *Soil Science Society of America Journal* 55, 438–442.

Jackson, M. L., 1958. Soil Chemical Analysis. Prentice Hall, Englewood Cliffs, N.J.

Jackson, M. L., 1969. Soil Chemical Analysis—Advanced Course. 2nd ed. Published by the author, Department of Soils, University of Wisconsin, Madison.

Jacob, J. S., 1995a. Ancient Maya wetland agricultural fields at Cobweb Swamp, Belize: Construction, chronology, and possible function. *Journal of Field Archaeology* 22, 175–190.

Jacob, J. S., 1995b. Archaeological pedology in the Maya lowlands. *In* Pedological Perspectives in Archaeological Research (M. E. Collins, B. J. Carter, B. G. Gladfelter, and R. J. Southard, Eds.), pp. 51–80. Special Publication no. 44. Soil Science Society of America, Madison, Wisc.

Jacobs, P. M., 1998. Influence of parent material grain size on genesis of the Sangamon Geosol in south-central Indiana. *Quaternary International* 51/52, 127–132.

James, L. A., 1988. Rates of organic carbon accumulation in young mineral soils near Burroughs glacier, Glacier Bay, Alaska. *Physical Geography* 9, 50–70.

James, P., D. Chester, and A. Duncan, 2000. Volcanic soils: Their nature and significance for archaeology. *In* The Archaeology of Geological Catastrophes (W. G. McGuire, D. R. Griffiths, P. L. Hancock, and I. S. Stewart, Eds.), pp. 317–338. Special Publications no. 171. Geological Society, London.

Janak, E. F., and W. A. Martin, 1987. Pedology: Implications for archaeological interpretation. *In* The Bird Point Island and Adams Ranch Sites, Richland Creek, Texas (J. E. Bruseth and W. A. Martin, Eds.), pp. 11–15. Technical Series Vol. III, Archaeological Research program, Institute for the Study of Earth and Man. Southern Methodist University, Dallas, Tex.

Janitzky, P., 1986a. Organic carbon (Walkley-Black method). *In* Field and Laboratory Procedures used in a Soil Chronosequence Study (M. J. Singer and P. Janitzky, Eds.), pp. 34–36. Bulletin no. 1648. U.S. Geological Survey, Washington, D.C.

Janitzky, P., 1986b. Particle-size analysis. *In* Field and Laboratory Procedures used in a Soil Chronosequence Study (M. J. Singer and P. Janitzky, Eds.), pp. 11–16. Bulletin no. 1648. U.S. Geological Survey, Washington, D.C.

Jenkins, D., 1985. Chemical and mineralogical composition in the identification of paleosols. *In* Soils and Quaternary Landscape Evolution (J. Boardman, Ed.), pp. 23–43. Wiley, Chichester.

Jenny, H., 1941. Factors of Soil Formation. McGraw-Hill, New York.

Jenny, H., 1980. The Soil Resource. Springer, New York.

Jenny, H., and C. D. Leonard, 1934. Functional relationships between soil properties and rainfall. *Soil Science* 38, 363–381.

Jia, L. P., 1980. Early Man in China. Foreign Language Press, Beijing.

Jing, Z., G. Rapp, and T. Gao, 1995. Holocene landscape evolution and its impact on the Neolithic and Bronze age sites in the Shangqui area, northern China. *Geoarchaeology* 10, 481–513.

Johnson, D. L., 1983. Geomorphic setting. *In* Bug Hill: Excavation of a Multicomponent Midden Mound in the Jackfork Valley, Southeast Oklahoma (J. H. Altschul), pp. 17–31. Report of Investigation no. 81–1. New World Research, Pollock, La.

Johnson, D. L., 1989. Subsurface stone lines, stone zones, artifact-manuport layers, and biomantles produced by bioturbation via pocket gophers (*Thomomys botae*). *American Antiquity* 54, 370–389.

Johnson, D. L., 1990. Biomantle evolution and the redistribution of earth materials and artifacts. *Soil Science* 149, 84–102.

Johnson, D. L., 1993. Dynamic denudation evolution of tropical, subtropical and temperate landscapes with three tiered soils: Toward a general theory of landscape evolution. *Quaternary International* 17, 67–78.

Johnson, D. L., 1994a. Biomechanical processes and the Gaia paradigm in a pedogeomorphic and pedo-archaeologic framework: Dynamic denudation. *In* Proceedings of the First International Conference on Pedo-Archaeology (J. E. Foss, M. W. Morris, and M. E. Timpson, Eds.), pp. 41–68. Special Publication, Agricultural Experiment Station. University of Tennessee, Knoxville.

Johnson, D. L., 1994b. Reassessment of early and modern soil horizon designation frameworks and associated pedogenetic processes: Are midlatitude A-E-B-C horizons equivalent to tropical M S W horizons? *Soil Science (Trends in Agricultural Science)* 2, 77–91.

Johnson, D. L., 1998. Paleosols are buried soils. *Quaternary International* 51/52, 7.

Johnson, D. L., 2000. Soils and soil-geomorphology theories and models: The Macquarie connection. Book review forum on "Soils: A New Global View." *Annals of the Association of American Geographers* 90, 775–782.

Johnson, D. L., 2002. Darwin would be proud: Bioturbation, dynamic denudation, and the power of theory in science. *Geoarchaeology* 17, 7–40.

Johnson, D. L., and K. L. Hansen, 1974. The effects of frost-heaving on objects in soils. *Plains Anthropologist* 19, 81–98.

Johnson, D. L., and N. C. Hester, 1972. Origin of stone pavements on Pleistocene marine terraces in California. *Proceedings, Association of American Geographers* 4, 50–53.

Johnson, D. L., and F. D. Hole, 1994. Soil formation theory: A summary of its principal impacts on geography, geomorphology, soil-geomorphology, Quaternary geology, and paleopedology. *In* Factors of Soil Formation: A Fiftieth Anniversary Retrospective (R. Amundson, Ed.), pp. 111–126. Special Publication no. 33. Soil Science Society of America, Madison, Wisc.

Johnson, D. L., E. A. Keller, and T. K. Rockwell, 1990. Dynamic pedogenesis: New views on some key soil concepts, and a model for interpreting Quaternary soils. *Quaternary Research* 33, 306–319.

Johnson, D. L., D. R. Muhs, and M. L. Barnhardt, 1977. The effects of frost-heaving on objects in soils; 2, Laboratory experiments. *Plains Anthropologist* 22, 133–147.

Johnson, D. L., and D. Watson-Stegner, 1987. Evolution model of pedogenesis. *Soil Science* 143, 349–366.

Johnson, D. L., and D. Watson-Stegner, 1990. The soil-evolution model as a framework for evaluating pedoturbation in archaeological site formation. *In* Archaeological Geology of North America (N. P. Lasca and J. Donahue, Eds.), pp. 541–560. Centennial Volume no. 4. Geological Society of America, Boulder, Colo.

Johnson, D. L., D. Watson-Stegner, D. N. Johnson, and R. J. Schaetzl, 1987. Proisotropic and proanisotropic processes of pedoturbation. *Soil Science* 143, 278–292.

Johnson, E., 1986. Late Pleistocene and early Holocene paleoenvironments on the Southern High Plains (USA). *Géographie Physique et Quaternaire* 40, 249–261.

Johnson, E., 1987a. Cultural activities and interactions. Lubbock Lake: Late Quaternary Studies on the Southern High Plains (E. Johnson, Ed.) pp. 120–158. Texas A&M University Press, College Station.

Johnson, E., Ed., 1987b. Lubbock Lake: Late Quaternary Studies on the Southern High Plains. Texas A&M University Press, College Station.

Johnson, E., 1987c. Paleoenvironmental overview. *In* Lubbock Lake: Late Quaternary Studies on the Southern High Plains (E. Johnson, Ed.), pp. 90–99. Texas A&M University Press, College Station.

Johnson, E., 1991. Late Pleistocene cultural occupation on the Southern Plains. *In* Clovis: Origins and Adaptations (R. Bonnichsen and K. L. Turnmire, Eds.), pp. 215–236. Center for the Study of the First Americans, Orono, Me.

Johnson, E., and V. T. Holliday, 1986. The Archaic record at Lubbock Lake. *Plains Anthropologist Memoir* 21, 7–54.

Johnson, E., and V. T. Holliday, 1989. Lubbock Lake: Late Quaternary cultural and environmental change on the Southern High Plains, USA. *Journal of Quaternary Science* 4, 145–165.

Johnson, E., and V. T. Holliday, 1995. Archaeology and Late Quaternary Environments of the Southern High Plains. *Bulletin of the Texas Archeological Society* 66, 519–540.

Johnson, G. D., 1977. Paleopedology of *Ramapithecus*-bearing sediments, North India. *Geologische Rundschau* 66, 192–216.

Johnson, G. D., P. H. Rey, R. H. Ardrey, C. F. Visser, N. D. Opdyke, and R. A. Khan Tahirkheli, 1981. Paleoenvironments of the Siwalik Group, Pakistan and India. *In* Hominid Sites: Their Geologic Settings (G. Rapp, Jr. and C. F. Vondra, Eds.), pp. 197–254. Selected Symposium no. 63, American Association for the Advancement of Science, Washington, D.C.

Johnson, W. C., and B. Logan, 1990. Geoarchaeology of the Kansas River Basin, central Great Plains. *In* Archaeological Geology of North America (N. P. Lasca and J.

Donahue, Eds.), pp. 267–299. Centennial Volume no. 4. Geological Society of America, Boulder, Colo.

Johnson, W. C., and C. W. Martin, 1987. Holocene alluvial-stratigraphic studies from Kansas and adjoining states of the east-central Plains. *In* Quaternary Environments of Kansas (W. C. Johnson, Ed.), pp. 109–122. Guidebook Series no. 5. Kansas Geological Survey, Lawrence.

Johnson, W. H., A. K. Hansel, E. A. Bettis, III, P. F. Karrow, G. J. Larson, T. V. Lowell, and A. F. Schneider, 1997. Late Quaternary temporal and event classifications, Great Lakes region, North America. *Quaternary Research* 47, 1–12.

Jones, K. L., 1990. Settlement chronology on river terrace land forms: A New Zealand case study. *Geoarchaeology* 5, 255–273.

Jorgensen, D. W., 1992. Use of soils to differentiate dune age and to document spatial variation in eolian activity, northeast Colorado, U.S.A. *Journal of Arid Environments* 23, 19–34.

Jorstad, T., T. East, J. M. Adovasio, J. Donohue, and R. Stuckenrath, 1986. Paleosols and prehistoric populations in the High Plains. *Geoarchaeology* 2, 163–181.

Josenhans, H. W., D. W. Fedje, K. W. Conway, and J. V. Barrie, 1995. Post glacial sea levels on the Western Canadian continental shelf: Evidence for rapid change, extensive subaerial exposure, and early human habitation. *Marine Geology* 125, 73–94.

Jungerius, P. D., 1969. Soil evidence of postglacial tree line fluctuations in the Cypress Hills area, Alberta, Canada. *Arctic and Alpine Research* 1, 235–245.

Jungerius, P. D., Ed., 1985. Soils and Geomorphology. Catena Verlag, Catena Supplement 6. Cremlingen, Germany.

Kamprath, E. J., and M. E. Watson, 1980. Conventional soil and tissue tests for assessing the phosphorus status of soils. *In* The Role of Phosphorus in Agriculture (F. E. Khasawneh, E. C. Sample, and E. J. Kamprath, Eds.), pp. 433–469. Soil Science Society of America, Madison, Wisc.

Karlstrom, E. T., 1988. Rates of soil formation on Black Mesa, northeast Arizona: A chronosequence in late Quaternary alluvium. *Physical Geography* 9, 301–327.

Kato, Y., and T. Matsui, 1979. Some applications of paleopedology in Japan. *Geoderma* 22, 45–60.

Kaufman, I. R., and B. R. James, 1991. Anthropic epipedons in oyster shell middens of Maryland. *Soil Science Society of America Journal* 55, 1191–1193.

Keeley, H. C. M., 1981. Recent work using soil phosphorus analysis in archaeological prospection. *Revue d'Archéométrie* 5, 89–95.

Keen, D. H., 1995. Molluscan assemblages from the loess of north central China. *Quaternary Science Reviews* 14, 699–706.

Kemp, R. A., 1985a. The cause of redness in some buried and non-buried soils in eastern England. *Journal of Soil Science* 36, 329–334.

Kemp, R. A., 1985b. The decalcified Lower Loam at Swanscombe, Kent: A buried Quaternary soil. *Proceedings of the Geological Association* 96, 343–355.

Kemp, R. A., 1985c. Soil Micromorphology and the Quaternary. Quaternary Research Association Technical Guide no. 2. Quaternary Research Association of Britain, Cambridge.

Kemp, R. A., 1998. Role of micromorphology in paleopedological research. *Quaternary International* 51/52, 133–141.

Kemp, R. A., 1999. Micromorphology of loess-paleosol sequences: A record of paleoenvironmental change. *Catena* 35, 181–198.

Kemp, R. A., 2001. Pedogenic modification of loess: Significance for palaeoclimatic reconstructions. *Earth Science Reviews* 54, 145–156.

Kemp, R. A., and E. Derbyshire, 1998. The loess soils of China as record of climatic change. *European Journal of Soil Science* 49, 525–539.

Kemp, R. A., E. Derbyshire, F.-A. Chen, and H.-Z. Ma, 1996. Pedosedimentary development and paleoenvironmental significance of the S1 palaeosol on the northeastern margin of the Qinghai-Xizang (Tibetan) Plateau. *Journal of Quaternary Science* 11, 95–106.

Kemp, R. A., E. Derbyshire, and X.-M. Meng, 1997. Micromorphological variation of the S1 paleosol across northwest China. *Catena* 31, 77–90.

Kemp, R. A., H. Jerz, W. Grottenthaler, and R. C. Preece, 1994. Pedosedimentary fabrics of soils within loess and colluvium in southern England and Germany. *In* Soil Micromorphology (A. Ringrose-Voase and G. Humphries, Eds.), pp. 207–219. Elsevier, Amsterdam.

Kerr, J. P., 1995. Phosphate imprinting within Mound A at the Huntsville site. *In* Pedological Perspectives in Archaeological Research (M. E. Collins, B. J. Carter, B. G. Gladfelter, and R. J. Southard, Eds.), pp. 133–149. Special Publication no. 44. Soil Science Society of America, Madison, Wisc.

Kesel, R. H., and B. E. Spicer, 1985. Geomorphologic relationships and ages of soils on alluvial fans in the Rio General valley, Costa Rica. *Catena* 12, 149–166.

Kimbel, W. H., R. C. Walter, D. C. Johanson, K. E. Reed, J. L. Aronson, Z. Assefa, C. W. Marean, G. G. Eck, R. Bobe, E. Hovers, Y. Rak, C. Vondra, T. Yemane, D. York, Y. Chen, N. M. Evensen, and P. E. Smith, 1996. Late Pliocene *Homo* and Oldowan tools from the Hadar Formation, Ethiopia. *Journal of Human Evolution* 31, 549–561.

King, R. H., 1983. Soils and archaeological surveys: The case of the Canadian Palaipaphos survey project. *In* Archaeological Survey in the Mediterranean Area (D. R. Keller and D. W. Rupp, Eds.), pp. 101–107. BAR International Series no. 155, Oxford.

Kingston, J. D., B. D. Marino, and A. Hill, 1994. Isotopic evidence for Neogene Hominid paleoenvironments in the Kenya Rift Valley. *Science* 264, 955–959.

Klappa, C. F., 1983. A process-response model for the formation of pedogenic calcretes. *In* Residual Deposits: Surface Related Weathering Processes and Materials (R. C. L. Wilson, Ed.), pp. 211–220. Geological Society of London, London.

Kleber, M., J. Rößner, C. Chenu, B. Glaser, H. Knicker, and R. Jahn, 2003. Prehistoric alteration of soil properties in a central German chernozemic soil: In search of pedologic indicators for prehistoric activity. *Soil Science* 168, 292–306.

Klein, R. G., 1969. Man and Culture in the Late Pleistocene: A Case Study. Chandler Publishing, San Francisco.

Klein, R. G., 1973. Ice-Age Hunters of the Ukraine. The University of Chicago Press, Chicago.

Klute, A. A., Ed., 1986. Methods of Soil Analysis, Part 1—Physical and Mineralogical Methods. 2nd ed. Monograph no. 9. American Society of Agronomy, Madison, Wisc.

Knapp, A. B., and W. Ashmore, 1999. Archaeological landscapes. *In* Archaeologies of Landscape (W. Ashmore and A. B. Knapp, Eds.), pp. 1–30. Blackwell, Malden, Mass.

Knapp, G., 1985. Review of "Advances in Abandoned Settlement Analysis" by R. C. Eidt. *Geographical Review* 75, 370–372.

Knox, E. G., 1965. Soil individuals and soil classification. *Soil Science Society of America Proceedings* 29, 79–84.

Knuepfer, P. L. K., and L. D. McFadden, Eds., 1990. Soils and Landscape Evolution: Proceedings of the 21st Binghamton Symposium, pp. 197–575. Elsevier, New York.

Kolb, M. F., 1996. Part III: Geomorphology. *In* Archaeology, Geomorphology, and Land Use History at Wright-Patterson Air Force Base, Ohio (J. R. Haas, M. F. Kolb, and J. D. Richards), pp. 105–152. Reports of Investigations no. 389. Great Lakes Archaeological Research Center, Milwaukee, Wisc.

Kolb, M. F., 2000. Geomorphological and geoarchaeological investigations. *In* The Lake Onamia-Trunk Highway 169 Data Recovery Project, Mille Lacs County, Minnesota, volume 1: Final Technical Report (D. Mather and E. J. Abel, Eds.), pp. 13.1–13.51. Loucks Project Report no. 96506-1. Loucks Associates, Minneapolis.

Kolb, M. F., and J. A. Homburg, 1991. Soil and sediment analyses. *In* The Lower Verde Archaeological Project: Laboratory Manual (C. J. Ellick and S. M. Whittlesey, Eds.), pp. 43–62. Statistical Research, Tucson, Ariz.

Kolb, M. F., N. P. Lasca, and L. C. Goldstein, 1990. A soil-geomorphic analysis of the midden deposits at the Aztalan site, Wisconsin. *In* Archaeological Geology of North America (N. P. Lasca and J. Donahue, Eds.), pp. 199–218. Centennial Volume no. 4. Geological Society of America, Boulder, Colo.

Konrad, V. A., R. Bonnichsen, and V. Clay, 1983. Soil chemical identification of ten thousand years of prehistoric human activity areas at the Munsungun lake thoroughfares, Maine. *Journal of Archaeological Science* 10, 13–28.

Kraus, M. J., 1999. Paleosols in clastic sedimentary rocks: Their geologic applications. *Earth-Science Reviews* 47, 41–70.

Kraus, M. J., and A. Aslan, 1999. Palaeosol sequences in floodplain environments: A hierarchical approach. *In* Palaeoweathering, Palaeosurfaces and Related Continental Deposits (M. Thiry and R. Simon-Coinçon, Eds.), pp. 303–321. International Association of Sedimentologists, Special Publication no. 27. Oxford University Press, Malden, Mass.

Kraus, M. J., and T. M. Bown, 1986. Paleosols and time resolution in alluvial stratigraphy. *In* Paleosols: Their Recognition and Interpretation (V. P. Wright, Ed.), pp. 180–207. Princeton University Press, Princeton, N.J.

Kraus, M. J., and T. M. Bown, 1988. Pedofacies analysis: A new approach to reconstructing ancient fluvial sequences. *In* Paleosols and Weathering Through Geologic Time: Principles and Applications (J. Reinhardt and W. R. Sigleo, Eds.), pp. 143–152. Special Paper no. 216. Geological Society of America, Boulder, Colo.

Ku, T.-L., 2000. Uranium-series methods. *In* Quaternary Geochronology: Methods and Applications (J. S. Noller, J. M. Sowers, and W. R. Lettis, Eds.), pp. 101–114. AGU Reference Shelf no. 4. American Geophysical Union, Washington, D.C.

Ku, T.-L., W. B. Bull, S. T. Freeman, and K. G. Knauss, 1979. Th^{230}-U^{234} dating of pedogenic carbonates in gravelly desert soils of Vidal Valley, southeastern California. *Geological Society of America Bulletin* 90, 1063–1073.

Ku, T.-L., and Z.-C. Liang, 1984. The dating of impure carbonates with decay-series isotopes. *Nuclear Instruments and Methods in Physics Research* 223, 563–571.

Kukla, J., 1970. Correlations between loesses and deep-sea sediments. *Geologiska Föreningens I Stockholm Förhandlingar* 92, 148–180.

Kukla, G. J., 1975. Loess stratigraphy of Central Europe. *In* After the Australopithecines (K. W. Butzer and G. L. Issac, Eds.), pp. 99–188. Mouton, The Hague.

Kukla, G. J., 1977. Pleistocene land-sea correlations. I: Europe. *Earth Science Reviews* 13, 307–374.

Kukla, G., 1987. Loess stratigraphy in Central China. *Quaternary Science Reviews* 6, 191–219.

Kukla, G., and Z.-S. An, 1989. Loess stratigraphy in Central China. *Palaeogeography, Palaeoclimatology, Palaeoecology* 72, 203–225.

Kumar, S., B. Parkash, M. L. Manchanda, A. K. Singhvi, and P. Srivastava, 1996. Holocene landform and soil evolution of the western Gangetic Plains: Implications for neotectonics and climate. *Zeitschrift für Geomorphologie* supplementband 116, 283–312.

Kuo, S., 1996. Phosphorus. *In* Methods of Soil Analysis, Part 3—Chemical Methods (D. L. Sparks, A. L. Page, P. A. Helmke, R. H. Loeppert, P. N. Soltanpour, M. A. Tabatabai, C. T. Johnston, and M. E. Sumner, Eds.), pp. 869–919. Book Series no. 5. Soil Science Society of America, Madison, Wisc.

Kurenkova, E. I., Y. N. Gribchenko, and L. V. Grekhova, 1995. Paleolithic archaeology and the paleogeographical study of loess-paleosol sequences in eastern Europe. *GeoJournal* 36, 229–235.

Kusumgar, S., D. P. Argawal, and B. S. Kotlia, 1985. Magnetic stratigraphy of the Karewas of the Kashmir Valley. *In* Climate and Geology of Kasmir and Central Asia (D. P. Agrawal, S. Kusumgar, and R. V. Krishnamurthy, Eds.), pp. 13–17. Current Trends in Geology, v. VI. Today and Tomorrow's Printers and Publishers, New Delhi.

LaBelle, J., V. T. Holliday, and D. J. Meltzer, 2003. Early Holocene Paleoindian deposits at Nall playa, Oklahoma Panhandle. *Geoarchaeology* 18, 5–34.

Lancaster, N., 1995. Geomorphology of Desert Dunes. Routledge, New York.

Lang, A., H.-P. Niller, and M. M. Rind, 2003. Land degradation in Bronze Age Germany: Archaeological, pedological, and chronometric evidence from a hilltop settlement on the Frauenberg, Niederbayern. *Geoarchaeology* 18, 757–778.

Langley-Turnbaugh, S. J., and J. G. Bockheim, 1997. Time-dependent changes in pedogenic processes on marine terraces in coastal Oregon. *Soil Science Society of America Journal* 61, 1428–1440.

Langmaid, K. K., 1964. Some effects of earthworm invasion in virgin podzols. *Canadian Journal of Soil Science* 44, 34–37.

Larsen, C. E., and J. Schuldenrein, 1990. Depositional history of an archaeologically dated floodplain, Haw River, North Carolina. *In* Archaeological Geology of North America (N. P. Lasca and J. Donahue, Eds.), pp. 161–181. Centennial Special Volume no. 4. Geological Society of America, Boulder, Colo.

Lasca, N. P., and J. Donahue, Eds., 1990. Archaeological Geology of North America. Centennial Volume no. 4. Geological Society of America, Boulder, Colo.

Laury, R. L., and C. C. Albritton, Jr., 1975. Geology of middle Stone Age archaeological sites in the main Ethiopian Rift Valley. *Bulletin of the Geological Society of America* 86, 999–1011.

Lazarenko, A. A., 1984. The loess of Central Asia. *In* Late Quaternary Environments of the Soviet Union (A. A. Velichko, Ed.), pp. 125–131. University of Minnesota Press, Minneapolis.

Leamy, M. L., J. D. G. Milne, W. A. Pullar, and J. G. Bruce, 1973. Paleopedology and soil stratigraphy in the New Zealand Quaternary succession. *New Zealand Journal of Geology and Geophysics* 16, 723–744.

Lehmann, J., D. C. Kern, B. Glaser, and W. I. Woods, Eds., 2004. Amazonian Dark Earths: Origin, Properties, and Management. Kluwer, Dordrecht, The Netherlands.

Leigh, D. S., 1996. Soil chronosequence of Brasstown Creek, Blue Ridge Mountains, USA. *Catena* 26, 99–114.

Leigh, D. S., 1998. Evaluating artifact burial by eolian versus bioturbation processes, South Carolina Sandhills, USA. *Geoarchaeology* 13, 309–330.

Leigh, D. S., 2001. Buried artifacts in sandy soils: Techniques for evaluating pedoturbation versus sedimentation. *In* Earth Sciences and Archaeology (P. Goldberg, V. T. Holliday, and C. R. Ferring, Eds.), pp. 269–293. Kluwer Academic/Plenum Publishers, New York.

Leigh, D. S., and J. S. Cable, 1997. Stratigraphy and soil chronosequence of the Brasstown sites: A model for age assessment of alluvium in the southern Blue Ridge, U.S.A. *In* Proceedings of the Second International Conference on Pedo-Archaeology (A. C. Goodyear, J. E. Foss, and K. E. Sassaman, Eds.), pp. 1–9. Occasional Papers of the South Carolina Institute of Archaeology and Anthropology. University of South Carolina, Columbia.

Leighton, M. M., 1934. Some Observations on the Antiquity of Man in Illinois. *Transactions of the Illinois State Academy of Sciences* 25, Springfield. 83 pp.

Leighton, M. M., 1936. Geological Aspects of the Findings of Primitive Man, near Abilene, Texas. *Medallion Papers* 24, 1–44.

Leighton, M. M., 1937. The significance of profiles of weathering in stratigraphic archaeology. *In* Early Man (G. G. MacCurdy, Ed.), pp. 163–172. Lippincott, New York.

Leighton, M. M., 1958. Principles and viewpoints in formulating the stratigraphic classifications of the Pleistocene. *Journal of Geology* 66, 700–709.

Leonardi, G., M. Miglavacca, and S. Nardi, 1999. Soil phosphorus analysis as an integrative tool for recognizing ancient ploughsoils. *Journal of Archaeological Science* 26, 343–352.

Leopold, L. B., and J. P. Miller, 1954. A Post-glacial Chronology for some Alluvial Valleys in Wyoming. U.S. Geological Survey Water Supply Paper no. 1261. Government Printing Office, Washington, D.C.

Leverett, F., 1898. The weathered zone (Yarmouth) between the Illinoian and Kansan till sheets. *Journal of Geology* 6, 238–243.

Lewis, D. R., 1978. Use of phosphate analysis for determining land use patterns. *Bulletin of the Texas Archeological Society* 49, 309–317.

Lewis, R. J., J. E. Foss, M. W. Morris, M. E. Timpson, and C. A. Stiles, 1994. Trace-element analysis in pedo-archaeology studies. *In* Proceedings of the First International

Conference on Pedo-Archaeology (J. E. Foss, M. E. Timpson, and M. W. Morris, Eds.), pp. 81–88. The University of Tennessee Agricultural Experiment Station, Knoxville.

Lillios, K. T., 1992. Phosphate fractionation of soils at Agroal, Portugal. *American Antiquity* 57, 495–506.

Lima, H. N., C. E. R. Schaefer, J. W. V. Mello, R. J. Gilkes, and J. C. Ker, 2002. Pedogenesis and pre-Colombian land use of "Terra Preta Anthrosols" ("Indian black earths") of western Amazonia. *Geoderma* 110, 1–17.

Limbrey, S., 1975. Soil Science in Archaeology. Academic Press, London.

Limbrey, S., 1978. Changes in quality and distribution of the soils of lowland Britain. *In* The Effect of Man on the Landscape: The Lowland Zone (S. Limbrey and J. G. Evans, Eds.), pp. 21–27. Research Report no. 21. Council for British Archaeology, London.

Limbrey, S., and J. G. Evans, Eds., 1978. The Effect of Man on the Landscape: The Lowland Zone. Research Report no. 21. Council for British Archaeology, London.

Linderholm, J., and E. Lundberg, 1994. Chemical characterization of various archaeological soil samples using main and trace elements determined by inductively coupled plasma emission spectrometry. *Journal of Archaeological Science* 21, 303–314.

Lindholm, R. C., 1993. Soil maps as an aid to making geologic maps. *Journal of Geologic Education* 41, 352–357.

Lindholm, R. C., 1994a. Information derived from soil maps. *Environmental Geology* 23, 271–275.

Lindholm, R. C., 1994b. The value of soil maps to geologists: An acknowledgment. *Soil Survey Horizons* 35, 40–48.

Lippi, R. D., 1988. Paleotopography and phosphate analysis of a buried jungle site in Ecuador. *Journal of Field Archaeology* 15, 85–97.

Lisitsina, G. N., 1976. Arid soils—The source of archaeological information. *Journal of Archaeological Science* 3, 55–60.

Litwinionek, L., E. Johnson, and V. T. Holliday, 2003. The playas of the Southern High Plains: An archipelago of human occupation for 12,000 years on the North American grasslands. *In* Islands on the Plains: Ecological, Social, and Ritual Use of Landscapes (M. Kornfeld and A. J. Osborn, Eds.), pp. 21–43. The University of Utah Press, Salt Lake City.

Liu, T. K., R. T. Odell, W. C. Etter, and T. H. Thornburn, 1966. Comparison of clay contents determined by hydrometer and pipette methods using reduced major axis analysis. *Soil Science Society of America Proceedings* 30, 665–669.

Liu, T.-S., Ed., 1987. Aspects of Loess Research. China Ocean Press, Beijing.

Liu, T.-S., 1988. Loess in China. 2nd ed. China Ocean Press/Springer, Beijing.

Liu, T.-S., Z.-S. An, B.-Y. Yuan, and J.-M. Han, 1985. The loess-paleosol sequence in China and climatic history. *Episodes* 8, 21–28.

Liu, T.-S., Z.-L. Ding, and Z.-T. Guo, Eds.,1991. Loess, Environment, and Global Change. Science Press, Beijing.

Liu, T.-Z., W. S. Broecker, J. W. Bell, and C. W. Mandeville, 2000. Terminal Pleistocene wet event recorded in rock varnish from Las Vegas Valley, southern Nevada. *Palaeogeography, Palaeoclimatology, Palaeoecology* 161, 423–433.

Liu, X.-M., T. Rolph, J. Bloemendal, J. Shaw, and T.-S. Liu, 1995. Quantitative estimates of palaeoprecipitation at Xifeng, in the Loess Plateau of China. *Palaeogeography, Palaeoclimatology, Palaeoecology* 113, 243–248.

Lof, P., 1987. Soils of the World (chart). Elsevier, Amsterdam.

Lomov, S. P., and V. A. Ranov, 1984. Distribution of Palaeolithic tools in the buried soils of Tadzhikistan. *Soviet Soil Science* 16, 18–28.

Lomov, S. P., and V. A. Ranov, 1985. The peculiarities of the Pleistocene palaeosol formations and distribution of embedded Palaeolithic tools. *In* Climate and Geology of Kasmir and Central Asia (D. P. Agrawal, S. Kusumgar, and R. V. Krishnamurthy, Eds.), pp. 227–240. Current Trends in Geology, v. VI. Today and Tomorrow's Printers and Publishers, New Delhi.

Lopinot, N. H., J. H. Ray, and M. D. Conner, 1998. The 1997 Excavations at the Big Eddy Site (23CE426) in Southwest Missouri. Center for Archaeological Research, Special Publication no. 2. Southwest Missouri State University, Springfield, Mo.

Lorch, W., 1930. Neue methoden der siedlungsgeshichte. *Geographisches Zeitschrift* 45, 294–305.

Lorch, W., 1939. Methodische Untersuchungen zur Wüstungsforschung. Arbeiten zur Landesund Volksforschung, Band 4. Anstalt für geschichtlich Landeskunde an der Friedrish-Schiller Universittät, Jena.

Lorch, W., 1940. Die siedlungsgeographische phosphatmethode. *Die Naturwissenschaften* 28, 633–640.

Lorch, W., 1954. Die anthropogenen bodenphosphate des hohenstaufen-gipfels. *Jahrbücher für Statistik and Landeskunde von Baden-Württemberg* 1, 367–375.

Lovis, W. A., Jr., 1976. Quarter sections and forests: An example of probability sampling in the northeastern Woodlands. *American Antiquity* 41, 364–372.

Lowe, D. J., 1986. Controls on the rates of weathering and clay mineral genesis in airfall tephras: A review and New Zealand case study. *In* Rates of Chemical Weathering of Rocks and Minerals (S. M. Colman and D. P. Dethier, Eds.), pp. 265–330. Academic Press, Orlando, Fla.

Lowe, D. J., Ed., 1996. Tephra, Loess, and Paleosols—An Integration. *Quaternary International* 34–36, 1–261.

Lowe, J. J., and M. J. C. Walker, 1997. Reconstructing Quaternary Environments. 2nd ed. Longman, Essex.

Lozek, V., 1990. Molluscs in loess: Their paleoecological significance and role in geochronology—Principles and methods. *Quaternary International* 7/8, 71–79.

Luff, R., and P. Rowley-Conwy, 1994. The (dis)integration of environmental archaeology. *In* Whither Environmental Archaeology (R. Luff and P. Rowley-Conwy, Eds.), pp. 1–3. Oxbow Books, Oxford.

Luternauer, J. L., J. J. Clague, K. W. Conway, J. V. Barrie, B. Blaise, and R. W. Mathewes, 1989. Late Pleistocene terrestrial deposits on the continental shelf of western Canada: Evidence for rapid sea-level change at the end of the last glaciation. *Geology* 17, 357–360.

Machette, M. N., 1975. Geologic Map of the Lafayette Quadrangle, Adams, Boulder, and Jefferson Counties, Colorado. U.S. Geological Survey Map MF-656. U.S. Government Print. Office, Washington, D.C.

Machette, M. N., 1985. Calcic soils of the southwestern United States. *In* Soils and Quaternary Geology of the Southwestern United States (D. L. Weide, Ed.), pp. 1–21. Special Paper no. 203. Geological Society of America, Boulder, Colo.

Machette, M. N., 1986. Calcium and magnesium carbonates. *In* Field and Laboratory Procedures Used in a Soil Chronosequence Study (M. J. Singer and P. Janitzky, Eds.), pp. 30–33. Bulletin no. 1648. U.S. Geological Survey, Washington, D.C.

Mack, G. H., W. C. James, and H. C. Monger, 1993. Classification of paleosols. *Geological Society of America Bulletin* 105, 129–136.

Mackay, J. R., W. H. Mathews, and R. S. MacNeish, 1961. Geology of the Engigstciak archaeological site, Yukon Territory. *Arctic* 14, 25–52.

Macphail, R. I., 1981. Soil and botanical studies of the "Dark Earth." *In* The Environment of Man: The Iron Age to the Anglo-Saxon Period (M. Jones and G. W. Dimbleby, Eds.), pp. 309–331. BAR British Series no. 87, Oxford.

Macphail, R. I., 1983. The micromorphology of dark earth from Gloucester, London and Norwich: An analysis of urban anthropogenic deposits from the Late Roman to Early Medieval periods in England. *In* Soil Micromorphology, vol. 1: Techniques and Applications (P. Bullock and C. P. Murphy, Eds.), pp. 245–252. A B Academic Publishers, Berkhamsted.

Macphail, R. I., 1986. Paleosols in archaeology: Their role in understanding Flandrian pedogenesis. *In* Paleosols: Their Recognition and Interpretation (V. P. Wright, Ed.), pp. 263–290. Princeton University Press, Princeton, N.J.

Macphail R. I., 1987. A review of soil science in archaeology in England. *In* Environmental Archaeology: A Regional Review (H. C. M. Keeley, Ed.), pp. 332–379. Occasional Paper no. 1. English Heritage, London.

Macphail, R. I., 1991. The archaeological soils and sediments. *In* Maiden Castle: Excavations and Field Survey 1985–6 (N. M. Sharples, Ed.), pp. 106–118. English Heritage, London.

Macphail, R. I., 1992. Soil micromorphological evidence for ancient soil erosion. *In* Past and Present Soil Erosion: Archaeological and Geographical Perspectives (M. Bell and J. Boardman, Eds.), pp. 197–216. Oxbow Monographs no. 22. Oxbow Books, Oxford.

Macphail, R. I., 1994. The reworking of urban stratigraphy by human and natural processes. *In* Urban-Rural Connexions: Perspectives from Environmental Archaeology (A. R. Hall and H. K. Kenward, Eds.), pp. 13–43. Oxbow Books, Oxford.

Macphail, R. I., 1998. A reply to Carter and Davidson's "An evaluation of the contribution of soil micromorphology to the study of ancient arable agriculture." *Geoarchaeology* 13, 549–564.

Macphail, R. I., and M.-A. Courty, 1985. Interpretation and significance of urban deposits. *In* Proceedings of the Third Nordic Conference on the Application of Scientific Methods in Archaeology (T. Edgren and H. Jungner, Eds.), pp. 71–83. The Finnish Antiquarian Society, Helsinki.

Macphail, R. I., and G. M. Cruise, 2000. Rescuing our urban archaeological soil heritage: A multidiciplinary microstratigraphical approach. *In* Proceedings, First International Conference of Urban, Industrial, Traffic and Mining Areas (W. Burghardt and C. Dornauf, Eds.), pp. 5–10. University of Essen, Essen.

Macphail, R. I., and G. M. Cruise, 2001. The soil micromorphologist as a team player: A multianalytical approach to the study of European microstratigraphy. *In* Earth Sciences and Archaeology (P. Goldberg, V. T. Holliday, and C. R. Ferring, Eds.), pp. 241–267. Kluwer Academic/Plenum Publishers, New York.

Macphail, R. I., and P. Goldberg, 1990. The micromorphology of tree subsoil hollows: Their significance to soil science and archaeology. *In* Soil Micromorphology: A Basic and Applied Science (L. A. Douglas, Ed.), pp. 425–429. Elsevier, New York.

Macphail, R. I., and P. Goldberg, 1995. Recent advances in micromorphological interpretations of soils and sediments from archaeological sites. *In* Archaeological Sediments and Soils: Analysis, Interpretation and Management (A. J. Barham and R. I. Macphail, Eds.), pp. 1–24. Institute of Archaeology, London.

Macphail, R. I., M.-A. Courty, and A. Gebhardt, 1990a. Soil micromorphological evidence of early agriculture in Northern Europe. *World Archaeology* 22, 53–69.

Macphail, R. I., M.-A. Courty, and P. Goldberg, 1990b. Soil micromorphology in archaeology. *Endeavor* (new series) 14, 163–171.

Macphail, R. I., G. Cruise, R. Engelmark, and J. Linderholm, 2000. Integrating soil micromorphology and rapid chemical survey methods: New developments in reconstructing past rural settlement and landscape organization. *In* Interpreting Stratigraphy: Site Evaluation, Recording Procedures and Stratigraphic Analysis (S. Roskams, Ed.), pp. 71–80. BAR International Series no. 910, Oxford.

Macphail, R. I., J. C. C. Romans, and L. Robertson, 1987. The application of micromorphology to the understanding of Holocene soil development in the British Isles; with special reference to Holocene cultivation. *In* Soil Micromorphology (N. Fedoroff, L. M. Bresson, and M.-A. Courty, Eds.), pp. 647–656. Association Française pour l'Etude du Sol, Plaisir.

Madsen, D. B., J.-Z. Li, R. G. Elston, X. Cheng, R. L. Bettinger, G. Kan, P. J. Brantingham, and Z. Kan, 1998. The loess/paleosol record and the nature of the Younger Dryas climate in central China. *Geoarchaeology* 13, 847–869.

Maher, B. A., 1986. Characterization of soils by mineral magnetic measurements. *Physics of the Earth and Planetary Interiors* 42, 76–92.

Maher, B. A., A. Alekseev, and T. Alekseeva, 2002. Variation of soil magnetism across the Russian steppe: Its significance for use of soil magnetism as a paleorainfall proxy. *Quaternary Science Reviews* 21, 1571–1576.

Maher, B. A., and R. Thompson, 1991. Mineral magnetic record of the Chinese loess and paleosols. *Geology* 19, 3–6.

Maher, B. A., and R. Thompson, 1992. Paleoclimatic significance of the mineral magnetic record of the Chinese loess and paleosols. *Quaternary Research* 37, 155–170.

Maher, B. A., and R. Thompson, 1995. Paleorainfall reconstructions from pedogenic magnetic susceptibility variations in the Chinese loess and paleosols. *Quaternary Research* 44, 383–391.

Maher, B. A., and R. Thompson, 1999. Palaeomonsoons I: The magnetic record of palaeoclimate in the terrestrial loess and palaeosol sequences. *In* Quaternary Climates, Environments and Magnetism (B. A. Maher and R. Thompson, Eds.), pp. 81–125. Cambridge University Press, Cambridge.

Major, J., 1951. A functional, factorial approach to plant ecology. *Ecology* 32, 392–412.

Malde, H. E., 1964. Environment and man in arid America. *Science* 145, 123–129.

Mandel, R. D., 1992. Soils and Holocene landscape evolution in central and southwestern Kansas: Implications for archaeological research. *In* Soils and Landscape Evolution (V. T. Holliday, Ed.), pp. 41–100. Smithsonian Institution Press, Washington, D.C.

Mandel, R. D., 1994. Holocene Landscape Evolution in the Pawnee River Valley, Southwestern Kansas. Bulletin no. 236. Kansas Geological Survey, Topeka.

Mandel, R. D., 1995. Geomorphic controls of the Archaic record in the Central Plains of the United States. *In* Archaeological Geology of the Archaic Period in North America (E. A. Bettis, III, Ed.), pp. 37–66. Special Paper no. 297. Geological Society of America, Boulder, Colo.

Mandel, R. D., Ed., 2000. Geoarchaeology of the Great Plains: A Historical Perspective. University of Kansas Press, Lawrence.

Mandel, R. D., and E. A. Bettis, III, 2001a. Late Quaternary Landscape Evolution in the South Fork of the Big Nemaha River Valley, Southeastern Nebraska and Northeastern Kansas. Guidebook no. 11, Conservation and Survey Division, University of Nebraska-Lincoln, Lincoln.

Mandel, R. D., and E. A. Bettis, III, 2001b. Use and analysis of soils by archaeologists and geoscientists: A North American perspective. *In* Earth Sciences and Archaeology (P. Goldberg, V. T. Holliday, and C. R. Ferring, Eds.), pp. 173–204. Kluwer Academic/Plenum Publishers, New York.

Mandryk, C. A. S., H. Josenhans, D. W. Fedje, and R. W. Mathewes, 2001. Late Quaternary paleoenvironments of northwestern North America: Implications for inland versus coastal migration routes. *Quaternary Science Reviews* 20, 301–314.

Manzanilla, L., 1996a. Corporate groups and domestic activities at Teotihuacan. *Latin American Antiquity* 7, 228–246.

Manzanilla, L., 1996b. Soil analyses to identify ancient human activities. *Canadian Journal of Soil Science* 76, 107–108.

Markewich, H. W., and M. J. Pavich, 1991. Soil chronosequence studies in temperate to subtropical, low-latitude, low-relief terrain with data from the eastern United States. *Geoderma* 51, 213–239.

Martin, C. W., 1993. Radiocarbon ages on late Pleistocene loess stratigraphy of Nebraska and Kansas, Central Great Plains, U.S.A. *Quaternary Science Reviews* 12, 179–188.

Martin, C. W., and W. C. Johnson, 1995. Variation in radiocarbon ages of soil organic matter fractions from late Quaternary buried soils. *Quaternary Research* 43, 232–237.

Martini, I. P., and W. Chesworth, Eds., 1992. Weathering, Soils and Paleosols. Elsevier, New York.

Mason, J. A., and M. S. Kuzilla, 2000. Episodic Holocene loess deposition in central Nebraska. *Quaternary International* 67, 119–131.

Mason, O. K., and S. C. Gerlach, 1995. Chukchi hot spots, paleo-polynyas, and caribou crashes: Climatic and ecological dimensions of north Alaska prehistory. *Arctic Anthropology* 32, 101–130.

Matthews, J. A., 1980, Problems and implications of 14C dates from a Podzol buried beneath an end moraine at Haugabreen, southern Norway. *Geografiska Annaler* 62-A, 185–208.

Matthews, J. A., 1985. Radiocarbon dating of surface and buried soils: principles, problems and prospects. *In* Geomorphology and Soils (K. S. Richards, R. R. Arnett, and S. Ellis, Eds.), pp. 269–288. Allen and Unwin, London.

Mattingly, G. E. G., and R. J. B. Williams, 1962. A note on the chemical analysis of a soil buried since Roman times. *Journal of Soil Science* 13, 253–258.

Mausbach, M. J., and L. P. Wilding, Eds., 1991. Spatial Variabilities of Soils and Landforms. Special Publication no. 28. Soil Science Society of America, Madison, Wisc.

Maxwell, T. A., and C. V. Haynes, Jr., 2001. Sand sheet dynamics and Quaternary landscape evolution of the Selima Sand Sheet, southern Egypt. *Quaternary Science Reviews* 20, 1623–1647.

May, D. W., 1989. Holocene alluvial fills in the South Loup Valley, Nebraska. *Quaternary Research* 32, 117–120.

May, D. W., 1992. Late Holocene valley-bottom aggradation and erosion in the South Loup River Valley, Nebraska. *Physical Geography* 13, 115–132.

May, D. W., 2002. Stratigraphic studies at Paleoindian sites around Medicine Creek Reservoir. *In* Medicine Creek: Seventy Years of Archaeological Investigations (D. C. Roper, Ed.), pp. 37–53. University of Alabama Press, Tuscaloosa.

McBrearty, S., 1990. Consider the humble termite: Termites as agents of post-depositional disturbance at African archaeological sites. *Journal of Archaeological Science* 17, 111–143.

McDonald, E. V., and A. J. Busacca, 1992. Late Quaternary stratigraphy of loess in the Channeled Scabland and Palouse regions of Washington State. *Quaternary Research* 38, 141–156.

McDonald, E. V., and A. J. Busacca, 1998. Unusual timing of regional loess sedimentation triggered by glacial outburst flooding in the Pacific Northwest US. *In* Dust Aerosols, Loess Soils and Global Change (A. J. Busacca, Ed.), pp. 163–166. Washington State University, Pullman.

McDonald, R. C., R. F. Isbell, J. G. Speight, J. Walker, and M. S. Hopkins, 1998. Australian Soil and Land Survey Field Handbook. 2nd ed. Department of Primary Industries and Energy and CSIRO173 Australia, Canberra.

McDonald, W. A., and G. R Rapp, Jr., Eds., 1972. The Minnesota Messinia Expedition: Reconstructing a Bronze Age Regional Environment. University of Minnesota Press, Minneapolis.

McDowell, P. F., 1988. Chemical enrichment of soils at archaeological sites: Some Oregon case studies. *Physical Geography* 9, 247–262.

McFadden, L. D., 1988. Climatic influences on rates and processes of soil development in Quarternary deposits of southern California. *In* Paleosols and Weathering through Geologic Time: Principles and Applications (J. Reinhardt and W. R. Sigleo, Eds.), pp. 153–177. Special Paper 216. Geological Society of America, Boulder, Colo.

McFadden, L. D., and D. M. Hendricks, 1985. Changes in the content and composition of pedogenic iron oxyhydroxides in a chronosequence of soils in southern California. *Quaternary Research* 23, 189–204.

McFadden, L. D., and E. V. McDonald, 1998. Paleopedology, paradigms, and paleosols: A comment on the issue of disciplinary status. *Quaternary International* 51/52, 23–26.

McFadden, L. D., and J. C. Tinsley, 1985. Rate and depth of pedogenic-carbonate accumulation in soils: Formulation and testing of a compartment model. *In* Soils and Quaternary Geology of the Southwestern United States (D. L. Weide, Ed.), pp. 23–41. Special Paper no. 203. Geological Society of America, Boulder, Colo.

McFadden, L. D., and R. J. Weldon, 1987. Rates and processes of soil development on Quaternary terraces in Cajon Pass, California. *Geological Society of America Bulletin* 98, 280–293.

McFadden, L. D., S. G. Wells, and J. C. Dohrenwend, 1986. Influences of Quaternary climate changes on processes of soil development on desert loess deposits of the Cima Volcanic field, California. *Catena* 13, 361–389.

McFadgen, B. G., 1980a. Age relationship between a Maori plaggen soil and Moa-hunter sites on the west Wellington coast (New Zealand). *New Zealand Journal of Geology and Geophysics* 23, 249–256.

McFadgen, B. G., 1980b. Maori Plaggen soils in New Zealand, their origin and properties. *Journal of the Royal Society of New Zealand* 10, 3–19.

McFaul, M., K. L. Traugh, G. D. Smith, and W. Doering, 1994. Geoarchaeologic analysis of South Platte River terraces: Kersey, Colorado. *Geoarchaeology* 9, 345–374.

McHugh, W. P., J. F. McCauley, C. V. Haynes, C. S. Breed, and G. G. Schaber, 1988. Paleorivers and geoarchaeology in the southern Egyptian Sahara. *Geoarchaeology* 3, 1–40.

McKeague, J. A., 1978. Manual on Soil Sampling and Methods of Analysis. 2nd ed. Canadian Society of Soil Science, Ottawa.

McKeague, J. A., F. DeConinck, and D. P. Franzmeier, 1983. Spodosols. *In* Pedogenesis and Soil Taxonomy, part 2, The Soil Orders (L. P. Wilding, N. E. Smeck, and G. F. Hall, Eds.), pp. 217–252. Developments in Soil Science 11B. Elsevier, Amsterdam.

McKennan, R. A., and J. P. Cook, 1968. Prehistory of Healy Lake, Alaska. *8th International Congress of Anthropological and Ethnological Sciences Proceedings* 3, 182–184.

McManamon, F. P., 1984. Discovering sites unseen. *Advances in Archaeological Method and Theory* 7, 223–292.

Mehlich, A., 1978. New extractant for soil test evaluation of phosphorus, potassium, magnesium, calcium, sodium, manganese, and zinc. *Communications in Soil Science and Plant Analysis* 9, 477–492.

Mehlich, A., 1984. Mehlich 3 soil test extractant: A modification of Mehlich 2 extractant. *Communications in Soil Science and Plant Analysis* 15, 1409–1416.

Meixner, R., 1986a. Total phosphorus (extraction). *In* Field and Laboratory Procedures Used in a Soil Chronosequence Study (M. J. Singer and P. Janitzky, Eds.) pp. 43–44. Bulletin no. 1648, U.S. Geological Survey, Washington, D.C.

Mello Araujo, A. G., and J. C. Marcelino, 2003. The role of armadillos in the movement of archaeological materials: An experimental approach. *Geoarchaeology* 18, 433–460.

Melson, W. G., 1994. The eruption of 1968 and tephra stratigraphy of Arenal volcano. *In* Archaeology, Volcanism, and Remote Sensing in the Arenal Region, Costa Rica (P. D. Sheets and B. R. McKee, Eds.), pp. 24–45. University of Texas Press, Austin.

Meng, X.-M., E. Derbyshire, and R. A. Kemp, 1997. Origin of the magnetic susceptibility signal in Chinese loess. *Quaternary Science Reviews* 16, 833–839.

Mercader, J., R. Martí, J. L. Martínez, and A. Brooks, 2002. The nature of "stone-lines" in the African Quaternary record: Archaeological resolution at the rainforest site of Mosumu, equitorial Guinea. *Quaternary International* 89, 71–96.

Michie, J., 1990. Bioturbation and gravity as a potential site formation process: The Open Area site, 38GE261, Georgetown County, South Carolina. *South Carolina Antiquities* 22, 27–46.

Michlovic, M. G., D. G. Hopkins, and J. L. Richardson, 1988. An interdisciplinary procedure for the identification and study of archaeological sites in sedimentary contexts. *Soil Survey Horizons* 29, 3–8.

Middleton, W. D., and T. D. Price, 1996. Identification of activity areas by multi-element characterization of sediments from modern and archaeological house floors using inductively coupled plasma-atomic emission spectroscopy. *Journal of Archaeological Science* 23, 673–687.

Millard, A., 2001. The deterioration of bone. *In* Handbook of Archaeological Sciences (D. R. Brothwell and A. M. Pollard, Eds.) pp. 637–647. Wiley, New York.

Miller, N. F., and K. L. Gleason, Eds., 1994a. The Archaeology of Garden and Field. University of Pennsylvania Press, Philadelphia.

Miller, N. F., and K. L. Gleason, 1994b. Fertilizer in the identification and analysis of cultivated soil. *In* The Archaeology of Garden and Field (N. F. Miller and K. L. Gleason, Eds.), pp. 25–43. University of Pennsylvania Press, Philadelphia.

Milne, G., 1935a. Composite units for the mapping of complex soil associations. *Transactions of the Third International Congress of Soil Science* 1, 345–347.

Milne G., 1935b. Some suggested units for classification and mapping, particularly for East African soils. *Soils Research* 4, 183–198.

Monger, H. C., 1995. Pedology in arid lands archaeological research: An example from southern New Mexico-western Texas. *In* Pedological Perspectives in Archaeological Research (M. E. Collins, B. J. Carter, B. G. Gladfelter, and R. J. Southard, Eds.), pp. 35–50. Special Publication no. 44. Soil Science Society of America, Madison, Wisc.

Moody, U. L., and W. Dort, Jr., 1990. Microstratigraphic analysis of sediments and soils: Wasden archaeological site, eastern Snake River Plain, Idaho. *In* Archaeological Geology of North America (N. P. Lasca and J. Donahue, Eds.), pp. 361–382. Centennial Volume no. 4. Geological Society of America, Boulder, Colo.

Moore, P. D., 1990. Soils and ecology: Temperate wetlands. *In* Wetlands: A Threatened Landscape (M. Williams, Ed.), pp. 95–114. Basil Blackwell, Oxford.

Moore, T. R., and D. Denton, 1988. The role of soils in the interpretation of archaeological sites in northern Quebec. *In* Conceptual Issues in Environmental Archaeology (J. L. Bintliff, D. A. Davidson, and E. G. Grant, Eds.), pp. 25–37. University Press, Edinburgh.

Mora, C., L. F. Herrera, I. Cavelier, and C. Rodríguez, 1991. Cultivars, Anthropic Soils and Stability: A Preliminary Report of Archaeological Research in Araracuara, Colombia Amazonia. University of Pittsburgh Latin American Archaeology Reports no. 2. University of Pittsburgh, Pittsburgh.

Morey, D. F., and G. M. Crothers, 1998. Clearing up clouded waters: Palaeoenvironmental analysis of freshwater mussel assemblages from the Green River shell middens, western Kentucky. *Journal of Archaeological Science* 25, 907–926.

Morgan, M. F., 1941. Chemical soil diagnosis by the universal testing system. *Connecticut Agricultural Experiment Station Bulletin* 450, 579–628.

Morris, M. W., J. T. Ammons, and P. Santas, 1997. Evidence for subsurface translocation of ceramic artifacts in a Vertisol in eastern Crete, Greece. *In* Proceedings of the Second International Conference on Pedo-Archaeology (A. C. Goodyear, J. E. Foss, and K. E. Sassaman, Eds.), pp. 41–51. Occasional Papers of the South Carolina Institute of Archaeology and Anthropology. University of South Carolina, Columbia.

Morrison, R. B., 1967. Principles of Quaternary soil stratigraphy. *In* Quaternary Soils (R. B. Morrison and H. E. Wright, Jr., Eds.), pp. 1–69. University of Nevada Desert Research Institute, Center for Water Resources Research, Reno.

Morrison, R. B., 1978. Quaternary soil stratigraphy-concepts, methods, and problems. *In* Quaternary Soils (W. C. Mahaney, Ed.), pp. 77–108. Geo Abstracts, Norwich.

Morrison, R. B., and H. E. Wright, Jr., Eds., 1967. Quaternary Soils. Desert Research Institute, University of Nevada, Reno.

Movius, H. L., Jr., 1944. Early Man and Pleistocene Stratigraphy in Southern and Eastern Asia. Harvard University Peabody Museum of American Archaeology and Ethnology 19(3), 125.

Mücher, H. J., R. T. Slotboom, and W. J. Ten Veen, 1990. Palynology and micromorphology of a man-made soil: A reconstruction of the agricultural history since late-medieval times of the Posteles in The Netherlands. *Catena* 17, 55–67.

Muhs, D. R., 1984. Intrinsic thresholds in soil systems. *Physical Geography* 5, 99–110.

Muhs, D. R., 1985. Age and paleoclimatic significance of Holocene sand dunes in northeastern Colorado. *Annals of the Association of American Geographers* 75, 566–582.

Muhs, D. R., R. R. Kautz, and J. J. MacKinnon, 1985. Soils and the location of cacao orchards at a Maya site in western Belize. *Journal of Archaeological Science* 12, 121–137.

Muhs, D. R., T. W. Stafford, J. Been, S. A. Mahan, J. Burdett, G. Skipp, and Z. M. Rowland, 1997a. Holocene eolian activity in the Minot dune field, North Dakota. *Canadian Journal of Earth Science* 34, 1442–1459.

Muhs, D. R., T. W. Stafford, S. D. Cowherd, S. A. Mahan, R. Kihl, P. B. Maat, C. A. Bush, and J. Nehring, 1996. Origin of the late Quaternary dune fields of northeastern Colorado. *Geomorphology* 17, 129–149.

Muhs, D. R., T. W. Stafford, J. B. Swinehart, S. D. Cowherd, S. A. Mahan, C. A. Bush, R. F. Madole, and P. B. Maat, 1997b. Late Holocene eolian activity in the mineralogically mature Nebraska Sand Hills. *Quaternary Research* 48, 162–176.

Murphy, C. P., 1986. Thin Section Preparation of Soils and Sediments. A B Academic Press, Berkhamsted.

Murphy, J., and J. P. Riley, 1962. A modified single solution method for determination of phosphate in natural waters. *Analytica Chimica Acta* 27, 31–36.

Nash, D. T., and M. D. Petraglia, 1987. Natural Formation Processes and the Archaeological Record. BAR International Series no. 352, Oxford.

Neal, V. E., and I. K. Paintin, 1986. Rates of weathering of ^{14}C-dated late Quaternary volcaniclastic deposits in the western United States. *In* Rates of Chemical Weathering of Rocks and Minerals (S. M. Colman and D. P. Dethier, Eds.), pp. 331–350. Academic Press, Orlando, Fla.

Nelson, D. A., and N. J. Sauer, 1984. An evaluation of postdepositional changes in the trace element content of human bone. *American Antiquity* 49, 141–147.

Nelson, D. W., and L. E. Sommers, 1982. Total carbon, organic carbon, and organic matter. *In* Methods of Soil Analysis, Part 2—Chemical and Microbiological Properties. 2nd ed. (A. L. Page, R. H. Miller, and D. R. Keeney, Eds.), pp. 539–579. Monograph no. 9. American Society of Agronomy, Madison, Wisc.

Nesbitt, H. W., and G. M. Young, 1989. Formation and diagenesis of weathering profiles. *The Journal of Geology* 97, 129–147.

Nettleton, W. D., B. R. Brasher, and R. J. Ahrens, 1998. A classification system for buried paleosols. *Quaternary International* 51/52, 175–183.

Nettleton, W. D., C. G. Olson, and D. A. Wysocki, 2000. Paleosol classification: Problems and solutions. *Catena* 41, 61–92.

Neumann, T. W., 1978. A model for the vertical distribution of flotation-size particles. *Plains Anthropologist* 23, 85–101.

Nikiforoff, C. C., 1949. Weathering and soil evolution. *Soil Science* 67, 219–230.

Nordt, L. C., 2001. Stable carbon and oxygen isotopes in soils: Applications for archaeological research. *In* Earth Sciences and Archaeology (P. Goldberg, V. T. Holliday, and C. R. Ferring, Eds.), pp. 419–448. Kluwer Academic/Plenum Publishers, New York.

Nordt, L. C., T. W. Boutton, C. T. Hallmark, and M. R. Waters, 1994. Late Quaternary vegetation and climate changes in central Texas based on the isotopic composition of organic carbon. *Quaternary Research* 41, 109–120.

Nordt, L. C., T. W. Boutton, J. S. Jacob, and R. D. Mandel, 2002. C_4 plant productivity and climate-CO_2 variations in south-central Texas during the late Quaternary. *Quaternary Research* 58, 182–188.

Nordt, L. C., E. F. Kelly, T. W. Boutton, and O. A. Chadwick, Eds., 1998. Biogeochemistry of isotopes in soil environments: Theory and application. *Geoderma* (special issue) 82, 1–293.

North American Commission on Stratigraphic Nomenclature (NACOSN), 1983. North American stratigraphic code. *American Association of Petroleum Geologists Bulletin* 67, 841–875.

Nunez, M., 1990. Phosphate survey. *Norwegian Archaeological Review* 23, 120–127.

Nunez, M., and A. Vinberg, 1990. Determination of anthropic soil phosphate on Aland. *Norwegian Archaeological Review* 23, 93–104.

Oakes, H., and J. Thorp, 1951. Dark clay soils of warm regions variously called *Rendzina*, Black Cotton Soils, Regur and Tirs. *Soil Science Society of America Proceedings* 15, 347–354.

O'Brien, M. J., R. E. Warren, and D. E. Lewarch, Eds., 1982. The Cannon Reservoir Human Ecology Project. Academic Press, New York.

O'Brien, P., 1990. An experimental study of the effects of salt erosion on pottery. *Journal of Archaeological Science* 17, 393–401.

Olsen, S. R., and L. A. Dean, 1965. Phosphorus. *In* Methods of Soil Analysis, Part 2—Chemical and Microbiological Properties (C. A. Black, Ed.), pp. 1035–1049. Monograph no. 9. American Society of Agronomy, Madison, Wisc.

Olsen, S. R., and L. E. Sommers, 1982. Phosphorus. *In* Methods of Soil Analysis, Part 2—Chemical and Microbiological Properties. 2nd ed. (A. L. Page, R. H. Miller, and D. R. Keeney, Eds.), pp. 403–430. Monograph no. 9. American Society of Agronomy, Madison, Wisc.

Olson, C. G., and W. D. Nettleton, 1998. Paleosols and the effects of alteration. *Quaternary International* 51/52, 185–194.

Olson, G. W., 1981. Soils and the Environment: A Guide to Soil Surveys and their Applications. Chapman and Hall, New York.

Olson, G. W., 1983. An evaluation of soil properties and potentials in different volcanic deposits. *In* Archaeology and Volcanism in Central America: The Zapotitan Valley of El Salvador (P. D. Sheets, Ed.), pp. 52–56. University of Texas Press, Austin.

Orme, A. R., 1990. Wetland morphology, hydrodynamics and sedimentation. *In* Wetlands: A Threatened Landscape (M. Williams, Ed.), pp. 42–94. Basil Blackwell, Oxford.

Ottaway, J. H., 1984. Persistence of organic phosphate in buried soils. *Nature* 307, 257–259.

Overstreet, D. F., 1998. Late Pleistocene geochronology and the Paleoindian penetration of the southwestern Lake Michigan basin. *The Wisconsin Archeologist* 79, 28–52.

Overstreet, W. C., R. C. Eidt, R. Brinkmann, D. E. Detra, and T. T. Chao, 1988. Orientation survey using phosphate fractionation to distinguish anthrosols from native sediments in the Wadi al-Jubah archaeological area, Yemen Arab Republic. *In* Geological and Archaeological Reconnaissance in the Yemen Arab Republic, 1985 (W. C. Overstreet, M. J. Grolier, and M. R. Toplyn, Eds.), pp. 121–154. American Foundation for the Study of Man, Washington, D.C.

Overstreet, W. C., and M. J. Grolier, 1988. Reconnaissance geology of the al-Jubah quadrangle, Yemen Arab Republic. *In* Geological and Archaeological Reconnaissance in the Yemen Arab Republic, 1985 (W. C. Overstreet, M. J. Grolier, and M. R. Toplyn, Eds.), pp. 155–288. American Foundation for the Study of Man, Washington, D.C.

Overstreet, W. C., and M. J. Grolier, 1996. Summary of the environmental background for the human occupation of the al-Jadidah Basin in Wadi al-Jubah, Yemen Arab Republic. *In* Environmental Research in Support of Archaeological Investigations in the Yemen Arab Republic, 1982–1987 (M. J. Grolier, R. Brinkmann, and J. A. Blakely, Eds.), pp. 337–429. American Foundation for the Study of Man, Washington, D.C.

Page, A. L., R. H. Miller, and D. R. Keeney, Eds., 1982. Methods of Soil Analysis, Part 2—Chemical and Microbiological Properties. 2nd ed. Monograph no. 9. American Society of Agronomy, Madison, Wisc. 1159 pp.

Palmer, A. S., and B. J. Pillans, 1996. Record of climatic fluctuations from ca. 500 ka loess deposits and paleosols near Wanganui, New Zealand. *Quaternary International* 34–36, 155–162.

Panfil, M. S., T. W. Gardner, and K. G. Hirth, 1999. Late Holocene stratigraphy of the Tetimpa archaeological sites, northeast flank of Popcatépetl volcano, central Mexico. *Geological Society of America Bulletin* 111, 204–218.

Pape, J. C., 1970. Plaggen soils in The Netherlands. *Geoderma* 4, 229–255.

Pappu, S., 1999. A study of natural site formation processes in the Kortallayar Basin, Tamil Nadu, South India. *Geoarchaeology* 14, 127–150.

Parnell, J. J., R. E. Terry, and C. Golden, 2001. Using in-field phosphate testing to rapidly identify middens at Piedras Negras, Guatemala. *Geoarchaeology* 16, 855–873.

Parnell, J. J., R. E. Terry, and Z. Nelson, 2002. Soil chemical analysis applied as an interpretive tool for ancient human activities in Piedras Negras, Guatemala. *Journal of Archaeological Science* 29, 379–404.

Parsons, R. B., C. A. Balster, and O. A. Ness, 1970. Soil Development and geomorphic surfaces, Willamette Valley, Oregon. *Soil Science Society of America Proceedings* 34, 485–491.

Parsons, R. B., W. H. Scholtes, and F. F. Riecken, 1962. Soils of Indian mounds in northeastern Iowa as benchmarks for studies of soil genesis. *Soil Science Society of America Proceedings* 26, 491–496.

Pasini, G., and M. L. Colalongo, 1997. The Pliocene-Pleistocene boundary-stratotype at Vrica, Italy. *In* The Pleistocene Boundary and the Beginning of the Quaternary (J. A. Van Couvering, Ed.), pp. 15–45. Cambridge University Press, Cambridge.

Paton, T. R., and G. Humphreys, 2000. Reply. Book review forum on "Soils: A New Global View." *Annals of the Association of American Geographers* 90, 787–789.

Paton, T. R., G. S. Humphreys, and P. B. Mitchell, 1995. Soils: A New Global View. Yale University Press, New Haven, Conn.

Paul, E. A., 1969. Characterization and turnover rate of soil humic constituents. *In* Pedology and Quaternary Research (S. Pawluk, Ed.) pp. 63–76. The University of Alberta Printing Department, Edmonton.

Pawluk, S., 1978. The pedogenic profile in the stratigraphic section. *In* Quaternary Soils (W. C. Mahaney, Ed.), pp. 61–75. Geo Abstracts, Norwich.

Pearsall, D. M., 1989. Paleoethnobotany: A Handbook of Procedures. Academic Press, San Diego.

Pearsall, D. M., and D. R. Piperno, Eds., 1993. Current Research in Phytolith Analysis: Applications in Archaeology and Paleoecology. The University Museum of Archaeology and Anthropology, MASCA Research Papers in Science and Archaeology, v. 10. University of Pennsylvania, Philadelphia.

Pécsi, M., Ed., 1984. Lithology and Stratigraphy of Loess and Paleosols. Geographical Research Institute, Hungarian Academy of Sciences, Budapest.

Pécsi, M., Ed., 1987. Loess. *GeoJournal* 15, 133–208.

Pécsi, M., Ed., 1991. Distribution and chronological problems of loess. *GeoJournal* 24, 139–220.

Pécsi, M., and G. Richter, 1996. Löss. *Zeitschrift für Geomorphologie* supplementband 98, 1–391.

Pécsi, M., and A. A. Velichko, Eds., 1995. Loess-paleosol and paleoclimatic investigations. *GeoJournal* 36, 115–294.

Pendall, E. G., and R. Amundson, 1990. Soil/landform relationships surrounding the Harappa archaeological site, Pakistan. *Geoarchaeology* 5, 301–322.

Pendall, E. G., J. W. Harden, S. E. Trumbore, and O. A. Chadwick, 1994. Isotopic approach to soil carbonate dynamics and implications for paleoclimatic interpretations. *Quaternary Research* 42, 60–71

Pettry, D. E., and J. A. Bense, 1989. Anthropic epipedons in the Tombigbee Valley of Mississippi. *Soil Science Society of America Journal* 53, 505–511.

Phillips, J., 1997. Simplexity and the reinvention of equifinality. *Geographical Analysis* 29, 1–15.

Pierce, C., 1992. Effects of pocket gopher burrowing on archaeological deposits: A simulation approach. *Geoarchaeology* 7, 185–208.

Pierzynski, G. M., Ed., 2000. Methods of Phosphorous Analysis for Soils, Sediments, Residuals, and Waters. Southern Cooperative Series Bulletin no. 396. U.S. Department of Agriculture and North Carolina State University, Southern Extension/Research Activity-Information Exchange Group 17, Raleigh. Available: http://www.soil.ncsu.edu/sera17/ publications/sera17-2/pm_cover.htm.

Pillans, B. J., 1988. Loess chronology in Wanganui, New Zealand. *In* Loess: Its Distribution, Geology and Soils (D. N. Eden and R. J. Furkert, Eds.), pp. 175–191. A. A. Balkema Publishers, Rotterdam.

Pincha, P. R., and M. L. Gregg, 1993. Chronostratigraphy of Upper James River floodplain sediments: Implications for southeastern North Dakota archaeology. *Geoarchaeology* 8, 203–215.

Piperno, D. R., 1988. Phytolith Analysis: An Archaeological and Geological Perspective. Academic Press, San Diego.

Plaisance, G., and A. Cailleux, 1981. Dictionary of Soils. Amerind Publishing, New Delhi.

Plog, F., 1971. Some operational considerations. *In* The Distribution of Prehistoric Population Aggregates (G. J. Gumerman, Ed.), pp. 45–54. Anthropological Reports no. 1. Prescott College, Prescott, Ariz.

Pohl, M. D., Ed., 1990. Ancient Maya Wetland Agriculture: Excavations on Albion Island, Northern Belize. Westview Special Studies in Archaeological Research, Boulder, Colo.

Pohl, M. D., P. R. Bloom, and K. O. Pope, 1990. Interpretation of wetland farming in northern Belize: Excavations at San Antonio Rio Hondo. *In* Ancient Maya Wetland Agriculture: Excavations on Albion Island, Northern Belize (M. D. Pohl, Ed.), pp. 187–254. Westview Special Studies in Archaeological Research, Boulder, Colo.

Pope, G. A., and R. Rubenstein, 1999. Anthroweathering: Theoretical framework and case study for human-impacted weathering. *Geoarchaeology* 14, 247–264.

Pope, K. O., and T. J. Van Andel, 1984. Late Quaternary alluviation and soil formation in the southern Argolid: Its history, causes and archaeological implications. *Journal of Archaeological Science* 11, 281–306.

Pope, K. O., C. N. Runnels, and T. L. Ku, 1984. Dating middle Paleolithic red beds in southern Greece. *Nature* 312, 264–266.

Porat, N., L.-P. Zhou, M. Chazan, T. Noy, and L. K. Horwitz, 1999. Dating the Lower Paleolithic open-air site of Holon, Israel by luminescence and ESR techniques. *Quaternary Research* 51, 328–341.

Pote, D. H., and T. C. Daniel, 2000. Analyzing for total phosphorus and total dissolved phosphorus in water samples. *In* Methods of Phosphorous Analysis for Soils, Sediments, Residuals, and Waters (G. M. Pierzynski, Ed.), pp. 94–97. Southern Cooperative Series Bulletin no. 396. Southern Extension/Research Activity-Information Exchange Group 17, U.S. Department of Agriculture and North Carolina State University, Raleigh.

Prewitt, E. R., 1981. Cultural chronology in Central Texas. *Bulletin of the Texas Archeological Society* 52, 65–89.

Prewitt, E. R., 1983. From Circleville to Toyah: Comments on Central Texas chronology. *Bulletin of the Texas Archeological Society* 54, 201–238.

Prosch-Danielsen, L., and A. Simonsen, 1988. Principle component analysis of pollen, charcoal and soil phosphate data as a tool in prehistoric land-use investigation at Forsandmoen, southwest Norway. *Norwegian Archaeological Review* 21, 85–102.

Proudfoot, B., 1976, The analysis and interpretation of soil phosphorus in archaeological contexts. *In* Geoarchaeology (D. A. Davidson and M. L. Shackley, Eds.), pp. 93–113. Westview Press, Boulder, Colo.

Provan, D. M., 1971. Soil phosphate analysis as a tool in archaeology. *Norwegian Archaeological Review* 4, 37–50.

Prusinkiewicz, Z., R. Bednarek, A. Kosko, and M. Szmyt, 1998. Paleopedological studies of the age and properties of illuvial bands at an archaeological site. *Quaternary International* 51/52, 195–201.

Purdy, B. A., Ed., 1988. Wet Site Archaeology. Telford Press, Caldwell, N.J.

Purdy, B. A., 1991. The Art and Archaeology of Florida's Wetlands. CRC Press, Boca Raton, Fla.

Purdy, B. A., Ed., 2001. Enduring Records: The Environmental and Cultural Heritage of Wetlands. Oxbow Books, Oxford.

Purdy, B. A., and D. E. Clark, 1987. Weathering of inorganic materials: Dating and other applications. *Advances in Archaeological Method and Theory* 11, 211–253.

Pyddoke, E., 1961. Stratification for the Archaeologist. Phoenix House, London.

Pye, K., 1987. Aeolian Dust and Dust Deposits. Academic Press, New York.

Quade, J., R. M. Forester, W. L. Pratt, and C. Carter, 1998. Black Mats, spring-fed streams, and late-glacial-age recharge in the southern Great Basin. *Quaternary Research* 49, 129–148.

Quade, J., M. L. John, T. P. Cater, J. A. Ojha, and T. M. Harrison, 1995. Late Miocene environmental change in Nepal and the northern Indian subcontinent: Stable isotopic evidence from paleosols. *Geological Society of America Bulletin* 107, 1381–1397.

Quine, T., 1995. Soil analysis and archaeological site formation studies. *In* Archaeological Sediments and Soils: Analysis, Interpretation and Management (A. J. Barham and

R. I. Macphail, Eds.), pp. 77–98. Institute of Archaeology, University College, London.

Rabenhorst, M. C., and D. Swanson, 2000. Histosols. *In* Handbook of Soil Science (M. E. Sumner, Ed.), pp. E183–E209. CRC Press, Boca Raton, Fla.

Raiswell, R., 2001. Defining the burial environment. *In* Handbook of Archaeological Sciences (D. R. Brothwell and A. M. Pollard, Eds.), pp. 595–603. Wiley, New York.

Ranov, V., 1995. The "Loessic Palaeolithic" in south Tadjikistan, Central Asia: Its industries, chronology, and correlation. *Quaternary Science Reviews* 14, 731–745.

Rapp, G., Jr., 1975. The archaeological field staff: The geologist. *Journal of Field Archaeology* 2, 229–237.

Rapp, G., Jr., and S. E. Aschenbrenner, Eds., 1978. Excavations at Nichoria in Southwest Greece. Volume 1, Site Environs, and Techniques. University of Minnesota Press, Minneapolis.

Rapp, G., Jr., R. Bullard, and C. A. Albritton, 1974. Geoarchaeology? The geologist. *Newsletter of the Geological Society of America* 9, 1.

Rapp, G., Jr., and C. L. Hill, 1998. Geoarchaeology: The Earth-Science Approach to Archaeological Interpretation. Yale University Press, New Haven, Conn.

Rawling, J. E., 2000. A review of lamellae. *Geomorphology* 35, 1–9.

Ray, J. H., 1998. Cultural components. *In* The 1997 Excavations at the Big Eddy Site (23CE426) in Southwest Missouri (N. H. Lopinot, J. H. Ray, and M. D. Conner, Eds.), pp. 111–220. Center for Archaeological Research, Special Publication no. 2. Southwest Missouri State University, Springfield, Mo.

Reed, S., N. Bailey, and O. Onokpise, 2000. Soil Science for Archaeologists. Florida A&M University and Southeast Archaeological Center, National Park Service, Washington, D.C. 32 pp. Available: http://www.cr.nps.gov/seac/soils-index/htm.

Reeves, B. O. K., and J. F. Dormaar, 1972. A partial Holocene pedological and archaeological record for the southern Alberta Rocky Mountains. *Arctic and Alpine Research* 4, 325–336.

Reheis, M. C., J. G. Goodmacher, J. W. Harden, L. D. McFadden, T. K. Rockwell, R. R. Shroba, J. M. Sowers, and E. M. Taylor, 1995. Quaternary soils and dust deposition in southern Nevada and California. *Geological Society of America Bulletin* 107, 1003–1022.

Reheis, M. C., J. W. Harden, L. D. McFadden, and R. R. Shroba, 1989. Development rates of late Quaternary soils, Silver Lake Playa, California. *Soil Science Society of America Journal* 53, 1127–1140.

Reid, K. C., 1984. Fire and ice: New evidence for the production and preservation of late Archaic fiber-tempered pottery in the middle-latitude lowlands. *American Antiquity* 49, 55–76.

Reider, R. G., 1975. Morphology and genesis of soils on the Prairie Divide Deposit (pre-Wisconsin), Front Range, Colorado. *Arctic and Alpine Research* 7, 353–372.

Reider, R. G., 1980. Late Pleistocene and Holocene soils of the Carter/Kerr-McGee archeological site, Powder River basin, Wyoming. *Catena* 7, 301–315.

Reider, R. G., 1982a. Soil development and paleoenvironments. *In* The Agate Basin Site (G. C. Frison and D. Stanford, Eds.), pp. 331–344. Academic Press, New York.

Reider, R. G., 1982b. The soil of Clovis age at the Sheaman archaeological site, eastern Wyoming. *The University of Wyoming, Contributions to Geology* 21, 195–200.

Reider, R. G., 1987. Soil formation and paleoenvironmental interpretation at the Horner site, Park County, Wyoming. *In* The Horner Site: The Type Site of the Cody Cultural Complex (G. C. Frison and L. C. Todd, Eds.), pp. 347–360. Academic Press, New York.

Reider, R. G., 1990. Late Pleistocene and Holocene pedogenic and environmental trends at archaeological sites in plains and mountain areas of Colorado and Wyoming. *In* Archaeological Geology of North America (N. P. Lasca and J. Donahue, Eds.), pp. 335–360. Centennial Volume no. 4. Geological Society of America, Boulder, Colo.

Reider, R. G., G. A. Huckleberry, and G. C. Frison, 1988. Soil evidence for postglacial forest-grassland fluctuations in the Absaroka Mountains of northwestern Wyoming, U.S.A. *Arctic and Alpine Research* 20, 188–198.

Reider, R. G., and E. T. Karlstrom, 1987. Soils and stratigraphy of the Laddie Creek site (48BH345), an Altithermal-age occupation in the Big Horn Mountains, Wyoming. *Geoarchaeology* 2, 29–47.

Reider, R. G., N. J. Kuniansky, D. M. Stiller, and P. J. Uhl, 1974. Preliminary investigation of comparative soil development on Pleistocene and Holocene geomorphic surfaces of the Laramie basin, Wyoming. *In* Applied Geology and Archaeology: The Holocene History of Wyoming (M. Wilson, Ed.), pp. 27–33. Report of Investigation no. 10. Geological Survey of Wyoming, Laramie.

Reinhardt, J., and W. R. Sigleo, Eds., 1988. Paleosols and Weathering through Geologic Time: Principles and Applications. Special Paper no. 216. Geological Society of America, Boulder, Colo.

Reitz, E. J., L. A. Newsom, and S. J. Scudder, 1996. Issues in environmental archaeology. *In* Case Studies in Environmental Archaeology (E. J. Reitz, L. A. Newsom, and S. J. Scudder, Eds.), pp. 3–16. Plenum Press, New York.

Renfrew, C., 1976. Archaeology and the earth sciences. *In* Geoarchaeology (D. A. Davidson and M. L. Shackley, Eds.), pp. 1–5. Westview Press, Boulder, Colo.

Retallack, G. J., 1983. Late Eocene and Oligocene Paleosols from Badlands National Park, South Dakota. Special Paper no. 193. Geological Society of America, Boulder, Colo.

Retallack, G. J., 1986. The fossil record of soils. *In* Paleosols: Their Recognition and Interpretation. (V. P. Wright, Ed.), pp. 1–57. Princeton University Press, Princeton, N.J.

Retallack, G. J., 1988. Field recognition of paleosols. *In* Paleosols and Weathering through Geologic Time: Principles and Applications (J. Reinhardt and W. Sigleo, Eds.), pp. 1–20. Special Paper no. 216. Geological Society of America, Boulder, Colo.

Retallack, G. J., 1990. Soils of the Past: An Introduction to Paleopedology. Unwin and Hyman, New York.

Retallack, G. J., 1991a. Miocene Paleosols and Ape Habitats of Pakistan and Kenya. Oxford University Press, New York.

Retallack, G. J., 1991b. Untangling the effects of burial alteration and ancient soil formation. *Annual Review of Earth and Planetary Sciences* 19, 183–206.

Retallack, G. J., 1993. Classification of paleosols: Discussion. *Geological Society of America Bulletin* 105, 1635–1636.

Retallack, G. J., 1994. The environmental factor approach to the interpretation of paleosols. *In* Factors of Soil Formation: A Fiftieth Anniversary Retrospective (R. Amundson, Ed.), pp. 31–64. Special Publication no. 33. Soil Science Society of America, Madison, Wisc.

Retallack, G. J., 1997a. A Colour Guide to Paleosols. Wiley, New York.

Retallack, G. J., 1997b. Neogene expansion of North American prairie. *Palaios* 12, 380–390.

Retallack, G. J., 1998. Core concepts of paleopedology. *Quaternary International* 51/52, 203–212.

Retallack, G. J., 2001. Soils of the Past: An Introduction to Paleopedology. 2nd ed. Blackwell Science, Oxford.

Retallack, G. J., D. P. Dugas, and E. A. Bestland, 1990. Fossil soils and grasses of a middle Miocene east African grassland. *Science* 247, 1325–1328.

Richards, K. S., R. R. Arnett, and S. Ellis, Eds., 1985. Geomorphology and Soils. Allen and Unwin, London.

Roberts, B. K., 1987. Landscape archaeology. *In* Landscape and Culture: Geographical and Archaeological Perspectives (J. M. Wagstaff, Ed.), pp. 77–95. Basil Blackwell, New York.

Robertson-Rintoul, M. S. E., 1986. A quantitative soil stratigraphic approach to the correlation and dating of post-glacial river terraces in Glen Feshie, western Cairngorms. *Earth Surface Processes and Land Forms* 11, 605–617.

Robinson, W. O., 1927. The determination of organic matter in soils by means of hydrogen peroxide. *Journal of Agricultural Research* 34, 339–356.

Rockwell, T. K., D. L. Johnson, E. A. Keller, and G. R. Dembroff, 1985. A late Pleistocene-Holocene soil chronosequence in the Ventura Basin, southern California, USA. *In*

Geomorphology and Soils (K. S. Richards, R. R. Arnett, and J. Ellis, Eds.), pp. 309–327. Allen and Unwin, London.

Ronen, A., 1975. The Paleolithic archaeology and chronology of Israel. *In* Problems in Prehistory: North Africa and the Levant (F. Wendorf and A. E. Marks, Eds.), pp. 229–248. Southern Methodist University Press, Dallas, Tex.

Ronen, A., 1977. Mousterian sites in red loam in the coastal plain of Mount Carmel. *Eretz Israel* 13, 183–190.

Ronen, A., 1991. The lower Paleolithic site Evron-Quarry in western Galilee, Israel. *Sonderveroffentlichungen, Geologisches Institut der Universitat zu Koln* 82, 187–212.

Ronen, A., A. Tsatskin, and S. A. Laukhin, 1999. The genesis and age of Mousterian paleosols in the Carmel Coastal Plain, Israel. *In* Dorothy Garrod and the Progress of the Palaeolithic (W. Davies and R. Charles, Eds.), pp. 135–151. Oxbow Books, Oxford.

Rossignol, J., and L. Wandsnider, Eds., 1992. Space, Time, and Archaeological Landscapes. Plenum Publishing, New York.

Rousseau, D.-D., 1990. Statistical analyses of loess molluscs for paleoecological reconstructions. *Quaternary International* 7/8, 81–89.

Rousseau, D.-D., N. Gerasimenko, Z. Matviischina, and G. Kukla, 2001. Late Pleistocene environments of the Central Ukraine. *Quaternary Research* 56, 349–356.

Rovner, I., 1988. Macro- and micro-ecological reconstruction using plant opal phytolith data from archaeological sediments. *Geoarchaeology* 3, 155–163.

Royer, D. L., 1999. Depth to pedogenic carbonate horizon as a paleoprecipitation indicator? *Geology* 27, 1123–1126.

Ruellan, A., 1971. The history of soils: Some problems of definition and interpretation. *In* Paleopedology: Origin, Nature and Dating of Paleosols (D. H. Yaalon, Ed.), pp. 3–13. University of Israel Press, Jerusalem.

Ruhe, R. V., 1956. Geomorphic surfaces and the nature of soils. *Soil Science* 82, 441–455.

Ruhe, R. V., 1965. Quaternary paleopedology. *In* The Quaternary of the United States (H. E. Wright and D. G. Frey, Eds.), pp. 755–764. Princeton University Press, Princeton, N.J.

Ruhe, R. V., 1970. Soils, paleosols, and environment. *In* Pleistocene and Recent Environments of the Central Great Plains (W. Dort and J. K. Jones, Eds.), pp. 37–52. University of Kansas Press, Lawrence.

Ruhe, R. V., and J. G. Cady, 1967. The relation of Pleistocene geology and soils between Bentley and Adair in southwestern Iowa. *In* Landscape Evolution and Soil Formation in Southwestern Iowa. (R. V. Ruhe, R. B. Daniels, and J. G. Cady, Eds.), pp. 1–92. Technical Bulletin no. 1349. U.S. Department of Agriculture, Washington, D.C.

Ruhe, R. V., R. B. Daniels, and J. G. Cady, Eds., 1967. Landscape Evolution and Soil Formation in Southwestern Iowa. Technical Bulletin no. 1349. U.S. Department of Agriculture, Washington, D.C.

Ruhe, R. V., and C. G. Olson, 1980. Soil welding. *Soil Science* 130, 132–139.

Runnels, C. N., 1988. A prehistoric survey of Thessaly: New light on the Greek middle Paleolithic. *Journal of Field Archaeology* 15, 277–290.

Runnels, C. N., and T. H. Van Andel, 1993a. A handaxe from Kokkinopilos, Epirus, and its implications for the Paleolithic of Greece. *Journal of Field Archaeology* 20, 191–203.

Runnels, C. N., and T. H. Van Andel, 1993b. The lower and middle Paleolithic of Thessaly, Greece. *Journal of Field Archaeology* 20, 299–317.

Running, G. L., IV, 1995. Archaeological geology of the Rustad Quarry site (32RI775): An Early Archaic site in southeastern North Dakota. *Geoarchaeology* 10, 183–204.

Rutter, N. W., 1978. Soils in archaeology. 1977 Annual Report and Review of Soil Science, Geological Survey of Canada Paper no. 78-6, Geosciences in Canada, Ottawa. 1 p.

Rye, O. S., 1981. Pottery Technology: Principles and Reconstruction. Taraxacum, Washington, D.C.

Salvador, A., Ed., 1994. International Stratigraphic Guide: A Guide to Stratigraphic Classification, Terminology, and Procedure. 2nd ed. Geological Society of America, Boulder, Colo.

Sanborn, P., and S. Pawluk, 1980. An early Holocene Solodic paleosol near Priddis, Alberta. *Canadian Journal of Earth Science* 17, 1323–1330.
Sanchez, A., M. L. Canabate, and R. Lizcano, 1996. Phosphorus analysis at archaeological sites: An optimization of the method and interpretation of the results. *Archaeometry* 38, 151–164.
Sandor, J. A., 1987. Initial investigation of soils in agricultural terraces in the Colca Valley, Peru. *In* Pre-Hispanic Agricultural Terraces in the Andean Region (W. M. Denevan, K. Mathewson, and G. Knapp, Eds.), pp. 163–192. BAR International Series no. 359, Oxford.
Sandor, J. A., 1992. Long-term effects of prehistoric agriculture on soils: Examples from New Mexico and Peru. *In* Soils and Landscape Evolution (V. T. Holliday, Ed.), pp. 217–245. Smithsonian Institution Press, Washington, D.C.
Sandor, J. A., 1995. Searching soil for clues about southwest prehistoric agriculture. *In* Soil, Water, Biology, and Belief in Prehistoric and Traditional Southwestern Agriculture (H. W. Toll, Ed.), pp. 119–137. Special Publication no. 2. New Mexico Archaeological Council, Albuquerque.
Sandor, J. A., and N. S. Eash, 1995. Ancient agricultural soils in the Andes of Southern Peru. *Soil Science Society of America Journal* 59, 170–179.
Sandor, J. A., P. L. Gersper, and J. W. Hawley, 1986. Soils at prehistoric agricultural terracing sites in New Mexico (3 parts). *Soil Science Society of America Journal* 50, 166–180.
Saucier, R. T., 1966. Soil-survey reports and archaeological investigations. *American Antiquity* 31, 419–422.
Sawbridge, D. F., and M. A. Bell, 1972. Vegetation and soils of shell middens on the coast of British Columbia. *Ecology* 53, 840–849.
Schaetzl, R. J., 1986. Complete soil profile inversion by tree uprooting. *Physical Geography* 7, 181–189.
Schaetzl, R. J., 1996. Spodosol-Alfisol intergrades: Bisequal soils in NE Michigan, USA. *Geoderma* 74, 23–47.
Schaetzl, R. J., 2000. Shock the world (and then some). Book review forum on "Soils: A New Global View." *Annals of the Association of American Geographers* 90, 772–774.
Schaetzl, R. J., and S. A. Isard, 1996. Regional-scale relationships between climate and strength of podzolization in the Great Lakes region, North America. *Catena* 28, 47–69.
Schaetzl, R. J., and D. L. Mokma, 1988. A numerical index of Podzol and Podzolic soil development. *Physical Geography* 9, 232–264.
Schaetzl, R. J., and C. J. Sorenson, 1987. The concept of "buried" versus "isolated" paleosols: Examples from northeastern Kansas. *Soil Science* 143, 426–435.
Schaetzl, R. J., L. R. Barrett, and J. A. Winkler, 1994. Choosing models for soil chronofunctions and fitting them to data. *European Journal of Soil Science* 45, 219–232.
Scharpenseel, H. W., 1971a. Radiocarbon dating of soils. *Soviet Soil Science* 3, 76–83.
Scharpenseel, H. W., 1971b. Radiocarbon dating of soils—Problems, troubles, hopes. *In* Paleopedology: Origin, Nature and Dating of Paleosols (D. H. Yaalon, Ed.), pp. 77–87. University of Israel Press, Jerusalem.
Schelling, J., 1970. Soil genesis, soil classification and soil survey. *Geoderma* 4, 165–193.
Schiegl, S., P. Goldberg, O. Bar-Yosef, and S. Weiner, 1996. Ash deposits in Hayonim and Kebara Caves, Israel: Macroscopic, microscopic and mineralogical observations, and their archaeological implications. *Journal of Archaeological Science* 23, 763–781.
Schiffer, M. B., 1983. Toward the identification of formation processes. *American Antiquity* 48, 675–706.
Schiffer, M. B., 1987. Formation Processes of the Archaeological Record. University of New Mexico Press, Albuquerque.
Schlezinger, D. R., and B. L. Howes, 2000. Organic phosphorus and elemental ratios as indicators of prehistoric human occupation. *Journal of Archaeological Science* 27, 479–492.
Schoeneberger, P. J., D. A. Wysocki, E. C. Benham, and W. D. Broderson, 1998. Field Book for Describing and Sampling Soils. National Soil Survey Center, National Resource Conservation Service, U.S. Department of Agriculture, Lincoln, Ne.

Schuldenrein, J., 1991. Coring and the identity of cultural-resource environments: A comment on Stein. *American Antiquity* 56, 131–137.

Schuldenrein, J., 1995. Geochemistry, phosphate fractionation, and the detection of activity areas at prehistoric North American sites. *In* Pedological Perspectives in Archaeological Research (M. E. Collins, B. J. Carter, B. G. Gladfelter, and R. J. Southard, Eds.), pp. 107–132. Special Publication no. 44. Soil Science Society of America, Madison, Wisc.

Schultz, C. B., and J. C. Frye, Eds., 1968. Loess and Related Eolian Deposits of the World. University of Nebraska Press, Lincoln.

Schwarz, G. T., 1967. A simplified chemical test for archaeological field work. *Archaeometry* 10, 57–63.

Schweger, C., 1985. Geoarchaeology of northern regions: Lessons from cryoturbation at Onion Portage, Alaska. *In* Archaeological Sediments in Context (J. K. Stein and W. R. Farrand, Eds.), pp. 127–141. Center for the Study of Early Man, Peopling of the Americas Edited Volume Series no. 1. University of Maine, Orono.

Schwertmann, U., 1993. Relations between iron oxides, soil color, and soil formation. *In* Soil Color (J. M. Bigham and E. J. Ciolkosz, Eds.), pp. 51–69. Special Publication no. 31. Soil Science Society of America, Madison, Wisc.

Schwertmann, U., E. Murad, and D. G. Schultze, 1982. Is there Holocene reddening (hematite formation) in soils of axeric temperate areas? *Geoderma* 27, 209–223.

Scott, W. E., 1977. Quaternary glaciation and volcanism, Metolius River area, Oregon. *Geological Society of America Bulletin* 88, 113–124.

Scott, G. R., 1963. Quaternary Geology and Geomorphic History of the Kassler Quadrangle, Colorado. U.S. Geological Survey Professional Paper no. 421-A. Government Printing Office, Washington, D.C.

Scudder, S. J., 1996. Human influence on pedogenesis: Midden soils on a southwest Florida Pleistocene island. *In* Case Studies in Environmental Archaeology (E. J. Reitz, L. A. Newsom, and S. J. Scudder, Eds.), pp. 55–69. Plenum Press, New York.

Scudder, S. J., J. E. Foss, and M. E. Collins, 1996. Soil science and archaeology. *Advances in Agronomy* 57, 1–76.

Scully, R. W., and R. W. Arnold, 1981. Holocene alluvial stratigraphy in the upper Susquehanna River basin, New York. *Quaternary Research* 15, 327–344.

Segovia, A. V., 1997. Considerations about fragipans of the eastern United States. *In* Proceedings of the Second International Conference on Pedo-Archaeology (A. C. Goodyear, J. E. Foss, and K. E. Sassaman, Eds.), pp. 121–123. Occasional Papers of the South Carolina Institute of Archaeology and Anthropology. University of South Carolina, Columbia.

Self-Davis, M. L., P. A. Moore, Jr., and B. C. Joern, 2000. Determination of Water- and/or dilute salt-extractable phosphorus. *In* Methods of Phosphorous Analysis for Soils, Sediments, Residuals, and Waters (G. M. Pierzynski, Ed.), pp. 24–26. Southern Cooperative Series Bulletin no. 396. Southern Extension/Research Activity-Information Exchange Group 17, U.S. Department of Agriculture and North Carolina State University, Raleigh.

Sellards, E. H., 1952. Early Man in America. University of Texas Press, Austin.

Sellards, E. H., and G. L. Evans, 1960. The Paleo-Indian cultural succession in the Central High Plains of Texas and New Mexico. *In* Men and Cultures (A. F. C. Wallace, Ed.), pp. 639–649. University of Pennsylvania Press, Philadelphia.

Semmel, A., 1989. Paleopedology and geomorphology: Examples from the western part of Central Europe. *In* Paleopedology: Nature and Applications of Paleosols (A. Bronger and J. Catt, Eds.), pp. 143–162. Catena Supplement 16. Catena Verlag, Cremlingen, Germany.

Shackley, M. L., 1975. Archaeological Sediments: A Survey of Analytical Methods. Butterworths, London.

Shackley, M. L., 1981. Environmental Archaeology. Allen and Unwin, London.

Shaler, N. S., 1891. The origin and nature of soils. 12th Annual Report 1890–1891, part 1, pp. 213–345. U. S. Geological Survey, Washington, D.C.

Sharpley, A. N., H. Tiessen, and C. V. Cole, 1987. Soil phosphorus forms extracted by soil tests as a function of pedogenesis. *Soil Science Society of America Journal* 51, 362–365.

Sheets, P. D., Ed., 1983. Archaeology and Volcanism in Central America: The Zapotitan Valley of El Salvador. University of Texas Press, Austin.

Sheets, P. D., and D. K. Grayson, Eds., 1979. Volcanic Activity and Human Ecology. Academic Press, New York.

Sheets, P. D., and B. R. McKee, Eds., 1994. Archaeology, Volcanism, and Remote Sensing in the Arenal Region, Costa Rica. University of Texas Press, Austin.

Shipman, J. A., 1997. Soil Survey of the Lincoln County Area, Oregon. U.S. Department of Agriculture, Soil Conservation Service, Washington, D.C.

Shlemon, R. J., 1978. Quaternary soil-geomorphic relationships, southeastern Mojave desert, California and Arizona. Quaternary Soils (W. C. Mahaney, Ed.), pp. 187–207. Geo Abstracts, Norwich.

Shlemon, R. J., and F. E. Budinger, 1990. The archaeological geology of the Calico site, Mojave Desert, California. *In* Archaeological Geology of North America (N. P. Lasca and J. Donahue, Eds.), pp. 301–314. Centennial Volume no. 4. Geological Society of America, Boulder, Colo.

Shoji, S., M. Nanzyo, and R. Dahlgren, 1993. Volcanic Ash Soils: Genesis, Properties and Utilization. Elsevier, Amsterdam.

Shotyk, W., 1992. Organic soils. *In* Weathering, Soils, and Palaeosols (I. P. Martini and W. Chesworth, Eds.), pp. 203–224. Elsevier, New York.

Shroba, R. R., and P. W. Birkeland, 1983. Trends in late-Quaternary soil development in the Rocky Mountains and Sierra Nevada of the western United States. *In* Late-Quaternary Environments of the United States, vol. 1 (H. E. Wright, Jr., Ed.), pp. 145–146. University of Minnesota Press, Minneapolis.

Sieveking, G. De G., I. H. Longworth, M. J. Hughes, A. J. Clark, and A. Millett, 1973. A new survey of Grime's Graves—First report. *Proceedings of the Prehistoric Society* 39, 182–218.

Simonson, R. W., 1959. Outline of a generalized theory of soil genesis. *Soil Science Society of America Proceedings* 23, 152–156.

Simonson, R. W., 1978. Multiple process model of soil genesis. *In* Quaternary Soils (W. C. Mahaney, Ed.), pp. 1–25. Geo Abstracts, Norwich.

Simonson, R. W., 1987. Historical Aspects of Soil Survey and Soil Classification. Soil Science Society of America (reprinted from *Soil Survey Horizons*), Madison, Wisc.

Simonson, R. W., 1995. Airborne dust and its significance to soils. *Geoderma* 65, 1–43.

Simonson, R. W., 1997. Evolution of soil series and type concepts in the United States. *In* History of Soil Science (D. H. Yaalon and S. Berkowicz, Eds.), pp. 79–108. Advances in GeoEcology, vol. 29. Catena Verlag, Reiskirchen, Germany.

Simpson, I. A., 1985. Stable carbon isotope analysis of anthropogenic soils and sediments in Orkney. *In* Palaeoenvironmental Investigations: Research Design, Methods and Data Analysis (N. R. J. Fieller, D. D. Gilbertson, and N. G. A. Ralph, Eds.), pp. 55–79. BAR International Series no. 258, London.

Simpson, I. A., 1997. Relict properties of anthropogenic deep top soils as indicators of infield management in Marwick, West Mainland, Orkney. *Journal of Archaeological Science* 24, 365–380.

Simpson, I. A., R. G. Bryant, and U. Tveraabak, 1998. Relict soils and early arable land management in Lofoten, Norway. *Journal of Archaeological Science* 25, 1185–1198.

Sims, J. T., 2000a. Soil test phosphorus: Bray and Kurtz P-1. *In* Methods of Phosphorous Analysis for Soils, Sediments, Residuals, and Waters (G. M. Pierzynski, Ed.), pp. 13–14. Southern Cooperative Series Bulletin no. 396. Southern Extension/Research Activity-Information Exchange Group 17, U.S. Department of Agriculture and North Carolina State University, Raleigh.

Sims, J. T., 2000b. Soil test phosphorus: Mehlich 1. *In* Methods of Phosphorous Analysis for Soils, Sediments, Residuals, and Waters (G. M. Pierzynski, Ed.), pp. 15–16. Southern Cooperative Series Bulletin no. 396. Southern Extension/Research

Activity-Information Exchange Group 17, U.S. Department of Agriculture and North Carolina State University, Raleigh.

Sims, J. T., 2000c. Soil test phosphorus: Mehlich 3. *In* Methods of Phosphorous Analysis for Soils, Sediments, Residuals, and Waters (G. M. Pierzynski, Ed.) pp. 17–19. Southern Cooperative Series Bulletin no. 396. Southern Extension/Research Activity-Information Exchange Group 17, U.S. Department of Agriculture and North Carolina State University, Raleigh.

Sims, J. T., 2000d. Soil test phosphorus: Olsen P. *In* Methods of Phosphorous Analysis for Soils, Sediments, Residuals, and Waters (G. M. Pierzynski, Ed.), pp. 20–21. Southern Cooperative Series Bulletin no. 396. Southern Extension/Research Activity-Information Exchange Group 17, U.S. Department of Agriculture and North Carolina State University, Raleigh.

Singer, M. J., P. Fine, K. L. Verosub, and O. A. Chadwick, 1992. Time dependence of magnetic susceptibility of soil chronosequences on the California Coast. *Quaternary Research* 37, 323–332.

Singer, M. J., and P. Janitzky, Eds., 1986. Field and Laboratory Procedures used in a Soil Chronosequence Study. Bulletin 1648. U.S. Geological Survey, Washington, D.C.

Sinitsyn, A. A., 1996. Kostenki 14 (Markina Gora): Data, problems, and perspectives. *Préhistoire Européene* 9, 273–313.

Sivan, D., G. Gvirtzman, and E. Sass, 1999. Quaternary stratigraphy and paleogeography of the Galilee coastal plain, Israel. *Quaternary Research* 51, 280–294.

Sjöberg, A., 1976. Phosphate analysis of anthropic soils. *Journal of Field Archaeology* 3, 447–454.

Skibo, J. M., 1992. Pottery Function: A Use-Alteration Perspective. Plenum Press, New York.

Skibo, J. M., and M. B. Schiffer, 1987. The effects of water on processes of ceramic abrasion. *Journal of Archaeological Science* 14, 83–96.

Skibo, J. M., M. B. Schiffer, and K. C. Reid, 1989. Organic-tempered pottery: An experimental study. *American Antiquity* 54, 122–146.

Skinner, S. M., 1986. Phosphorous as an anthrosol indicator. *Midcontinental Journal of Archaeology* 11, 51–78.

Slate, J. L., W. B. Bull, T.-L. Ku, M. Shafiqullah, D. J. Lynch, and Y.-P. Huang, 1991. Soil-carbonate genesis in the Pinacate volcanic field, northwestern Mexico. *Quaternary Research* 35, 400–416.

Smalley, I. J., I. F. Jefferson, T. A. Dijkstra, and E. Derbyshire, 2001. Some major events in the development of the scientific study of loess. *Earth Science Reviews* (special issue) 54, 5–18.

Smeck, N. E., 1973. Phosphorus: An indicator of pedogenetic weathering processes. *Soil Science* 115, 199–206.

Smeck, N. E., 1985. Phosphorus dynamics in soils and landscapes. *Geoderma* 36, 185–199.

Smeck, N. E., E. C. A. Runge, and E. E. Mackintosh, 1983. Dynamics and genetic modeling of soil systems. *In* Pedogenesis and Soil Taxonomy, part 1, Concepts and Interactions (L. P. Wilding, N. E. Smeck, and G. F. Hall, Eds.), pp. 51–81. Developments in Soil Science 11A. Elsevier, Amsterdam.

Smith, G. D., and M. McFaul, 1997. Paleoenvironmental and geoarchaeologic implications of late Quaternary sediments and paleosols: North-central to southwestern San Juan Basin, New Mexico. *Geomorphology* 21, 107–138.

Smith, N. J. H., 1980. Anthrosols and human carrying capacity in Amazonia. *Annals of the Association of American Geographers* 70, 29–55.

Soil Classification Working Group, 1998. The Canadian System of Soil Classification, 3rd ed. Agriculture and Agri-Food Canada Publication no. 1646. NRC Research Press, Ottawa.

Soil Science Society of America, 1987. Glossary of Soil Science Terms. Soil Science Society of America, Madison, Wisc.

Soil Survey Division Staff, 1993. Soil Survey Manual (revised). Agriculture Handbook no. 18. U.S. Department of Agriculture, Washington, D.C.

Soil Survey Laboratory Staff, 1992. Soil Survey Laboratory Methods Manual. Version 2.0. National Soil Survey Center Soil Survey Investigations Report no. 42. U.S. Department of Agriculture, Lincoln, Ne.

Soil Survey Laboratory Staff, 1996. Soil Survey Laboratory Methods Manual. Version 3.0. National Soil Survey Center Soil Survey Investigations Report no. 42. U.S. Department of Agriculture, Lincoln, Ne.

Soil Survey for Scotland, 1981. 1:50,000 Soil Maps of Orkney: Orkney Mainland, Orkney Northern Isles and Orkney Hoy. Macaulay Institute for Soil Research, Aberdeen.

Soil Survey Staff, 1951. Soil Survey Manual. Agriculture Handbook no. 18. U.S. Department of Agriculture, Washington D.C.

Soil Survey Staff, 1960. Soil Classification, A Comprehensive System, 7th Approximation. U.S. Department of Agriculture, Washington, D.C.

Soil Survey Staff, 1972. Soil Survey Laboratory Methods and Procedures for Collecting Soil Samples. Soil Conservation Service, Soil Survey Investigations Report no. 1. U.S. Department of Agriculture, Washington, D.C.

Soil Survey Staff, 1975. Soil Taxonomy. Agriculture Handbook no. 436. U.S. Department of Agriculture, Washington, D.C.

Soil Survey Staff, 1984. Procedures for Collecting Soil Samples and Methods of Analysis for Soil Survey. Soil Conservation Service Soil Survey Investigations Report no. 1. U.S. Department of Agriculture, Washington, D.C.

Soil Survey Staff, 1999. Soil Taxonomy. 2nd ed., Agriculture Handbook no. 436. U.S. Department of Agriculture, Washington, D.C.

Solecki, R. S., 1951. Notes on soil analysis and archaeology. *American Antiquity* 16, 254–256.

Sombroek, W. G., 1966. Amazon Soils: A Reconnaissance of the Soils of the Brazilian Amazon Region. Centre for Agricultural Publications and Documentation, Wageningen.

Sombroek, W., D. Kern, T. Rodriques, M. da S. Cravo, T. C. Jarbas, W. Woods, and B. Glaser, 2002. Terra Preta and Terra Mulata: Pre-Columbian Amazon kitchen middens and agricultural fields, their sustainability and their replication. *In* Transactions, 17th World Congress of Soil Science, Symposium no. 18, pp. 1935-1–1935-9). International Society of Soil Science, Bangkok, Thailand.

Sorenson, C. J., 1977. Reconstructed Holocene bioclimates. *Annals of the Association of American Geographers* 67, 214–222.

Sorenson, C. J., and J. C. Knox, 1973. Paleosols and paleoclimates related to late Holocene forest/tundra border migrations: Mackenzie and Keewatin, N.W.T., Canada. *In* International Conference on the Prehistory and Paleoecology of Western North American Arctic and Subarctic (S. Raymons and P. Schlederman, Eds.), pp. 187–204. University of Calgary Archaeological Association, The University of Alberta Printing Department, Edmonton.

Sorenson, C. J., J. C. Knox, J. A. Larsen, and R. A. Bryson, 1971. Paleosols and the forest border in Keewatin, N.W.T. *Quaternary Research* 1, 468–473.

Sorenson, C. J., K. H. Sallee, and R. D. Mandel, 1987. Holocene and Pleistocene soils and geomorphic surfaces of the Kansas River valley. *In* Quaternary Environments of Kansas (W. C. Johnson, Ed.), pp. 93–102. Guidebook Series no. 5. Kansas Geological Survey, Lawrence.

Sowers, J. L., B. Szabo, T. Jull, T.-L. Ku, M. C. Reheis, O. A. Chadwick, S. W. Robinson, and R. G. Amundson, 1988. Age data for Kyle Canyon soils and deposits. *In* This Extended Land, Geological Journeys in the Southern Basin and Range (D. L. Weide and M. L. Faber, Eds.), pp. 141–142. Geological Society of America, Cordilleran Section, Field Trip Guidebook, Boulder, Colo.

Sparks, D. L., A. L. Page, P. A. Helmke, R. H. Loeppert, P. N. Soltanpour, M. A. Tabatabai, C. T. Johnson, and M. E. Sumner, Eds., 1996. Methods of Soil Analysis: Part 3—Chemical Methods. Book Series 5. Soil Science Society of America, Madison, Wisc.

Stafford, C. R., 1995. Geoarchaeological perspectives on paleolandscapes and regional subsurface archaeology. *Journal of Archaeological Method and Theory* 2, 69–104.

Stafford, C. R., and S. D. Creasman, 2002. The hidden record: Late Holocene landscapes and settlement archaeology in the Lower Ohio River Valley. *Geoarchaeology* 17, 117–140.

Stafford, C. R., D. S. Leigh, and D. L. Asch, 1992. Prehistoric settlement and landscape change on alluvial fans in the Upper Mississippi River Valley. *Geoarchaeology* 7, 287–314.

Stafford, T. W., Jr., 1981. Alluvial geology and archaeological potential of the Texas Southern High Plains. *American Antiquity* 46, 548–565.

Steen-McIntyre, V., 1985. Tephrochronology and its application to archaeology. *In* Archaeological Geology (G. Rapp and J. A. Gifford, Eds.), pp. 265–302. Yale University Press, New Haven, Conn.

Stein, J. K., 1982. Geologic analysis of the Green River shell middens. *Southeastern Archaeology* 1, 22–39.

Stein, J. K., 1983. Earthworm activity: A source of potential disturbance of archaeological sediments. *American Antiquity* 48, 277–289.

Stein, J. K., 1984. Organic matter and carbonates in archaeological sites. *Journal of Field Archaeology* 11, 239–346.

Stein, J. K., 1985. Interpreting sediments in cultural settings. *In* Archaeological Sediments in Context (J. K. Stein and W. R. Farrand, Eds.), pp. 5–19. Center for the Study of Early Man, Peopling of the Americas, Edited Volume Series 1. University of Maine, Orono.

Stein, J. K., 1986. Coring archaeological sites. *American Antiquity* 51, 505–527.

Stein, J. K., 1987. Deposits for archaeologists. *Advances in Archaeological Method and Theory* 11, 337–395.

Stein, J. K., 1991. Coring in CRM and archaeology: A reminder. *American Antiquity* 56, 138–142.

Stein, J. K., 1992a. The analysis of shell middens. *In* Deciphering a Shell Midden (J. K. Stein, Ed.), pp. 1–24. Academic Press, New York.

Stein, J. K., Ed., 1992b. Deciphering a Shell Midden. Academic Press, New York.

Stein, J. K., 1992c. Organic matter in archaeological contexts. *In* Soils and Landscape Evolution (V. T. Holliday, Ed.), pp. 193–216. Smithsonian Institution Press, Washington, D.C.

Stein, J. K., 2000. Stratigraphy and archaeological dating. *In* It's About Time: A History of Archaeological Dating In North America (S. E. Nash, Ed.), pp. 14–40. University of Utah Press, Salt Lake City.

Stein, J. K., 2001a. Archaeological sediments in cultural environments. *In* Sediments in Archaeological Context (J. K. Stein and W. R. Farrand, Eds.), pp. 1–28. University of Utah Press, Salt Lake City.

Stein, J. K., 2001b. A review of site formation processes and their relevance to geoarchaeology. *In* Earth Sciences and Archaeology (P. Goldberg, V. T. Holliday, and C. R. Ferring, Eds.), pp. 77–106. Kluwer Academic/Plenum Publishers, New York.

Stein, J. K., and W. R. Farrand, Eds., 2001. Sediments in Archaeological Context. University of Utah Press, Salt Lake City.

Sternberg, R. W., and J. S. Creager, 1961. Comparative efficiencies of size analysis by hydrometer and pipette methods. *Journal of Sedimentary Petrology* 31, 96–100.

Stevens, P. R., and T. W. Walker, 1970. The chronosequence concept and soil formation. *Quarterly Review of Biology* 45, 333–350.

Stevenson, F. J., 1969. Pedohumus: Accumulation and diagenesis during the Quaternary. *Soil Science* 107, 470–479.

Stevenson, F. J., and M. A. Cole, 1999. Cycles of Soil: Carbon, Nitrogen, Phosphorus, Sulfur, Micronutrients. 2nd ed. Wiley, New York.

Stevenson, M. G., 1985. The formation of artifact assemblages at workshop/habitation sites: Models from Peace Point in northern Alberta. *American Antiquity* 50, 63–81.

Stewart, M., J. Custer, and D. Kline, 1991. A deeply stratified archaeological and sedimentary sequence in the Delaware River Valley of the Middle Atlantic region. *Geoarchaeology* 6, 169–182.

Stimmel, C. A., R. G. V. Hancock, and A. M. Davis, 1984. Chemical analysis of archaeological soils from Yagi site, Japan. *In* Archaeological Chemistry—III (J. B. Lambert, Ed.), pp. 79–96. Advances in Chemistry Series no. 205. American Chemical Society, Washington, D.C.

Stremme, H. E., 1989. Thermoluminescence dating of the pedostratigraphy of the Quaternary Period in NW Germany. *Geoderma* 45, 185–195.

Styles, T. R., 1985. Holocene and Late Pleistocene Geology of the Napoleon Hollow Site in the lower Illinois Valley. Center for American Archaeology Research Series no. 5. Kampsville Archeological Center, Kampsville, Ill.

Sullivan, A. P., III. 2000. Effects of small-scale prehistoric runoff agriculture on soil fertility: The developing picture from upland terraces in the American Southwest. *Geoarchaeology* 15, 291–313.

Swanson, D. K., 1990a. Landscape classes: Higher-level map units for soil survey. *Soil Survey Horizons* 31, 52–54.

Swanson, D. K., 1990b. Soil landform units for soil survey. *Soil Survey Horizons* 31, 17–21.

Swanson, D. K., 1993. Comments on "The Soil Survey as Paradigm-Based Science." *Soil Science Society of America Journal* 57, 1164.

Swart, P. K., K. C. Lohmann, J. McKenzie, and S. Sarin, Eds., 1993. Climate Change in Continental Isotopic Records. Geophysical Monograph no. 78. American Geophysical Union, Washington, D.C.

Syers, J. K., M. L. Jackson, V. E. Berkheiser, R. N. Clayton, and R. W. Rex, 1969. Eolian sediment influence on pedogenesis during the Quaternary. *Soil Science* 107, 421–427.

Syers, J. K., G. W. Smillie, and J. D. H. Williams, 1972. Calcium fluoride formation during extraction of calcareous soils with fluoride: I. Implications to inorganic P fractionation schemes. *Soil Science Society of America Proceedings* 36, 20–30.

Tabatabai, M. A., 1996. Soil organic matter testing: An overview. *In* Soil Organic Matter: Analysis and Interpretation (F. R. Magdoff, M. A. Tabatabai, and E. A. Hanlon, Eds.), pp. 1–9. Special Publication no. 46. Soil Science Society of America, Madison, Wisc.

Tamplin, M. J., 1969. The application of pedology to archaeological research. *In* Pedology and Quaternary research (S. Pawluk, Ed.), pp. 153–161. The University of Alberta Printing Department, Edmonton.

Tandarich, J. P., 1998a. Agricultural geology: Disciplinary history. *In* Science of the Earth: An Encyclopedia of Events, People, and Phenomena (G. A. Good, Ed.), pp. 23–29. Garland Publishing, New York.

Tandarich, J. P., 1998b. Pedology: Disciplinary history. *In* Science of the Earth: An Encyclopedia of Events, People, and Phenomena (G. A. Good, Ed.), pp. 666–670. Garland Publishing, New York.

Tandarich, J. P., and S. W. Sprecher, 1994. The intellectual background for the factors of soil formation. *In* Factors of Soil Formation: A Fiftieth Anniversary Retrospective (R. Amundson, Ed.), pp. 1–13. Special Publication 33. Soil Science Society of America, Madison, Wisc.

Tandon, S. K., and S. Kumar, 1999. Semi-arid/arid zone calcretes: A review. *In* Paleoenvironmental Reconstruction in Arid Lands (A. K. Singhvi and E. Derbyshire, Eds.), pp. 109–152. A. A. Balkema, Rotterdam.

Tarnocai, C., and K. W. G. Valentine, 1989. Relict soil properties of the Arctic and Subarctic regions of Canada. *In* Paleopedology: Nature and Applications of Paleosols (A. Bronger and J. Catt, Eds.), pp. 9–39. Catena Supplement 16. Catena Verlag, Cremlingen, Germany.

Taylor, R. E., 1987. Radiocarbon Dating: An Archaeological Perspective. Academic Press, New York.

Taylor, R. E., A. Long, and R. S. Kra, Eds., 1992. Radiocarbon After Four Decades: An Interdisciplinary Perspective. Springer, New York.

Tchernov, E., L. K. Horwitz, A. Ronen, and A. Lister, 1994. The faunal remains from Evron Quarry in relation to other Lower Paleolithic hominid sites in the southern Levant. *Quaternary Research* 42, 328–339.

Terry, R. E., P. J. Hardin, S. D. Houston, S. D. Nelson, M. W. Jackson, J. Carr, and J. Parnell, 2000. Quantitative phosphorus measurement: A field test procedure for archaeological site analysis at Piedras Negras, Guatemala. *Geoarchaeology* 15, 151–166.

Thornes, J. B., 1987. The palaeo-ecology of erosion. *In* Landscape and Culture: Geographical and Archaeological Perspectives (J. M. Wagstaff, Ed.), pp. 37–55. Basil Blackwell, Oxford.

Thorp, J., 1941. The influence of environment on soil formation. *Soil Science Society of America Proceedings* 6, 39–46.

Thorp, J., 1949. Interrelations of Pleistocene geology and soil science. *Geological Society of America Bulletin* 60, 1517–1526.

Thorp, J., W. M. Johnson, and E. C. Reed, 1951. Some post-Pleistocene buried soils of central United States. *Journal of Soil Science* 2, 1–19.

Thorson, R. M., 1990. Geologic contexts of archaeological sites in Beringia. *In* Archaeological Geology of North America (N. P. Lasca and J. Donahue, Eds.), pp. 399–420. Centennial Volume no. 4. Geological Society of America, Boulder, Colo.

Thorson, R. M., and T. D. Hamilton, 1977. Geology of the Dry Creek site: A stratified early man site in interior Alaska. *Quaternary Research* 7, 172–188.

Thorson, R. M., and V. T. Holliday, 1990. Just what is geoarchaeology? *Geotimes* 35, 19–20.

Thurston, T. L., 2001. Landscapes of Power, Landscapes of Conflict: Site Formation in the South Scandinavian Iron Age. Kluwer Academic/Plenum Publishers, New York.

Tiffany, J. A., and L. R. Abbott, 1982. Site-catchment analysis: Applications to Iowa archaeology. *Journal of Field Archaeology* 9, 313–322.

Timpson, M. E., and J. E. Foss, 1994. The use of particle-size analysis as a tool in pedological investigations of archaeological sites. *In* Proceedings of the First International Conference on Pedo-Archaeology (J. E. Foss, M. E. Timpson, and M. W. Morris, Eds.), pp. 69–80. The University of Tennessee Agricultural Experiment Station, Knoxville.

Truog, E., 1930. Determination of the readily available phosphorus in soils. *Journal of the American Society of Agronomy* 22, 874–882.

Tsatskin, A., and A. Ronen, 1999. Micromorphology of a Mousterian Paleosol in aeolianites at the site Habonim, Israel. *Catena* 34, 365–384.

Tsatskin, A., M. Weinstein-Evron, and A. Ronen, 1995. Weathering and pedogenesis of wind-blown sediments in the Mt. Carmel caves, Israel *Quaternary Proceedings* 4, 83–93.

Turner, B. L., II, and P. D. Harrison, 1983. Pulltrouser Swamp: Ancient Maya Habitat, Agriculture, and Settlement in Northern Belize. University of Texas Press, Austin.

Turner, E. S., and T. R. Hester, 1993. A Field Guide to Stone Artifacts of Texas Indians. 2nd ed. Texas Monthly Press, Austin.

Turner, W. B., J. L. Hofman, and G. R. Brakenridge, 1982. Technique to aid in recording and field interpretation of stratigraphic sections in archaeological deposits. *Journal of Field Archaeology* 9, 133–136.

Ugolini, F. C., and R. J. Zasoski, 1979. Soils derived from tephra. *In* Volcanic Activity and Human Ecology (P. D. Sheets and D. K. Grayson, Eds.), pp. 83–124. Academic Press, New York.

U.S. Salinity Laboratory Staff, 1954. Diagnosis and Improvement of Saline and Alkali Soils. U.S. Department of Agriculture Handbook no. 60. Government Printing Office, Washington D.C.

Valentine, K. W. G., and J. B. Dalrymple, 1975. The identification, lateral variation and chronology of two buried paleocatenas at Woodhall Spa and West Runton, England. *Quaternary Research* 4, 551–590.

Valentine, K. W. G., and J. B. Dalrymple, 1976. Quaternary buried paleosols: A critical review: *Quaternary Research* 6, 209–222.

Valentine, K. W. G., K. R. Fladmark, and B. E. Spurling, 1980. The description, chronology and correlation of buried soils and cultural layers in a terrace section, Peace River Valley, British Columbia. *Canadian Journal of Soil Science* 60, 185–197.

Valentine, K. W. G., R. H. King, J. F. Dormaar, W. J. Vreeken, C. Tarnocai, C. R. De Kimpe, and S. A. Harris, 1987. Some aspects of Quaternary soils in Canada. *Canadian Journal of Soil Science* 67, 221–247.

Van Andel, T. H., 1994. Geo-archaeology and archaeological science—A personal view. *In* Beyond the Site: Regional Studies in the Aegean Area (P. N. Kardulias, Ed.), pp. 25–44. University Press of America, New York.

Van Andel, T. H., 1998. Paleosols, red sediments, and the Old Stone Age in Greece. *Geoarchaeology* 13, 361–390.

Van Andel, T. H., C. N. Runnels, and K. O. Pope, 1986. Five thousand years of land use and abuse in the southern Argolid. *Hesperia* 55, 103–128.

Van Andel, T. H., and E. Zangger, 1990. Landscape stability and destabilization in the prehistory of Greece. *In* Man's Role in the Shaping of the Eastern Mediterranean Landscape (S. Bottema, G. Entjes-Nieborg, and W. Van Zeist, Eds.), pp. 139–157. A. A. Balkema Publishers, Rotterdam.

Van Andel, T. H., E. Zangger, and A. Demitrack, 1990. Land use and soil erosion in prehistoric and historical Greece. *Journal of Field Archaeology* 17, 379–396.

Van Breemen, N., and P. Buurman, 2002. Soil Formation. 2nd ed. Kluwer Academic Publishers, Boston.

Van de Westeringh, W., 1988. Man-made soils in the Netherlands, especially in sandy areas ("plaggen soils"). *In* Man-Made Soils (W. Groennmann-van Waateringe and M. Robinson, Eds.), pp. 5–19. BAR International Series no. 410, Oxford.

Van der Merwe, N. J., and P. H. Stein, 1972. Soil chemistry of post molds and rodent burrows: Identification without excavation. *American Antiquity* 37, 245–254.

Van Dijk, D. C., A. M. H. Riddler, and R. K. Rowe, 1968. Criteria and problems in ground-surface correlations with reference to a regional correlation in southeastern Australia. *Proceedings of the 9th International Congress of Soil Science, Adelaide* 4, 131–138.

Van Heeringen, R., and L. Theunissen, 2001. Repeated water table lowering in the Dutch delta: A major challenge to the archaeological heritage management of pre- and protohistoric wetlands. *In* Enduring Records: The Environmental and Cultural Heritage of Wetlands (B. A. Purdy, Ed.), pp. 271–276. Oxbow Books, Oxford.

Van Nest, J., 2002. The good earthworm: How natural processes preserve upland Archaic archaeological sites of western Illinois. *Geoarchaeology* 17, 53–90.

Van Nest, J., D. K. Charles, J. E. Buikstra, and D. L. Asch, 2001. Sod blocks in Illinois Hopewell mounds. *American Antiquity* 66, 633–650.

Van Vliet-Lanoë, B., 1985. Frost effects in soils. *In* Soils and Quaternary Landscape Evolution (J. Boardman, Ed.), pp. 117–158. Wiley, Chichester.

Van Vliet-Lanoë, B., 1998. Frost and soils: Implications for paleosols, paleoclimates and stratigraphy. *Catena* 34, 157–183.

Velichko, A. A., 1990. Loess-paleosol formation on the Russian Plain. *Quaternary International* 7/8, 103–114.

Vepraskas, M. J., 1992. Redoximorphic Features for Identifying Aquic Conditions. North Carolina Agricultural Research Service Technical Bulletin no. 301. North Carolina State University, Raleigh.

Verosub, K. L., P. Fine, M. J. Singer, and J. TenPas, 1993. Pedogenesis and paleoclimate: Interpretation of the magnetic susceptibility record of Chinese loess-paleosol sequences. *Geology* 21, 1011–1014.

Vidic, N., M. Pavich, and F. Lonbik, 1991. Statistical analysis of soil properties on a Quaternary terrace sequence in the upper Sava River valley, Slovenia, Yugoslavia. *Geoderma* 51, 189–212.

Vita-Finzi, C., 1969. The Mediterranean Valleys: Geological Changes in Historical Times. Cambridge University Press, Cambridge.

Vizcaíno, A. S., and M. L. Cañabate, 1999. Identification of activity areas by soil phosphorus and organic matter analysis in two rooms of the Iberian sanctuary "Cerro El Pajarillo." *Geoarchaeology* 14, 47–62.

Vogel, G., 2002. A Handbook of Soil Description for Archeologists. Technical Paper no. 11. Arkansas Archeological Survey, Fayetteville.

Voight, E. E., and M. J. O'Brien, 1981. The use and misuse of soils-related data in mapping and modeling past environments: An example from the central Mississippi River valley. *Contract Abstracts and CRM Archeology* 2, 22–35.

Vreeken, W. J., 1975. Principal kinds of chronosequences and their significance in soil history. *Journal of Soil Science* 26, 378–394.

Wagstaff, J. M., Ed., 1987. Landscape and Culture: Geographical and Archaeological Perspectives. Basil Blackwell, New York.

Walker, T. W., 1964. The significance of phosphorous in pedogenesis. *In* Experimental Pedology (E. G. Hallsworth and D. V. Crawford, Eds.), pp. 295–316. Butterworths, London.

Walker, T. W., and J. K. Syers, 1976. The fate of phosphorus during pedogenesis. *Geoderma* 15, 1–19.

Walter, N. F., G. R. Hallberg, and T. E. Fenton, 1978. Particle size analysis by Iowa State University soil survey laboratory. *In* Standard Procedures for Evaluation of Quaternary Materials in Iowa (G. R. Hallberg, Ed.), pp. 61–74. Technical Information Series no. 8. Iowa Geological Survey, Iowa City.

Wandsnider, L., 1988. Experimental investigation of the effect of dune processes on archaeological remains. *American Archeology* 7, 18–29.

Wang, H., S. H. Ambrose, C. L. J. Liu, and L. R. Follmer, 1997. Paleosol stable isotope evidence for early Hominid occupation of east Asian temperate environments. *Quaternary Research* 48, 228–238.

Wang, Y., R. Amundson, and S. Trumbore, 1996. Radiocarbon dating of soil organic matter. *Quaternary Research* 45, 282–288.

Warren, R. E., 1982a. The historical setting. *In* The Cannon Reservoir Human Ecology Project (M. J. O'Brien, R. E. Warren, and D. E. Lewarch, Eds.), pp. 337–368. Academic Press, New York.

Warren, R. E., 1982b. Prehistoric settlement patterns. *In* The Cannon Reservoir Human Ecology Project (M. J. O'Brien, R. E. Warren, and D. E. Lewarch, Eds.), pp. 29–70. Academic Press, New York.

Warren, R. E., C. K. McDonnell, and M. J. O'Brien, 1981. Soils and settlement in the Southern Prairie Peninsula. *Contract Abstracts and CRM Archeology* 2, 36–49.

Warren, R. E., and M. J. O'Brien, 1981. Regional sample stratification: The drainage class technique. *Plains Anthropologist* 26, 213–227.

Warren, R. E., and M. J. O'Brien, 1982. Holocene dynamics. *In* The Cannon Reservoir Human Ecology Project (M. J. O'Brien, R. E. Warren, and D. E. Lewarch, Eds.), pp. 71–84. Academic Press, New York.

Wasson, R. J., Ed., 1982. Quaternary Dust Mantles of China, New Zealand, and Australia. Australia National University, Canberra.

Waters, M. R., 1985. Late Quaternary alluvial stratigraphy of Whitewater Draw, Arizona: Implications for regional correlation of fluvial deposits in the American Southwest. *Geology* 13, 705–708.

Waters, M. R., 1992. Principles of Geoarchaeology: A North American Perspective. The University of Arizona Press, Tucson. 398 pp.

Waters, M. R., 2000. Alluvial stratigraphy and geoarchaeology in the American Southwest. *Geoarchaeology* 15, 537–557.

Waters, M. R., and C. V. Haynes, 2001. Late Quarternary arroyo formation and climate change in the American southwest. *Geology* 29, 399–402.

Waters, M. R., and D. D. Kuehn, 1996. The geoarchaeology of place: The effect of geological processes on the preservation and interpretation of the archaeological record. *American Antiquity* 61, 483–497.

Waters, P. L., and N. W. Rutter, 1984. Utilizing paleosols and volcanic ash in correlating Holocene deposits in southern Alberta. *In* Correlation of Quaternary Chronologies (W. C. Mahaney, Ed.), pp. 203–223. Geo Books, Norwich.

Weide, D. L., 1966. Soil pH as a guide to archaeological investigations. University of California (Los Angeles) Archaeological Survey Annual Report no. 8, pp. 155–163. University of California, Los Angeles.

Weider, M., and G. Gvirtzman, 1999. Micromorphological indications on the nature of the Late Quaternary paleosols in the southern coastal plain of Israel. *Catena* 35, 219–237.

Weiner, S., Q.-Q. Xu, P. Goldberg, J.-Y. Liu, and O. Bar-Yosef, 1998. Evidence for the use of fire at Zhoukoudian, China. *Science* 281, 251–253.

Weisler, M. I., 1999. The antiquity of aroid pit agriculture and significance of buried A horizons on Pacific atolls. *Geoarchaeology* 14, 621–654.

Wells, E. C., R. E. Terry, J. J. Parnell, P. J. Hardin, M. W. Jackson, and S. D. Houston, 2000. Chemical analyses of ancient anthrosols in residential areas at Piedras Negras, Guatemala. *Journal of Archaeological Science* 27, 449–462.

Wells, L. E., 2001. A geomorphological approach to reconstructing archaeological settlement patterns based on surficial artifact distribution: Replacing humans on the landscape. *In* Earth Sciences and Archaeology (P. Goldberg, V. T. Holliday, and C. R. Ferring, Eds.), pp. 107–141. Kluwer Academic/Plenum Publishers, New York.

Wells, S. D., L. D. McFaden, and J. D. Schultz, 1990. Eolian landscape evolution and soil formation in the Chaco dune field, southern Colorado Plateau, New Mexico. *In* Soils and Landscape Evolution (P. L. K. Knuepfer and L. D. McFadden, Eds.), pp. 517–546. Elsevier, New York.

Wendorf, F., Ed., 1961. Paleoecology of the Llano Estacado. Publication of the Fort Burgwin Research Center no. 1, The Museum of New Mexico Press, Santa Fe.

Wendorf, F., 1970. The Lubbock Subpluvial. *In* Pleistocene and Recent Environments of the Central Great Plains (W. Dort and J. K. Jones, Eds.), pp. 23–36. The University of Kansas Press, Lawrence.

Wendorf, F., and J. J. Hester, Eds., 1975. Late Pleistocene Environments of the Southern High Plains. Publication of the Fort Burgwin Research Center no. 9. Taos, N.M.

Wendorf, F., and R. Schild, Eds., 1974a. A Middle Stone Age Sequence from the Central Rift Valley, Ethiopia. Polska Akademia Nauk Instytut Historii Kultury Materialnej, Warsaw.

Wendorf, F., and R. Schild, 1974b. Stratigraphy and lithology. *In* A Middle Stone Age Sequence from the Central Rift Valley, Ethiopia (F. Wendorf and R. Schild, Eds.), pp. 28–55. Polska Akademia Nauk Instytut Historii Kultury Materialnej, Warsaw.

Weston, D., 1996. Soil science and the interpretation of archaeological sites: A soil survey and magnetic susceptibility analysis of Altofts 'Henge,' Normanton, West Yorkshire. *Archaeological Prospection* 3, 39–50.

Wheat, J. B., 1974. First excavations at the Lubbock Lake site. *In* History and Prehistory of the Lubbock Lake Site (C. A. Black, Ed.), pp. 15–42. The Museum Journal. West Texas Museum Association, Lubbock.

White, E. M., 1978. Cautionary note on soil phosphate data interpretation for archaeology. *American Antiquity* 43, 507–508.

Wiant, M. D., E. R. Hajic, and T. R. Styles, 1983. Napoleon Hollow and Koster site stratigraphy: Implications for Holocene landscape evolution and studies of Archaic period settlement patterns in the lower Illinois valley. *In* Archaic Hunters and Gatherers in the American Midwest (J. L. Phillips and J. A. Brown, Eds.), pp. 147–164. Academic Press, New York.

Wilding, L. P., and L. R. Drees, 1983. Spatial variability and pedology. *In* Pedogenesis and Soil Taxonomy, part 1, Concepts and Interactions (L. P. Wilding, N. E. Smeck, and G. F. Hall, Eds.), pp. 83–116. Developments in Soil Science 11A. Elsevier, Amsterdam.

Wilding, L. P., and K. W. Flach, 1985. Micropedology and soil taxonomy. *In* Soil Micromorphology and Soil Classification (L. A. Douglas and M. L. Thompson, Eds.), pp. 1–16. Special Publication no. 15. Soil Science Society of America, Madison, Wisc.

Wilding, L. P., N. E. Smeck, and G. F. Hall, 1983a. Glossary of horizon designations. *In* Pedogenesis and Soil Taxonomy, part 2, The Soil Orders (L. P. Wilding, N. E. Smeck and G. F. Hall, Eds.), pp. 383–387. Developments in Soil Science 11B. Elsevier, Amsterdam.

Wilding, L. P., N. E. Smeck, and G. F. Hall, Eds, 1983b. Pedogenesis and Soil Taxonomy, part 2, The Soil Orders. Developments in Soil Science 11B. Elsevier, Amsterdam.

Wilding, L. P., and D. Tessier, 1988. Genesis of Vertisols: Shrink-swell phenomena. *In* Vertisols: Their Distribution, Properties, Classification and Management (L. P. Wilding and R. Puentes, Eds.), pp. 55–81. U.S. Department of Agriculture, Soil Management

Support Services Technical Monograph no. 18. Texas A&M University Printing Center, College Station, Tex.

Wilkinson, T. J., 1976. Soil and sediment structures as an aid to archaeological interpretation: Sediments at Dibsi Faraj, Syria. *In* Geoarchaeology (D. A. Davidson and M. L. Shackley, Eds.) pp. 277–287. Westview Press, Boulder, Colo.

Wilkinson, T. J., 1988. The archaeological component of agricultural soils in the Middle East: The effects of manuring in antiquity. *In* Man-Made Soils (W. Groennmann-van Waateringe and M. Robinson, Eds.), pp. 93–114. BAR International Series no. 410, Oxford.

Wilkinson, T. J., 1990. Soil development and early land use in the Jazira region, Upper Mesopotamia. *World Archaeology* 22, 87–103.

Wilkinson, T. J., 1997. Holocene environments of the High Plateau, Yemen. Recent geoarchaeological investigations. *Geoarchaeology*, 12, 833–864.

Williams, G. E., and H. A. Polach, 1971. Radiocarbon dating of arid-zone calcareous paleosols. *Geological Society of America Bulletin* 82, 3069–3086.

Williams, J. D. H., J. K. Syers, R. F. Harris, and D. E. Armstrong, 1971. Fractionation of inorganic phosphate in calcareous lake sediments. *Soil Science Society of America Proceedings* 35, 250–255.

Williams, J. D. H., J. K. Syers, and T. W. Walker, 1967. Fractionation of soil inorganic phosphate by a modification of Chang and Jackson's procedure. *Soil Science Society of America Proceedings* 31, 736–739.

Wilmsen, E. M., and F. H. H. Roberts, 1978. Lindenmeier, 1934–1974: Concluding Report of Investigations. Smithsonian Contributions to Anthropology no. 24. Smithsonian Institution, Washington, D.C.

Wilson, M. C., 1990. Archaeological geology in western Canada: Techniques, approaches, and integrative themes. *In* Archaeological Geology of North America (N. P. Lasca and J. Donahue, Eds.), pp. 61–86. Centennial Volume 4. Geological Society of America, Boulder, Colo.

Winograd, I. J., J. M. Landwehr, K. R. Ludwig, T. B. Coplen, and A. C. Riggs, 1997. Duration and structure of the past four interglaciations. *Quaternary Research* 48, 141–154.

Wintle, A. G., 1990. A review of current research on TL dating of loess. *Quaternary Science Reviews* 9, 385–397.

Witty, J. E., and E. G. Knox, 1989. Identification, role in Soil Taxonomy, and worldwide distribution of fragipans. *In* Fragipans: Their Occurrence, Classification, and Genesis (N. E. Smeck and E. J. Ciolkosz, Eds.), pp. 1–9. Special Publication no. 24. Soil Science Society of America, Madison, Wisc.

WoldeGabriel, G., G. Heiken, T. D. White, B. Asfaw, W. K. Hart, and P. R. Renne, 2000. Volcanism, tectonism, and the paleoanthropological record in the Ethiopian Rift system. *In* Volcanic Hazards and Disasters in Human Antiquity (F. W. McCoy and G. Heiken, Eds.), pp. 83–99. Special Paper no. 345. Geological Society of America Boulder, Colo.

Wood, W. R., and D. L. Johnson, 1978. A survey of disturbance processes in archaeological site formation. *Advances in Archaeological Method and Theory* 1, 315–381.

Woods, W. I., 1975. The analysis of abandoned settlements by a new phosphate field test method. *The Chesopiean, A Journal of North American Archaeology* 13, 1–45.

Woods, W. I., 1977. The quantitative analysis of soil phosphate. *American Antiquity* 42, 248–251.

Woods, W. I., 1984. Soil chemical investigations in Illinois archaeology: Two example studies. *In* Archaeological Chemistry—III (J. B. Lambert, Ed.), pp. 67–77. Advances in Chemistry Series no. 205. American Chemical Society, Washington, D.C.

Woods, W. I., 1995. Comments on the Black Earths of Amazonia. *Papers and Proceedings of Applied Geography Conferences (State University of New York at Binghamton)* 18, 159–165.

Woods, W. I., 2004. Development of anthrosol research. *In* Amazonian Dark Earths: Origin, Properties, and Management (J. Lehmann, D. C. Kern, B. Glaser, and W. I. Woods, Eds.), pp. 3–14. Kluwer, Dordrecht, The Netherlands.

Woods, W. I., and J. M. McCann, 1999. The anthropogenic origin and persistence of Amazonian dark earths. *The Yearbook of the Conference of Latin American Geographers* 25, 7–14.

Woodward, J. C., M. G. Macklin, and J. Lewin, 1994. Pedogenic weathering and relative-age dating of Quaternary alluvial sediments in the Pindus Mountains of northwest Greece. *In* Rock Weathering and Landform Evolution (D. A. Robinson and R. B. G. Williams, Eds.), pp. 259–283. John Wiley and Sons, Chichester.

Wright, H. T., and W. B. Roosa, 1966. The Barnes site: A fluted point assemblage from the Great Lakes region. *American Antiquity* 31, 850–860.

Wright, V. P., Ed., 1986. Paleosols: Their Recognition and Interpretation. Princeton University Press, Princeton, N.J.

Wymer, J., 1985. The Palaeolithic Sites of East Anglia. Geo Books, Norwich.

Yaalon, D. H., 1971a. Criteria for the recognition and classification of paleosols. *In* Paleopedology: Origin, Nature and Dating of Paleosols (D. H. Yaalon, Ed.), pp. 153–158. University of Israel Press, Jerusalem.

Yaalon, D. H., Ed., 1971b. Paleopedology: Origin, Nature and Dating of Paleosols. University of Israel Press, Jerusalem.

Yaalon, D. H., 1971c. Soil-forming processes in time and space. *In* Paleopedology: Origin, Nature and Dating of Paleosols (D. H. Yaalon, Ed.), pp. 29–40. University of Israel Press, Jerusalem.

Yaalon, D. H., 1975. Conceptual models in pedogenesis: Can soil-forming functions be solved? *Geoderma* 14, 189–205.

Yaalon, D. H., 1983. Climate, time and soil development. *In* Pedogenesis and Soil Taxonomy, part 1, Concepts and Interactions (L. P. Wilding, N. E. Smeck, and G. F. Hall, Eds.), pp. 233–251. Developments in Soil Science 11A. Elsevier, New York.

Yaalon, D. H., 1997. History of soil science in context: International perspective. *In* History of Soil Science (D. H. Yaalon and S. Berkowicz, Eds.), pp. 1–13. *Advances in GeoEcology* 29, Catena Verlag, Reiskirchen, Germany.

Yaalon, D. H., and S. Berkowicz, Eds., 1997. History of Soil Science: International Perspectives. *Advances in GeoEcology* 29, Catena Verlag, Reiskirchen, Germany.

Yaalon, D. H., and B. Yaron, 1966. Framework for man-made soil changes: An outline of metapedogenesis. *Soil Science* 102, 272–277.

Yakimenko, E. Y., 1995. Pleistocene paleosols in the loess and loess-like sediments of the central part of the Russian Plain. *Quaternary Science Reviews* 14, 747–753.

Yassoglou, N. J., and C. F. Haidouti, 1978. Soil formation. *In* Excavations at Nichoria in Southwest Greece. Volume 1, Site, Environs, and Techniques (G. Rapp and S. E. Aschenbrenner, Eds.), pp. 31–40. University of Minnesota Press, Minneapolis.

Yassoglou, N. J., and C. Nobeli, 1972. Soil studies. *In* The Minnesota Messinia Expedition: Reconstructing a Bronze Age Regional Environment (W. A. McDonald and G. R. Rapp, Jr., Eds.), pp. 171–176. University of Minnesota Press, Minneapolis.

Yule, B., 1990. The "dark earth" and late Roman London. *Antiquity* 64, 620–628.

Zangger, E., 1993. The Geoarchaeology of the Argolid. Argolis Zweiter Band, Deutsches Archäologisches Institut Athen, Gebr. Mann, Berlin.

Zárate, M., N. Flegenheimer, and M. Pécsi, Eds., 1993. Properties, chronology and paleoclimatic significance of loess. *Quaternary International* 17, 3–110.

Zech, W., L. Haumaier, and R. Hempfling, 1990. Ecological aspects of soil organic matter in tropical land use. *In* Humic Substances in Soil and Crop Sciences: Selected Readings (P. MacCarthy, C. E. Clapp, R. L. Malcolm, and P. R. Bloom, Eds.), pp. 187–202. American Society of Agronomy and Soil Science Society of America, Madison, Wisc.

Zeuner, F. E., 1959. The Pleistocene Period: Its Climate, Chronology and Faunal Successions. Hutchinson and Company, London.

Zhang, G.-L., and Z.-T. Gong, 2003. Peogenic evolution of paddy soils in different soil landscapes. *Geoderma* 115, 15–29.

Zhang, M., L. Ma, W. Li, B. Chen, and J. Jia, 2003. Genetic characteristics and taxonomic classification of Fimic Anthrosols in China. *Geoderma* 115, 31–44.

Zhang, Z.-H., Z.-Y. Zhang, and Y.-S. Wang, 1991. Loess in China. *In* The Quaternary of China (Z.-H. Zhang and S.-X. Shao, Eds.), pp. 459–508. China Ocean Press, Beijing.

Zhou, L.-P., T. H. Van Andel, and A. Lang, 2000. A luminescence dating study of open-air Paleolithic sites in western Epirus, Greece. *Journal of Archaeological Science* 27, 609–620.

Zoltai, S. C., 1995. Permafrost distribution in peatlands of west-central Canada during the Holocene warm period 6000 years BP. *Géographie Physique et Quaternaire* 49, 45–54.

Zvelebil, M., 1987. Wetland settlements in eastern Europe. *In* European Wetlands in Prehistory (J. M. Coles and A. J. Lawson, Eds.), pp. 94–116. Clarendon Press, Oxford.

Index